THE CYBERCITIES READER

For too long Information and Communications Technologies (or ICTS) have been lazily portrayed as means to simply escape into a parallel world – to withdraw from the body, or the city, in some utopian, or dystopian, stampede online. Such perspectives deny the fact that the so-called 'information society' is an increasingly urban society. They ignore the ways in which new technologies now mediate every aspect of everyday urban life. And they obscure a key question: *how do the multifaceted realities of city regions interrelate in practice with new technologies in different ways in different places?*

The Cybercities Reader explores this question. It is the most comprehensive, international and interdisciplinary analysis yet of the relationships between cities, urban life and new technologies. The book incorporates:

- Thirty-one of the best published writings in the field with thirty-two specially commissioned pieces.
- The work of writers from thirteen nations and twelve disciplines.
- Detailed discussions of cybercity history, theory, economic processes, mobilities, physical forms, social and cultural worlds, digital divides, public domains, strategies, politics and futures.
- Analyses of postmodern technoculture, virtual reality and the body, global city economies, urban surveillance, 'intelligent' transportation, e-commerce, teleworking, community informatics, digital architecture, urban technology strategies, and the role of cities and new technologies in the 'war on terrorism'.
- Detailed case studies of 'virtual cities' in Amsterdam, Internet cabins in Lima, back offices in Jamaica, 'smart' highways in Melbourne, technopoles in New York, mobiles in Helsinki, e-commerce convenience stores in Tokyo, high-tech business parks in Bangalore, public spaces in Mexico City, and urban ICT strategies in Kuala Lumpur, California and Singapore.
- Editor's introductions, guides to further reading for each theme and piece.
- Over fifty plates, tables, and diagrams.
- A dedicated web site at **http://www.geographyarena.com/geographyarena/cybercities** providing annotated links, using the same structure as the book, to a wide range of relevant resources on the Internet.

The Cybercities Reader is essential reading for anyone interested in how cities and new technologies are helping to remake each other at the start of this quintessentially urban digital century.

Stephen Graham is Professor of Urban Technology at the Global Urban Research Unit in Newcastle University's School of Architecture, Planning and Landscape. He is co-author of *Telecommunications and the City* and *Splintering Urbanism*, both published by Routledge, and co-editor of *Managing Cities*.

THE ROUTLEDGE URBAN READER SERIES

Series editors

Richard T. LeGates
Professor of Urban Studies, San Francisco State University

Frederic Stout
Lecturer in Urban Studies, Stanford University

The Routledge Urban Reader Series responds to the need for comprehensive coverage of the classic and essential texts that form the basis of intellectual work in the various academic disciplines and professional fields concerned with cities.

The readers focus on the key topics encountered by undergraduates, graduates and scholars in urban studies and allied fields. They discuss the contributions of major theoreticians and practitioners and other individuals, groups, and organizations that study the city or practise in a field that directly affects the city.

As well as drawing together the best of classic and contemporary writings on the city, each reader features extensive general, section and selection introductions prepared by the volume editors to place the selections in context, illustrate relations among topics, provide information on the author and point readers towards additional related bibliographic material.

Each reader will contain:

- Approximately thirty-six *selections* divided into approximately six sections. Almost all of the selections will be previously published works that have appeared as journal articles or portions of books.
- A *general introduction* describing the nature and purpose of the reader.
- Two- to three-page *section introductions* for each section of the reader to place the readings in context.
- A one-page *selection introduction* for each selection describing the author, the intellectual background of the selection, competing views of the subject matter of the selection and bibliographic references to other readings by the same author and other readings related to the topic.
- A plate section with twelve to fifteen plates and illustrations at the beginning of each section.
- An index.

The types of readers and forthcoming titles are as follows:

THE CITY READER

The City Reader: third edition – an interdisciplinary urban reader aimed at urban studies, urban planning, urban geography and urban sociology courses – will be the *anchor urban reader*. Routledge published a first edition of *The City Reader* in 1996 and a second edition in 2000. *The City Reader* has become one of the most widely used anthologies in urban studies, urban geography, urban sociology and urban planning courses in the world.

URBAN DISCIPLINARY READERS

The series will contain *urban disciplinary readers* organized around social science disciplines. The urban disciplinary readers will include both classic writings and recent, cutting-edge contributions to the respective disciplines. They will be lively, high-quality, competitively priced readers which faculty can adopt as course texts and which will also appeal to a wider audience.

TOPICAL URBAN ANTHOLOGIES

The urban series will also include *topical urban readers* intended both as primary and supplemental course texts and for the trade and professional market.

INTERDISCIPLINARY ANCHOR TITLE

The City Reader: third edition
Richard T. LeGates and Frederic Stout (eds)

URBAN DISCIPLINARY READERS

The Urban Geography Reader
Nick Fyfe and Judith Kenny (eds)

The Urban Sociology Reader
Jan Lin and Christopher Mele (eds)

The Urban Politics Reader
Elizabeth Strom and John Mollenkopf (eds)

The Urban and Regional Planning Reader
Eugenie Birch (ed.)

TOPICAL URBAN READERS

The City Cultures Reader: second edition
Malcolm Miles and Tim Hall, with Iain Borden (eds)

The Cybercities Reader
Stephen Graham (ed.)

The Sustainable Urban Development Reader
Stephen M. Wheeler and Timothy Beatley (eds)

The Global Cities Reader
Neil Brenner and Roger Keil (eds)

For further information on The Routledge Urban Reader Series
please visit our web site:

www.geographyarena.com/geographyarena/urbanreaderseries

or contact:

Andrew Mould
Routledge
11 New Fetter Lane
London EC4P 4EE
England
andrew.mould@routledge.co.uk

Richard T. LeGates
Urban Studies Program
San Francisco State University
1600 Holloway Avenue
San Francisco, California 94132
(415) 338-2875
dlegates@sfsu.edu

Frederic Stout
Urban Studies Program
Stanford University
Stanford, California 94305-6050
(650) 725-6321
fstout@stanford.edu

The Cybercities Reader

Edited by

Stephen Graham

Routledge
Taylor & Francis Group

LONDON AND NEW YORK

First published 2004 by Routledge
11 New Fetter Lane, London EC4P 4EE

Simultaneously published in the USA and Canada
by Routledge
29 West 35th Street, New York, NY 10001

Routledge is an imprint of the Taylor & Francis Group

Designed and typeset in Amasis and Akzidenz Grotesk by
Keystroke, Jacaranda Lodge, Wolverhampton
Printed and bound in Great Britain by
Bell & Bain Ltd, Glasgow

British Library Cataloguing in Publication Data
A catalogue record for this book is available from the British Library

Library of Congress Cataloging in Publication Data
The cybercities reader/[edited by] Stephen Graham.
     p. cm. – (Routledge urban reader series)
   Includes bibliographical references and index.
   1. Information technology–Social aspects. 2. Telecommunication–Social aspects.
   3. Cities and towns. 4. Information society. I. Title: Cybercities reader.
   II. Graham, Stephen, 1965– III. Series.
   T14.5.C93 2004
   303.48'3–dc21   2003010584

ISBN 0–415–27955–0 (hbk)
ISBN 0–415–27956–9 (pbk)

For Ben and Oliver

Plate 1 Mobile citizenship: commuters text as they walk in Tokyo's Shinjuku district, June 2003. (Photograph by Stephen Graham.)

Contents

Illustrations

PLATES

FIGURES

TABLES

Contributors

John Adams is Professor of Geography at University College London, UK.

Philip Agre is Professor at the Department of Information Studies, University of California, Los Angeles, USA.

Yuko Aoyama is an Assistant Professor and Henry J. Leir Faculty Fellow of Geography at Clark University, Worcester, Massachusetts, USA.

Nick Barley is an architectural writer from London, UK.

Anne Beamish is Assistant Professor in the Community and Regional Planning Program in the School of Architecture, University of Texas at Austin, USA.

The **Benton Foundation** (http://www.benton.org) is a research and pressure group based in Washington, DC, whose mission is to to 'articulate a public interest vision for the digital age and to demonstrate the value of communications for solving social problems'.

The late **Deirdre Boden** was Professor at the University of Copenhagen Business School, Denmark, and, prior to that, on the Sociology Faculty at Lancaster University, UK.

Stefano Boeri is an architect and teaches Urban Design at the Universities of Venice and Mendrisio, Italy.

Andreas Broeckmann is the Director of the Transmedia ars festival, Berlin, Germany.

Tim Bunnell is Assistant Professor at the Department of Geography in the National University of Singapore.

Thomas J. Campanella is Assistant Professor in the Department of City and Regional Planning at the University of North Carolina, Chapel Hill, USA.

Zac Carey is a Diploma School student at the Architectural Association School of Architecture, London, UK.

Manuel Castells is Professor of City and Regional Planning, and Professor of Sociology at the University of California, Berkeley, as well as Research Professor at the Internet Interdisciplinary Institute, Universitat Oberta de Catalunya, Barcelona, Spain.

Lieven de Cauter is a philosopher and art historian who teaches at the Departement Architectuur, Heverlee, Belgium.

Neil Coe teaches at Manchester University's School of Geography, UK.

Mike Crang is Reader in the Geography Department at Durham University, UK.

Susan Davis is Professor of Communication at the University of California at San Diego, USA.

Gilles Deleuze (1925–1994) was a French Philosopher whose last position was as Chair at Université de Paris VIII, France.

Jean-Michel Dewailly works at the Département de Géographie, Université Lumière, Lyon II, France.

Fred Dewey is a public space activist and writer on culture and politics who lives in Los Angeles, USA.

Martin Dodge is a Lecturer at the Centre for Advanced Spatial Analysis, University College London, UK.

Keller Easterling is an author and Associate Professor at Yale's School of Architecture, USA.

Ana María Fernández-Maldonado is an urban researcher working at the Section of Urbanism of the Faculty of Architecture of the Delft University of Technology in the Netherlands.

Andrew Gillespie is Director of in the Centre for Urban and Regional Development Studies (CURDS) in Newcastle University, UK.

Mark Gottdeiner is Professor of Sociology at the University of Buffalo, USA.

Stephen Graham is a Professor in the Global Urban Research Unit (GURU) at Newcastle University's School of Architecture, Planning and Landscape in the UK.

Keith Hampton is Assistant Professor in MIT's Department of Urban Studies and Planning, USA.

Ken Hillis is Assistant Professor of Communication Studies and Adjunct Professor of Geography at the University of North Carolina at Chapel Hill, USA.

David Holmes lectures in Media and Communication at Monash University in Australia.

Thomas Horan is Executive Director of the Claremont Information Technology Institute, and Associate Professor in the School of Information Science at the Claremont Graduate University, USA.

Peter Huber is a Senior Fellow with the Manhattan Institute of Policy Research in New York, and a Partner of Digital Power Capital.

Timo Kopomaa is a Senior Researcher in Helsinki University of Technology, Centre for Urban and Regional Studies, Finland.

Danny Kruger is a researcher with the Demos think-tank in London, UK.

Geert Lovink is a media theorist and Internet critic, holding a fellowship at the Centre for Critical and Cultural Studies, University of Queensland, Brisbane, Australia.

Robert Luke is a Graduate Fellow of the Knowledge Media Design Institute at the University of Toronto, Canada.

Timothy Luke is Professor of Political Science at Virginia Polytechnic Institute and State University, USA.

David Lyon is Professor of Sociology at Queen's University, Canada.

Shirin Madon teaches in Information and Communication Studies at the London School of Economics, UK.

Simon Marvin is Professor at Salford University's Centre for Sustainable Urban and Regional Futures (SURF) in the UK.

Mark Mills is co-author of the *Digital Power* report investment newsletter, and a Partner of Digital Power Capital.

William Mitchell is Dean of the School of Architecture and Planning at MIT in the USA.

Harvey Molotch is Professor of Sociology and Metropolitan Studies at New York University.

David Morley is a Professor of Media and Communications at Goldsmith's College, University of London, UK.

Vincent Mosco is Professor of Communication at Carleton University, Ottawa, Canada.

Andrew Murphy is a Lecturer in the School of Geography, Earth and Environmental Sciences at the University of Birmingham, UK.

Samuel Nunn is a Professor at the School of Public and Environmental Affairs, Indiana University–Purdue University, Indianapolis, USA.

Martin Pawley is an architectural critic working in London, UK.

Pnina Ohana Plaut is Senior Lecturer in Town Planning at the Faculty of Architecture and Town Planning, Technion – Israel Institute of Technology, Haifa, Israel.

Ranald Richardson is a researcher at the Centre for Urban Regional Development Studies at Newcastle University in the UK.

Saskia Sassen is Professor of Sociology at the University of Chicago, USA.

Richard Sclove is the founder and former executive director of the Loka institute (http://www.loka.org).

Susan Shwartenberg is an urban archaeologist and artist who lives in San Francisco, USA.

Mimi Sheller is a Lecturer in Sociology at Lancaster University in the UK.

Walter Siembab is Director of Siembab Planning Associates, a firm specialising in planning and the implementation of regional cyber-strategies.

Ewart Skinner is an Assistant Professor in the Department of Telecommunications at Bowling Green State University, USA.

The late **Ithiel de Sola Pool** (1917–1984) was Professor of Political Science at MIT, USA.

Rebecca Solnit writes about public space, landscape and environmental issues and lives in San Francisco, USA.

Thomas Streeter is Professor of Communications at the University of Wisconsin, USA.

Joel Tarr is Professor of Urban History at Carnegie Mellon University in Pittsburg in the USA.

Nigel Thrift is Professor of Geography at Bristol University in the UK.

Anthony Townsend teaches and researches on technology and urban development at New York University in the USA.

John Urry is Professor in Sociology at Lancaster University in the UK.

Paul Virilio is an urbanist, writer and technology critic who lives in Paris, France.

Nina Wakeford works at the Digital World Research centre at the University of Surrey in the UK.

Colin Warren is an independent scholar who lives in the USA.

Robert Warren is Professor at the School of Urban Affairs and Public Policy, University of Delaware in the USA

Stacy Warren works at the Department of Geography and Anthropology, Eastern Washington University, USA.

Melvyn Webber is Professor Emeritus of Planning, University of California, Berkeley, USA.

Henry Wai-chung Yeung teaches and researches at the Geography Department of the National University of Singapore.

Matthew Zook is Assistant Professor of Geography at the University of Kentucky, USA.

Acknowledgements

The idea for this book first developed when Andrew Gillespie, Simon Marvin and I hosted a conference titled 'Cities in the Global Information Society: International Perspectives' in November 1999 in Newcastle, UK. The range and quality of the speakers at that conference, many of whom have also contributed to this book, made one thing very clear to us: studies of the relations between cities and Information and Communications Technologies (ICTs) were at a 'take-off' stage in a wide variety of disciplines and contexts around the world.

For a variety of reasons, our efforts to assemble some of these papers and put them into an edited book failed to achieve success. But it quickly became apparent in the process that there was a real need for a major, crosscutting and synthesising book which could do more than an edited collection of chapters. So was born the idea of a major reader on cities and ICTs – a book that would attempt to take stock of, and help provide shape to, the loosely affiliated sub-discipline which is crystallising around the study of how cities and ICTs interrelate. Whilst neither Simon nor Andrew have been involved in the actual production of the book, I owe them a considerable debt as excellent and supportive colleagues and for their help in generating the ideas for this work.

The second debt I owe is to colleagues and students in Newcastle University's School of Architecture, Planning and Landscape. Over the years, students in my 'Cities and New Technologies' option have provided great feedback and critical commentary on the use of many of the sources in this book. Rodrigo Firmino provided excellent support in the earlier stages of identifying and copying possible readings. Elizabeth Storey's efforts in contacting copyright holders was dogged, determined and, ultimately, extremely successful. Thanks are due, also, to colleagues in the School's new Global Urban Research Unit. Completing this book, as it grew progressively in complexity and size, took far too much of my time away from working on the launch of the Unit. Thanks for your support!

My third debt lies with the authors included in this text. I am particularly grateful to the forty or so people who gladly contributed specially produced or modified readings, often at very short notice. Many thanks for the excellent quality of your work, and for being so accommodating and supportive in editing the initial drafts down for publication.

Fourth, I owe thanks to the staff at Routledge – especially Andrew Mould, Ann Michael and Melanie Attridge – who provided excellent support for the book even through considerable difficulties and delay.

Finally, to Annette – my greatest support of all – what can I say? Thanks for putting up with this ever-expanding beast! I suspect you will be as happy as I to see it published.

Stephen Graham
Newcastle, UK
April 2003

INTRODUCTION

■ From dreams of
transcendence to
the remediation
of urban life

Plate 2 New media complex in the new Sony Plaza at Potsdamer Platz, Berlin, 2002.
(Photograph by Stephen Graham.)

The so-called 'information society' is an increasingly urban society. The 'digital age' is an age which is dominated by cities and metropolitan regions to an extent that is unprecedented in human history. This situation raises a critical question: What is the intersection between digital technologies and urban life? The aim of this book is to address this question. It seeks to explore the crucial relationship between two of the most significant processes of contemporary social, economic, geographical, political and cultural change across the world: the intensifying urbanisation of the planet, and the rapid acceleration in the use and capability of digital information and communications technologies (ICTs).

Up until the late 1990s, the complex links between cities and electronic communications generated a curiously scarce literature. Since their inception, urban studies, policy and planning had tended to neglect electronic means of communication because of their relative invisibility compared to physical communications systems (i.e. transportation) (Mandlebaum, 1986). Communications studies disciplines, meanwhile, had traditionally tended to neglect the dominant role of modern cities as crucibles of innovation in electronic communication and the organisation of information, knowledge, and electronic flows (Jowett, 1993). With a few notable exceptions (e.g. Sola-Pool, 1997; Gottmann, 1990; Meier, 1962; Abler, 1974) the best analyses of the intersections of electronic technologies and the cultures and processes of urban life had not focused on the late twentieth century. Instead, they had centred on the ways in which the electro-mechanical technologies of speed, light and power – from film, electricity, the telegraph and the telephone to radio and the railways – had helped to shape the great modernisation, and acceleration, of western urban life in the late nineteenth and early twentieth centuries (e.g. Simmel, 1964; Banham, 1980; Kern, 1986; see Thrift, 1996).

Thankfully, this situation has now changed dramatically. Since the mid-1990s, high quality theoretical, empirical and policy research on the links between ICTs and the changing nature of cities and urban life has rapidly emerged in many disciplines across the world. Specialist researchers on cities and ICTs now exist in urban studies, anthropology, geography, planning, art, sociology, architecture, cultural studies, transport studies and information, communication and media studies.

Together, this work is starting to constitute a sub-discipline which we might label urban ICT studies. It has begun to demonstrate how the urban dominance in the use of new ICTs like the Internet, mobile phones, Geographical Information Systems (GISs), and new media like Virtual Reality (VR) significantly influence how these media are shaped and used. It has also started to show how extraordinary current advances in new media and ICT applications and technologies are starting to influence the forms, processes, experiences and ideas of urban life in a wide variety of contexts across the world.

As a result, it is now startlingly clear that global urbanisation trends, and the intensifying use of computers, the Internet, telephones and digital media in social, economic and cultural life, are actually closely interrelated processes of change. Against the widespread assumption between the 1960s and late 1990s that electronic communications would necessarily work to *undermine* the large metropolitan region, all evidence suggests that the two are actually supporting each other. Both, in fact, are constitutive elements of broader processes of modernisation, internationalisation, industrialisation, restructuring, and cultural change.

The application of ICTs within and between cities – whilst an intensely uneven process – thus constitutes a critical and strategic nexus which underpins the development of human societies, settlements, and, indeed, civilisations. But why are ICTs helping to facilitate processes of intensifying global urbanisation? Three main reasons can be highlighted.

First, ICTs allow specialist urban centres, with their high value-added services, manufacturing, and cultural and knowledge industries, to extend their powers, markets and control over ever-more distant regional, national, international and even global hinterlands. ICTs support the accelerating and spiralling contacts, transactions, communications flows and interactions that help bind, integrate and add economic dynamism to the vast, extended and multi-centred urban settlements, corridors and regions of our age.

Second, in an intensely volatile global economy, the growing speed, complexity, reflexiveness, and riskiness of innovation in all sectors − even those that can theoretically be pushed entirely online − seems to demand a parallel concentration in those cities with the assets and 'innovative milieux' to sustain ongoing competitiveness. This explains why the greatest planning problems in many emerging 'multimedia clusters' and digital growth centres in the urban regions of the north and south surround transportation and car parking. Workers still need to move their physical bodies to be 'in the thick of' the innovation process, even though their products can often then be instantaneously sent online to distant markets and users.

Finally, demand for ICTs − mobile and land line phones, satellite and cable TVs, computer networks, electronic commerce, Internet services, and all the rest − is overwhelmingly driven by the growth of metropolitan markets. Large 'global' cities, especially, are of disproportionate importance in driving all aspects of ICT investment and innovation. This is because of the speed, complexity, dynamism, mobility (and, in the case of mobile phones, sometimes immobility!), of cities and urban settlements. It is also because of cities' cultures of modernisation, their concentrations of capital, their relatively high disposable incomes, their cosmopolitan and multicultural social worlds, and their high concentrations of internationally oriented firms, institutions and people. As much of the economic product base of cities becomes mediated by flows of electronic information and symbols, with the progressive digitisation of money, services, media, information, education and logistics, this critical nexus between cities and ICTs will only strengthen.

The anything–anywhere–anytime dream: fantasies of transcendence on an urbanising planet

And yet we face a curious paradox here. For, despite the critical and strategic importance of city–ICT relations, the emerging research on them still often remains relatively peripheral in the traditional disciplines within which it takes place. Debates about both the implications of ICTs for cities and urban life, and the implications of the cities and urban life for ICTs, have yet to move centre stage within urban and social research.

Why is this the case? One explanation is the justified scepticism amongst many social scientists about the hype which surrounds new technologies. Another is the legacy of an influential discourse − especially strong between the 1960s and late 1990s − which implied that the emerging stream of innovations such as fax, satellite, computer networks, and Virtual Reality (VR) would actually threaten the very existence of cities. It is worth exploring this discourse in a little more detail. If we are to understand the complex realities of city–ICT relations we will need to analyse why this discourse fails so completely to describe the emerging relationships between cities, ICTs and urban life.

Fantasies of transcendence: the origins

Between the 1960s and late 1990s a wide range of influential commentators, business writers, futurists, novelists, media theorists, architects and social scientists portrayed ICTs as being effectively at war with any activity which generated the need, or desire, for geographical concentration in cities and urban regions. Physical transportation flows, many such commentators also implied, might similarly be gradually replaced by growing flows and capabilities of electronic communications. This wave of excited rhetoric drew on the earlier 'disurbanist' schemes of a wide range of urban commentators, planners and utopianists since the end of the nineteenth century (Wigley, 2002). Flows of 'bits', these commentators suggested, might simply grow in scale to substitute for flows of 'atoms'. This, it was alleged, might even lead to the gradual 'dematerialisation' of advanced societies (Negroponte, 1995; Marvin, 1997).

Such assumptions were especially dominant in the Anglo-Saxon world. Here, media theorists such as Marshall McLuhan (1964), futurists such as Alvin Toffler (1980), and technology gurus such as Nicholas Negroponte (1995) and Bill Gates (1995), published a stream of provocative, and very influential, accounts. Building on long-established suggestions that new technologies would overcome the need for spatial proximity, these authors implied that the spread of speed-of-light, digital exchanges, and the explosion of digital domains, would, necessarily, imply some catastrophic collapse for cities. It is worth looking back at some examples of such predictions:

- *1964*: The media theorist Marshall McLuhan suggests that the coming of the 'global village' will inevitably mean that 'the city as a form of major dimensions must inevitably dissolve like the fading shot in a movie'.
- *1964*: The urban geographer Melvin Webber argues that, in the near future, 'for the first time in history, it might be possible to locate on a mountain top and to maintain intimate, real-time and realistic contact with business and other societies. All persons tapped into the global communications network would have ties approximating those used in a given metropolitan region'.
- *1987*: In an article titled 'The vanishing city', the geographer Anthony Pascal suggests that 'the era of the computer and the communication satellite is inhospitable to the high density city. With the passage of time [will come] spatial regularity; the urban system converges on, even if never quite attains, complete areal uniformity'.
- *1991*: The US futurists Naisbitt and Aburdene state that 'in many ways, if cities did not exist, it now would not be necessary to invent them'.
- *1993*: The influential French philosopher Paul Virilio observes that the diffusion of ICTs into cities will mean that 'the city of the past slowly becomes a paradoxical agglomeration in which relations of immediate proximity give way to interrelationships over distance'.
- *1995*: Nicholas Negroponte, Director of MIT's Media Lab, asserts that 'digital living will include less and less dependence upon being in a specific place at a specific time, and the *transmission of place* itself will start to become possible' (emphasis added). Soon, he believed, homes would be equipped with 'electronic windows' allowing him, from Boston, to 'see the Alps, hear the cowbells, and smell the (digital) manure in summer'. Thus, whilst being in Boston, he would, in a way, be 'very much in Switzerland'.
- *1996*: The architecture critic Shafraaz Kaba poses the question: 'Why would you want to drive for an hour, get stuck in traffic, and be scolded by your boss, when work is a few keystrokes away from the comfort of your home-office?' 'Why', he asked, 'even *build* in reality? If the sensations of space provokes the same emotion in virtual reality as they do in real life, why go through the expense of construction, building and maintenance? Nothing degrades in cyberspace!' (original emphasis).
- *1997*: Martin Pawley, an influential British architecture critic, pronounces that, 'in urban terms, once time has become instantaneous, space becomes unnecessary. In a spaceless city, the whole population might require no more than the 30-atom diameter light beam of an optical computer system'.
- *1999*: The US Romm report, an influential exploration of the links between the Internet and global warming, argues that the Internet has 'the ability to turn retail buildings into web sites and to turn warehouses into better supply chain software, to dematerialize paper and CDs into electrons and trucks into fibre' (9).
- *2000*: The British writer Charles Leadbeater pronounces that 'at the moment it is as if we occupy two worlds at once, especially in our congested cities: the physical world of clogged roads, which is inefficient, slow moving, rigid, and the immaterial world of computers and communications, in which we can work at the touch of a button. The ease of use and responsiveness of the new, immaterial economy will make us increasingly frustrated with our experience of the cumbersome old economy of physical machines and roads' (7).

The implications of these predictions seemed depressing for those who cherished and valued vibrant cities in all their physical juxtaposition, conflict, unpredictability, and social and cultural diversity. If such scenarios were to be believed, human reliance on urban places, urban infrastructures, transport flows, and even the corporeal presence of human bodies could, and would, ultimately be transcended. After 8,000 years, urban life, with all its messy, material ambivalence, its clashing territorialities, and its associated congestion, poverty, conflict, violence, pollution and social problems, was on the way out. Crudely put, the pure, clean, and limitless realm of (postmodern) 'cyberspace' would rescue us from the dirty, polluted, contested, finite, (modern) city. To believe such authors was to believe that a new, post-urban age was being ushered in which would finally allow the ruralised utopias of the twentieth-century architects like Frank Lloyd Wright to be realised.

'The dazzling light': elements of the post-urban fantasy

This post-urban fantasy had four main elements. These are worth exploring in some depth.

The 'theology of "Cyberspace": Casting Away the "Ballast of Materiality"'

The first element was the discourse surrounding 'cyberspace'. Replacing physical movement in and around the city, according to this discourse, would be the limitless domains of dematerialised cyberspace – the digital domain that results from the convergence of computers with digital telecommunications and new media technologies. Since William Gibson coined the term in his 1984 cyberpunk novel *Neuromancer*, cyberspace has been defined as 'a parallel universe' (Benedikt, 1991) or 'a new kind of space, invisible to our direct senses, a space which might become more important than physical space itself [and which is] layered on top of, within and between the fabric of traditional geographical space' (Batty, 1993, 615–16). Gibson himself famously depicted it as 'a consensual hallucination, a graphic representation of data abstracted from the banks of every computer in the human system' (Gibson, 1994).

This stream of excited, romantic, and often utopian accounts of the ICT-based demise of cities stressed the apparently logical technological basis for this post-urban societal revolution (see Coyne, 1999). Rather then being seen as technologies to be adopted, and shaped, within the fine-grained practices of everyday life, ICTs, and cyberspatial technologies, were cast here as a 'dazzling light' 'shining above everyday concerns' (Haythornthwaite and Wellman, 2002, 4).

This was especially common during the massive boom of 'information age' utopianism that paralleled the rapid growth of the Internet and world wide web between the early 1990s and the year 2000. Here prevailed what Jay Bolter and Richard Grusin have called the 'theology' of cyberspace. This was based on a master narrative which suggested that ICTs transform 'information from something separate and contained within computers to a space we can inhabit' (2000, 180). The world of cyberspace, invariably, was pictured in this discourse as 'a world that exists in the future . . . a hope, an expectation, for future fulfilment' (Coyne, 1995, 154).

A key part of this dream was the desire of commentators like Michael Benedikt who yearned for a future 'with the ballast of materiality cast away' because humans could 'inhabit' a virtual world of pure information (1991, 123). 'Reality will be replaced gradually, piece by piece', predicted the software engineer David Gelerntner in 1991. Instead of 'reality' there would be 'a software imitation' (3). Crucially, the purity of software-constructed 'virtual' reality, and the perceived contamination of the 'real', material, reality of cities, were cast in this discourse as two completely separate realms. Because it was immaterial, infinitely extendable, and an absolute substitution for all forms of physical mobility and corporeal presence, 'cyberspace' was thus seen here as a force for:

> decontaminating the natural and urban landscapes, redeeming them, saving them from the chain-dragging bulldozers of the paper industry, from the diesel smoke of courier and post-office trucks, from jet fuel flames and clogged airports, from billboards, trashy and pretentious architecture, hour-long freeway communities, ticket lines, choked subways . . . from all the inefficiencies, pollution

(chemical and informational), and corruptions attendant to moving information attached to things
. . . across, over and under the vast and bumpy surface of the earth rather than letting it fly free in
the soft hail of electrons that is cyberspace (Benedikt, 1991, 3).

The sense of a technological revolution of astonishing speed only added to this feeling of euphoria,
utopianism, and a pervasive 'technological sublime' (Coyne, 1999). Between the 1980s and late 1990s
computing power was doubling every two years. Microchips were starting to saturate every domain of life.
Soon, it seemed, everything from cars, TVs, cameras, video cameras, to music machines, games machines,
phones, watches, diaries, domestic appliances, toys, clothes and even human bodies would be filled with
ICTs and turned into globally connected digital appliances or terminals. Linking this great web of terminals
together, broadband telecommunications services over fibre optics and global satellite links promised to open
up limitless, transcontinental communications spaces or bandwidths (at negligible costs that were
increasingly independent of distance).

From the late 1990s the explosion of wireless infrastructures, moreover, promised to unhitch computing
and communications from fixed locations. Beyond all this, by the late 1990s the Internet – harnessing and
integrating the power of all three of these phenomena – was the fastest growing communications medium
in human history. It brought in its wake an apparently anti-spatial world where digital streams of information,
data, images and video – manipulated and processed through an infinitely complex global skein of computer
networks which pervaded every domain of contemporary society – seemed to operate like some giant
'nervous system' for the planet. Such technologies were portrayed within end of city discourses as devices
through which human beings could 'have total control or omnipotence, play God, by simulating, mastering,
redefining, manipulating, and controlling space, time, community, thought, and life' (Coyne, 1999, 4).

*From space and time to real-time: the 'death of distance' and the utopianism
of a neoliberal global economy*

The second element of the post-urban dream involved the assumption that ICTs would entail little less than
a generalised collapse of geographical constraints which would undermine cities and urban life. With these
apparent revolutions in the technological basis of society, a second broad group of commentators and
analysts perceived it to be both logical and obvious that the technologies supporting what *The Economist*
(1995) called the 'death of distance' would also threaten the city. The chain of logic through which this would
occur seemed simple – even self-evident. One newer human invention designed to facilitate communication
– ICTs – used speed-of-light flows of information to allow 'real time' interactions to occur. This minimised
the time constraints which had traditionally inhibited human interaction. In so doing, ICTs thus allowed space
constraints to be annihilated.

Meanwhile, another human invention that had developed partly to facilitate communication – the city –
had evolved through for 8,000 years of urban history by concentrating human life in geographical territory.
Through minimising space constraints cities had allowed the time constraints on physical human interaction
to be reduced: even in major metropolitan regions physical, face-to-face contact could be achieved within
a reasonable time (Graham and Marvin, 1996).

With an exponential growth in the capability of ICTs, the former, it was implied, would simply grow in scale
and sophistication to such an extent that they would directly displace, and substitute for, the latter. It followed
that, in the future, the key relationships in the 'digital age' or the 'information society' would not link a person
or a group and their home, their neighbourhood, their city, or their nation. Rather, they would continuously
telescope between the individual and the integrated, planet-wide, digital (and capitalist) civilisation as a
whole.

The finitudes and congestions of the city, it seemed, would only slow down and impede the perfect
transactional fluency of what Bill Gates (1995) termed 'friction-free capitalism' – a global neoliberal
marketplace where individuals engaged seamlessly as consumers with the entire planet based on globally
stretched ICTs (see Herman and Sloop, 2000). Such an assertion was closely integrated into the widespread

neoliberal economic triumphalism and utopianism that accompanied the collapse of communism. Here right-wing economic commentators asserted that ICTs helped to 'open up' the post-communist and developing worlds, so allowing their 'development', based on 'free' trade, 'market forces', and foreign direct investment from the advanced capitalist countries.

Cyberlibertarianism: 'information infrastructures' as an inherently democratic transformation

The third element of the post-urban discourse involved utopian assumptions that ICTs could meld communities, and mediate democratic processes, in ways that would completely replace the political, social, and anthropological roles of bodily interaction in urban places. Such was the mystique attached to new digital technologies between the 1960s and 1990s that purported ICT-mediated shifts away from place and city-based lives were often uncritically assumed to automatically also involve shifts towards more democratic, egalitarian, decentralised and ecologically sensitive societies. That central purveyor of cyberspace rhetoric, *Wired* magazine, for example, proclaimed in their 1996 *Manifesto for the Digital Society* that:

> the Digital Revolution that is sweeping across society is actually a communications revolution which is transforming society. When used by people who understand it, digital technology allows information to be transmitted and transmuted in fundamentally limitless ways. This ability is the basis of economic success around the world. But it offers more than that. It offers the priceless intangibles of friendship, community and understanding. It offers a new democracy dominated neither by vested interests of political parties nor the mob's baying howl. It can narrow the gap that separates capital from labour; it can deepen the bonds between people and planet (1996, 43–4).

From 'virtual communities' of various scales working to reduce the need for face-to-face encounters, through to national and global 'information infrastructures' or 'superhighways', ICTs were widely cast here as tools to (re)install some idealised Athenian or Jeffersonian democracy. With the use of ICTs to support bi-directional and decentralised communications, communities and democratic links could be forged irrespective of geography, resuscitating democracy itself. Michael Heim (1991, 73) even argued that 'the computer network appears as a godsend in providing forums for people in surprising personal proximity . . . without the physical limitations of geography, time zones, or conspicuous social status'. In 1994 the then US Vice President, Al Gore, argued that ICTs would fuse into a 'global information infrastructure'. This, he believed, would 'circle the globe with information superhighways on which all people can travel . . . [It] will be a metaphor for democracy itself'. Gore predicted 'a new Athenian age of democracy forged in the fora the GII will create' (cited in Mattelaart, 1999, 189).

Cyborg bodies: terminal citizens, telepresence and the 'infinite city'

A final component of the post-urban discourse involved a range of libertarian and critical writers who argued that the progressive blending of ICTs with human bodies was revolutionising citizenship and so transcending the human reliance on place. To believe these accounts was to assume that, as society effectively dematerialised and urban activities effectively migrated into the extending domains of virtual interaction, so cities would grow increasingly anachronistic as political and cultural spaces of expression. Instead, they would gradually be complemented, and even replaced, by a fast-expanding universe of virtual realities, virtual communities, digital economies, and unlimited, electronically constructed domains accessible from anywhere at any time.

In a direct challenge to the human experience of place, such domains would increasingly evolve to be experienced immersively through Virtual Reality (VR) technologies. These fuse microprocessors and digital media ever-more seamlessly with the human bodies and nervous systems of their cyborg (cybernetic-organism) hosts. Such 'terminal citizens' could, in a sense, 'leave' their physical bodies ('meatspace'), and their local geographies, 'behind' the arrays of technology they used to create, and inhabit, these simulated, replaceable, and infinitely extensible, virtual worlds. The suggestion of this element of the discourse was that 'in a virtual world you can have instant access to any coordinate in data space. You can be here, there or everywhere, unlike the limited spatially constrained world we usually experience. [We] can immerse ourselves in such spaces so that we are there in ways that mean as much as, and possibly more than, being in a physical place' (Coyne, 1995, 4–6).

From their technological launch-pads into cyberspace and VR, humans could thus be telepresent within infinite arrays of electronic domains, housed within data infrastructures strung out across the planet and through its airwaves. Moreover, through real-time flows of bits and bytes, and blizzards of photons and electrons, they could actually access these domains with near perfect (hyper)mobility and virtually zero friction. Such omnipotent subjects could thus be 'telepresent' on the other side of the planet – engaging in virtual trading, virtual games, virtual tourism, virtual consumption, even virtual sex – more easily than they could physically move their real bodies to the other side of the room, the local shop, or the other side of the local city (see Robins, 1995).

In reviewing such technological potential the architecture critic Shafraaz Kaba, writing in 1996, argued that these logics would inevitably mean that 'the future [would] be a war between the city of bits and the city of atoms' (5). In an economy, culture and society of generalised, online, and speed-of-light interactivity, any activity would be possible from any place and at any time. Everything would be 'one click away' (*ibid.*, 7). In a long-standing dream to escape from the perpetual transport crises afflicting modern urban society, Kaba believed that, eventually, little or no physical movement would be required at all. Here he echoed the futurist James Martin who had even proclaimed in the 1970s energy crisis that telecommunications was a 'substitute for gasoline' (1978, 193).

What future for face-to-face stock exchanges, economies, retailing, socialising, romance, political activity, protest, working, sport, tourism or leisure within and between the physical city, in a world of what BT's futurist Peter Cochrane (1998, 4) termed 'universal information access through the infinite city' made possible through ICTs?

The common theme: assumptions of substitution and technological determinism

Readers should note that this brief summary is a simplified account of thirty years' worth of depictions that ICTs and cities were effectively polar opposite domains 'at war' with each other. Far from being a single, unified discourse, these accounts varied significantly. They derived from a wide range of sources: futurism, urban studies, cyberpunk science fiction, media and critical theory, and the business press. They drew on a complex range of philosophical and theoretical positions and antecedents (see Coyne, 1999). Most were positive, excited and breathless, fuelled by a negative attitude to the supposed perils of the industrial city and a positive, pastoral idea of how technology could fuel ruralisation, decentralisation and the empowerment of people and social groups. Others (notably the writings of Paul Virilio and cyberpunk science fiction writers) were bleak and pessimistic, and involved critiques of the direction of global political-economic and technological change.

Crude fantasies of a complete substitution of corporeal (i.e. bodily) presence and movement within and between urban places are now less common in the west than during their 1960s–1990s heyday. Nevertheless, excited hyperbole stressing the transcendent mobilities possible through ICTs is still a dominant part of the discourses about 'information' or 'network societies'. 'At the turn of the twenty-first century', the anthropologist Caren Kaplan has argued, it is still the case that:

> the rhetoric of cyberspace and information technologies relies heavily on a hyperbole of unlimited power through disembodied mobility. Whether we read theorizations of new cityscapes published by university presses or advertisements in magazines for wireless Internet connections, references

to boundless space, unfettered mobility, and speedy transfers abound . . . More and more in this context, the concept of a person or of human beings appears to depend on the attenuated possibilities of cyberspace (2002, 34).

All the depictions analysed above had one thing in common: they were based on a general, and uncritical, use of the metaphor that cities would simply be 'impacted' by new communications technologies in the same way as planets are impacted by asteroids. In all the above accounts ICT technologies were portrayed as arriving from 'out there', as a transformative 'force' or 'shock' hitting the fabric of urban society.

These accounts reflected the long-standing tendencies within technological culture, especially in the west, to regard new technologies and infrastructures as 'largely unproblematic and even autonomous in shaping the life and form of urban areas' (Aibar and Bijker, 1997). In all of these depictions ICTs were portrayed in general as some disembodied, external 'wave' of change which, quite literally, transformed cities and urban life single-handed. Alvin Toffler (1980) even used the metaphor 'Third Wave' to capture the latest ICT-based societal revolution, following on from the First (agricultural) and Second (industrial) revolutions.

This view reflects the classic, deterministic view of the role of new communications and transport innovations in which 'changes in technology lead inexorably to changes in urban form' and urban life (Hodge, 1990, 87). In this view, new networked infrastructures like the Internet become little more than 'progenitor[s] of new urban geometries' (*ibid.*, 87). In such rhetoric, a simple, linear, cause-and-effect chain is assumed where the technology itself is seen as the direct causal agent of urban change.

Commonly, this intellectual device is quickly translated into the broader use of technological and infrastructural depictions of historical urban 'ages': from the 'hydraulic civilisations' of the first urban centres in Mesopotamia, through to the 'steam', 'electric', 'auto', the 'nuclear', and 'information age' metropolises of the past three centuries.

Beyond the 'dazzling light': from dreams of transcendence to analysing the remediation of everyday urban life

The problem with such approaches, however, is that they tend to portray technologies as having overwhelming power in ushering in simple and discrete societal shifts which seem to amount to some naturalistic process of urban, and societal, evolution. In the process, the parallels between historical periods tend to be underplayed (or ignored completely). The tendencies of newer technologies to overlay, and subtly combine with (rather than replace) earlier ones is often forgotten. And the forms and processes of city life tend to be simply read off as the deterministic result of the intrinsic nature of the new generation of technology.

Within deterministic end of city discourses the urban 'impacts' of cyberspace and ICTs were assumed to follow inevitably from the way such technologies compressed the time and space constraints that had previously shackled human life. Because they operated at the general, speculative level, many of these scenarios also implied that all cities (say London, Lima, Hong Kong, Johannesburg, Amsterdam, Bangalore and Seattle) would somehow be 'impacted' by ICTs in similar ways, irrespective of their starkly contrasting geographical positions, sociologies, economies, cultures, political situations and histories.

Such so-called 'technological determinism' – the reading off of the universal effects of technologies from their intrinsic properties – is attractive. It helps to create powerful scenarios, clear stories, and tangible predictions. It makes good copy for the media and creates glamorous notions of a new electronic 'frontier' awaiting colonisation by those hardy pioneers that are up to the task. Such rhetoric seems to tally well with the early periods of the diffusion of a new set of technologies and practices. This is because, at this stage, people literally see the social world, and the urban landscape around them, gradually filling up over time with new equipment, new artefacts, new practices, and new ways of organising – as if from nowhere. Above all, technological determinism and the metaphors of impact accord well with the dominant experience in the west, where, as Stephen Hill (1988) put it, 'the pervasive experience of technology is one of apparent inevitability'.

This pervasive reliance on technological determinism and cartoonish end of city visions has actually worked to obscure the complex relationships between new communications and information technologies and cities and urban life that have emerged as ICTs have diffused to be embedded in real everyday lives and practices. Simple, binary allegations that new technologies are ushering in a single, separate, virtual domain – which challenges the future of the city – have tended to undermine more subtle and sophisticated thinking about the ways in which ICTs actually relate to the spaces and places bound up with human territorial life.

There is thus a desperate need, as sociologists Haythornthwaite and Wellman (2002, 4–5) suggest, to move beyond generalised and deterministic discourses about the 'impacts' of 'cyberspace' on society to look in rich empirical detail at the complex ways in which ICT technologies are being used in real ways, in real urban regions, in the real world. As they suggest 'the reality of the Internet is more important than the dazzle' (*ibid.*). Indeed, some now argue that many ICT-related shifts in everyday urban life are actually more intriguing than could possibly have been predicted in the generalised scenarios pumped out between the 1960s and 1990s. The architecture critic Keller Easterling (2003, 3), for example, argues that 'the explosion of changes to the world's markets, cities, and means of shipping and communication [are] far more strange and unpredictable than any of the swaggering futurology scenarios'.

The media theorists Bolter and Grusin (1999) have shown that the whole raft of current ICT innovations are not being used in ways that are divorced from the use of existing media, means of communication, and material practices in places. Rather, they are allowing for the subtle 'remediation' of TV, newspapers, magazines, radio, telephones, publishing, books, art, video, photography, face-to-face communication, and the social and anthropological experience, and construction, of place. This is happening as established practices subtly combine with, rather than disappear through, socially constructed technological potentials.

ICTs, far from being a complete and revolutionary break with the past, thus maintain many intimate connections with old media, old technologies, old practices, and old (electromechanical) infrastructures and spaces (telephone systems, broadcasting systems, electricity systems, highway systems, streets, airline systems, logistics systems). The so-called 'information age', then, is best considered not as a revolution but as a complex and subtle amalgam of new technologies and media fused on to, and 'remediating' old ones (Bolter and Grusin, 2000, 183). There are a great many more continuities and synergies than many would have us believe. We are not experiencing some wholesale, discrete, break with the urban past ushered in by the 'impacts' of new technology. Rather, we are experiencing a complex and infinitely diverse range of transformations where new and old practices and media technologies become mutually linked and fused in an ongoing blizzard of change. As Bolter and Grusin (2000, 183) suggest, 'cyberspace'

is very much a part of our contemporary world and . . . it is constituted through a series of remediations. As a digital network, cyberspace remediates the electric communications networks of the past 150 years, the telegraph and the telephone; as virtual reality, it remediates the visual space of painting, film, and television; and as social space, it remediates such historical places as cities and parks and such 'nonplaces' as theme parks and shopping malls. Like other contemporary telemediated spaces, cyberspace refashions and extends earlier media, which are themselves embedded in material and social environments.

The weaknesses of the anything–anywhere–anytime dream

Within the overall failure of resorting to technological determinism it is possible to identify six major weaknesses of ICT-based end of city discourses. It is worth exploring these in a little more detail here. We need to do this because addressing these failings provides the foundation for the comprehensive analyses of the relationships between cities and ICTs that follow in the rest of this book.

Ignoring global urbanisation and mobility trends

The first problem is obvious: deterministic end of city accounts are simply, empirically wrong. As Table 1 demonstrates, these accounts fly in the face of a contemporary reality marked by the greatest processes of urbanisation in human history. They fail to explain the rapid rates of economic, demographic and physical growth not just of the 'megacities' of the global south but of many high-tech and old industrial cities of the global north, many of which have experienced a startling economic renaissance in the last twenty or so years (at least in certain districts).

A few figures give a flavour of the scale of contemporary urbanisation processes on our planet. Between 1900 and 2000 the world's urban population grew from 10 per cent to 50 per cent of the global population. Bombay's population has quadrupled in thirty years (1972–2002). In one hour sixty people move in to Manila, forty-seven move to Delhi, twenty-one to Lagos, twelve to London, nine to New York and two to Paris. By 2025 there are likely to be 5 billion urban dwellers on Earth (two-thirds of whom will live in 'developing' nations) (all figures from Koolhaas *et al.*, 2001, 1–6).

End of city discourses studiously ignore the fact that urbanisation and growing ICT use are actually going hand in hand (see Table 1). City spaces are dominant hubs which shape and configure all aspects of global ICT infrastructure investment and global Internet and telephone traffic. Cities and urban regions are actually massively dominant in driving demand for land and mobile telephony, the Internet, and new media technologies. This dominance shows no sign whatsoever of slackening. Again, a few figures are salutary here. Eighty per cent of telecommunications investment within France goes into Paris (Graham and Marvin, 1996). In 1998, 25 per cent of all of the UK's international telecommunications traffic was funnelled into a single optic fibre network in central London run by WorldCom that was only 230 kilometers long (Graham, 1999). As Anthony Townsend reports (p. 144 of this book) more optic fibre underlies the island of Manhattan than is threaded across the whole of Africa. And, in 2003, the Internet geographer Matthew Zook showed that only five metropolitan regions – New York, Los Angeles, San Francisco, London and Washington DC – accounted for 17.7 per cent of the world's total Internet domains. These are the familiar .coms, .nets, .orgs etc. that are widely portrayed as being 'placeless' and 'without a geography' (see http://www.zookNIC.com/Domains/index.html). It is, in short, no exaggeration to say that cities of all types across the world are the powerhouses for all aspects of ICT innovation, research, application, and representation.

End of city perspectives also ignore the rapid rise in all forms of physical mobility at all geographical scales. These have actually grown in *parallel* with the application of ICTs; they have not been replaced by ICTs. Evidence for this is difficult to avoid: the growing gridlock of city streets; the staggering rise of global automobile ownership; the exponential growth of airline travel; intensifying levels of consumer and business tourism; growing energy consumption; and a general acceleration of flows of goods, commodities and raw materials at all scales across the world.

Ours is certainly *not* a world of reducing physical mobility! Between 1950 and 1980 global car ownership rose from 50 to 350 million. By 2001 there were 500 million cars; by 2030 there are likely to be over 1 billion (Bell, 2001). Between 1800 and 2000 transport and electronic communications flows in France both rose *in parallel* by a factor of more than a thousand (Graham and Marvin, 1996, 22). In the 1990s international telephone traffic and airline traffic grew in parallel (in the EU by 10 per cent and 3–4 per cent respectively per year). Finally, both fast transport and fast telecommunications networks are tending to extend their reach in parallel, to interconnect the same cities and parts of cities. For example, in the 1990s alone the motorway network in the EU grew by more than 25 per cent. At the same time Europe's most economically dynamic cities, and the corridors between them, were invisibly filled by millions of kilometers of new optic fibre, handling unprecedented information exchanges.

This is not to deny that some real substitution of face-to-face contact and physical movement by ICT-based exchange can occur. This happens with online and telephone banking, some forms of teleworking, and the use of DVDs in place of cinema visits, for example (although DVDs still need to be made, shipped and bought). But any relative substitution has tended to be dwarfed by a much deeper, absolute intensification of physical mobilities. And much of this physical mobility is actually generated, accommodated, managed, or made more bearable by ICT-based systems and connections. ICTs allow many more aircraft, trains and cars to be squeezed into the available physical spaces of hard infrastructure networks. They stimulate the growth of tourist and business trips by extending people's and organisations' perceptual horizons. And the saturation of the world's cities and urban corridors by gridlocked traffic would be a great deal more problematic without mobile phones!

So we can reject the idea of some ICT-mediated death of cities at a simple, empirical level. At the same time, however, cities are being changed radically by current processes of economic, political, social, and technological change. Global urbanisation and social, economic, demographic, and ICT and mobility trends are intimately involved in the complex remaking of urban areas. Technologies like the Internet, mobile phones, webcams and video surveillance are, along with massive demographic, structural, economic and cultural shifts, helping to make cities look – and feel – less and less like the classical western idea of what a 'city' might actually be. Today's multicentred megacities and international urban corridors – which can house 30–60 million people across hundred-mile stretches of apparently chaotic territory – have little resemblance to the classical cities of medieval Europe that were so often idealised in urban thinking and urban theory up until the 1960s. Some consequently label our age one of urbanisation without cities – at least in the classical sense (Castells, 1996; see Manuel Castells, p. 82). 'Even as the data gather and diversify', writes French philosopher Nadia Tazi:

> as the statistics run wild in the face of the flight from the countryside, as certain metropolitan centres grow and others become more complex, the notion of the city itself empties out and dilutes . . . It is still accepted as a noun, a substance, although it no longer designate a merely physical territory, nor only the indefinite play of virtualities and digital variabilities.

In an internationally integrating, urban, and increasingly ICT-mediated society, it is therefore now clear that the word 'city' becomes more diffuse just as the condition of all spaces within societies becomes more generally urban.

Debunking the 'virtual' myth: ignoring the material geographies of ICTs

The second problem with ICT-based end of city visions is that they ignore the very material realities that make the supposedly 'virtual' realms of 'cyberspace' possible. 'Cyberspaces' do not exist on their own; the many supposedly 'virtual' domains and worlds are brought into existence, and constantly facilitated, by massive, globally extended sets of material systems and infrastructures. In their obsession with the ethereal worlds of cyberspace – with the blizzards of electrons, photons, and bits and bytes on screens – end of city commentators have consistently ignored the fact that it is real wires, real fibres, real ducts, real leeways, real satellite stations, real mobile towers, real web servers, and – not to be ignored – real electricity systems that make all of this possible. All these are physically embedded and located in real places. They are expensive. They are profoundly material. They sharply condition the functionality of digitally mediated encounters (contrast an always-on broadband Internet computer with the scratchy and slow service available by dialing up over a phone line). Finally, ICTs have very real geographies in the traditional sense (i.e. they are in some places and not in others).

Because the material bases for cyberspace are usually invisible they tend only to be noticed when they collapse or fail through wars, terrorist attack, natural disasters, or technical failure. Even when they are visible – as with mobile phone masts and telecom towers – they are often disguised so as not to be 'unsightly' (in the UK many mobile towers are camouflaged as fake trees!). Thus, in contrast to the vast and land-hungry infrastructural edifices that sustain transport, electricity and water flows, the myth that cyberspace is an ethereal and immaterial realm continues to retain power.

It is also now very very clear that the geographical patterns of the material bases and investment patterns of cyberspace are not spreading across the world equally. They are extremely, perhaps increasingly, uneven. In a world of increasing economic and infrastructural liberalisation, the giant transnational media and

Table 1 The urban-digital century: indicative statistics, 2000–2030

Area (with population of biggest city, 2000/2015, in millions)	Urbanisation trends			Landline telephone access trends	Internet trends					Mobile phone trends	Intl phone traffic
	% population urban, 2000	% population urban, 2015 (est.)	% population urban, 2030 (est.)	Phone lines per 1,000 people, 1991/1998	% population using the Internet, 2001	Number of people using the Internet, 2001 (month)	Number of people using the Internet, 2002 (month – where available)	% growth in international Internet bandwidth, 2000–1	Location of top 5 hub cities for international Internet traffic (rank and millions of bits per second Mbps), 2001	Number of mobile phones in use (Jan. 2003)	Annual rate of growth in intl phone traffic (%)
Africa (Cairo 10.5/13.7)	37.9	46.5	54.5	–	–	6,310,000 (Dec.)	–	89.6	–	34,000,000	10.1
Kenya (Nairobi 2.3/3.7)	33.1	44.5	54.0	8/12	1.61	500,000 (Dec.)	–	–	–	–	–
S. Africa (Cape Town 2.9/3.4)	50.4	56.3	64.3	87/115	7.03	3,068,000 (Dec.)	–	–	–	–	–
Asia/Pacific (Tokyo 26.4/26.4)	36.7	44.7	53.4	–	–	187,240,000 (Dec.)	–	129.3	–	310,000,000	14.1
India (Mumbai/Bombay 18/26.1)	28.4	35.9	45.8	6/22	0.49	5,000,000 (Dec.)	–	–	–	5,700,000	–
China (Shanghai 12.8/14.5)	32.1	40.7	50.3	6/70	2.08	26,500,000 (Dec.)	45,800,000 (July)	–	–	200,000,000	–
Japan (Tokyo 26.4/26.4)	78.8	81.5	84.8	441/503	37.2	47,080,000 (Dec.)	56,000,000 (June)	–	–	–	–
Iran (Teheran 7.2/8.7)	61.6	68.8	74.6	40/112	0.38	250,000 (Dec.)	–	–	–	–	–
Australia (Sydney 3.6/3.9)	84.7	86.0	88.5	456/512	52.49	10,060,000 (Aug.)	–	–	–	–	–

Region (city size)										
Europe (Paris 9.6/9.7)	74.8	78.6	82.6	–	–	190,910,000 (Dec.)	–	109.8	1 London (237,389) 2 Paris (179,000) 4 Amsterdam (170,000) 5 Frankfurt (160,000)	320,000,000	11.6
Bulgaria (Sofia 1.2/1.2)	69.6	74.5	79.3	242/329	7.59	585,000 (April)	–	–	–	–	–
UK (London 7.64/7.64)	89.5	90.8	92.4	441/557	55.32	33,000,000 (June)	34,300,000 (Sept.)	–	–	–	–
Latin America (São Paulo 17.7/20.4)	75.3	79.9	83.2	–	–	33,350,000	–	479.2	–	–	10.1
Brazil (São Paulo 17.7/20.4)	81.3	86.5	88.9	121/185	6.84	11,940,000 (Sept.)	13,980,000 (Sept.)	–	–	–	–
Argentina (Buenos Aires 12.6/14.07)	89.9	92.6	93.9	95/203	10.38	3,880,000 (July)	–	–	–	–	–
N. America (New York 16.6/17.4)	77.2	80.9	84.4	–	–	182,670,000 (Dec.)	–	144.3	3 New York (173,000)	–	–
USA (New York 16.6/17.4)	77.2	81.0	84.5	545/661	59.75	166,140,000 (Aug.)	165,750,000 (April)	–	–	140,000,000	0.8
Canada (Toronto 4.7/5.3)	77.1	79.9	83.6	565/634	45.71	14,440,000 (July)	16,990,000 (Feb)	–	–	–	–
World Tokyo (26.4/26.4)	47	53.4	60.3	–	–	605,600,000	–	174	–	1,300,000,000	–

Sources: Urban, city size and telepehone access data from UNCHS (2001).
Internet data from Nua: http://www.nua.ie/surveys/how_many_online/index.html
Mobile phone data from GSM group http://www.gsmgroup.net/mobile-phone-statistics.htm
Telephone traffic and internet bandwidth and hub city data from Telegeography Inc. http://www.telegeography.com/press/releases/2002/19-nov-2002.html

infrastructure firms that build and control the material bases for cyberspace tend to concentrate their investments where the main markets are – in major cities, urban regions, and metropolitan corridors. For example, even now, large swathes of the world's poorest countries have little telecommunications infrastructure to speak of. A third of the world's population has yet to make a telephone call (let alone log on to the net). And, even in advanced industrial cities, the spaces where one is able to access the new premium ICT services – such as broadband, third-generation mobile, or wireless Internet – are often still limited to the 'premium bubbles' of connectivity located in downtown cores, affluent suburbs, airports or university campuses (Graham, 2000). In the UK, for example, the new 'Wi-Fi' services, which allow people with laptops to connect wirelessly to the Internet at fast speeds, are, initially at least, only going to be available in airport lounges and selected chains of branded coffee shops and petrol stations. To further highlight how local these geographies of premium access can be a new breed of Internet 'crackers' now walk the streets of the world's finance capitals, laptops in hand, to find points from which they can enter the local, broadband wireless networks of corporations. Once connected these street-level activists mark up their boundaries with chalk so that these hidden infrastructures can be publicly consumed.

All this is the result of simple economics. In liberalised marketplaces infrastructure firms, not surprisingly, funnel their investment into 'cherry picking' high demand, low risk, and low cost areas to try to maximise profits. Indeed, in these post-dot.com 'crunch' days, many just try to pay off the huge debts necessary to build private infrastructure in the first place, without falling into bankruptcy.

In general, the result of these processes is that cities, and especially the global and internationally oriented high-tech spaces within the urban regions of the global north and south, benefit from enormous, and intensifying, concentrations of ICT infrastructures. They enjoy cut-throat competition between multiple providers. And they have cutting-edge access to whole suites of the latest highly advanced and cost-effective services which are always at least a few steps ahead of more remote areas (as well as poorer, adjacent neighbourhoods).

These advantages add significantly to the other benefits that such cities enjoy in services, advanced skills, concentrations of market, and the availability of property, other infrastructure and 'soft' cultural and lifestyle attractions. It also limits the degree to which industries that rely heavily of ICTs can simply upsticks and shift to remote areas (which often suffer from poor, outdated, expensive, and monopolistic ICT services and infrastructures as well as poor physical transportation and small pools of labour).

As a result, even when firms do decentralise from major cities – to exploit the capacity of ICTs to mediate exchanges through the relocation of call centres for example – they tend to move to other, more peripheral, cities, rather than to remote or rural areas. This is in order to access large pools of labour, infrastructure, specialised services, cultural attractions, transportation and logistics links, and property. So the spread of activities from major metropolitan regions to other urban regions tends to be much more significant than the growth of bucolic 'electronic cottages' in truly remote environments.

Above all, we should remember that capitalism, even when mediated by ICTs, remains an intrinsically material mode of production. Contemporary capitalism is based on extended geographies of physical flow and complex geographical divisions of labour which are stretched out across the world. The 'dematerialisation' thesis is, therefore, a dangerous red herring (see Bernadini and Galli, 1993; Marvin, 1997). Even relatively prosaic examples reveal its lie. When offices give email to workers, for example, on average their paper consumption goes *up* by 40 per cent as people print more and more to keep track of the accelerated flows of information (Brooke, 2001). 'New technologies appear to promise ever-increasing degrees of disembodiment or detachment', suggests Caren Kaplan (2002, 34). But, in reality:

> They are as embedded in material relations as any other practices. They require hard industries as well as light ones. In addition to the bright and mobile world of designers and users, human hands build the machines in factories that are located in specific places that are regulated by particular political and economic practices. Thus, in the production of the machinery and materials of cyberspace, another form of mobility can be discerned, that of labor in this moment of globalisation.

Less obviously, we should also note that the frenzied cycle of innovation and obsolescence in electronic technologies – designed to be consumed and rapidly disposed of in the world's high-tech urban regions – produces its own (usually ignored) material mobilities and divisions of labour. A continuous and massive transfer is occurring of unwanted and often virtually new electronic equipment which flows 'downhill on an economic path of least resistance' from the major cities of the global north to the processing spaces of Asia, Africa and Latin America (Shabi, 2002, 36). In the semiurban area of Guiyu, in the Guandong province of China, for example, over 100,000 men, women and children earn US$1.50 a day breaking discarded servers, computers, mobiles and other electronic equipment by hand to extract valuable steel, aluminium, copper aluminium, manganese and gold. As Rachel Shabi suggests, these hidden urban spaces are 'grotesque sci-fi fusions of technology and deprivation'. In such places terrible health and contamination problems reveal the true extent of the lies which surround the dreams of dematerialisation, a global 'Athenian' democracy, and egalitarian ICT-based mobilities. 'This', she writes, 'is the gloomy underside of our glorious technology' (*ibid.*).

Overgeneralisation and the weaknesses of 'impact' metaphors

The third problem with the utopian and anti-urban predictions of end of city visionaries is that they massively over-generalise. They imply that all experiences are the same anywhere and that ICTs relate to all cities in the same way at all times. This ignores an increasingly sophisticated body of theoretical and empirical research which suggests that, whilst ICTs do have important implications for cities, the relationships are much more subtle, complex and contingent than that endlessly repeated world of simple, deterministic, substitution, total dematerialisation, and a wholesale stampede of urban life into the clean and infinitely extendable domains of 'cyberspace'.

ICTs are helping to facilitate significant reconfigurations in the geography, mobility patterns and social, economic and cultural dynamics of cities, and in the ways in which urban life is represented, lived and managed. But these changes are subtle. They are often counterintuitive. And they tend to involve many other processes of change.

In short, ICTs do not have 'impacts' on cities in and of themselves. Rather, complex social processes and practices occur which shape the nature, use and application of ICTs in urban (and non-urban) contexts in a wide variety of contingent ways. ICT use subtly combines with, rather than replaces, the traditional urban experience which centres on the placement and movement of the human body (or bodies) in the room, the building, the street, the transport network, the neighbourhood or the city.

The use and experience of ICTs is therefore associated with a myriad of urban changes in different spaces, times and contexts. Indeed, one ICT artefact – say an Internet computer – can itself be used to sustain a wide range of uses by a range of different people at different times of the day and in different physical situations. Each may entail different relations between ICT-based exchange and the spaces, times and social worlds of the places that provide the context for its use.

All this means that generalisation about ICTs and cities is hazardous to say the least (see Munt, 2001). The ways in which places become enmeshed into geographically and temporally stretched electronic networks like the Internet form an extraordinarily diverse, contingent process. And, whilst there certainly are a growing range of transnational and even 'global' interactions on the Internet, we must also remember that many such relationships are profoundly local.

A wide range of relations are thus likely to exist between ICTs and urban structures, forms, landscapes, experiences and the cultural particularities of different urban spaces and times. The communications scholar Bolter speculates, for example, that:

> perhaps the Japanese will construct cyberspace as an extension of their dense urban corridors. On the other hand, people can live in the suburbs and participate in cyberspace from their homes, as many Americans do now. Or, as Americans do, they can commute between one cyberspace location in the workplace (a corporate communications system) and another in their homes (America

Online). Thus cyberspace can be a reflection of the American suburbs and exurbs, the Japanese megacities, or the European combination of large and medium-sized cities. Cyberspace need not be the uniform entity suggested by the current metaphor popular in the United States, the information superhighway (Bolter, 1995, 2).

Theoretical laxity and the dangers of binary thinking: overemphasising ICTs, underemphasising place

The fourth problem with deterministic end of city predictions is that they tend to dramatically overestimate the capabilities of ICTs to mediate human relationships. At the same time, they dramatically underestimate the complexity, richness, and the continuing anthropological and cultural power generated by co-present human bodies in places. The communications theorist Sawhney criticises 'the very transmission-oriented view of human communication' in ICT and city debates. Here 'the purpose of human communication is reduced to transfer of information and the coordination of human activity. The ritual or the communal aspect of human communication is almost totally neglected' (Sawhney, 1996, 309).

In such transmission-oriented accounts more information or more bandwidth is always equated with more knowledge, more mutual understanding and more wisdom. Squeezing more bits down a wire, or over a wireless system, is assumed to lead to deeper, more satisfying relationships. And the crucial difference between formal knowledge – which tends to be more easily accessible through ICT networks – and tacit knowledge – which is often developed and divulged much more in trustful, face-to-face encounters – tends to be ignored or underplayed.

Above all, whilst there is no doubt that ICTs can act as 'prostheses' to extend human actions, identities and communities in time and space, it does not follow that the human self is 'released from the fixed location of the body, built environment or nation'. Rather, 'the self is always somewhere, always located in some sense in some place, and cannot be totally unhoused' (Kaplan, 2002, 34).

Technologically determinist commentators thus fail to appreciate the complex social, cultural and economic ecologies, and the sheer resilience, of the corporeal lives of people in places. They fail to see that many of these cannot simply be substituted by ICTs no matter how broadband, 3D, or immersive the substitutes. Quite the reverse, in fact. For the social construction and experience of the body and space and place actually ground and contextualise applications and uses of new technologies. 'The urban world networked by [Bill] Gates' technologies strung out on the wire', suggests cultural geographer Denis Cosgrove, 'is not disconnected, abstract, inhuman; it is bound in the places and times of actual lives, into human existences that are as connected, sensuous and personal as they ever have been' (Cosgrove, 1996, 1495).

In fact, as ICTs diffuse more widely and become more taken for granted and ubiquitous it is increasingly apparent, at least in richer cities, that they are being used to subtly reconfigure the place-based worlds and mobilities of everyday urban life (see Mulder, 2002). ICT interactions have now moved from the status of novelty to rapidly diffuse into all walks of life. In many contexts they are now increasingly ubiquitous – even banal. In a sense, then, ICTs have now 'produced the ordinary' in the sense that they are woven so completely into the fabric of everyday urban life that they become more and more ignored (Amin and Thrift, 2002, 103, original emphasis).

Rather than treat them as meteoric impactors, then, it is becoming more and more common to simply take ICTs for granted as part of the fabric of urban life. Today, ICT traffic and infrastructure *is* the contemporary city; nowadays, much of the city *is* ICTs (see Latour and Hermant, 1998). Separating them today makes no sense. One is not 'virtual' whilst the other is 'real'. Rather, cities, bodies, physical flows and ICT exchanges are socially shaped in combination, in parallel – together. They are combined ensembles. They recursively interact and mutually constitute each other. This mutual constitution is always provisional and contingent; it always occurs dynamically, in practice, over time, in a vast variety of ways. The British geographer Kevin Robins believes that 'through the development of new technologies, we are, indeed, more and more open to experiences of de-realization and de-localization'. 'But', he argues, 'we continue to have physical and localized existences.' We must therefore consider what Kevin Robins calls our fused 'state of suspension' between these two conditions (Robins, 1995, 153).

Cities and ICTs are thus fused into 'socio-technical' and 'hybrid' complexes. Many examples are relevant here. Vast and unknowable domains of software now mediate an increasing proportion of people's interactions with the city, with each other, and with all types of services. Internet sites called 'virtual cities' gather local services and information to help people articulate more effectively with the places in which they live. Tiny 'smart' radio-linked microchips are now being integrated into even low value products as means of controlling logistics flows, supporting surveillance, and preventing shoplifting. Countless millions of webcams allow an extending galaxy of places to be represented and 'consumed' from afar at all geographical scales in (near) real time. E-commerce sites, as well as 'stretching' economic transactions to transnational scales, are being used to change the ways in which people shop for food and consume local products and services. (This reflects the fact that, even in the USA – that most mobile of societies – urban economies are still overwhelmingly local and 80 per cent of transactions still occur within twenty miles of people's homes.)

In addition, ICTs are being used to reconfigure local public and private transport systems (as in the electronic road pricing of city centres). They are supporting a reconfiguration of automobiles and road systems. Mobile phones and Personal Digital Assistants (PDAs) are being used to remodel users' experiences of city services and urban public spaces. New third-generation mobiles (G3) even mean that users can receive running commentaries on local services and spaces, as they move around cities, through what are known as 'location based services' (clearly this also has the potential to clog the systems with masses of unwanted spam). Technologies like closed circuit TV and local radio are allowing for subtle changes in the regulation and management of specific public spaces in cities. ICTs, meanwhile, are facilitating radical changes in the delivery of local public services and in the organisation of local democracy.

It is also clear that on- and off-line social interactions and information exchanges now blend together in very complex ways to subtly change cultural geographies of community, friendship, identity and kinship. As with the use of ICTs to sustain a myriad of cultural diasporas, such complex and multi-scaled social connections can be both local and transnational at the same time. ICTs may subtly alter the ways in which local social relations knit together to shape places. A social experiment in Helsinki in 2002 known as FLIRT, for example, allowed single people to leave text messages about themselves in the local 'cells' that make mobile phone systems work. Other single people wandering into these cells received any messages available. 'Participants would disseminate information about themselves gradually, with messages becoming more personal and revealing as time went by. For those involved, the experiment turned Helsinki into a citywide chat room' (Wilson, 2003, 65).

New ICT innovations are also changing the way cities are being managed and built. Geographically referenced ICTs like Global Positioning Systems (GPSs) and Geographical Information Systems (GISs) are supporting a reordering of time and space within and between cities. This is happening as flows and territory are reorganised at different geographical scales, from the precise mapping of social geographies to support location decision-making and 'red lining' amongst banks, insurers and retailers, through to new bar-coded logistics systems and new techniques of urban killing and warfare (see Pickles, 1995). Such ICT-based logistics systems 'are a new infrastructure of the city, providing unprecedented synchronization and organisation in seeming formlessness'. They 'reorganise the pattern of the city and allow its destabilisation. [They] are the mechanism by which the virtual establishes the logic of the real' (Hosaya and Schaefer, 2001, 157).

Finally, despite massive changes in the nature of cities and urban life, it is important to stress that the very idea of the 'city' is proving a resilient one in ICT-mediated culture and everyday life (Bleecker, 1994). As games like 'SimCity' and the proliferation of 'city' or 'town' style interfaces on web sites demonstrate, the idea of the city acts as a powerful metaphor for organising the experience and use of the new media themselves. ICTs thus change our idea of the city just as our idea of the city changes the way we shape and consume new media.

The 'dot.con': glossy ideologies of the 'information age' as a camouflage for neoliberalism and uneven development

The penultimate problem with end of the city scenarios is that they – at least in their utopian incarnations – tend to promote simplistic, biased and glossy ideologies of the so-called 'information age'. Many of these have been relentlessly used to improve the public relations profiles of governments and digital media and telecommunications firms. Over the past three decades, upbeat depictions of ICTs have also been commonly used to generate public funding and subsidies for media transnationals. This has been done as local, national and transnational policy-makers struggle to make their jurisdictions 'competitive' and symbolically 'high-tech' in the 'global information society'.

But such utopian ICT discourses now ring decidedly hollow. The collective image of ICTs now appears ambivalent at best, and deeply problematic at worst. In the last decade the hyperbole has now been tarnished by a succession of major societal crises as ICT-mediated models of capitalist restructuring have led to severe problems. Consider the Asian financial crisis of 1998–9 (partly induced by electronic currency speculations); the spectacular collapse of the dot.com stock market 'bubble' between 2000 and 2002; the numerous fraud scandals that have bankrupted many of the supposed 'stars' of the new telecom world since 2000; and the financial meltdown in Argentina in the same period.

From the perspective of the international recession that has afflicted large parts of the world since the dot.com crunch, it is increasingly clear that the late 1990s was a period within which ICT industries, with their carefully fuelled utopian discourses, perpetuated little less than a giant 'dot.con' (see Cassidy, 2002). The huge speculative bubble of 1997–2001 showed powerfully how deep the connections were between utopian ideologies of ICTs and the 'information age' and the accelerating swings of speculation and depression in financial and neoliberal capitalism. As Steven Poole (2003, 18) writes of the height of the boom, 'how could priceline.com, a web site that sold airline tickets, come to be valued more than the entire US airline industry? Why did people buy shares in webvan.com, an online grocery-delivery service that lost $35m on sales of $395,000 in its first six months?' (see Cassidy, 2002).

This is a telling example of how, over the years, the endlessly repeated rose-tinted tales of the use of ICTs, to sublimely and unproblematically liberate us from all the ills of industrial and urban civilisation, have done little but camouflage the social, political, environmental or economic conflicts which continuously emerge at the interplay of cities and new media within capitalism (see Carrier and Miller, 1998). John Cassidy (2002) argues that in the dot.com bubble of the 1990s technological utopianianism allowed corporate and finance capital to fleece investors, generate public subsidies and political favouritism, and deflect attention from unprecedented corporate fraud, greed and plunder. In addition, such discourses were carefully constructed to ignore stark digital divides which condition highly uneven geographical and social access to ICTs at all geographical scales. Finally, in their suggestions of an automatically egalitarian digital future based on non-stop stock-market growth, utopian ICT discourses ignored the ways in which ICTs are being structured, and used, to add power to certain groups in society, often at the direct (relative or absolute) expense of other people or other places (see Cassidy, 2002).

To their critics, then, utopian ICT discourses, and the 'virtual' realities of neoliberal economic thought to which they have been so closely tied, are in fact nothing but camouflage screens (Carrier and Miller, 1998). They mask the roles ICTs have played, materially and discursively, in facilitating the intensifying corporate control of cities, economies, infrastructures and, indeed, the international economic system. Utopian discourses and ideologies of the 'information age' have also largely masked the deepening environmental crises caused by contemporary capitalism (see the Guandong case above). They have shifted attention from the deepening social, economic, and cultural inequalities that are emerging across our urbanising planet. Above all, such depictions have deflected attention away from the roles of ICT innovations and are allowing corporate capital to turn more and more domains (land, labour, space, time, identity, access, infrastructure, mobility, even genetic code) into commodities to be patented, sold and traded through ICT-based financial, transaction, and surveillance systems. These systems are strung out through the neoliberalising world economy allowing it to function (Dyer-Witheford, 1999; Schiller, 1999; Slack and Fejes, 1987; Brenner and Theodore, 2002).

To these critics it is therefore a myth to assume that all social interests can benefit from new ICT technologies and the practices they support. Rather, the construction of new ICT-based systems of mobility, flow and representation for some will always involve the construction of barriers for others. Experiences of ICTs are therefore highly contingent and relative. 'One person's infrastructure is another's difficulty' (Leigh-Star, 1999, 380).

Urban dimensions are an important part of the construction, and perpetuation, of glossy and benign ideologies of the information age. Reifying the need for cities to somehow act as single agents so that they can 'compete', be on the 'information superhighway', be 'switched on', and be technologically modern within the 'digital economy' or 'information society', is a virtually ubiquitous discourse in contemporary urban politics and planning. Such depictions provide the platform for massive financial and urban subsidies to new high-tech development spaces and infrastructures. These benefit certain interests whilst damaging others.

The symbolic power of information infrastructures provides many opportunities for the commodification and re-branding of urban places through marketing campaigns. These invariably involve the labelling of a city, development or urban district using the prefix 'silicon . . .', 'cyber . . .', 'information . . .', 'techno . . .', 'e- . . .', 'intelligent . . .' or 'multimedia . . .' (I am sure readers can add more). Jessop and Sum (2000) call this process the 'siliconisation' of urban entrepreneurialism.

In certain instances, the symbolic power of ICTs can actually become more important than their substantive application. This point is important when one considers the relationships between ICTs and the political economies of the deepening interconnection of the world (a process known as 'globalisation'). Rather than miraculously ushering in some global democratic utopia, the best global ICT connections are in the hands of the corporate and military worlds. Instead of somehow equalising geography, corporate ICT systems, in particular, are being carefully designed to *exploit* differences between places within carefully configured global divisions of labour. In so doing such mobility and logistics systems allow transnational corporations to place their different functions in the widely varying urban regions of the world whilst maintaining remote control, in real time, in order to try to maximise profitability. As the French urbanist Jean-Marc Offner puts it, rather than somehow causing 'territory to disappear', as in the immaterial dream of so many information technology utopianists, 'it is precisely *the fact that a multitude of places exist* which creates the need for exchange based on infrastructure networks' (emphasis added) (1996, 26).

Thus, it is preposterous to argue that cities are being made obsolete in an ICT-mediated world capitalist economy. Rather, their elements and parts are being reconfigured within corporately dominated geometries of changed mobility. This is being done as efforts intensify to deepen the geographical divisions of labour that are associated with neoliberal capitalism on a transplanetary scale. Within this process cities (and parts of cities) fare extremely unevenly. Some become more important as global powerhouses and switching centres of digital creativity, finance, control, governance and capital. Others benefit as research and development or innovation centres. Still others might be lucky enough to receive call centres, chip manufacturers, web servers or the co-location sites of online gambling, sex or data storage companies. But vast swathes of our urbanising planet – especially in the global South – are actually being marginalised and by-passed by the global ICT-mediated economy altogether. Here tiny elites may use ICTs as ways of maintaining their privilege in (and protection from) the city. But the vast majority of urban poor will remain totally disconnected from them. 'A virtual relation is about the differences between places and what passes between them', writes McKenzie Wark (2002, 401). 'It is about how places differ without forgetting that they are connected, and how they are connected without forgetting that they differ.'

Assumptions of technical rationality and the denial of politics: leaving no space for social and policy innovation

The final problem with end of city visions is that, in their depoliticised depictions of cities being 'impacted' by waves of autonomous, future technology, which seem to arrive from elsewhere, they imply that there is little or no space for policy innovation in cities, urban regions, or nations and supranational blocs. They suggest that there are few opportunities to harness ICTs, and to use policies and strategies to mediate how they articulate with individual cities, individual communities, and individual places. And they imply that the transformation of cityscapes and urban ways of life through ICTs is more a technical matter than a political one.

Such effects have stunted the development of reflective ICT policy debates amongst media industries, urban planners, urban policy-makers, developers, community development coalitions and governance partnerships in cities. As the urban researcher Robert Warren (1989, 14) argues, 'benign projections give little indication that there are significant policy issues which should be on the public agenda'.

Deterministic and anti-urban ideas about telecommunications provide an extremely simplistic set of mistaken and mythical ideas. These severely hamper our ability to think critically about using the new technologies to support innovations in urban policies and strategy-making. The simple attractiveness of these ideas compounds the many existing problems facing policy-makers at the urban level when attempting to intervene in the fast-moving and arcane world of telecommunications and ICTs. After all, this is an area where, in contrast to transportation, they have traditionally had very little power, knowledge, or experience.

Above all, in failing to stress the ways in which successful ICT innovations must necessarily occur through social practice and reflexive linkages between people, organisations, practices and machines – which need continuous work – deterministic accounts have led indirectly to countless technological failures. On numerous occasions city governments, firms, universities, and other institutions have decided to buy sparkling new ICT systems 'off the shelf'. Many of these have been expected to have their miraculous, deterministic, 'impacts' in and of themselves. Thus, there has often been insufficient attention to the social processes of innovation that might allow positive effects to be realised in practice.

Despite these problems, the imperatives of trying to shape ICTs to support creative policy solutions to the challenges of community regeneration, local government restructuring, urban economic development, city planning, and social activism are so pressing that they are now leading to a range of much more mature and sophisticated urban ICT policy solutions (see, for example, Keeble and Loader, 2001).

Imagining cybercities: subtle articulations of the urban and the digital

Our extended critique of end of city scenarios is now complete. As a result, the starting position for this book becomes very clear. This is the irrefutable reality that the twenty-first century will be a century marked by *both* the deepening urbanisation of all parts of our planet *and* a growing reliance on fast-advancing information and communications technologies. Drawing on the work of Christine Boyer (1992, 1995, 1996), the hybrid concept of the 'cybercity' is used in this book to denote the inseparable fusion of relations that are mediated by ICTs with those that are mediated between human presence, and movement, within and between urban places. The concept underlines that new media practices and technologies do not substitute for the city or the body in any simple and direct sense. It is utilised as the central organising idea through this book because it:

- Captures the hybrid material and 'socio-technical' interconnections of the new media, and the spatialities and mobilities of contemporary urban physical, social, economic and cultural life.
- Allows attention to fall in parallel on the changing materialities of urban and mediated life, the changing social relations that surround these shifts, and the ways in which ideas, and representations, of the city and new media are being changed together in different ways in different places.
- Emphasises the ways in which the creation, use and experience of knowledge and technology – including cybernetic systems utilising digital microprocessing and feedback control – now blur seamlessly into the political economies, and experiences, of place in an internationalising capitalist society (see Wiener, 1948; Beniger, 1986).
- Supports an interdisciplinary, critical, multidimensional, international, and parallel investigation. On the one hand, this can encompass the importance of the city for the development, and imagination, of new media. On the other, it can also involve an analysis of the importance of ICTs, and the new media, for

changing the form, shape, nature and experience of cities, and for influencing our ideas as to what a 'city' might actually be these days (see Boyer, 1996).

The ease with which the cybercity concept sustains a multidimensional perspective makes it particularly useful as the basis for an extremely broad book such as this. This work seeks to analyse, understand and theorise the complex and multifaceted interconnections between cities and ICTs. Such a task presents a rather daunting prospect. But the wide range of high quality scholarship that has emerged over the last decade or so helps considerably in this task. This has emerged in urban studies, planning, geography, sociology, anthropology, architecture, urban design, transport studies, cultural studies, philosophy, art, and media, information, and communication studies. Unfortunately, however, the very spread of this work, in both geographical and disciplinary terms, means that it is very difficult to track down, both in hard copy form and on the web. This presents problems to both researchers and students and inhibits the emergence of urban ICT studies as a coherent sub-discipline in its own right within the above disciplines.

The aims of this book

In this context, the *Cybercities Reader* has been designed to be the first book that systematically brings together the very best research into the ways in which cities, urban life and ICTs interrelate from the perspectives of a full range of disciplines, scales, and geographical contexts. The book therefore provides the first accessible and comprehensive introduction to the wide range of interactions between cities and digital media technologies.

The book has three aims. First, it seeks to bring together and introduce the best work to date which addresses the complex interrelationships between the use and application of electronic technologies and practices and processes of contemporary metropolitan life and urban development.

Second, the *Cybercities Reader* adopts a uniquely international perspective which aims to transcend the Anglo-American domination of recent English-language debates on ICTs and cities. In particular, the book draws together for the first time discussion of recent experiences of innovations in digital technologies from the cities of continental Europe, South and East Asia, Australasia, and Latin America, as well as North America and the UK.

Finally, the book engages with the proliferating range of polices, strategies and initiatives which have recently emerged which attempt to shape creatively the articulations between cities and ICTs. The book provides a state-of-the-art snapshot of some of the ways in which policies for cities and those for ICTs are being combined across the world, as efforts proliferate to harness digital technologies to urban development, revitalisation, and cultural policy objectives.

How to use this book

The *Cybercities Reader* is split into three parts and nine sections. Each section includes an Editor's Introduction to the section as a whole and a series of selected Readings. Each Reading is also preceded by a brief Editor's Introduction. Most are followed by lists of the references cited by the Reading's author(s). Following this, a list of the references cited in the Editor's introduction to the Reading is provided, along with his suggestions for further reading. These have been designed to allow readers to explore quickly literature relevant to each entry in the book. Each section of Readings has been designed to be read on its own or, for those using the book to accompany a course, each can form the basis for a teaching or discussion session. The nine sections are also used to structure the links at a companion web site for the book which can be viewed http://www. geographyarena.com/geographyarena/cybercities.

Part 1 of the book is entitled 'Understanding cybercities'. This brings together three sections which together seek to place contemporary city–ICT interactions in their historical, theoretical, and geographical contexts. Section I, 'Cybercity archaeologies', looks back to the experiences of urban media and communications improvements in the nineteenth and twentieth centuries. It explores how technologies like the telegraph, telephone, and cable TV were represented and used within changing structures and experiences of urban life between the 1850s and 1970s. Section II, 'Theorising cybercities', brings together state-of-the-art theorisations of the relationships between ICTs and cities from a range of perspectives. The final section of Part 1, Section III, is titled 'Cybercities: hybrid forms and recombinant spaces'. This includes a variety of reflections of the ways in which ICTs are being fused with cityspaces and cityscapes to create 'hybrid' forms which amount to complex recombinations of the urban and the digital which can only be understood as whole.

Part 2 of the book, 'Cybercity dimensions', builds on the contextual, historical and theoretical insights in Part 1. It includes five sections which together amount to an in-depth, interdisciplinary, and international reflection on the substantive realities of cybercities in a wide range of contexts across the world. These sections address, in turn:

IV Cybercity mobilities: the complex relationships between electronic and virtual mobilities and the physical flows and mobilities of people, goods and freight on urban streets and the physical transportation systems within and between cities (roads, highways, airline networks, logistic systems etc.).

V Cybercity economies: the relations between online economic activities and the construction and reconstruction of place-based urban economies.

VI Social and cultural worlds of cybercities: The articulations between social and cultural uses of ICTs and the social and cultural processes of change in the built, and represented, city.

VII Cybercity public domains and digital divides: The connections between the public worlds, and politics of cities and those mediated by ICTs and the links between urban social inequality and digital divides perpetuated through uneven access to ICTs or social surveillance or sorting through ICTs.

Finally, Part 3 of the book is titled 'Shaping cybercities?'. This reflects on the politics of policy, planning, prescription and prediction in the shaping of city–ICT relations and draws together some speculations on the medium and long-term futures of cybercities. It has two sections. Section VIII, 'Cybercity strategy and politics', collects some of the latest discussions of the scope for, and nature of, policies, initiatives, strategies and interventions that strive to shape cities and ICTs in parallel. These analyse a variety of cases in North and Latin America, Asia and Europe. The last section of the book, 'Cybercity futures', brings together some of the most powerful and thought-provoking reflections on the futures of cybercities.

References

Abler, R. (1974), 'The geography of communications'. In M. Eliot Hurst (ed.), *Transportation Geography*, New York: McGraw Hill, 327–46.

Aibar, E. and Bijker, W. (1997), 'Constructing a city: the Cerdà plan for the extension of Barcelona', *Science, Technology and Human Values*, 22(1), 3–30.

Amin, A. and Thrift, N. (2002), *Cities: Reimagining the Urban*, Cambridge: Polity.

Banham, R. (1980), *Theory and Design in the First Machine Age*, Cambridge, MA: MIT Press.

Batty, M. (1993), 'The geography of cyberspace', *Environment and Planning B: Planning and Design*, 20, 615–16.

Bell, J. (ed.) (2001), *Carchitecture: When the Car and the City Collide*, Basel: Birkhäuser.

Benedikt, M. (1991), 'Introduction'. In M. Benedikt (ed.), *Cyberspace: First Steps*, Cambridge, MA: MIT Press.

Beniger, J. (1986), *The Control Revolution: Technological and Economic Origins of the Information Society*, Cambridge, MA: Harvard University Press.

Bernardini, O. and Galli, R. (1993), 'Dematerialization: long term trends in the intensity of use of materials and energy, *Futures*, May, 431–48.

Bleecker, J. (1994), 'Urban crisis: past, present and virtual', *Socialist Review*, 24, 189–221.

Bolter, D. (1995), 'The social construction of telepolis'. *Mimeo*.

Bolter, J. and Grusin, R. (2000), *Remediation: Understanding New Media*, Cambridge, MA: MIT Press.

Boyer, C. (1992), 'The imaginary real world of cybercities', *Assemblage*, 18, 115–27.

Boyer, C. (1995), 'The great frame up: fantastic appearances in contemporary spatial politics'. In H. Liggett and D. Perry (eds), *Spatial Practices*. London: Sage, 81–109.

Boyer, C. (1996), *Cybercities: Visual Perception in the Age of Electronic Communication*, New York: Princeton University Press.

Brenner, N. and Theodore, N. (2002), *Spaces of Neoliberalism*, Oxford: Blackwell.

Brooke, J. (2001), 'Paper factories boom despite computer era', *Herald Tribune*, April 23, 15.

Carrier, J. and Miller, D. (eds) (1998), *Virtualism: A New Political Economy*, London: Berg.

Cassidy, J. (2002), *Dot.Con: The Real Story of Why the Internet Bubble Burst*, London: Penguin.

Cochrane, P. (1998), 'The infinite city'. Mimeo.

Cosgrove, D. (1996), 'Windows on the city', *Urban Studies*, 33(8), 1495–8.

Coyne, R. (1995), *Designing Information Technology in the Postmodern Age*, Cambridge, MA: MIT Press.

Coyne, R. (1999), *Technoromanticism: Digital Narrative, Holism, and the Romance of the Real*, Cambridge, MA: MIT Press.

Dyer-Witheford, N. (1999), *Cyber-Marx: Cycles and Circuits of Struggle in High-Technology Capitalism*, Chicago: University of Illinois Press.

Easterling, K. (2002), 'A-ware', *Journal of Architectural Education*, 56(2), 3–4.

Economist (1995), *The Death of Distance*, Telecommunications Survey, 30 September–6 October.

Gates, W. (1995), *The Road Ahead*, London: Hodder and Stoughton.

Gelerntner, D. (1991), *Mirror Worlds*, New York: Oxford University Press.

Gibson, W. (1994), *Neuromancer*, London: Voyager.

Gottmann, J. (1990), *Since Megalopolis: The Urban Writings of Jean Gottmann*, Baltimore: Johns Hopkins University Press.

Graham, S. (1999), 'Global grids of glass: on global cities, telecommunications and planetary urban networks', *Urban Studies*, May, 36 (5–6), 929–49.

Graham, S. (2000), 'Constructing premium networked spaces: reflections on infrastructure networks and contemporary urban development', *International Journal of Urban and Regional Research*, 24(1), 183–200.

Graham, S. and Marvin, S. (1996), *Telecommunications and the City: Electronic Spaces, Urban Places*, London: Routledge.

Haythornthwaite, C. and Wellman, B. (2002), 'Moving the Internet out of cyberspace'. In B. Wellman and C. Haythornthwaite (eds), *The Internet and Everyday Life*, Oxford: Blackwell, 3–44.

Heim, M. (1991), 'The erotic ontology of cyberspace'. In M. Benedikt (ed.), *Cyberspace: First Steps*, Cambridge, MA: MIT Press, 59–80.

Hill, S. (1988), *The Tragedy of Technology*, London: Pluto.

Hodge, D. (1990), 'Geography and the political economy of urban transportation', *Urban Geography*, 11(1), 87–100.

Hosaya, H. and Chaefer, M. (2001), 'Bit structures'. In C. Chung, J. Inaba, R., Koolhaas and S. Tsung Leong (eds), *Harvard Design School Guide to Shopping*, Cologne: Taschen, 157–63.

Jarvis, H., Pratt, A. and Cheng-Chong, P. (2000), *The Secret Life of Cities: Social Reproduction of Everyday Life*, New York: Pearson Education.

Jessop, B. and Sum, N.-L. (2000), 'An entrepreneurial city in action: Hong Kong's emerging strategies in and for (inter)urban competition', *Urban Studies*, (37), 2287–313.

Jowett, G. (1993), 'Urban communication: the city, media and communications policy'. In P. Gaunt (ed.), *Beyond Agendas: New Directions in Communications Research*, Westpoint, CT: Greenwood Press, 41–56.

Kaba, S. (1996), 'Building the future: an architectural manifesto for the next Millennium', *Web Architecture Magazine*, April.

Kaplan, C. (2002), 'Transporting the subject: technologies of mobility and location in an era of globalization', *Proceedings of the Modern Language Association of America*, 117(1), 32–42.

Keeble, L. and Loader, B. (eds) (2001), *Community Informatics: Shaping Computer-Mediated Social Relations*, London: Routledge.

Latour, B. and Hermant, E. (1998), *Paris ville invisible*, Paris: La Decouverte.

Leadbetter, C. (2000), *Living on Thin Air: The New Economy*, London: Penguin.

Leigh-Star, S. (1999), 'The ethnography of infrastructure', *American Behavioral Scientist*, 43(3), 377–91.

McLuhan, H. (1964), *Understanding Media: The Extension of Man*, London: Sphere.

Mandlebaum, S. (1986), 'Cities and communication: the limits of community', *Telecommunications Policy*, June, 132–40.

Marshall, T. (1997), 'Futures, foresight and forward looks', *Town Planning Review*, 68(1), 31–50.

Masser, I., Sviden, O. and Wegener, M. (1992), *The Geography of Europe's Futures*, London: Belhaven.

Martin, J. (1978), *The Wired Society*, Englewood Cliffs, NJ: Prentice Hall.

Marvin, S. (1997), 'Environmental flows: telecommunications and the dematerialisation of cities', *Futures*, 29 (1).

Mattelaart, A. (1999), 'Mapping modernity: Utopia and communications networks'. In D. Cosgrove (ed.), *Mappings*, London: Reaktion Books, 169–92.

Meier, R. (1962), *A Communications Theory of Urban Growth*, Cambridge, MA: MIT Press.

Mulder, A. (ed.) (2002), *Transurbanism*, Rotterdam: NAi Publishers.

Munt, S. (2001), *Technospaces: Inside the New Media*, London: Continuum.

Naisbitt, J. and Aburdene, P. (1991), *Megatrends 2000: Ten Directions for the 1990s*, New York: Avon.

Negroponte, N. (1995), *Being Digital*, London: Hodder and Stoughton.

Offner, J. (1996), '"Reseaux" et "Large technical System": concepts complémentaires ou concurrent?', *Flux*, 26, October–December, 17–30.

Pascal, A. (1987), 'The vanishing city', *Urban Studies*, 24, 597–603.

Pawley, M. (1997), *Terminal Architecture*, London: Reaktion Books.

Pickles, J. (ed.) (1995), *Ground Truth: The Social Implication of Geographic Information Systems*, New York: Guilford.

Poole, S. (2003), 'Review of Dot.Con, by John Cassidy', *Guardian Review*, 25 January, 18.

Robins, K. (1995), 'Cyberspace and the world we live in'. In M. Featherstone and R. Burrows (eds), *Cyberspace/Cyberbodies/Cyberpunk*, London: Sage, 135–56.

Robins, K. and Hepworth, M. (1988), 'Electronic spaces: new technologies and the future of cities', *Futures*, April, 155–76.

Romm, J. (1999), *The Internet Economy and Global Warming: A Scenario of the Impact of E-Commerce on Energy and the Environment*, Washington, DC: Center for Energy and Climate Solutions.

Sawhney, H. (1996), 'Information superhighway: metaphors as midwives', *Media, Culture and Society*, 18, 291–314.

Schiller, D. (1999), *Digital Capitalism: Networking the Global Market System*, Cambridge, MA: MIT Press.

Shabi, R. (2002), 'The E-wasteland', *The Guardian Weekend*, November 30, 36–43.

Simmel, G. (1964), 'The metropolis and mental life'. In K. Wolff (ed.), *The Sociology of George Simmel*, New York: Free Press, 409–24.

Slack, J. D. and Fejes, F. (eds) (1987), *The Ideology of the Information Age*, Norwood, NJ: Ibex.

Sola Pool, I. de (ed.) (1976), *The Social Impact of the Telephone*, Cambridge, MA: MIT Press.

Tazi, N. (2001), 'Fragments of net-theory'. In R. Koolhaas, S. Boeri and S. Kwinter (eds), *Mutations*, Bordeaux: ACTAR, 42–50.

Thrift, N. (1995), 'Inhuman geographies: landscapes of speed, light and power'. In N. Thrift, *Spatial Formations*, London: Routledge, 256–310.

Toffler, A. (1980), *The Third Wave*, New York: Morrow.

United Nations Centre for Human Settlements (2001), *Cities in a Globalizing World: Global Report on Human Settlements*, London: Earthscan.

Virilio, P. (1993), 'The third interval: a critical transition'. In V. Andermatt-Conley (ed.), *Rethinking Technologies*, Minneapolis: University of Minnesota Press, 3–10.

Wark, M. (2002), 'To the vector the spoils'. In T. Levin, U. Frohne and P. Weibel (eds), *Ctrl Space: Rhetorics of Surveillance from Bentham to Big Brother*, Cambridge, MA: MIT Press.

Warren, R. (1989), 'Telematics and urban life', *Journal of Urban Affairs*, 11(4), 339–46.

Webber, M. (1964), 'The urban place and the non-place urban realm'. In M. Webber, J. Dyckman, D. Foley, A. Guttenberg, W. Wheaton and C. Whurster (eds), *Explorations Into Urban Structure*, Philadelphia: University of Pennsylvania Press, 79–153.

Wiener, N. (1948), *Cybernetics; or, Control and Communication in the Animal and the Machine*, Cambridge, MA: MIT Press.

Wigley, M. (2002), 'Resisting the city'. In A. Mulder (ed.), *Transurbanism*, Rotterdam: NAi Publishers, 103–22.

Wilson, H. (2003), 'We just don't click anymore', *Observer Magazine*, 30 March, 64–5.

Wired Magazine (1996), 'The *Wired* manifesto', October, 42–6.

General suggestions for further reading

General overviews of cities and ICTs

Borja, J. and Castells, M. (1997), *Local and Global: Managing Cities in the Information Age*, London: Earthscan.

Brotchie, J., Newton, P., Hall, P. and Nijkamp, P. (eds) (1985), *The Future of Urban Form: The Impact of New Technology*, London: Croom Helm and Nichols.

Brotchie, J. , Hall, P. and Newton, P. (eds) (1987), *The Spatial Impact of Technological Change*, London: Croom Helm.

Brotchie, J., Batty, M., Hall, P. and Newton, P. (eds) (1991), *Cities of the 21st Century,* London: Halsted.

Brunn, S. and Leinbach, T. (eds) (1991), *Collapsing Space and Time: Geographic Aspects of Communications and Information*, London: Harper.

Castells, M. (1996), *The Rise of the Network Society*, Oxford: Blackwell.

Castells, M. (2001), *The Internet Galaxy*, Oxford: Oxford University Press.

De Meyer, D., Versluys, K., Borret, K., Eeckhout, B., Jacobs, S. and Keunen, B. (1999), *The Urban Condition: Space, Community and Identity in the Contemporary Metropolis*, Rotterdam: 010 Publishers.

Dodge, M. and Kitchin, R. (2001), *Mapping Cyberspace*, London: Routledge.

Downey, J. and McGuigen, J. (eds) (1999), *Technocities*, London: Sage.

Droege, P. (ed.) (1997), *Intelligent Environments*, Amsterdam: North Holland.

Dutton, W. (1996), *Information and Communications Technologies: Visions and Realities*, Oxford: Oxford University Press.

Dutton, W. (1999), *Society on the Line: Information Politics in the Digital Age*, Oxford: Oxford University Press.

Graham, S. and Marvin, S. (1996), *Telecommunications and the City: Electronic Spaces, Urban Places*, London: Routledge.

Graham, S. and Marvin, S. (2000), *Splintering Urbanism: Networked Infrastructures, Technological Mobilities and the Urban Condition*, London: Routledge.

Kitchen, R. (1998), *Cyberspace*, London: Wiley.

Koolhaas, R., Boeri, S. and Kwinter, S. (2000), *Mutations*, Bordeaux: ACTAR.

Lash, S. and Urry, J. (1994), *Economies of Signs and Space*, London: Sage.

Wheeler, J., Aoyama, Y. and Warf, B. (eds) (2000), *Cities in the Telecommunications Age: The Fracturing of Geographies*, London and New York: Routledge.

Shields, R. (1996), *Cultures of Internet: Virtual Spaces, Real Histories, Living Bodies*, London: Sage.

United Nations Centre for Human Settlements (2001), *Cities in a Globalizing World: Global Report on Human Settlements*, London: Earthscan.

Wellman, B. and Haythornthwaite, C. (eds) (2002), *The Internet and Everyday Life*, Oxford: Blackwell.

Wilson, M. and Corey, K. (2000), *Information Tectonics: Space, Place and Technology in an Electronic Age*, London: Wiley.

Fantasies of transcendence: the anything–anywhere–anytime dream

Positive and negative discussions emphasising transcendence of the body and the city via ICTs. These cover 'terminal citizens', cyberlibertarianism, technological utopianism and end-of-city scenarios:

Benedikt, M. (ed.) (1992), *Cyberspace: First Steps*, Cambridge, MA: MIT Press.

Brook, J. and Boal, I. (1995), *Resisting the Virtual Life: The Culture and Politics of Information*, San Francisco: City Lights.

Coyne, R. (1999), *Technoromanticism: Digital Narrative, Holism, and the Romance of the Real*, Cambridge, MA: MIT Press.

Gates, W. (1996), *The Road Ahead*, London: Hodder and Stoughton.

Gelerntner, D. (1991), *Mirror Worlds*, New York: Oxford University Press.

Hayles, N. K. (1999), *How We Became Posthuman: Virtual Bodies in Cybernetics, Literature, and Informatics*, Chicago: University of Chicago Press.

Herman, A. and Sloop, J. (2000), '"Red alert!" Rhetorics of the World Wide Web and "friction free" capitalism'. In A. Herman and T. Swiss (eds), *The World Wide Web and Contemporary Cultural Theory*, New York: Routledge, 77–98.

Keleman, M. and Smith, W. (2001), 'Community and its "virtual" promises: a critique of cyberlibertarian rhetoric', *Information, Communications and Society*, 4(3), 370–87.

Leadbetter, C. (2000), *Living on Thin Air: The New Economy*, London: Penguin.

Martin, J. (1978), *The Wired Society*, London: Prentice Hall.

Mattelaart, A. (1999), 'Mapping modernity: Utopia and communications networks'. In D. Cosgrove (ed.), *Mappings*, London: Reaktion Books, 169–92.

McBeath, G. and Webb, S. (2000), 'On the nature of future worlds? Considerations of virtuality and utopia', *Information, Communications and Society*, 3(1), 1–16.

McLuhan, M. (1964), *Understanding Media*, London: Routledge.

Negroponte, N. (1995), *Being Digital*, London: Hodder and Stoughton.

Pascal, A. (1987), 'The vanishing city', *Urban Studies*, 24, 597–603.

Pawley, M. (1997), *Terminal Architecture*, London: Reaktion Books.

Robins, K. and Hepworth, M. (1988), 'Electronic spaces: new technologies and the future of cities', *Futures*, April, 155–76.

Toffler, A. (1980), *The Third Wave*, New York: Morrow.

Virilio, P. (1997), *Open Sky*, London: Verso.

Warf, B. (2000), 'Compromising positions: the body in cyberspace'. In J. Wheeler, Y. Aoyama and B. Warf (eds), *Cities in the Telecommunications Age: The Fracturing of Geographies*, London and New York: Routledge, 54–70.

Webber, M. (1968), 'The post-city age'. *Daedalus*, fall.

Winger, A. (1997), 'Finally: a withering away of cities?', *Futures*, 29(3), 251–6.

Zylinska, Z. (2002) *The Cyborg Experiments: The Extensions of the Body in the Media Age*, London: Continuum.

'Beyond the dazzling light': the 'remediation' of everyday urban life

Generally recent analyses emphasising the ways in which ICTs, now that they are increasingly ubiquitous, support subtle 'remediations' in everyday urban life.

Amin, A. and Thrift, N. (2002), *Cities: Reimagining the Urban*, Cambridge: Polity.

Armitage, J. and Roberts, J. (eds) (2002), *Living with Cyberspace: Technology and Society in the 21st Century*, London: Continuum.

Bolter, J. and Grusin, R. (1999) *Remediation: Understanding New Media*, Cambridge MA: MIT Press.

Brown, J.-S. and Duguid, P. (2000), *The Social Life of Information*, Cambridge, MA: Harvard Business School Press.

Cosgrove, D. (1996), 'Windows on the city', *Urban Studies*, 33(8), 1495–8.

Jarvis, H., Pratt, A. and Cheng-Chong, P. (2000), *The Secret Life of Cities: Social Reproduction of Everyday Life*, London: Prentice Hall.

Kaplan, C. (2002), 'Transporting the subject: technologies of mobility and location in an era of globalization', *Proceedings of the Modern Language Association of America* (PMLA), 117(1), 32–42.

Latour, B. and Hermant, E. (1998), *Paris ville invisible*, Paris: La Decouverte.

Munt, S. (2001), *Technospaces: Inside the New Media*, London: Continuum.

Robins, K. (1995), 'Cyberspace and the world we live in'. In Mike Featherstone and Roger Burrows (eds), *Cyberspace/Cyberbodies/Cyberpunk*, London: Sage, 135–56.

Thrift, N. (1996), 'New urban eras and old technological fears: reconfiguring the goodwill of electronic things', *Urban Studies*, 33(8), 1463–93.

Thrift, N. (1997), 'Cities without modernity, cites with magic', *Scottish Geographical Magazine*, 113(3), 138–49.

Wellman, B. and Haythornthwaite, C. (eds), *The Internet and Everyday Life*, Oxford: Blackwell, 3–44.

Woolgar, S. (ed.) (2002), *Virtual Society? Get Real!* Oxford: Oxford University Press.

PART ONE

Understanding Cybercities

Plate 3 Domestic satellite TV installations in Dubai, United Arab Emirates. (Photograph by Stephen Graham.)

I CYBERCITY ARCHAEOLOGIES

Plate 4 Blizzard of 1888, New Street, New York, showing masses of telegraph and telephone poles weighed down by snow. (Source: Museum of the City of New York.)

INTRODUCTION

Discussions about ICTs and cities tend to deny continuities and emphasise revolutionary transformations. New media technologies and practices are often abstracted from their historical contexts and presented simplistically as meteoric impactors arriving as if from nowhere to revolutionise pre-existing places and mobility practices. The subtle interrelations between old and new, non-mediated and mediated, that occur within the fine-grained fabric of cities and everyday life, tend to be glossed over in such hyperbolic discourse.

And yet, over the past 8,000 years, cities have always been places where innovations in transport, communications media, printing, publishing, the processing information and the creation of knowledge have been concentrated. Infrastructures and techniques that facilitate connections to distant spaces and times have been central to the growth of cities throughout urban history. The most powerful cities have always gained wealth through concentrating military, economic and political control over far-off people and places. Technologies of information processing and 'remote control' are thus intrinsic to any notion of the 'urban'. Innovations in information processing and communication – from the invention of writing to modern bureaucracies and the software-controlled spaces of the Internet – have also been central to facilitating more and more complex urban societies and more mobile urban cultures.

Thus, the leading roles that contemporary cities have as concentrations of Internet activity and as centres of innovation in online, mobile, and e-commerce services, in teleworking practices, and as sites for the leading-edge application of a whole range of ICT services, reflect strong historical continuities with earlier periods.

Cybercities thus have complex archaeologies which are the concern of this first part of the book. Such archaeologies are expressed in the close continuities between the urban dominance of older, electromechanical and electronic media (the telegraph, the telephone, publishing, electricity generation, TV and radio broadcasting) and the contemporary worlds of the Internet, mobile phones, broadband cable and multimedia. They are also reflected in the material reliance of Internet and mobile phone infrastructures upon massive, embedded concentrations of older telephone, media, energy, street and transport infrastructures within and between cities. Very often, for example, the world's most capable telecommunications infrastructures are actually physically located in centuries-old canals, hydraulic power systems, streets, sewers, subway tunnels and railway channels. These, quite literally, form the root systems of the world's major cities.

Finally, the archaeologies of cybercities are marked by the strong continuities in the ways in which various generations of new media have been represented, idealised and lambasted as their meanings and uses are constructed – and contested – within a wide variety of urban cultures. For example, the notion that space and distance-transcending technologies can overcome the need for spatial concentration in cities – because of the quality of interaction across distance that they bring – has accompanied the introduction of all new communications media since the late nineteenth century.

References and suggestions for further reading

Abbate, J. (2000), *Inventing the Internet*, Cambridge, MA: MIT Press.

Banham, R. (1980), *Theory and Design in the First Machine Age*, Cambridge, MA: MIT Press.

Beniger, J. (1986), *The Control Revolution: Technological and Economic Origins of the Information Society*, Cambridge, MA: Harvard University Press.

Carey, J. (1989), *Communication as Culture: Essays on Media and Society*, London: Routledge.

Corn, J. and Horrigan, B. (1984), *Yesterday's Tomorrows: Past Visions of the American Future*, Baltimore: Johns Hopkins University Press.

De Landa, M. (1997), *A Thousand Years of Nonlinear History*, New York: Swerve.

Downey, G. (2002), *Telegraph Messenger Boys: Labor, Communication and Technology, 1850–1950*, New York: Routledge.

Donald, J. (1999), *Imagining the Modern City*, London: Athlone Press.

Dupuy, G (1991), *L'urbanisme des réseaux: théories et méthodes*, Paris: Armand Colin.

Fischer, C. (1992), *America Calling: A Social History of the Telephone to 1940*, Oxford: University of California Press.

Giddens, A. (1990), *The Consequences of Modernity*, Oxford: Polity.

Goodman, D. and Chant, C. (1999), *European Cities and Technology: Industrial to Post-Industrial Cities*, London: Routledge.

Gottmann, J. (1990), *Since Megalopolis: The Urban Writings of Jean Gottmann*, Baltimore: Johns Hopkins University Press.

Graham, S. and Marvin, S. (2001) *Splintering Urbanism: Networked Infrastructures, Technological Mobilities, and the Urban Condition*, London: Routledge, Chapter 2.

Hall, P. (1988), *Cities of Tomorrow*, Oxford: Blackwell.

Hall, P. (1999), *Cities and Civilization*, London: Weidenfeld and Nicolson.

Hall, P. and Preston, P. (1988), *The Carrier Wave: New Information Technology and the Geography of Innovation, 1846–2003*, London: Unwin.

Hughes, T. (1983), *Networks of Power: Electrification of Western Society, 1880–1930*, London and Baltimore: Johns Hopkins University Press.

Kern, S. (1986), *The Culture of Time and Space, 1880–1918*, Cambridge, MA: Harvard University Press.

Macphee, G. (2002), *The Architecture of the Visible: Technology and Urban Visual Culture*, London: Continuum.

Marvin, C. (1988), *When Old Technologies Were New: Thinking about Electric Communication in the Late Nineteenth Century*, Oxford: Oxford University Press.

Mattelaart, A. (1996), *The Invention of Communication*, Minneapolis, MN: University of Minnesota Press.

Mattelaart, A. (1994), *Mapping World Communication: War, Progress, Culture*, Minneapolis, MN: University of Minnesota Press.

Mattelaart, A. (1999), 'Mapping modernity: Utopia and communications networks'. In D. Cosgrove (ed.), *Mappings*, London: Reaktion Books, 169–92.

Meier, R. (1962), *A Communications Theory of Urban Growth*, Cambridge, MA: MIT Press.

Mumford, L. (1934), *Technics and Civilisation*, London: Routledge and Kegan Paul.

Mumford, L. (1938), *The Culture of Cities*, New York: Secker and Warburg.

Mumford, L. (1961), *The City in History*, New York: MJF Books.

Nye, D. (1997), *Narratives and Spaces: Technology and the Construction of American Culture*, Exeter: University of Exeter Press.

Roberts, G. and Steadman, P. (eds) (1999), *American Cities and Technology: Wilderness to Wired City*, London: Routledge.

Sennett, R. (1994), *Flesh and Stone: The Body and the City in Western Civilisation*, London: Faber.

Sola Pool, I. de (ed) (1976), *The Social Impact of the Telephone*, Cambridge, MA: MIT Press.

Standage, T. (1998), *The Victorian Internet*, New York: Berkeley.

Tarr J. and Dupuy G. (eds) (1988), *Technology and the Rise of the Networked City in Europe and North America*, Philadelphia: Temple University Press.

Shiel, M. and Fitzmaurice, T. (eds) (2001), *Cinema and the City: Film and Urban Societies in a Global Context*, Oxford: Blackwell.

Standage, T. (1998), *The Victorian Internet*, New York: Berkeley.

Thrift, N. (1996), *Spatial Formations*, London: Sage.

Wiener, N. (1948), *Cybernetics; or, Control and Communication in the Animal and the Machine*, Cambridge, MA: MIT Press.

Williams, R. (1973), *The Country and the City*, London: Hogarth Press.

O
N
E

'Inhuman Geographies: Landscapes of Speed, Light and Power'

from *Spatial Formations* (1996)

Nigel Thrift

Editor's Introduction

In this reading Nigel Thrift, a Professor of Geography at Bristol University, analyses in detail how innovations in technologies of speed, light and power combined and interrelated in nineteenth-century urban societies in the west to form an integrated 'machinic complex'. Rather than 'impacting' on cities, Thrift argues that this complex of telegraph, bicycle, cinema, telephone, mail, rail, light, gas, and electricity innovations was socially produced together in complex ways to help reshape the spaces, times, rhythms, experiences and representations of urban life.

As Thrift discusses, as an interrelated complex, these technologies helped to reshape the consciousness of time and space by linking widely separated geographical places with increasing speed. They were involved in changes in the feelings urban citizens had about the nature of their lives, their bodies, their cities, their cultures, their desires, and their futures. Together, their widening use brought the sense that urban life, and urban culture, were undergoing some 'great acceleration'. Such technologies also supported – but did not simply determine – complex sociological and geographical shifts: the colonisation of the urban night, the geographical dispersal of the metropolis, and the introduction of new spaces, practices and spectacles of production, distribution and consumption across the city. Finally, this machinic complex also helped to meld modern cities, industries and urban societies into giant technological systems. These were orchestrated and engineered at wider and wider geographical scales (the national, the transnational, even the global).

In helping to produce the extended, and speeded up, 'wired-tracked and piped' metropolis – with its regular temporal rhythms and spatial flows – this machinic complex thus laid the foundations for the mass production, mass distribution, mass consumption and mass media cities of the twentieth century (Tarr and Dupuy, 1988). It allowed such complex and distanciated societies based on new magnitudes and scales of speed, flow and connection to be managed and coordinated (see Beniger, 1986). And it laid the foundations for the application of cybernetic and computerised innovations in cities in the second half of the twentieth century (Wiener, 1948).

Thrift's analysis is notable because it is one of the few that does not privilege one particular innovation, technology or infrastructure within the construction of the broad machinic complex of speed, light and power that he focuses on. He usefully demonstrates the necessity of adopting sophisticated perspectives of how innovations in ICTs in contemporary cities closely interlink with transport, energy, water, media and other technologies, systems and techniques, to form the machinic complexes which underpin the development of contemporary cybercities.

THE NINETEENTH CENTURY

Speed

In the nineteenth century the technology of speed broke through the limits of walking and the horse into a period of progressively accelerating transport network technology – the stage coach and the horse-drawn tram, the railway and the electric train, and the bicycle. Thus in Britain by 1820 it was often quicker to travel by stage coach than on horseback. By 1830 movement between the major towns was some four or five times faster than in 1750. This increase in the speed of the stage coach was paralleled by an increase in both frequency of operation and in the number of destinations. The subsequent growth of the railway network made for even more dramatic leaps in speed, frequency and access. It is no surprise that 'the anni-hilation of space by time' was a favourite meditation for the Victorian writer. The effects were all the more arresting because they came to be experienced by so many people. By 1870, 336.5 million journeys were made by rail, the vast bulk of them by third-class passengers. In the growing cities a parallel process of democratisation was taking place measured out by the advent of the horse-drawn tram, the underground and, latterly, electric tramways. But perhaps the most dramatic change in travel was the invention of the bicycle. By 1855 there were already 400,000 cyclists in Britain and the 1890s saw the peak of this simple machine's popularity. The bicycle, which started as a piece of fun for young swells, foreshadowed the automobile in providing immediate, democratic access to speed.

The nineteenth century also saw the beginnings of new networks of communication which began to displace face-to-face communication, especially a rapid and efficient mail service and, latterly, the tele-graph and mass circulation newspapers. In Britain, the mail service expanded massively in the nineteenth century, as did new communication innovations like the postcard, Valentine cards, Christmas cards, and so on. By 1890, the Post Office was carrying 1,706 million letters a year. The telegraph was first used in 1839. By 1863 nearly 22,000 miles of line had been set up, transmitting over six million messages a year, from 3,381 points. However, it was not until the last quarter of the nineteenth century that the telegraph became an institution of communication genuinely used by the mass of the population. However, neither of these means of communication was instantaneous. The mail service still depended upon the velocity of a set of different means of transport while the telegraph service had to be actually reached (usually in a post office) before it could be used.

The exact social and cultural effects of this 'great acceleration' have certainly been disputed. But, amongst the changes attendant on a world of traffic flowing through multiple networks at least four might be counted as being significant. The first of these was a change in the consciousness of time and space. For example, so far as time consciousness was concerned, it seems clear that the population began to pay more attention to smaller distinctions in time. Thus, in the last decade of the nineteenth century watches became more popular. Again, it seems clear that a sense of an enlarged, simultaneous presence became more common, especially as a result of the telegraph. This was not just a temporal but also a spatial sense. In one sense space had been shrunk by the new simultaneity engendered by the shrinkage of travel and communica-tion times, and by new social practices like travelling to work and tourism. In another sense it was much enlarged. *The Times* of 1858 (quoted in Briggs, 1989, p. 29) wrote proudly of 'the vast enlargement . . . given to the sphere of human activity' by the telegraph and the press that now fed upon its pulses. It is also possible to identify, rather more tentatively, elements of a change in the perception of landscape. For example, Schivelsbuch (1986) has argued for the development of a 'panoramic perception' in which the world is presented as something seen from within a moving platform, as a passing, momentary spectacle to be glimpsed and consumed. A second change was the impact of this general speed-up on the texts of the period, whether in the form of enthusiastic paeans to machinery or in the form of counter-laments for slower, less mechanical and decidedly more authentic times gone by. [. . .]

A third change was in the nature of subjectivity. In particular, it is possible to note an increasing sense of the body as an anonymised parcel of flesh which is shunted from place to place, just like other goods. Each of these bodies passively avoided others, yet was still linked in distant events, a situation typified by Victorian vignettes of the railway passenger in greater communion with the newspaper than companions. [. . .]

Light

Another important part of the history of the machinic complex of mobility has been the history of the machinic complex of artificial light. In effect, this history dates from the end of the eighteenth century, when, for the first time, a technology that had not significantly altered for several hundred years began to change. Before this time, artificial light had been in short supply. To an extent, the work day had been emancipated from dependence on daylight by candles and oil lamps, but most households still used artificial light only very sparingly.

From the end of the eighteenth century through the early nineteenth century a whole series of inventions began to change this situation. The Argand oil-burning light (1783) was the first such invention. It was soon followed by the gaslight which was made possible by the invention of systems to produce gas from coal *c.* 1800. These were first used in factories and then, with the setting up of central gas supplies in cities (the first gasworks was set up in Britain in London in 1814), they spread to the household. The networks of gas mains prefigured railway tracks and electrical networks. By 1822 there were already 200 miles of gas main in London, and 53 English cities in all had gas companies. [. . .]

To some extent running in parallel with the development of gas lighting was the development of brighter and more spectacular electrical lighting. The arc light was invented in 1800 and was used in specific situations over long periods of time but was not put into general use in factories, shops and similar sites until the 1870s and 1880s. It was the invention of the electrical bulb in 1879 which heralded the widespread electrification of light and the decisive break between light and fire. The first central electricity-generating stations became operational in 1882 in New York and London. However, although London was amongst the first to have a fully functioning electric light system (around Holborn), subsequently, like Britain as a whole, it lagged in the adoption of electric light, partly because of economic conditions, partly because of legislation – which prevented the growth of electricity monopolies – and partly because of opposition from gas interests. Thus, it was not until the 1920s that electricity, and electrical light, was widely adopted in Britain. In contrast, by 1903 New York had 17,000 electric street lamps, 'while electric interior lighting and exterior displays had become "essential to competition" for downtown theatres, restaurants, hotels and department stores' (Nasaw, 1992, p. 274).

This history of an ever-expanding landscape of light which we now take so much for granted cannot be ignored. As a machinic complex it is particularly important because of five changes. The first of these is that it signifies the progressive colonisation of the night. In the towns and cities, at least, night had never been regarded as a period of general inactivity. The early pleasure gardens, for example, were famous for their displays of nightly social bustle amidst much remarked upon artificial light. [. . .]

A second change was that very gradually a separate set of human practices evolved which we now call 'night life'. In 1738 Hogarth had produced an engraving called *Night* as a part of a series, *The Four Times of Day*, which showed only lurching drunks and wandering vagrants inhabiting a night-time London street. But, it was an exaggeration. Lighted pleasure gardens and other places of entertainment had already started to produce a specific night social life, and through the eighteenth century this new kind of sociality was extended by the addition of theatres, shops (which used manufactured light to enhance their displays), cafés, and so on. It is no surprise, then, that by the late Victorian period London at night was seen by many foreign visitors as almost another country:

> *Great city of the midnight sun*
> *Whose day begins when day is done.*
> (Le Galliene, cited in Briggs, 1992, 24)

As the veil of night was lifted, so many effects were produced. Lighting removed some of the dangers that had once lurked in the dark. Again, the advent of lighting in the household produced new timings and forms of social interaction. Industrial production was also revolutionised as manufacturers began to work out shift systems which used light and night to enable continuous production.

A third change was that manufactured light was of central importance in the cultural construction of the 'dream spaces' of the nineteenth century: the department stores (with their illuminated display windows), the hotels, the theatres, the cafés, the world fairs, and so on. These urban spaces, which transformed consumption into a mode of being, depended on visual consumption, which, in turn, depended upon artificial lighting. Such spaces rapidly became *the* urban experience and their 'spectacular lighting . . . quickly became a central cultural practice' (Nye, 1990, 383) providing the opportunity for a new nocturnal round.

A fourth change was that manufactured light produced major new opportunities for surveillance. Foucault (1977, 93) has stressed the importance of light to practices of surveillance. There is no hiding in the light. For example, in a prison 'the strong light and the stare of a guard are better captors than the dark, which formerly was protective. The visibility is a trap.' With a light source, a focusing cell and a directed looking all becomes visible. Manufactured light, then, extended the state and industry's powers of surveillance. The unease that was current in the nineteenth century about streetlights that made the insides of houses visible, and the corresponding use of heavy curtains to block out this light (and the gaze of others), help to make this point.

A final change produced by manufactured light was a re-metaphorisation of texts and bodies as a result of the new perceptions. There were, to begin with, new perceptions of landscape: the night under artificial light was a new world of different colours and sensations, modulated by artificial lighting, which painters like Joseph Wright of Derby had been amongst the first to explore. Late Victorian London was often seen anew through the power of gaslight:

> *London, London, our delight*
> *Great flower that opens but at night.*
>
> (cited in Briggs, 1989, 25)

[. . .]

Power

A further important machinic complex is power. Electrical power is, of course, hardly a new invention. The invention of the Leyden jar sparked off the first electrical experiments in Germany and Switzerland in the 1740s. By the 1780s electricity had achieved some measure of recognition in medical circles, a popularity which it retained in the nineteenth century through devices like galvanic belts, electric baths, electric shock treatment and the like, for the treatment of physical and mental disorders. The telegraph reinforced this recognition by bringing about changes in 'the nature of language, of ordinary knowledge, of the very structures of awareness' (Carey, 1989, 202). It gave a particularly strong boost to a rapidly expanding 'rhetoric of the

electrical sublime' (Carey, 1989). This rhetoric was particularly marked in the United States, where it tended to be couched in the language of either religious aspiration or secular millenarianism. Henry Adams, for example, saw the telegraph as a demonic device displacing the Virgin with the Dynamo. But for Samuel Morse, the telegraph was an electrical nervous system able 'to diffuse with the speed of thought a knowledge of all that is occuring through the land' (cited in Carey, 1989, 207). In Britain, the rhetoric of the telegraph tended to be less extreme but, even there, it fired a powerful flow of thought, most especially when connected with matters of Empire, for which the telegraph was a powerful integrative metaphor.

But it was the promise of infinite light and power that made electricity into a household term. In particular, it was the advent in the 1880s of power stations which distributed electricity along a network that allowed electricity to come into its own, rather as gas had some decades earlier. Indeed, at first, these power stations were often depicted as electrical gasworks. But as it became clear that high voltage could be transported over long distances, electricity power stations were located at long distances away from where the electricity was actually used, thus giving the impression that electricity was a sourceless source, an absent presence. Further, unlike gas, electrical light was immediately accepted in drawing rooms as clean and odour-free. Electricity therefore reached more places more quickly and it quickly became a vast *collective* network of power.

Already in 1881 *Punch* was able to introduce into its pages a cartoon showing King Coal and King Steam watching the infant electricity grow and asking 'What will he grow in to?' Certainly, in the 1880s, it looked as though London might become an 'electropolis'. By 1891 London had 473,000 incandescent lamps, compared with Berlin's 75,000 and Paris's 67,600. [. . .]

Like speed and light, electrical power produced a number of important changes. Three of these stand out. The first change was the boost electrical power gave to new conceptions of time and space. Through the telegraph and the instant power provided by the light switch came ideas of absent presence, of a geography of communication that no longer depended upon actual transport. A second change was the re-metaphorisation of the body into the body electric. From the time of *Frankenstein*, electricity, energy and life [had been] synonymous' (Schivelsbuch, 1988, 71).

Now forces, engines, dynamos, discharges, currents and flow become a new linguistic currency of desire. More than even speed or light, electrical power:

> raised the possibility of an alteration in what Felix Guattari calls the 'social chemistry of desire', in the means through which physical desire, and motivation generally, could be represented, articulated and transmitted. If the social and the individual body are intimately related, then the massive programmes for the wiring of individual houses that went on throughout the late nineteenth century have a parallel in the wiring of individual bodies and in the way in which states of desire, animation, and incandescence are figured in literary and other texts (Armstrong, 1991, 307).

Thus a whole vocabulary grew up (live wires, human dynamos, electric performances, and so on [Nye, 1990]) that depended upon electricity, conceived as 'shadowy, mysterious, impalpable. It lives in the skies and seems to connect the spiritual and the material' (Czitrom, cited in Carey, 1989, 206).

The third and final change was electricity's critical role in the transition to a new form of more integrated capitalism. Society was tied together and made subservient to the new networks of power.

> the transformation of free competition into corporate monopoly capitalism confirmed in economic terms what electrification had confirmed technically; the end of individual enterprise and an autonomous energy supply. It is well known that the electrical industry was a significant factor in bringing

about these changes. An analogy between electrical power and finance capital springs to mind. The concentration and centralisation of energy in high capacity power stations corresponded to the concentration of economic power in the big banks (Schivelsbuch, 1988, 74).

REFERENCES FROM THE READING

Armstrong, T. (1991), 'The electrification of the body', *Textual Practice*, 8, 16–32.

Briggs, A. (1989), *Victorian Things*, London: Batsford.

Briggs, A. (1992), 'The later Victorian age'. In B. Ford (ed.), *Victorian Britain. The Cambridge Cultural History, Volume 7*, Cambridge: Cambridge University Press, 2–38.

Carey, J. W. (1989), *Communication as Culture: Essays on Media and Society*, London: Routledge.

Foucault, M. (1977), *Discipline and Punish: The Birth of the Prison*, London: Allen Lane.

Nasaw, D. (1992), 'Cities of light, landscapes of pleasure'. In D. Ward and O. Zunz (eds), *The Landscape of Modernity*, New York: Russell Sage Press, 273–86.

Nye, D. (1990), *Electrifying America: Social Meanings of New Technology*, Cambridge, MA: MIT Press.

Schivelsbuch, W. (1986), *The Railway Journey: The Industrialisation of Time and Space in the Nineteenth Century*, Berkeley: University of California Press.

Schivelsbuch, W. (1988), *Disenchanted Night: The Industrialisation of Light in the Nineteenth Century*, Berkeley: University of California Press.

Editor's references and suggestions for further reading

Beniger, J. (1986), *The Control Revolution: Technological and Economic Origins of the Information Society*, Cambridge, MA: Harvard University Press.

Dupuy, G. (1991), *L'urbanisme des réseaux: théories et méthodes*, Paris: Armand Colin.

Tarr, J. and Dupuy, G. (eds) (1988), *Technology and the Rise of the Networked City in Europe and North America*, Philadelphia: Temple University Press.

Thrift, N. (1990), 'Transport and communications 1730–1914'. In R. Dogshon and R. Butlin (eds), *An Historical Geography of England and Wales*, London: Academic Press, 453–84.

Thrift, N. (1995), 'Inhuman geographies: Landscapes of speed, light and power'. In N. Thrift, *Spatial Formations*, London: Routledge, 256–310.

Wiener, N. (1948), *Cybernetics; or, Control and Communication in the Animal and the Machine*, Cambridge, MA: MIT Press.

'The City and the Telegraph: Urban Telecommunications in the Pre-Telephone Era'

Journal of Urban History (1987)

Joel Tarr

Editor's Introduction

Before the advent of the telegraph all communication required either optical signals (by flags, fires or human hands) or physical movement (by people, pigeons, horses, ships or vehicles). As a result, even relatively small distances could take long periods (days, weeks, months, even years) for messages to move from sender to receiver. The telegraph was the first medium to use near speed of light flows of electricity (i.e. electrons) over wires to uncouple the communication of information from the need for physical movement. But how did the use and construction of the telegraph relate to the cities, and systems of cities, where it was first built and used? And how was telegraph use bound up with changes in urban life and structure?

In this reading Joel Tarr – a Professor of Urban History at Carnegie Mellon University in Pittsburg – analyses in detail how cities and the telegraph influenced each other during the period 1844–80. Telegraph networks transmitted electrical pulses, created using Morse code, along wires stretched across geographical space. They thus brought with them a remarkable increase in speed in communication between cities. Tarr shows how the first telegraphs were initially built between the main areas of potential demand – the centres of dominant business cities. The lines initially followed existing railways and related closely with railway operations. The dominant early users of the interurban telegraph systems, apart from the railways themselves (who used the telegraph to coordinate and control train traffic) were financial markets, brokers and commodity markets. In such markets the use of the telegraph allowed small time delays to be exploited, for the first time, for profit, by entrepreneurial capitalists. But the lack of telegraph links within cities led to the creation of large teams of messenger boys, who were employed by telegraph companies to physically carry telegrams to their destinations within the city.

The story of the telegraph and the city resonates strongly with the contemporary dynamics of cybercities in two important respects. First, in both eras, city cores maintained their advantages in controlling and hosting time-critical markets and central business activities through their early access to the best electronic communications media (as well as to their prestige value, unparalleled access to services, labour and cultural facilities, and unmatched physical transport infrastructures).

Second, in both periods, the explosion of electronic flows between cities actually brought about new means of methods of physically moving hardcopy material around dense city cores at high speed (see Downey, 2002). Today's equivalents of telegram message boys, of course, are bicycle couriers (Culley, 2001). These young men and women risk life and limb on a daily basis to move around the cores of the world's biggest and most congested cities. They deliver digital or non-digital products that are simply too important or sizeable to risk being pushed down a wire. Thus, as in the nineteenth century, the transnational flows of electronic signals help to stimulate the construction of social worlds in city cores in which the movement of bodies become the key factor.

In 1844, Samuel F. B. Morse and Alfred Vail first practically demonstrated the telegraph by sending the message, "What hath God wrought!" from Washington to Baltimore. This device worked by using a key to transmit coded electrical signals representing letters of the alphabet over wire to a receiver. The signals utilized in the United States were a series of dots and dashes known as the Morse code. The telegraph resulted in a remarkable increase in the speed of long-distance information transmission compared with the mails. Because of this advantage, businessmen and entrepreneurs were quick to adopt the technology. Within eight years of Morse's and Vail's demonstration, they had constructed 23,283 miles of wire connecting nearly 500 towns and cities. But while the great advantages of the telegraph for long-distance transmission of information were obvious, they were less clear in regard to intraurban communications. Distances in cities were much shorter, and advantages correspondingly smaller. Each sending and receiving instrument initially required a skilled operator, increasing operating costs and therefore limiting the dissemination of the technology among the set of potential users. These constraints resulted in development of the telegraph as a specialized rather than a universal communications technology.

In an attempt to escape these constraints, in the second half of the nineteenth century inventors and innovators developed a variety of automatic devices, signaling mechanisms and call boxes, indicators, and printing telegraphs. Some, such as the automatic telegraph, speeded up message transmission, while others, such as the call box, were limited to sending simple signals that could produce a response. Indicators and especially printing telegraphs handled more complex information and therefore had more potential uses. The indicators, for instance, would display information while the printing telegraphs could both send and receive messages and could translate coded messages into numbers or text at the receiving end. Although several of the nineteenth-century instruments invented were successful in specialized markets, it was not until the twentieth century that a telegraph printer suitable for general public use was available. The telegraph's technical characteristics, therefore, limited its intraurban effects to situations in which simple signals were sufficient to meet a need, or to those in which potential economic advantages justified the application of expensive automatic and printing equipment.

THE PRIVATE INTRAURBAN TELEGRAPH

[. . .]

The telegraph revolutionized business communications by providing almost instantaneous message transmission between connected points and by sharply reducing information and transactions costs. In these decades, as Allan Pred notes, "time was the great item with commercial men," and speed rather than low rates was the highest priority. The telegraph provided businessmen with immediate and extremely accurate information on market prices and quantities, thus reducing intermarket uncertainties. Mercantile and manufacturing businessmen in the nation's largest cities secured the most benefit from the telegraph. These advantages initially included preferentially low rates between major termini, the provision of multiple wire services, and priority for "through business" between major cities as opposed to messages between less important cities.

Because of its advantages over other existing forms of communications, brokers and speculators made "early and avid" use of the telegraph, thereby transforming market functioning by increasing the availability of price information. Traders formed eleven commodity exchanges in the larger cities between 1845 and 1871. These exchanges utilized telegraphic communications to buy and sell commodities while they were still being grown or transported, and thus "transformed the methods of financing and reduced the costs of the movement of American crops" (Chandler, 1983). The telegraph also played a crucial role in the centralization of the nation's securities markets in New York City between 1850 and 1880. By 1860 the Wall Street exchanges were connected with and set prices for every major city. Bankers, manufacturers, and wholesale and retail business people used the telegraph to negotiate back and forth throughout the day, and to buy and sell, thereby accelerating the rate of capital turnover and reinvestment.

But the increased volume of messages between cities also created a bottleneck in the distribution of information within the city. In the great financial centers such as Boston, New York City, and Philadelphia, young boys working as messengers provided the connections between exchanges, markets, and businesses. Larger brokerage houses often employed between twelve and fifteen messengers. Thus the speed of the short-distance transmission of information within the city was actually slower than long-distance transit time between cities, forming a blockage-of-flow problem within the communications system. Businessmen eagerly sought innovations that offered the potential of eliminating the bottlenecks, and hundreds of devices were patented. Attracted by the demand, inventors and entrepreneurs such as Thomas A. Edison and Moses G. Farmer gravitated to the big cities to set up workshops and to market their devices.

Although the telegraph had accentuated the problem of intraurban information flows by bringing additional messages into the city, adaptations of the basic technology also provided solutions. In 1866, S.S. Laws, a vice-president of the New York Gold Exchange with training in "electrical science," invented the gold indicator or gold reporting telegraph. The gold indicator was a device that received price information sent electrically over wires from the gold exchange and displayed it in the offices of subscribing brokers, eliminating the need both for a messenger and for a skilled operator.

REFERENCE FROM THE READING

Chandler, A. D. (1977), *The Visible Hand: The Managerial Revolution in American Business*, Cambridge, MA: MIT Press.

Editor's references and suggestions for further reading

Culley, T. (2001), *The Immortal Class: Bike Messengers and the Cult of Human Power*, New York: Villard.

Downey, G. (2002), *Telegraph Messenger Boys: Labor, Communication and Technology, 1850– 1950*, New York: Routledge.

Standage, T. (1998), *The Victorian Internet*, New York: Berkeley.

'The Structure of Cities'

from Ithiel de Sola Pool (ed.),
The Social Impact of the Telephone (1976)

Ithiel de Sola Pool

Editor's Introduction

By the late nineteenth century the telephone had started to displace the telegraph as the main means of electronic interpersonal and inter-business communication. Its diffusion through the metropolis was characterised by a paradox. This Reading draws on the work of one of the twentieth century's most influential communications scholars: Ithiel de Sola Pool (1917–1984). In it he argues that, at different times and places and in different ways, the telephone helped to facilitate both the dispersal and suburbanisation of US cities, and the intensifying concentration of city cores. As a classic example of the ways in which new communications media, and infra-structures, are socially produced in subtle combination with changes in the form and dynamics of cities, the telephone supported two apparently opposed effects.

On the one hand, as with the telegraph, the telephone supported control at a distance. Leading business people were freed from the need to supervise factories in person all the time through human surveillance. They could thus establish specialised headquarters in the place where urban centrality brought the best access to market information, infrastructure, specialised support services, and skilled labour. They therefore became increasingly concentrated in specialised head offices located in the central business districts of fast-growing major business cities. Once here, they could benefit from the parallel innovations in urban transport and building technologies. Production sites, in turn, could then be dispersed to lower cost, ex-urban locations, where improved transport and logistics innovations would allow them to link with supply and distribution chains easily (a process that now operates on an international scale as production is dispersed to cheap labour locations around the world whilst being controlled in real time through ICTs) (see Keller Easterling, p. 179).

Skyscrapers, in their own way, were thus made possible by the elevator and the telephone. Without the telephone, the skyscraper would have needed so many lifts to channel the number of messenger boys that would have been needed between its various floors and the outside world that the tiny amount of space left for occupants would have made them uneconomic to build.

On the other hand, telephones – along with widening access to street cars, metropolitan railways, automobiles, electricity, water and sewer systems, and the development of an increasingly home-centred urban culture – facilitated the dispersal of urban populations and the extension of suburbia. This was a particularly strong trend in the US. Telephones could help to overcome some of the negative aspects of geographical withdrawal from the city core. They could help people to continuously connect to families and friends, despite the increasingly wide geographical separations generated by urban sprawl and the growth of the region-wide metropolis or 'megalopolis'. Telephones could also allow consumption at a distance. And they could link dispersed suburbanites with a growing range of public, welfare, medical and educational services and bureaucracies.

Once again, the twentieth-century dynamics linking cities and telephones into machinic complexes resonate strongly with contemporary trends surrounding cybercities. Now, however, the remote-control capabilities of ICTs support the continued centralised power of the so-called 'global' cities that orchestrate the global economy (see Saskia Sassen, p. 195). At the same time, ICTs are supporting a partial decentralisation from city cores, as people who can telework play off the choice to live further from the city with the need to maintain the ability to physically travel less often to work (see Andrew Gillespie and Ranald Richardson, p. 212).

Similarly, ICTs are now used to string together production, management and distribution sites located in dozens of countries and cities across the world in real (or near real) time. Transnational ICT and logistics systems support a deepening, extending and lengthening division of labour and more and more complex choreographies of flow in data, products and assembled parts which integrate a myriad of urban spaces and times across the world (see Keller Easterling p. 179; Ewart Skinner, p. 218).

One of our working hypotheses as we began this study was that the automobile and the telephone – between them – were responsible for the vast growth of American suburbia and exurbia, and for the phenomenon of urban sprawl. There is some truth to that, but there is also truth to the reverse proposition that the telephone made possible the skyscraper and increased the congestion downtown.

The movement out to residential suburbs began in the decade before the telephone and long before the automobile . . . The streetcar was the key at the beginning. Today streetcars have vanished, and the automobile and the telephone do help make it possible for metropolitan regions to spread over thousands of square miles. But the impact of the phone today and its net impact seventy years ago are almost reverse. As John J. Carty tells it:

> It may sound ridiculous to say that Bell and his successors were the fathers of modern commercial architecture – of the skyscraper. But wait a minute. Take the Singer Building, the Flatiron, the Broad Exchange, the Trinity, or any of the giant office buildings. How many messages do you suppose go in and out of those buildings every day? Suppose there was no telephone and every message had to be carried by a personal messenger. How much room do you think the necessary elevators would leave for offices? Such structures would be an economic impossibility (cited in Mumford, 1908).

The prehistory of the skyscraper begins with the elevator in the 1850s; the first Otis elevator was installed in a New York City store in 1857, and with adaptation to electric power in the 1880s, it came into general use (Glaab and Brown, 1967). "The need to rebuild Chicago after the 1871 fire, rapid growth, and rising land values

encouraged experimentation in construction." In 1884, Jenney erected a ten-story building with a steel skeleton as a frame. The Woolworth Building with fifty-seven stories was opened in 1913. "By 1929 American cities had 377 skyscrapers of more than twenty stories" (Glaab and Brown, 1967, 280).

The telephone contributed to that development in several ways. We have already noted that human messengers would have required too many elevators at the core of the building to make it economic. Furthermore, telephones were useful in skyscraper construction; phones allowed the superintendent on the ground to keep in touch with the workers on the scaffolding. As the building went up, a line was dropped from the upper girders to the ground.

As the telephone broke down old business neighborhoods and made it possible to move to cheaper quarters, the telephone/tall-building combination offered an option of moving up instead of moving out. Before the telephone, businessmen needed to locate close to their business contacts. Every city had a furrier's neighborhood, a hatter's neighborhood, a wool neighborhood, a fishmarket, an egg market, a financial district, a shipper's district, and many others. Businessmen would pay mightily for an office within the few blocks of their trade center; they did business by walking up and down the block and dropping in on the places where they might buy or sell. For lunch or coffee, they might stop by the corner restaurant or tavern where their colleagues congregated.

Once the telephone was available, business could move to cheaper quarters and still keep in touch. A firm could move outward, as many businesses did, or move up to the tenth or twentieth story of one of the new tall buildings. Being up there without a telephone would have put an intolerable burden on communication.

The option of moving out from the core city and the resulting urban sprawl has been much discussed, but most observers have lost sight of the duality of the movement; the skyscraper slowed the spread. It helped keep many people downtown and intensified the downtown congestion. Contemporary observers noted this, but in recent decades we have tended to forget it. Burlingame, for example, said:

> It is evident that the skyscraper and all the vertical congestion of city business centers would have been impossible without the telephone. Whether, in the future, with its new capacities, it will move to destroy the city it helped to build is a question for prophets rather than historians (Burlingame, 1940, 96).

Burlingame, before World War II, already sensed that things were changing. The flight from downtown was already perceptible enough to be noted as a qualification to his description of the process of concentration; both processes have taken place at once throughout the era of the telephone. The telephone is a facilitator used by people with opposite purposes; so we saw it with crime, and so it is here, too. It served communication needs despite either the obstacle of congested verticality or the obstacle of distance; the magnitude of the opposed effects may differ from time to time, and with it the net effect. At an early stage the telephone helped dissolve the solid knots of traditional business neighborhoods and helped create the great new downtowns; but at a later stage, it helped disperse those downtowns to new suburban business and shopping centers.

The telephone contributed in some further ways to downtown concentration in the early years – we have forgotten how bad urban mail service was. The interurban mails worked reasonably well, but a letter across town might take a week to arrive. Given the miserable state of intracity communication, the telephone met a genuine need for those who conducted business within the city.

The telephone also contributed to urban concentration in the early days because the company was a supporter of zoning. The reasons were similar to ones motivating cablecasters today. Cablecasters are inclined to string cable TV through comfortable middle-class neighborhoods where houses are fairly close together but in which utility lines do not have to go underground; they get their highest rate of penetration at the lowest price that way. Only under the impact of regulation do they cover a city completely.

REFERENCES FROM THE READING

Burlingame, R. (1940), *Engines of Democracy*, New York: Scribner.

Glaab, C. N. and Brown, A. T. (1967), *A History of Urban America*, New York: Macmillan.

Mumford, J. K. (1908), 'This land of opportunity: the nerve center of business', *Harper's Weekly*, 52, August, 23.

Editor's references and suggestions for further reading

Fischer, C. (1992), *America Calling: A Social History of the Telephone*, Berkeley: University of California Press.

Katz, J. (1999), *Connections: Social and Cultural studies of the Telephone in American Life*, Brunswick, NJ: Transaction.

Poitras, C. (2000), *La cité au bout du fil: le téléphone à Montréal de 1879 à 1930*, Quebec: Les Presses de l'Université du Montréal.

Sola Pool, I. de (ed.) (1976), *The Social Impact of the Telephone*, Cambridge, MA: MIT Press.

'The Urban Place and the Non-Place Urban Realm'

from *Explorations Into Urban Structure* (1964)

Melvyn Webber

Editor's Introduction

One of the most powerful trends in modern societies has been the gradually uncoupling of the spaces of social, economic and cultural interaction, flow and mobility from the particularities of individual places (see Giddens, 1990). Urban commentators have long speculated on what lowering transport and communications costs, and the 'stretching' of relationships across territory (a process, in its ultimate form, which is known as 'globalisation') might mean for traditional face-to-face relationships and communities in cities.

This reading presents an extract from one of a series of extremely influential articles produced in the 1960s by the US urban geographer Melvyn Webber, who worked at the University of California at Berkeley. What, asked Webber, would the growth of long-distance connections, facilitated by telephones, cars, air travel and personal computers – what he termed the 'nonplace urban realm' – mean for the physical propinquity, and the urban place, of the traditional city ? Detailing the growth of professional and social interest communities across long distances, Webber argued that these presented a wide diversity which complemented, rather than destroyed, place-based communities in traditional urban communities.

In his later writings, including the article 'The post-city age' published in the journal *Daedalus* in the fall of 1968, Webber predicted a much more powerful growth of communications technologies. These, he believed, would lead to a non-place urban realm which would, in some cases, match the contact and information potential gained through living in a major metropolitan region.

Webber was one of the first urban scholars to seriously address the complex relationships between electronic media, growing mobility, and the nature and experience of late twentieth-century cities. To many, his concept of the non-place urban realm resonates strongly with the more widely used contemporary metaphor of cyberspace. Webber's work seems extremely prescient given the explosive growth of the Internet in the late 1990s.

THE INTEREST-COMMUNITIES

Both the idea of city and the idea of region have been traditionally tied to the idea of place. Whether conceived as physical objects, as interrelated systems of activities, as interacting populations, or as governmental dominions, a city or a region has been distinguishable from any other city or region by the fact of territorial separation.

The idea of community has similarly been tied to the idea of place. Although other conditions are associated with the community – including "sense of belonging," a body of shared values, a system of social organization, and interdependency – spatial proximity continues to be considered a necessary condition.

But it is now becoming apparent that it is the accessibility rather than the propinquity aspect of "place" that is the necessary condition. As accessibility becomes further freed from propinquity, cohabitation of a territorial place – whether it be a neighborhood, a suburb, a metropolis, a region, or a nation – is becoming less important to the maintenance of social communities.

The free-standing town of colonial America may have fixed the images that are still with us. At that time, when the difficulties and the costs of communicating and traveling were high, most of the townsman's associations were with other residents of the town he lived in. In turn, intratown intercourse tended to strengthen and stabilize the shared values and the systems of social organization; they must surely have reinforced the individual's "sense of belonging to his town." But even in colonial days when most men dealt solely with their neighbors, there were some who were simultaneously in close touch with others in distant towns, who were members of social communities that were not delimited to their home-town territories. As recently as two hundred years ago these long-distance communicators were uncommon, for the range of specialization and the range of interests among colonial populations were quite narrow. But today, the man who does not participate in such spatially extensive communities is the uncommon one.

Specialized professionals, particularly, now maintain webs of intimate contact with other professionals, wherever they may be. They share a particular body of values; their roles are defined by the organized structures of their groups; they undoubtedly have a sense of belonging to the groups; and, by the nature of the alliances, all share in a community of interests. Thus, these groups exhibit all the characteristics that we attribute to communities – except physical propinquity.

Spatial distribution is not the crucial determinant of membership in these professional societies, but interaction is. It is clearly no linguistic accident that "community" and "communication" share the Latin root *communis*, "in common." Communities comprise people with common interests who communicate with each other.

Though it is undoubtedly true that the most frequent interactions are among those members of professional communities who live and work within close distances of each other, the most productive contacts – those in which the content of the communication is the richest – are not necessarily the most frequent ones nor those that are made with associates who happen to be close

by. Although specialized professionals are a rather extreme example, we can observe the same kinds of associational patterns among participants in the various types of nonprofessional communities as well.

All this is generally recognized. There is no novelty in noting that members of occupational groups are members thereby of limited-interest fraternities. Similarly, it is commonly recognized that members of churches, clubs, political parties, unions, and business organizations, and that hobbyists, sportsmen, and consumers of literature and of the performing arts are thereby also members of limited-interest groups whose spatial domains extend beyond any given urban settlement. A "true" community, on the other hand, is usually seen as a multi-interest group, somewhat heterogeneous, whose unity comes from interdependencies that arise among groups when they pursue their various special group interests *at a common place.*

I do not challenge the utility of this idea for certain purposes. A metropolis is indeed a complex system in which interrelated and interdependent groups support each other by producing and distributing a wide assortment of services, goods, information, friendships, and funds. Groups located in a given metropolis, where they carry out activities of all sorts, certainly do thereby create a systematic structure through which they work; they do indeed form a community whose common interests lie in continuing the operations of the place-based metropolitan system.

Nevertheless, the place-community represents only a limited and special case of the larger genus of communities, deriving its basis from the common interests that attach to propinquity alone. Those who live near each other share an interest in lowering the social costs of doing so, and they share an interest in the quality of certain services and goods that can be supplied only locally. It is this thread of common interests in traffic flow on streets, garbage collection, facilities for child rearing, protection from miscreant neighbors and from the inhospitable elements, and the like, that furnishes the reason-for-being of municipal government. This thread is also the basis for certain business firms and voluntary institutions that supply other goods and services which inhabitants and firms in the place-community demand with frequent recurrence. But, over time, these place-related interests represent a decreasing proportion of the total bundle of interests that each of the participants holds.

With few exceptions, the adult American is increasingly able to maintain selected contacts with others on an interest basis, over increasingly great distances; and he is thus a member of an increasing number of interest-communities that are not territorially defined. Although it is clear that some proportion of his time is devoted to associations with others in his place-community, several long-run changes seem to be underway that are cutting this proportion back. I need mention only rising educational levels and hence greater access to information and ideas; increasing amounts of non-working time; rising income levels that make long-distance communication and travel feasible, and the technological changes that are making them possible; and the weakening barriers to association across ethnic, racial, and class lines. These changes are expanding the range of diversity in the average person's associations and are inducing a parallel reduction in the relative importance of place-related interests and associations.

Although each generation of Americans participates in more substantively and spatially diverse communities than did its predecessor, there is at any point in time a wide range of variation among contemporaries. Today there are people whose lives and interests are only slightly more diverse than those of their eighteenth-century counterparts; there are the notorious Brooklynites who, except for contacts through television, movies, and the other mass media, have never crossed the East River. But there are others who are at home throughout the world.

Editor's references and suggestions for further reading

Calhoun, C. (1986), 'Computer technology, large-scale social integration and the local community', *Urban Affairs Quarterly*, 22(2), 329–49.

Giddens, A. (1990), *The Consequences of Modernity*, Oxford: Polity.

Hall, P. (1996) 'Revising the nonplace urban realm: have we come full circle?', *International Planning Studies*, 1(1) 7–15.

Thompson, J. (1995), *The Media and Modernity: A Social Theory of the Media*, Cambridge: Polity.

Webber, M. (1964), 'The urban place and the non place urban realm'. In M. Webber, J. Dyckman, D. Foley, A. Guttenberg, W. Wheaton and C. Whurster (eds), *Explorations into Urban Structure*, Philadelphia: University of Pennsylvania Press, 79–153.

Webber, M. (1968), 'The post-city age', *Daedalus*, Fall, 107(4), 237–78.

'The Cable Fable Revisited: Discourse, Policy and the Making of Cable Television'

Critical Studies in Mass Communication (1987)

Thomas Streeter

Editor's Introduction

For over one and a half centuries electronic media have been accompanied by waves of hyperbolic and even messianic expectation. This has emanated from technology producers, advertisers, regulators, policy-makers, futurists, science fiction writers, novelists and all those set to benefit from uncritical support for new media innovations. In contemporary societies new media have few equals as symbols of progress, modernity and the transcendence of old shackles and constraints. As foci for utopian dreaming they are unmatched. And as centrepieces for efforts to whip consumers, public agencies and possible generators of profit into an expectant (and uncritical) frenzy of expectation, they have few equals.

All electronic media have generated their own particular discursive histories. All have perpetuated their own myths, fallacies and unrealistic expectations. Such discourses have had major influences on the shaping of adoption trends as processes of innovation, introduction, diffusion and, ultimately, 'banalisation' – the process through which a technical artefact becomes so banal and taken for granted that it ceases to have any novelty value – have played out over time.

In this reading the American communications scholar Thomas Streeter examines one such case: the fable that the introduction of cable TV to US cities in the 1970s would necessarily bring with it an urban democratic revolution. Streeter analyses the coalition of interests that participated in the widespread depiction of cable TV at that time as an inherently democratic medium which offered unlimited interactive potential and channel space which individuals and communities would be able to use and control. He documents how, in New York (as in many other US urban areas), the promise of the 'wired' or cabled city seemed to herald the possibility of a bi-directional and publicly organised and controlled information infrastructure. This would, it seemed, finally match the City's control of transport, water and other infrastructures. In so doing, the new two-way, community-level information infrastructure would work to counteract the overwhelmingly commercial, one-way, and passive nature of US terrestrial TV and radio broadcasting and newspaper publishing.

As it transpired, however, the democratic potential of cable TV has rarely been realised. Whilst there were many experiments in the 1980s to orchestrate wired cities through integrated cable and ICT networks organised at the local or city level (see the book *Wired Cities* by Dutton *et al.*, 1987), few of these were successful or sustainable.

Most cable networks are today small elements within the commercial portfolios of massive multimedia transnationals like AT&T and Time Warner. These behemoths have consolidated their control over the hard information infrastructures as well as the software and the content of so-called 'digital capitalism' (see Schiller, 2000). At best, current cable networks tend to offer local information and programming to be consumed in each city. Only a very small number offer support for local communities to make and broadcast their own programmes. At worst, today's digital and interactive cable systems merely pump hundreds of commercialised TV, pay-per-view and shopping channels into each home and attempt to squeeze income from these through individual surveillance and the tying-in of in-house online shopping and services promotions.

Relations with the particularities of place in contemporary cable networks are often arbitrary to say the least. Many offer no opportunity whatsoever for using the 'upstream' channel capacity to allow users to actually broadcast their own material to local communities: a key element of the democratic wired city visions of the 1960s and 1970s. In fact, most cable networks now are explicitly designed very 'asymmetrically'. That is, they have a very large 'downstream' capacity for consumers to enjoy corporate products, but very small (or non-existent) 'upstream' capacity for users' use. Even this is often limited to the signals generated by 'Press Now to Purchase' buttons.

The cable fable offers salutary lessons when it is set against the waves of utopian, deterministic and messianic discourses that have surrounded the growth of the Internet, multimedia and virtual realty technologies since the mid-1990s. These technologies, too, have been widely heralded as new media with intrinsic powers of democratisation, decentralisation and community building in cities.

The lessons from the cable experience are clear, however. These discourses tend to lead to assumptions that the new medium will have certain effects automatically, in isolation from the broader economic, political, social and regulatory contexts which actually shape their configuration, control and use. The production and serial repetition of utopian discourses about cities and ICTs are part of the wider production of mythological and symbolic representations which help shape new media in particular ways within contemporary capitalist society. And the messianic representation of new media usually has very little resemblance to what actually results once the media diffuse into the mainstream and become increasingly controlled by large, transnational, media corporations.

The imperative of maximising profitability, in particular, often tends to reduce the degree to which new ICT media are shaped to support democratic interactions and local community applications – even if this potential is significant (see Section VII). Explicit public, regulatory strategies and local and non-local social innovations and experiments are therefore required to hold large media transnationals to account and so open up ICTs to realise their full social potential (see Section VIII).

In 1974, Brenda Maddox introduced a book chapter titled "From CATV to Infinity" with the following passage:

An almost religious faith in cable television has sprung up in the United States. It has been taken up by organizations of blacks, of consumers and of educational broadcasters, by the Rand Corporation, the Ford Foundation, the American Civil Liberties Union, the electronics industry, the Americans for Democratic Action, the government of New York City, and – a tentative convert – the Federal Communications Commission. The faith is religious in that it begins with something that was once despised – a crude makeshift way of bringing television to remote areas – and sees it transformed over the opposition of powerful enemies into the cure for the ills of modern urban American society (145).

Today, it is hard not to think that this passage was written with irony. From the perspective of 1987, when for most people cable means MTV, HBO, and reruns, the notion that cable television could be a "cure for the ills of modern urban American society" seems an exaggeration to say the least. But the author, in 1974, was speaking with serious admiration: "The intriguing thing about cable television," she continued, "is that this faith may be in no way misplaced."

Maddox's passage is an example of what I call the "discourse of the new technologies," a pattern of talk common in the policy-making arena around 1970. [. . .]

The discourse had its effect because its structure – its particular mixture of themes, blind spots, and gaps – made possible an odd alliance between the community antenna television (CATV) industry, certain professional groups, and some liberal progressive organizations. The discourse created a terrain that helped make possible some major actions in the policy arena, actions that simple self-interest – of either a pluralist or Marxist, class variety – would not warrant. The goals, interests, and philosophies of the many contributors to the discourse of the new technologies were widely varied, sometimes to the point of being mutually antagonistic. Yet the participants did not understand the alliance as a compromise between groups with different but overlapping interests; rather, they saw it as a solid consensus, as what one policy activist dubbed "a great and growing body of competent, impartial opinion." The new discursive field helped create a *sense* of expert consensus, of unity and coherence where there actually was a variety of conflicting motivations, attitudes, and opinions.

Somewhat paradoxically, the gaps and contradictions in the new field of discourse, its lack of internal unity, made possible the sense of unity on the level of policy action. The gaps and contradictions played an important, double role. On the one hand, they allowed some to look out on the field and see a view very different from the view others saw; each group saw in the discourse a vision in the service of its goals and ignored crucial differences and potential conflicts with other groups. On the other hand, the gaps and contradictions allowed the discourse to unite all these diverse viewpoints in terms of a shared sense of awe and excitement. Maybe the "new technologies" of cable were good; maybe they were bad, but in any case they inspired a sense of urgency, of possibility, and of a need for action, for response [. . .]

In what Le Duc (1973, p. 5) described as "an ever expanding chorus of expert opinion," the new, hopeful view of cable television echoed throughout the policy arena in the late 1960s and early 1970s, appearing in numerous scholarly articles, studies, hearings, and journalistic publications. An important galvanizing force in this development was the Rand Corporation, which began research on cable television issues in 1969, with support from, among others, the Ford Foundation. Rand published more than a dozen reports on the topic over the next three years. The Alfred P. Sloan Foundation established the Commission on Cable Communications in the spring of 1970, which commissioned over 15 studies and produced a book length report (Sloan Foundation, 1971, p. vii). The fever went beyond the foundations, however. Articles appeared in *The New York Times* and *The Saturday Review*. The influential British weekly, *The Economist*, became a regular advocate of the new vision. A major article appeared in *The Nation* in the Spring of 1970, later to be published in expanded form as a book called *The Wired Nation*. Numerous progressive groups such as the ADA (Americans for Democratic Action) and

the ACLU (American Civil Liberties Union) became interested in and began making contributions to cable policy proceedings. While there are important differences in many of these texts, they all share a sense of urgency, a sense of activism, and a sense of working against stifling and powerful conservative forces. Cable had captured the imagination, not just of those traditionally concerned with television but of what seemed to be an entire cross section of the U.S. policymaking community.

Irving Kahn's confident reference to the existence of a "great and growing body of competent, impartial opinions" was not based on any explicit, thoroughly worked-out theory that can be located in a single statement or document. Rather, instances of the discourse typically were invoked in passing, as introductory or concluding passages to otherwise more concrete and specific arguments, policy recommendations, and research reports. For example, in 1968 the Advisory Task Force on CATV and Telecommunications for the City of New York published a report that was, for the most part, relatively brief and pragmatic. It recommended the introduction of state-of-the-art cable systems for each borough of the city, with rates and programming regulated, but not absolutely determined, by guidelines established by the city council. Most of its 50 pages referred to the specific details of the situation in New York. But the report concluded with the following passage:

> In conclusion the promise of cable television remains a glittering one . . . Those who own these electronic circuits will one day be the ones who will bring to the public much of its entertainment and news and information, and will supply the communications link for much of the city's banking, merchandising, and other commercial activities. With a proper master plan these conduits can at the same time be made to serve the City's social, cultural, and educational needs (New York City, 1968, v–vi).

It was this kind of passage that filtered most widely into policy discourse at large, not the report's data, analyses, and recommendations. The references to next generation high capacity, two-way cable systems, to satellites, to systems that combined voice, computer, and television signals all on the same wire, to the generally "glittering" promise of this new dazzling technology – these were the particulars of the New

York City report that found their way into discussions in the FCC (Federal Communications Commission), the Rand Corporation, and the elite popular press. Paradoxically, the specific details of the New York City report served largely as window dressing, while its vague speculations had a very concrete impact that went far beyond the borders of New York City. And this pattern was repeated in numerous other studies, books, and reports of the period.

REFERENCES FROM THE READING

Le Duc, D. R. (1973), *Cable Television and the FCC: A Crisis in Media Control*, Philadelphia: Temple University Press.

Maddox, B. (1974), *Beyond Babel: New Directions in Communications*, Boston: Beacon Press.

New York City, Mayor's Advisory Task Force on CATV and Telecommunications. (1968), *A Report on Cable Television and Cable Telecommunications in New York City.*

Sloan Foundation. (1971), *On the Cable: The Television of Abundance*, New York: McGraw-Hill.

Editor's references and suggestions for further reading

Dutton, W., Blumler, J. and Kraemer, K. (eds) (1987), *Wired Cities: Shaping the Future of Communications*, Washington, DC: Communications Library.

Keleman, M. and Smith, W. (2001), 'Community and its "virtual" promises: a critique of cyberlibertarian rhetoric', *Information, Communications and Society*, 4(3), 370–87.

Mattelaart, A. (1999), 'Mapping modernity: Utopia and communications networks'. In D. Cosgrove (ed.), *Mappings*, London: Reaktion Books, 169–92.

McBeath, G. and Webb, S. (2000), 'On the nature of future worlds? Considerations of virtuality and utopia', *Information, Communications and Society*, 3(1), 1–16.

Schiller, D. (2000), *Digital Capitalism: Networking the Global Market System*, Cambridge, MA: MIT Press.

Schuler, D. (1996), *New Community Networks: Wired For Change*, New York: Addison Wesley.

'Webcameras and the Telepresent Landscape'

Thomas J. Campanella

Editor's Introduction

One of the most notable recent developments bringing ICTs and cities into close articulation is the webcam phenomenon. As a perfect illustration of the ways in which ICTs and cities are today mutually constitutive, it is now possible, via the Internet, to access webcameras, as distant prostheses, to 'see' an expanding universe of urban (and non-urban) sites across the world on a computer screen. With similar ease one can watch the room next door or witness a sunset on the far side of the world.

But how does this explosion of telepresent, and instantly accessible, urban prosthetic eyes relate to constructions of technology, visibility, the urban spectacle, and cultures of looking, in previous eras? In this specially commissioned piece Thomas Campanella, an urbanist interested in the new media who works at the University of North Carolina, analyses these connections. Exploring the history and diffusion of webcams, Campanella traces their complex genealogical linkages to traditional post, telephone, telegraph, camera and television technologies. He also asserts that webcams are direct descendants of electromechanical attempts to project cityscapes onto screens for remote, real time, viewing, such as the *camera obscura* and *camera lucida*.

To Campanella, webcams herald a dramatic explosion in the availability and performance of remote prosthetic eyes for the viewing of distant places in (near) real time. They exemplify the ways in which cybercities are seamlessly fused into transglobal digital media systems, with many millions of terminals, which themselves act as cyborg-like prostheses of human senses and organs. And they intensify the logics of speed, distanciation, and the movement from asynchronous communications (which involve a delay between sending and receiving) to synchronous, or real time, ones (which have very little delay). What webcams do not do, however, is bring anything like the rich co-presence enjoyed when people are physically present in real places.

It was often claimed during the reign of Queen Victoria that the sun never set upon the British empire. While the age of global imperialism is long over, we have in recent years constructed an empire far greater than any in history. This vast and somewhat elusive realm is the cyberspatial empire of networked computers that stirred to life with the PRATT (developed by the Advanced Research Project Agency of the US Department of Defense) in the 1960s, and has since evolved into the Internet and world wide web. Today we are busily transposing into this new cultural space seemingly every detail of the human experience; cyberspace is, as David Gelernter, in his book *Mirror Worlds* (1991) memorably put it, a vast 'mirror world' reflecting reality itself.

In the early days of the digital revolution it was often heard that cyberspace threatened the robustness and vitality of the so-called 'real' world, that the two were opposing entities fated to an oil-and-water battle for supremacy. There were even predictions that the city itself, that paragon of physicality, was doomed. To these anti-urbanists, the modem's trill proclaimed a coming age of telecommuting from wired rural idylls. This vision, favored by Marshall McLuhan, Alvin Toffler, George Gilder and other futurists, was only the latest iteration of an age-old American scepticism about cities, something we inherited from the quixotic polymath depicted on our nickels.

But even in the early days of the web, there was evidence suggesting that people were using their net-worked computers to affirm and even celebrate the values of place that were supposedly heading the way of the dodo. For one, we started using computer-based communication to recreate the kinds of interaction once afforded by dense urban neighborhoods or small towns – the kinds of communal life celebrated by Jane Jacobs and Garrison Keillor. People who first met online found themselves arranging to meet in 'meat space.' And then a handful of hackers started patching old video cameras into their computers, apparently for no other reason than to share with others a glimpse of the physical environment from which they tapped into the global bitstream.

It was no great technological leap to realize that the digitized input from a camera could be distributed to a huge number of users via a computer network. This is just what a pair of Cambridge University computer scientists, Quentin Stafford-Fraser and Paul Jardetzky, did in 1991. After connecting a camera to a computer equipped with a video frame grabber, the pair chose as their subject a humble coffee pot next to a lab known as the Trojan Room. They wrote a simple client-server program that captured images from the camera every few minutes and distributed them on a local network, thus enabling people throughout the building to check if there was coffee before embarking on the long trek downstairs. Eventually Stafford-Fraser and Jardetzky began serving images over the web itself, in the process giving birth to the Internet's first 'webcamera.'

By the mid-1990s there were hundreds of webcams on the net. Following the geography of the net itself, the early units were all located in the United States, Europe or Japan. But they appeared rapidly in places farther off the digital mainline – from Mexico to Pakistan, South Africa to the Czech Republic. Today, most major cities in the world – with the exception of those in Africa, North Korea, and mainland China – can attest to at least one, and often scores, of webcams. The advent of this collective electronic retina brought new meaning to the old maxim about the sun and the British empire. With webcams, any modem jock

could prove that when the sun went down in Atlanta, Georgia, it was only just coming up over Tokyo bay. As the electro-optical matrix expanded around the globe, it became possible to simultaneously watch the sun rise and set at the same time – something that, as little as a decade ago, would have been the stuff of science fiction.

Webcams are liminal devices. They operate on the threshold of the physical and the cybernetic, like points of contact between reality and the virtual realm. The cameras grab data from the real world and translate it into machine-readable code. Those bits of digitized information may be no different from stored data on a hard drive; but because they are only briefly separated from the pulse and hue of life – for a time at least – they are infused with a touch of magic. Webcams begin to yield what Mark Stefik (1996, 263) has called 'a rich interaction that interweaves the images and agencies in the real world with those of the imagination and cyberspace.' Put another way, they 'map' reality onto cyberspace, and vice versa.

Contrary to projections that computer-based virtu-ality would abrogate traditional qualities of place and diminish our attachment to particular real-world environments, webcams prove that we are actually using networked computers to give real places new meaning in the cybernetic world. Webcams are espe-cially common means to allow iconic global tourist cities such as Florence to be visually 'consumed' at a distance (Plate 5). In major global cities like Paris whole suites of webcams are now marshalled to allow a comprehensive real-time picture of the elements of the subject city to be built up (Plate 6).

Few sites better illustrated the use of webcams to give urban places new meaning than the late, great, UpperWestSide Cam in New York City (Plate 7). Installed by David Spector in 1995, the webcamera was for a time among the most popular on the net. This was due in large part to the close-range view it afforded of a busy Manhattan street corner, and to the quality of the image served (details as fine as a person's face could be distinguished). The camera was mounted in a window above the intersection of 73rd Street and Columbus Avenue, where it took in an archetypal street corner in one of Manhattan's most vibrant neighbor-hoods.

As news of the webcam spread, its community of users expanded around the world. People held banners within its range, broadcasting messages to friends and relatives on the other side of the world. Old residents

Plate 5 The Florence Duomo Cupola webcam.

of the neighborhood used the webcam to glimpse back nostalgically. One couple planted themselves periodically in front of the camera to 'be' with their daughter in Sweden. More, genuine interest developed in the built environment captured by the lens; inquiries poured in about a particular restaurant in view, or a shop across the street, even about a certain delivery truck which continually appeared curbside. The UpperWestSide Cam Frequently Asked Questions (FAQ) filled with detailed information about a unique urban environment. The webcam generated real interest in a real place – a Manhattan street corner and a New York neighborhood. The camera added a whole new stratum of cultural space to the corner of 73rd and Columbus; now, it was not only a bustling New York intersection, but a street corner in cyberspace as well.

On a broader scale, webcams have also brought about something of revolution in our perceptions of time, space and global geography. Before this technology, the *synchronous* observation of distant places (those beyond, say, the reach of a telescope) was for all practical purposes an impossibility. To watch the sun set over Mount. Fuji in real time would have required either physically being present in Japan, or having access to a live television broadcast of the mountain at dawn. Webcams are byte-sized portals into distant worlds; they have, like the telephone and the Boeing jet, helped dwindle the one-time vastness of the earth.

Indeed, the story of technology is largely one of abnegating distance (time expressed in terms of space, or vice versa). For millennia, real-time communication was limited to the range of sound and sight. Signaling devices such as smoke, drums, cannon (or Paul Revere's belfry-borne lantern) were effective only as far as weather and topography permitted. Asynchronous communication, using drawn or written messages, was being carried around the globe by the time of the Renaissance, but the lag between dispatch and receipt could be years. In the middle ages the average person's world was small indeed – limited by the hills or walls surrounding their town. Travel was costly and dangerous; those who took to the roads were often criminals and outcasts from society. Indeed, the etymology of the word 'travel' is the Old French *travailler* or 'travail' – to toil and suffer hardship.

Plate 6 Webcam from the Place de La Bastille. (Source: Parispourvous.com)

Plate 7 The Upper West Side webcam in New York City.

It was not until about 1850 that technology began to profoundly expand the spatial parameters of daily life. The development of rail transport had the greatest impact. The locomotive destroyed the tyranny of distance, and ended what Stephen Kern (1983, 213) has called 'the sanctuary of remoteness.' Once-distant rural towns suddenly found themselves newly within reach of urban markets (if they were fortunate enough to be near a rail line; places bypassed by the train, such as many New England hill towns, often found themselves newly remote). Rail transport also brought about a new temporal order: an abundance of local time zones made the scheduling of trains a logistical nightmare, and eventually led to the adoption of a uniform time standard in the United States.

Subsequent advances in transportation technology – fast steamers, the Suez Canal, and eventually the airplane – compressed the great distances separating Europe, Asia, and America. Circumnavigation of the globe, a dream of ages, became reality not long after Jules Verne's *Around the World in Eighty Days* was published in 1873. Inspired by the novel, American journalist Nellie Bly became, in 1890, the first to circle the earth in less than the vaunted eighty days. In the following two decades, the scale of the globe itself progressively shrank. A journey to China had, by 1936, been reduced to a two-day flight by China Clipper. With the arrival of commercial jet aviation, traversing the earth was within a day's travel and a middle-class budget.

Innovations in electronics further annulled the old order of time and space. The first electric telegraph line, linking Baltimore and Washington, went into service in 1844; a mere twenty years later the first transatlantic cable was installed – the first segment of today's global telecom network. Marconi discovered that telegraphic signals could be transmitted via electromagnetic waves, and in 1902 he succeeded in sending the first transatlantic wireless message. The telephone, which spanned the United States by 1915, brought the power of distant synchronous communication into the kitchen and prompted predictions of home-based work as early as 1914.

The development of the digital computer – and especially the Internet – further neutralized the old order of geography. The net was conceived to do just that. PRATT, a Cold War-era defense initiative intended to create a multi-nodal knowledge-sharing infrastructure that could withstand a nuclear strike to one or more of its nodes, effectively made geography irrelevant. If any one part of the system was destroyed by a Soviet Intercontinental Ballistic Missile – New York or Washington, for instance – mission-critical data would simply re-route itself around the blockage.

While the development of the worldwide webcam network has helped span the gulf between place and cyberspace, it has also yielded something of a great grassroots telepresence project. *Telepresence* is a term applied to a wide range of phenomena, often inaccurately. It was coined in 1980 by MIT Artifical Intelligence pioneer Marvin Minsky, who applied it to tele-operation systems used in remote object-manipulation applications. One working definition of telepresence is 'the experience of presence in an environment by means of a communication medium.' Put another way, it is the mediated perception of a distant real environment.

The genealogy of telepresence begins with simple optical devices such as the telescope, microscope, *camera lucida* and *camera obscura*. Each of these ported an observer into a remote scene in real time; but it is the latter that is the webcam's most proximate antecedent. Though its optical principles were described by the fifth century BC, the camera obscura is generally attributed to the German astronomer Kepler, who designed a portable instrument for use in a tent. Guyot later described a camera obscura which projected an image upward onto a transparent tabletop, foreshadowing the desktop monitor. His device enabled viewing by those gathered around a table, and could also accommodate tea cups or a book.

Camera obscura technology scaled easily, and eventually room-size stations were built. One of these was at the Royal Observatory, Greenwich. Using mirrors and a lens, the camera image was projected on a concave plaster of Paris table, viewable by many people at once. Like so many webcams today, cameras obscura were often situated to enable a sweeping view of cityscape – the most famous of which was the Outlook Tower in Edinburgh, later acquired by the redoubtable town planner Sir Patrick Geddes.

Synchronous co-presence by electronic means remained a dream until relatively recently. The earliest depiction, published in an 1879 edition of *Punch*, imagined an 'Edison Telephonoscope' enabling family members in Ceylon to be telepresent in a Wilton Place villa. The first experiments in transmitting still images by telegraph took place in the 1840s. Twenty years later Abbé Caselli devised a system which used rotating cylinders wrapped with tin foil to transmit and receive photographs and handwritten notes. As early as the 1880s, photographs had been transmitted via radio signal in England; and by 1935, Wirephotos enabled the rapid transmission of photographs around the globe.

The electrical transmission of *live* images was first explored by the German physicist Paul Nipkow in the 1880s. Nipkow understood that the electrical conductivity of selenium changed with exposure to light, and that all images were essentially composed of patterns of light and dark. Based on this, he devised an apparatus to scan (using a rotating disk) a moving image into its light-and-dark components and convert this into electrical signals. The signals would then illuminate a distal set of lamps, projecting the scanned image on a screen. Nipkow's ideas, which remained theoretical, provided the basis for the early development of television.

Until the advent of the net, television remained the closest thing to telepresence most people would ever experience. And while videoconferencing technology has become more accessible in recent years, it has yet to move much further than the company boardroom. The arrival of the web, by providing inexpensive and ready access to a global computer network, made elementary telepresence a reality for anyone with a modem, PC, and an inexpensive camera such as the original Connectix QuickCam. The world wide web brought telepresence to the grassroots.

Admittedly, webcams hardly satisfy the more rigorous definitions of telepresence. David Zeltzer, for example, has argued that to achieve a true sense of 'being in and of the world' – real or virtual – requires full sensory immersion, a 'bath' of inputs, as he puts it. This is something that can be only be achieved with high-bandwidth, multisensory streams of data from the remote world. Held and Durlach have similarly argued that 'high telepresence' requires a high resolution image, a wide field of view, and a multiplicity of feedback channels – visual, aural, and tactile. The

system should also afford the user dexterity in manipulating the remote environment, where the user's movements are correlated to the actions of a remote 'slave robot.'

Obviously, there are few – if any – webcam sites on the net today that would meet such strict standards. Even with dramatic recent improvements in image quality and interactivity, webcams constitute only the most basic kind of telepresence. In the end, webcams afford what might be described as 'low telepresence' or 'popular telepresence.' But their limitations are, in this writer's view at least, compensated by the vast extent of the webcam network, which itself can be seen as offering remote-world mobility simply by enabling viewers to hop around the globe from one webcam to another.

'Telepresence' is an ambitious term in any case, and the virtual observation of any scene via cable and modem is but a pale cousin to the robustness of being there. No one sends postcards from cyberspace, not yet at least. Webcams will not cure seasonal affective disorder, and nobody in their right mind would turn down a junket to Bali for the BaliCam URL. Yet webcams retain a certain magical quality. That we can set our eyes on a sun-tossed Australian scene from the depths of a New England winter night is somehow reassuring, as if there on our desktop is proof positive that the home star is burning bright and heading toward our kitchen window.

REFERENCES FROM THE READING

Gelerntner, D. (1991), *Mirror Worlds*, New York: Oxford University Press.

Kern, S. (1983), *The Culture of Time and Space, 1880–1918*, Cambridge, MA: Harvard University Press.

Mark, S. (1996), *Internet Dreams: Archetypes, Myths, and Metaphors*, Cambridge, MA: MIT Press.

Stefik, M. (1996), *Internet Dreams: Archetypes, Myths and Metaphors*, Cambridge, MA: MIT Press.

Editor's references and suggested further reading

Campanella, T. (2001), 'Anti-urbanist city images and new media culture'. In L. Vale and S. Warner (eds), *Imaging the City*, New Brunswick: Rutgers Center for Urban Policy Research, 237–54.

Goldberg, K. (2001), *The Robot in the Garden: Telerobotics and Teleepistemology in the Age of the Internet*, Cambridge, MA: MIT Press.

O
N
E

II THEORISING CYBERCITIES

Plate 8 Web graffiti in Berlin, 2001. (Photograph by Stephen Graham.)

INTRODUCTION

All discussions about ICTs and cities are based, either explicitly or implicitly, on conceptual or theoretical assumptions about how place-based, face-to-face and mediated relationships interrelate. In this second part of the book we explore and illustrate the range of conceptual approaches to thinking about the relations between urban space, place and ICTs.

Broadly speaking, it is possible to identify three dominating types of conceptual and theoretical perspectives to analysing the interrelationships between cities and ICTs (see Graham, 1998). In the first, most familiar, position notions of *substitution* and *transcendence* dominate. The central idea here is that human territoriality, and the space and place-based dynamics of human life, can somehow be replaced using new ICT technologies. These conceptual frameworks were extensively analysed and criticised in the Introduction and so will not be discussed further here.

Second, there are a range of what might be termed *coevolution* perspectives. These argue that both electronic spaces and territorial spaces are necessarily produced *together*, as part of the ongoing restructuring of the capitalist political-economic system.

Finally, there are a set of *recombination* perspectives. These draw on recent work in actor-network theory and cyborg sociology. Here the argument is that a fully *relational* view of the links between technology, time, space and social life is necessary. To their proponents, such a perspective reveals how new technologies become enrolled into complex, contingent and subtle blendings of human actors and technical artefacts, to form actor-networks (which are socio-technical hybrids). Through these, social and spatial life become subtly and continuously recombined in complex combinations of new sets of spaces and times, which are always contingent and impossible to generalise. In what follows we will examine the last two perspectives in a little more detail.

Coevolution: urban places and electronic spaces produced together

Coevolution perspectives argue that complex articulations are emerging between interactions in geographical space and place, and the electronic domains accessible through new technologies. This state of suspension, or articulation, between place-based and electronically mediated realms is especially evident in the contemporary metropolis where a complex and multifaceted coevolution, articulation and synergy between place-based and telemediated exchange is emerging.

The usefulness of the coevolution perspective is that it underlines the fact that materially constructed urban places and telecommunications networks stand in a state of recursive interaction. They shape each other in complex ways. The coevolution perspective is therefore important for the stress it places on the parallel shaping and production of both cities and ICT systems. The approach emphasises that this is occurring within an overarching political and economic context marked by deepening corporate power, the use of ICTs to help commodify the essential facets of human life – time, space, human labour, and genetic code – and an intensifying global division of labour (Dyer-Witherford, 1999; Schiller, 1999).

Coevolution perspectives have been particularly associated with neo-Marxist political economic studies of the ways in which mobility systems within and between cities are made and remade within capitalism. The work of the geographer David Harvey is especially notable here. In 1989 Harvey suggested that the capabilities of ICTs are – along with increased transport, financial and capital flows – underpinning a process of what he termed 'time–space compression'. This expression captures the overcoming or reduction of time and space barriers (Harvey, 1989).

In Harvey's perspective, ICTs are supporting the exploration of new solutions to the tensions inherent within capitalism between what he calls 'fixity' and the demands within capitalism for 'motion', mobility and the circulation of information, money, capital, services, labour and commodities. Because it is driven by the search for new profits and 'capital accumulation', this perspective sees the capitalist mode of production as inevitably expansionary. This means that widely dispersed areas of production, consumption and exchange need to be integrated and coordinated. Geographic space needs to be 'commanded' and controlled by capitalist firms and logistics chains. Thus, new transport and telecommunications infrastructures must be built which accelerate the mobility of capital and money. This helps to overcome the spatial and temporal barriers that inhibit this expansion and coordination.

Both the new spatial structures that are built up for production and consumption – cities, industrial areas etc. – and the new telecommunications and transport networks, are fixed and embedded in geographic space (Swyngedouw, 1992). New telecommunications networks 'have to be immobilised in space, in order to facilitate greater movement for the remainder' (Harvey, 1985, 149). This makes them expensive, uncertain and risky to develop – especially for the profit-seeking firms that now tend to control them. According to Harvey, this inflexibility means that they then go on to present problems later to further 'rounds' of restructuring, as the geographical dynamism of capitalism play out. Harvey suggests that:

> the tension between fixity and mobility erupts into generalised crises, when the landscape shaped in relation to a certain phase of development . . . becomes a barrier to further [capital] accumulation. The landscape must then be reshaped around new transport and communications systems and physical infrastructures, new centres and styles of production and consumption, new agglomerations of labour power and modified social infrastructures (Harvey, 1993, 7).

Coevolution perspectives usefully reveal some of the political and economic power plays behind transformations in city–ICT relations. However, they must be used with care. This is because they can easily lead to over simplified and deterministic readings of complex socio-technical transformations. Harvey's account, for example, has been criticised as being technologically determinist (Smith, 2001). It certainly implies that city–ICT relations are little less than local manifestations of a globe-spanning struggle by dominant capitalist interests to install the mobility systems that the latest stages of capitalism functionally require. Coevolution perspectives can also fall into the trap of separating 'local' and 'global' scales. This risks reifying – i.e. according a simple and dominant form of agency to – 'globalisation' as some steam roller trundling over local places. It also fails to capture the ways in which social and economic life is continuously constructed to telescope dynamically between a whole variety of geographical scales.

Recombination: actor-network theory and cyborg urbanisation

In contrast, the third set of perspectives takes a much more fine-grained view of the ways in which urban places and technological systems are socially constructed in parallel. This view is anchored around the actor-network theories of French sociologist Michel Callon and Bruno Latour (see Law and Hassard, 1999). It also draws on recent theorisations of the American anthropologist of technology Donna Haraway (1991) on the emergence of blended human-technological 'cyborgs' (or cybernetic organisms). Utilising the work of these key theorists, a range of researchers from the sociology of science, science, technology and society, cultural anthropology, architecture, and, increasingly, geography, have argued in the past twenty or so years for a highly contingent, relational perspective of the linkage between technology and social worlds.

A particularly influential variant of this perspective is known as actor-network theory (ANT). ANT emphasises how particular social situations and human actors enrol pieces of technology, machines, as well as documents, texts and money, into actor-networks. It is fully 'relational' in that 'it is concerned with how all sorts of bits and pieces; bodies, machines, and buildings, as well as texts, are associated together in attempts to build order' (Bingham, 1996, 32).

Absolute spaces and times are meaningless in such a perspective. Agency is a purely relational process. Technologies only have contingent, and diverse, effects through the ways in which they become linked into specific social and cultural contexts by linked human and technological agency. What Pile and Thrift (1996, 37) call 'a vivid, moving, contingent and open-ended cosmology' emerges. The boundaries between humans and machines become ever-more blurred, permeable and cyborgian. And 'nothing *means* outside of its relations: it makes no more sense to talk of a machine in general than it does to talk of a human in general' (Bingham, 1996, 17, original emphasis). Nigel Thrift summarises the key argument of the approach which is that

> no technology is ever found working in splendid isolation as though it is the central node in the social universe. It is linked – by the social purposes to which it is put – to humans and other technologies of different kinds. It is linked to a chain of different activities involving other technologies. And it is heavily contextualised. Thus the telephone, say, at someone's place of work had (and has) different meanings from the telephone in, say, their bedroom, and is often used in quite different ways (Thrift, 1996, 1468).

This linkage of heterogeneous technological elements and actors, strung across distance via software and hardware, is thus seen as a difficult process requiring continuing effort. This effort is necessary to sustain relations which are necessarily both social and technical at the same time. The growing capabilities of telecommunications for supporting action at a distance and remote control, do not therefore negate the need for the human actors which use them to struggle to enrol technological agents into their efforts to attain real, meaningful remote control and action at a distance. 'Stories of remote control tend to tell of the sheer amount of work that needs to be performed before any sort of ordering through space becomes possible' (Bingham, 1996, 27). Such

> heterogeneous work involving programmers, silicon chips, international transmission protocols, users, telephones, institutions, computer languages, modems, lawyers, fibre-optic cables, and governments to name but a few, has had to be done to create envelopes stable enough to carry [electronic information] (Bingham, 1996, 31).

It follows that it is naive to assume there to be one single, unified 'cyberspace'. Rather, ANT suggests that there are multiple, heterogeneous networks. Within these, telecommunications and information technologies become closely enrolled with human actors, and with other technologies, into systems of socio-technical relations across space. As Nick Bingham (1996, 32) again argues, 'the real illusion is that cyberspace as a singular exists at all', rather than as an enormously varied 'skein of networks' (Latour, 1993, 120) straddling, linking, and weaving through different spaces. Thus, we need to consider the diverse, and interlinked, physical infrastructures of information technologies (cable, public switched telephone networks or PSTNs, satellite, mobile, microwave, Internet grids, transoceanic optic fibres etc.), and how they support the vast panoply of contingent actor-networks.

According to this approach, electronic domains, and cybercities, therefore need to be considered as fragmented, divided, and contested multiplicities of heterogeneous infrastructures and actor-networks. For example, there are tens of thousands of specialised corporate networks and intranets. The Internet provides the basis for countless Usenet groups, Listservers, corporate advertising sites, specialised web sites, multi-user dungeons (MUDs), corporate intranets, virtual communities and increasingly sophisticated flows of media and video. PSTNs, and the many competing telecoms infrastructures, support global systems of private automatic teller machine (ATM) networks, credit card and electronic clearing systems, as well as blossoming applications for CCTV, telehealth, teleshopping and telebanking, global logistics, remote monitoring, back office and telesales flows, electronic data interchange (EDI), electronic financial transactions and stock-market flows, as well as data and telephony flows. And specialised systems of satellite, broadband, cable and broadcasting networks support burgeoning arrays of television flow.

Each application has associated with it whole multiplicities of human actors and institutions, who must continually struggle to enrol and maintain the communications technologies, along with other technologies, money and texts, into producing some form of functioning social order. These, and the hundreds of other actor-networks, are always contingent. They are always constructed. They are never spatially universal. And they are always embedded in the micro-social worlds of individuals, groups and institutions. Such socio-technical networks 'always represent geographies of enablement and constraint' (Law and Bijker, 1992, 301); they always link the local and non-local in intimate relational, and reciprocal, connections.

Rather than hypothesising macro-level technological revolutions ANT stresses multiple, contingent worlds of social action. This underlines the difficulties involved in achieving social ordering at a distance through enrolling complex arrays of strung-out technological artefacts. In it humans emerge as more than just subjects whose lives are to be impacted; as more than bit-players within macro-level processes of global structural change. Actor-network theory underlines forcefully 'that living, breathing, corporeal human beings arrayed in various creatively improvised networks of relation still exist as something more than machine fodder' (Thrift, 1996, 1466).

Whilst they have many strengths, recombination and ANT perspectives have been criticised for ignoring the broader imbalances of political and social power in contemporary societies which provide the context for the construction of actor networks. It has also been suggested that, as a result of their strong emphasis on the social shaping of technology, they fail to take account of the very real ways in which technological systems and artefacts shape and influence the social world.

References and suggested further reading

Amin, A. and Thrift, N. (2001), *Cities: Reimagining the Urban*, Cambridge: Polity.

Bingham, N. (1996), 'Object-ions: from technological determinism towards geographies of relations', *Environment and Planning D: Society and Space*, 14, 635–57.

Castells, M. (1996), *The Rise of the Networked Society*, Oxford: Blackwell.

Cooper, S. (2002), *Technoculture and Critical Theory: In the Service of the Machine?*, London: Routledge.

Dyer-Witherford, N. (1999), *Cyber-Marx: Cycles and Circuits of Struggle in High-Technology Capitalism*, Chicago: University of Illinois Press.

Goldberg, K. (2001), *The Robot in the Garden: Telerobotics and Teleepistemology in the Age of the Internet*, Cambridge, MA: MIT Press.

Graham, S. (1998), 'The end of geography or the explosion of place? Conceptualising space, time and information technology', *Progress in Human Geography*, 22(2), 165–85.

Graham, S. and Healey, P. (1999), 'Relational concepts of space and place: Implications for planning theory and practice', *European Planning Studies*, 7(5), 623–46.

Graham, S. and Marvin, S. (1996), *Telecommunications and the City: Electronic Spaces, Urban Places*, London: Routledge, Chapter 3.

Graham, S. and Marvin, S. (2000), *Splintering Urbanism: Networked Infrastructures, Technological Mobilities and the Urban Condition*, London: Routledge, Chapter 5.

Gray, C. (ed.) (1995), *The Cyborg Handbook*, New York: Routledge.

Gray, C. H. (2001), *Cyborg Citizen*, New York: Routledge.

Haraway, D. (1991), 'A manifesto for cyborgs: science, technology, and socialist-feminism in the late twentieth century'. In D. Haraway (ed.), *Simians, Cyborgs and Women: The Reinvention of Nature*, New York: Routledge, 149–81.

Harvey, D. (1985), *The Urbanization of Capital*, Oxford: Blackwell.

Harvey, D. (1989), *The Condition of Postmodernity*, Oxford: Blackwell.

Harvey, D. (1993), 'From space to place and back again: reflections on the condition of postmodernity'. In J. Bird, B. Curtis, T. Putnam, G. Robertson and L. Tickner (eds), *Mapping the Futures: Local Cultures, Global Change*, London: Routledge, 3–29.

Latour, B. (1993), *We Have Never Been Modern*, London: Harvester and Wheatsheaf.

Law, J. and Bijker, W. (1992), 'Postscript: technology, stability and social theory'. In W. Bijker and J. Law (eds), *Shaping Technology, Building Society: Studies in Sociotechnical Change*, London: MIT Press, 290–308.

Law, J. and Hassard, J. (1999), *Actor-Network Theory and After*, Oxford: Blackwell.

Massey, D. (1993), 'Power-geometry and a progressive sense of place'. In J. Bird, B. Curtis, T. Putnam, G. Robertson and L. Tickner (eds), *Mapping the Futures: Local Cultures, Global Change*, London: Routledge, 59–69.

Mackenzie, A. (2002), *Transductions: Bodies and Machines at Speed*, London: Continuum.

Mitchell, W. (1995), *The City of Bits*, Cambridge, MA: MIT Press.

Mosco, V. (1996), *The Political Economy of Communication*, London: Sage.

Pile, S. and Thrift, N. (1996), 'Mapping the subject'. In S. Pile and N. Thrift (eds), *Mapping the Subject: Geographies of Cultural Transformation*, London: Routledge, 13–51.

Robins, R. (1995), 'Cyberspace and the world we live in'. In M. Featherstone and R. Burrows (eds), *Cyberpunk/Cyberspace/Cyberbodies*, London: Sage, 135–56.

Schiller, D. (1999), *Digital Capitalism: Networking the Global Market System*, Cambridge, MA: MIT Press.

Slack, S. and Fejes, F. (eds) (1987), *The Ideology of the Information Age*, Norwood, NJ: Ibex.

Smith, M. (2001), *Transnational Urbanism*, London: Sage.

Smith, M. and Marx, L. (eds) (1995), *Does Technology Drive History? The Dilemma of Technological Determinism*, Cambridge, MA: MIT Press.

Smith, R. G. (2003), 'World city-actor networks', *Progress in Human Geography*, 27(1), 25–44.

Sussman, G. (1997), *Communication, Technology and Politics in the Information Age*, London: Sage.

Swyngedouw, E. (1992), 'Communication, mobility and the struggle for power over space'. In G. Giannopoulos and A. Gillespie (eds), *Transport and Communications in the New Europe*, London: Belhaven, 305–25.

Thrift, N. (1996), 'New urban eras and old technological fears: reconfiguring the goodwill of electronic things', *Urban Studies*, 33(8), 1463–93.

Virilio, P. (1991), *Lost Dimension*, New York: Semiotext(e).

Webster, F. (1996), *Theories of the Information Society*, London: Routledge.

'Postscript on Societies of Control'

from *October* (1988)

Gilles Deleuze

Editor's Introduction

Our first theoretical reading comes from one of the most influential French philosophers of the late twentieth century: Gilles Deleuze (1925–1994). He takes as his starting point the suggestion by another French philosopher, Michel Foucault (1926–1984), that an intrinsic element of the evolution of modern societies in the eighteenth and nineteenth centuries was the construction of a whole variety of enclosed, disciplined spaces: homes, clinics, schools, hospitals, mad houses, and, most famously, prisons. To Foucault, the emergence of these was the result of a series of orchestrated strategies of attaining social control by modern bureaucratic institutions (Foucault, 1977). Relying on the ability of supervisors and professionals to physically view subservient, enclosed, populations, Foucault drew on the ideas of the English eighteenth-century utilitarian philosopher Jeremy Bentham to argue that these spaces achieved disciplinary control through 'panoptic' means. That is, they were explicitly designed so that their inhabitants always felt as though they were under the physical gaze of supervisors (even if this was not actually always the case). Foucault termed modern societies based on such notion of discipline 'disciplinary societies'.

In this famous reading, Deleuze argues powerfully that contemporary technological and social changes are fundamentally challenging such notions of enclosure and the attainment of control through these means. Increasingly, Deleuze argues, attempts at social control are now organised through extended webs of electronic monitoring and surveillance facilitated through ICTs (sensors, telecommunications systems, electronic transaction and tracking systems, databases, CCTV cameras, and so on).

The possibilities offered by these new techniques, Deleuze suggests, are leading to a crisis of all environments of enclosure. Attempts at social control float free from the disciplinary, enclosed spaces of modern societies, and modern cities, into an extending, virtually infinite, array of extended, and often invisible, ICT systems. These are strung out across the times and spaces of contemporary societies, from the body to global scales. Providing real time feedback and disciplinary power at a distance, these systems complement (and, in many cases, start to completely replace) the use of the visual gaze of human supervision over people, their bodies, their actions, and their consciousness, in particular spaces and places. Increasingly, then, people become 'digital subjects' which can be monitored, tracked, controlled and commodified, subjects who, in a sense, supervise themselves through their participation within extending webs of a myriad of electronic liaisons that always leave digital tracks.

The Australian cultural commentator McKenzie Wark (2002, 32) describes this transition another way. To him Foucault's 'disciplinary technologies' – such as panoptically designed prisons or clinics – 'enclose, they count and rank what they enclose'. By contrast, contemporary ICTs, and their antecedents – the telegraph and maritime navigation – are 'vectoral' rather than disciplinary technologies. These 'extend their grid out across open space, and project lines of movement across it'. They 'steadily make any resource something that can be connected to any other . . . In the periphery, vectoral power was always greater than the power of enclosure'. This is why ICTs

– from the telegraph to the Internet – have been such central supports to colonialism and globalisation for the past few centuries.

An early example of the recombination perspective, Deleuze's reading hints at the subtle ways in which ICT systems are mediating new hybrids of built space and continuous electronic surveillance and control (see David Lyon, p. 299). To Deleuze, such transformations herald a mutation of urban, capitalist society as a whole. This is a transformation he analysed in much more detail in his books with Felix Guattari, *Anti-Oedipus* and *A Thousand Plateaus*.

HISTORICAL

Foucault located the *disciplinary* societies in the eighteenth and nineteenth centuries; they reach their height at the outset of the twentieth. They initiate the organization of vast spaces of enclosure. The individual never ceases passing from one closed environment to another, each having its own laws: first, the family; then the school ("you are no longer in your family"); then the barracks ("you are no longer at school"); then the factory; from time to time the hospital; possibly the prison, the preeminent instance of the enclosed environment. It's the prison that serves as the analogical model: at the sight of some laborers, the heroine of Rossellini's *Europa '51* could exclaim, "I thought I was seeing convicts."

Foucault has brilliantly analyzed the ideal project of these environments of enclosure, particularly visible within the factory: to concentrate; to distribute in space; to order in time; to compose a productive force within the dimension of space-time whose effect will be greater than the sum of its component forces. But what Foucault recognized as well was the transience of this model: it succeeded that of the *societies of sovereignty*, the goal and functions of which were something quite different (to tax rather than to organize production, to rule on death rather than to administer life); the transition took place over time, and Napoleon seemed to effect the large-scale conversion from one society to the other. But in their turn the disciplines underwent a crisis to the benefit of new forces that were gradually instituted and which accelerated after World War II: a disciplinary society was what we already no longer were, what we had ceased to be.

We are in a generalized crisis in relation to all the environments of enclosure – prison, hospital, factory, school, family. The family is an "interior," in crisis like all other interiors – scholarly, professional, etc. The administrations in charge never cease announcing supposedly necessary reforms: to reform schools, to reform industries, hospitals, the armed forces, prisons. But everyone knows that these institutions are finished, whatever the length of their expiration periods. It's only a matter of administering their last rites and of keeping people employed until the installation of the new forces knocking at the door. These are the *societies of control*, which are in the process of replacing the disciplinary societies. "Control" is the name Burroughs proposes as a term for the new monster, one that Foucault recognizes as our immediate future. Paul Virilio (p. 78) also is continually analyzing the ultrarapid forms of free-floating control that replaced the old disciplines operating in the time frame of a closed system. There is no need here to invoke the extraordinary pharmaceutical productions, the molecular engineering, the genetic manipulations, although these are slated to enter into the new process. There is no need to ask which is the toughest or most tolerable regime, for it's within each of them that liberating and enslaving forces confront one another. For example, in the crisis of the hospital as environment of enclosure, neighborhood clinics, hospices, and day care could at first express new freedom, but they could participate as well in mechanisms of control that are equal to the harshest of confinements. There is no need to fear or hope, but only to look for new weapons.

LOGIC

The different internments or spaces of enclosure through which the individual passes are independent variables: each time one is supposed to start from zero, and although a common language for all these places exists, it is *analogical*. On the other hand, the different control mechanisms are inseparable variations, forming a system of variable geometry the language of which is *numerical* (which doesn't necessarily mean binary). Enclosures are *molds*, distinct castings, but controls are a *modulation*, like a self-deforming cast that will continuously change from one moment to the other, or like a sieve whose mesh will transmute from point to point.

This is obvious in the matter of salaries: the factory was a body that contained its internal forces at a level of equilibrium, the highest possible in terms of production, the lowest possible in terms of wages; but in a society of control, the corporation has replaced the factory, and the corporation is a spirit, a gas. Of course the factory was already familiar with the system of bonuses, but the corporation works more deeply to impose a modulation of each salary, in states of perpetual metastability that operate through challenges, contests, and highly comic group sessions. If the most idiotic television game shows are so successful, it's because they express the corporate situation with great precision. The factory constituted individuals as a single body to the double advantage of the boss who surveyed each element within the mass and the unions who mobilized a mass resistance; but the corporation constantly presents the brashest rivalry as a healthy form of emulation, an excellent motivational force that opposes individuals against one another and runs through each, dividing each within. The modulating principle of "salary according to merit" has not failed to tempt national education itself. Indeed, just as the corporation replaces the factory, *perpetual training* tends to replace the *school*, and continuous control to replace the examination. Which is the surest way of delivering the school over to the corporation.

In the disciplinary societies one was always starting again (from school to the barracks, from the barracks to the factory), while in the societies of control one is never finished with anything – the corporation, the educational system, the armed services being metastable states coexisting in one and the same modulation, like a universal system of deformation. In *The Trial*, Kafka, who had already placed himself at the pivotal point between two types of social formation, described the most fearsome of juridical forms. The *apparent acquittal* of the disciplinary societies (between two incarcerations); and the *limitless postponements* of the societies of control (in continuous variation) are two very different modes of juridical life, and if our law is hesitant, itself in crisis, it's because we are leaving one in order to enter into the other. The disciplinary societies have two poles: the signature that designates the *individual*, and the number or administrative numeration that indicates his or her position within a *mass*. This is because the disciplines never saw any incompatibility between these two, and because at the same time power individualizes and masses together, that is, constitutes those over whom it exercises power into a body and molds the individuality of each member of that body. (Foucault saw the origin of this double charge in the pastoral power of the priest – the flock and each of its animals – but civil power moves in turn and by other means to make itself lay "priest.") In the societies of control, on the other hand, what is important is no longer either a signature or a number, but a code: the code is a *password*, while on the other hand the disciplinary societies are regulated by *watchwords* (as much from the point of view of integration as from that of resistance). The numerical language of control is made of codes that mark access to information or reject it. We no longer find ourselves dealing with the mass/individual pair. Individuals have become "*dividuals*," and masses, samples, data, markets, or "*banks*." Perhaps it is money that expresses the distinction between the two societies best, since discipline always referred back to minted money that locks gold in as numerical standard, while control relates to floating rates of exchange, modulated according to a rate established by a set of standard currencies. The old monetary mole is the animal of the spaces of enclosure but the serpent is that of the societies of control. We have passed from one animal to the other, from the mole to the serpent, in the system under which we live but also in our manner of living and in our relations with others. The disciplinary man was a discontinuous producer of energy, but the man of control is undulatory, in orbit, in a continuous network. Everywhere *surfing* has already replaced the older *sports*.

Types of machines are easily matched with each type of society – not that machines are determining, but because they express those social forms capable of generating them and using them. The old societies of sovereignty made use of simple machines – levers, pulleys, clocks; but the recent disciplinary societies equipped themselves with machines involving energy, with the passive danger of entropy and the active danger of sabotage; the societies of control operate with machines of a third type, computers, whose passive danger is jamming and whose active one is piracy and the introduction of viruses. This technological evolution must be, even more profoundly, a mutation of capitalism, an already well-known or familiar mutation that can be summed up as follows: nineteenth-century capitalism is a capitalism of concentration, for production and for property. It therefore erects the factory as a space of enclosure, the capitalist being the owner of the means of production but also, progressively, the owner of other spaces conceived through analogy (the worker's familial house, the school). As for markets, they are conquered sometimes by specialization, sometimes by colonization, sometimes by lowering the costs of production. But, in the present situation, capitalism is no longer involved in production, which it often relegates to the Third World, even for the complex forms of textiles, metallurgy, or oil production. It's a capitalism of higher-order production. It no longer buys raw materials and no longer sells the finished products: it buys the finished products or assembles parts. What it wants to sell is services and what it wants to buy is stocks. This is no longer a capitalism for production but for the product, which is to say, for being sold or marketed. Thus it is essentially dispersive, and the factory has given way to the corporation. The family, the school, the army, the factory are no longer the distinct analogical spaces that converge towards an owner – state or private power – but coded figures – deformable and transformable – of a single corporation that now has only stockholders. Even art has left the spaces of enclosure in order to enter into the open circuits of the bank. The conquests of the market are made by grabbing control and no longer by disciplinary training, by fixing the exchange rate much more than by lowering costs, by transformation of the product more than by specialization of production. Corruption thereby gains a new power. Marketing has become the center or the "soul" of the corporation. We are taught that corporations have a soul, which is the most terrifying news in the world. The operation of markets is now the instrument of social control and forms the impudent breed of our masters. Control is short-term and of rapid rates of turnover, but also continuous and without limit, while discipline was of long duration, infinite and discontinuous. Man is no longer man enclosed, but man in debt. It is true that capitalism has retained as a constant the extreme poverty of three quarters of humanity, too poor for debt, too numerous for confinement: control will not only have to deal with erosions of frontiers but with the explosions within shanty towns or ghettos.

PROGRAM

The conception of a control mechanism, giving the position of any element within an open environment at any given instant (whether animal in a reserve or human in a corporation, as with an electronic collar), is not necessarily one of science fiction. Félix Guattari has imagined a city where one would be able to leave one's apartment, one's street, one's neighborhood, thanks to one's (dividual) electronic card that raises a given barrier; but the card could just as easily be rejected on a given day or between certain hours; what counts is not the barrier but the computer that tracks each person's position – licit or illicit – and effects a universal modulation.

The socio-technological study of the mechanisms of control, grasped at their inception, would have to be categorical and to describe what is already in the process of substitution for the disciplinary sites of enclosure, whose crisis is everywhere proclaimed. It may be that older methods, borrowed from the former societies of sovereignty, will return to the fore, but with the necessary modifications. What counts is that we are at the beginning of something. In the *prison system*: the attempt to find penalties of "substitution," at least for petty crimes, and the use of electronic collars that force the convicted person to stay at home during certain hours. For the *school system*: continuous forms of control, and the effect on the school of perpetual training, the corresponding abandonment of all university research, the introduction of the "corporation" at all levels of schooling. For the *hospital system*: the new medicine "without doctor or patient" that singles out potential sick people and subjects at risk, which in no way attests to individuation – as they say – but substitutes for the individual or numerical body the code of a "dividual" material to be controlled. In the *corporate system*: new ways of handling money, profits, and humans that no longer pass through the old factory form. These are very small examples, but ones that will allow for better understanding of what is meant by the crisis of the institutions, which is to say, the progressive and dispersed installation of a new system of domination. One of the most important questions will concern the ineptitude of the unions: tied to the whole of their history of struggle against the disciplines or within the spaces of enclosure, will they be able to adapt themselves or will they give way to new forms of resistance against the societies of control? Can we already grasp the rough outlines of these coming forms, capable of threatening the joys of marketing? Many young people strangely boast of being "motivated"; they re-request apprenticeships and permanent training. It's up to them to discover what they're being made to serve, just as their elders discovered, not without difficulty, the telos of the disciplines. The coils of a serpent are even more complex than the burrows of a molehill.

Editor's references and suggestions for further reading

Deleuze, G. and Guattari, F. (1988), *A Thousand Plateaus: Capitalism and Schizophrenia*, London: Athlone.

Deleuze, G. and Guattari, F. (1984), *Anti-Oedipus: Capitalism and Schizophrenia*, London: Athlone.

Foucault, M. (1977), *Discipline and Punish: The Birth of the Prison*, New York: Vintage.

Levin, T., Frohne, U. and Weibel, P. (eds) (2002), *Ctrl-Space: Rhetorics of Surveillance from Bentham to Big Brother*, Cambridge, MA: MIT Press.

Lyon, D. (1994), *The Electronic Eye: The Rise of Surveillance Society*, London: Polity.

Lyon, D. (2000), *Surveillance Society: Monitoring Everyday Life*, Buckingham: Open University Press.

Wark, M. (2002), 'Telegram from nowhere'. In R. Koolhaas, S. Boeri and S. Kwinter (eds), *Mutations*, Bordeaux: ACTAR, 30–9.

ONE

'The Third Interval'

from *Open Sky* (1997)

Paul Virilio

Editor's Introduction

Our second theoretical reading draws on the work of another influential French analyst of cities and ICTs: Paul Virilio. An architect and urbanist by training, Virilio is a leading critical theorist of the links between cities, speed, technology, war, space and time. Whilst his work is very diverse, in general, Virilio adopts an uncompromisingly substitutionist perspective to analysing the relations between ICTs and cities. He uses this to develop a series of critical and bleak portrayals of the nature of the current and future urban condition. In these he stresses the emergence of a world of hypermobility, electronic saturation, the computerisation of everything from war to consumption, and the digital manipulation of organisms, and natures, through genetic technologies.

In this reading Virilio argues that the city is threatened as its central raison d'être – the organisation of real space – becomes eclipsed by ICTs and their organisation of real time (based on speed of light electronic exchanges). Analysing the implications of this shift – at scales ranging from the human body and the city to the global capitalist economy – Virilio concludes that the future heralds a world where subjects will be universally telepresent anywhere at any time without physically moving their bodies in geographic space. The organisation of real time thus sits uncomfortably next to the challenges of organising real space that have long been the major concern of architecture and urban planning.

To Virilio, the saturation of societies and cities by extending arrays of ICTs thus means that arrival without departure now becomes possible. In fact, he argues that there is a kind of 'generalised arrival' as speed-of-light electronic signals mean that everything arrives without having to leave. Virilio speculates that the current growth of physical mobility will, in the future, shift to a growing inertia, as ICT use starts to substitute for the need for bodily presence and movement within and between cities.

In the end, certain elements of Virilio's critical analysis parallel some of the predictions of many of the cyber-utopian theorists analysed in the Introduction. Virilio finishes by arguing provocatively that the model for our future is that of the online disabled citizen: the human subject who is unable to physically move whilst being saturated by constant electronic mobilities. These are made possible through the integration of people's bodies and minds into sets of prosthetic devices which connect global universes of electronic space straight into users' senses and bodies. A further rendition of such a scenario, seen through a utopian rather than a critical lens, is delivered by the architectural commentator Martin Pawley (p. 401).

Without even leaving, we are already no longer there.
(Nikolai Gogol)

Critical *mass*, critical *moment*, critical *temperature*. You don't hear much about critical *space*, though. Why is this if not because we have not yet digested relativity, the very notion of space-time?

And yet critical space, and critical expanse, are now everywhere, due to the acceleration of communications tools that *obliterate the Atlantic* (Concorde), *reduce France to a square one and a half hours across* (Airbus) or *gain time over time* with the TGV, the various advertising slogans signalling perfectly the shrinking of geophysical space of which we are the beneficiaries but also, sometimes, the unwitting victims.

As for telecommunications tools, not content to limit extension, they are also eradicating all duration, any extension of time in the transmission of messages, images.

Mass transportation revolution of the nineteenth century, broadcasting revolution of the twentieth – a mutation and a commutation that affect both public and domestic space at the same time, to the point where we are left in some uncertainty as to their very reality, since the urbanization of *real space* is currently giving way to a preliminary urbanization of *real time*, with teleaction technologies coming on top of the technology of mere conventional television.

This abrupt transfer of technology, from the building of real-space infrastructures (ports, railway stations, airports) to the control of the real-time environment thanks to interactive teletechnologies (teleports), gives new life today to the critical dimension . . .

[. . .]

The urbanization of real time is in fact first the urbanization of *one's own body* plugged into various interfaces (keyboard, cathode screen, DataGlove or DataSuit), prostheses that make the super-equipped able-bodied person almost the exact equivalent of the motorized and wired disabled person.

If last century's revolution in transportation saw the emergence and gradual popularization of the dynamic motor vehicle (train, motorbike, car, plane), the current revolution in transmission leads in turn to the innovation of the ultimate vehicle: the static audiovisual vehicle, marking the advent of a behavioural inertia in the sender/receiver that moves us along from the celebrated *retinal persistence* which permits the optical illusion of cinematic projection to the *bodily persistence* of this 'terminal-man'; a prerequisite for the sudden

mobilization of the illusion of the world, of a *whole* world, telepresent at each moment, the witness's own body becoming the last urban frontier. Social organization and a kind of conditioning once limited to the space of the city and to the space of the family home finally closing in on the animal body.

This makes it easier to understand the decline in that unit of population, the family, initially extended then nuclearized, that is today becoming a single-parent family, individualism having little to do with the fact of a liberation of values and being more an effect of technological evolution in the development of public and private space, since the more the city expands and spreads its tentacles, the more the family unit dwindles and becomes a minority.

Recent *megalopolitan* hyperconcentration (Mexico City, Tokyo) being itself the result of the increased speed of exchanges, it looks as though we need to reconsider the importance of the notions of *acceleration* and *deceleration* (vector quantities with positive or negative velocities according to the physicists). But we also need to reconsider the less obvious notions of *true velocity* and *virtual velocity* – the speed of that which occurs unexpectedly: a crisis, for instance, an accident – properly to understand the importance of the 'critical transition' of which we are today helpless witnesses

[. . .]

So, politicians, just as much as urbanists, find themselves torn between the permanent requirements of organizing and constructing real space – with its land problems, the geometric and geographic constraints of the centre and the periphery – and the new requirements of managing the real time of immediacy and ubiquity, with its access protocols, its 'data packet transmissions' and its viruses, as well as the chronogeographic constraints of nodes and network interconnection. Long term for the topical and architectonic interval (the building); short, ultra-short – if not indeed non-existent – term for the teletopical interval (the network)

[. . .]

The question today posed by teletopical technologies is thus a major one for the planner, since the urbanization of real time permitted by the recent transmission revolution leads to a radical reversal in the order of the movement of displacement and of physical transportation. In fact, if operating remotely allows gradual elimination of the material infrastructures rigging out the territory in favour of the fundamentally immaterial wave trains of telesurveillance and instantaneous remote control, this is because the *journey* and its components are undergoing a veritable mutation-commutation. Where physical displacement from one point to another once supposed departure, a journey and arrival, the transport revolution of last century had already quietly begun to eliminate delay and change the nature of travel itself, arrival at one's destination remaining, however, a 'limited arrival' due to the very time it took to get there.

Currently, with the instantaneous broadcasting revolution, we are seeing the beginnings of a '*generalized arrival*' whereby everything arrives without having to leave, the nineteenth century's elimination of the journey (that is, of the space interval and of time) combining with the abolition of *departure* at the end of the twentieth, the journey thereby losing its successive components and being overtaken by *arrival* alone

[. . .]

Surely we cannot fail to foresee the future conditioning of the human environment behind this critical transition.

If last century's transport revolution already brought about a mutation in urban territory throughout the continent, the current revolution in (interactive) transmission is in turn provoking a commutation in the urban environment whereby the image prevails over the thing it is an image of; what was once a city becoming little by little a paradoxical agglomeration, relationships of immediate proximity giving way to remote interrelationships.

The paradoxes of acceleration are indeed numerous and disconcerting, in particular, the foremost among them: getting closer to the 'distant' takes you away proportionally from the 'near' (and dear) – the friend, the relative, the neighbour – thus making strangers, if not actual enemies, of all who are close at hand, whether they be family, workmates or neighbourhood acquaintances. This inversion of social practices, already evident in the development of communication equipment (ports, stations, airports), is further rein-

forced, radicalized, by the new telecommunications equipment (teleports).

Once more we are seeing a reversal in trends: where the motorization of transport and information once caused a *general mobilization* of populations, swept up into the exodus of work and then of leisure, instantaneous transmission tools cause the reverse: *a growing inertia*; television and especially remote control action no longer requiring people to be mobile, but merely to be mobile on the spot.

Home shopping, working from home, online apartments and buildings: 'cocooning', as they say. The urbanization of real space is thus being overtaken by this urbanization of real time which is, at the end of the day, the urbanization of the actual body of the city dweller, this *citizen-terminal* soon to be decked out to the eyeballs with interactive prostheses based on the pathological model of the 'spastic', wired to control his/her domestic environment without having physically to stir: the catastrophic figure of an individual who has lost the capacity for immediate intervention along with natural motricity and who abandons himself, for want of anything better, to the capabilities of captors, sensors and other remote control scanners that turn him into a being controlled by the machine with which, they say, he talks.

Service or servitude, that is the question. The old public services are in danger of being replaced by a domestic enslavement whose crowning glory would surely be home automation. Achieving a domiciliary inertia, the widespread use of techniques of *environmental control* will end in behavioural isolation, in intensifying the insularity that has always threatened the town, the difference between the (separate) 'block' and the (segregated) 'ghetto' remaining precarious

[. . .]

At the end of the century, there will not be much left of the expanse of a planet that is not only polluted but also shrunk, reduced to nothing, by the teletechnologies of generalized interactivity.

Editor's references and suggestions for further reading

Armitage, J. (2000), *Paul Virilio: From Modernism to Hypermodernism and Beyond*, London: Sage.
Cooper, S. (2002), *Technoculture and Critical Theory: In the Service of the Machine*?, London: Routledge.
Mackenzie, A. (2002), *Transductions: Bodies and Machines at Speed*, London: Continuum.

Virilio, P. (1987), 'The overexposed city', *Zone*, 1(2), 14–31.

Virilio, P. (1991), *The Lost Dimension*, New York: Semiotext(e).

Virilio, P. (1993), 'The third interval: a critical transition'. In V. Andermatt-Conley (ed.), *Rethinking Technologies*, London: University of Minnesota Press, 3–10.

Virilio, P. (1997), *Open Sky*, London: Verso.

Wark, M. (1998), 'On technological time, Virilio's overexposed city', *Arena*, 83, 1–21.

O
N
E

'Space of Flows, Space of Places: Materials for a Theory of Urbanism in the Information Age'

Manuel Castells

Editor's Introduction

One of the leading theorists of the changing nature of capitalist cities over the past thirty years has been the Catalan urban sociologist Manuel Castells. After pioneering Marxist analyses of cities and collective consumption in the 1960s and 1970s, Castells has, since the late 1980s, produced a range of highly influential analyses of the implications of ICTs, and the changing nature of capitalism, for cities, urbanisation, and social and cultural change. These have been published within his Blackwell books *The Informational City* (1989) and his trilogy of books on the *Information Age* (1996–8).

Adopting a broadly coevolutionist perspective to the relations between cities and ICTs, Castells centres his theorisation on the notion that cities are caught up in a complex interplay of what he calls the 'space of flows' – the accelerating domains of translocal and transnational technological movement and flow – and the 'space of places' – the geographic spaces and communities of everyday life in cities. His far-reaching analyses have addressed the implications of the tensions between these two domains for: the development of the networked economy; social and cultural struggles over resources and power; the changing nature of social movements and social identities; and the changing geographical structures of cities around the world.

In this extended reading Castells gives a comprehensive summary of his theorisation of how the interactions of the space of flows and the space of places combine to shape contemporary cities. Notably, unlike Deleuze or Virilio, Castells mobilises these concepts to sustain detailed empirical discussions of a range of transformations in a wide range of cities across the world.

In this summary Castells suggests that articulations between the space of flows and the space of places are leading to a wide variety of shifts of the ways in which function, form, and meaning are produced within contemporary cities. Using this perspective, Castells explores transformations in the economic dynamics of cities, urban physical form, the changing nature of the patriarchical family, the growing multiculturalism of cities, deepening patterns of social segregation, the growing influence of transnational organised crime, and challenges to the meaning and nature of urban public space. Prefiguring some of the reflections in Section VIII of this book, Castells also analyses the ways in which urban planning, design and governance practices might respond to these multiple transformations.

We have entered a new age, the Information Age. Spatial transformation is a fundamental dimension of the overall process of structural change. We need a new theory of spatial forms and processes, adapted to the new social, technological, and spatial context where we live. I will attempt here to propose some elements of this theory, a theory of urbanism in the information age. I will not develop the analysis of the meaning of the information age, taking the liberty to refer the reader to my trilogy on the matter (Castells, 1996–2000).

I will not build theory from other theories, but from the observation of social and spatial trends in the world at large. Thus, I will start with a summary characterization of the main spatial trends at the onset of the twenty-first century. Then I will propose a tentative theoretical interpretation of observed spatial trends. Subsequently I will highlight the main issues arising in cities in the information age, with particular emphasis on the crisis of the city as a socio-spatial system of cultural communication. I will conclude by drawing some of the implications of my analysis for planning, architecture and urban design.

THE TRANSFORMATION OF URBAN SPACE IN THE EARLY TWENTY-FIRST CENTURY

Spatial transformation must be understood in the broader context of social transformation: space does not reflect society, it expresses it, it is a fundamental dimension of society, inseparable from the overall process of social organisation and social change. Thus, the new urban world arises from within the process of formation of a new society, the network society, characteristic of the Information Age. The key developments in spatial patterns and urban processes associated with these macro-structural changes, can be summarized under the following headings (Scott, 2001):

■ Because commercial agriculture has been, by and large, automated, and a global economy has integrated productive networks throughout the planet, the majority of the world's population is already living in urban areas, and this will be increasingly the case: we are heading towards a largely urbanized world, which will comprise between two-thirds and three-quarters of the total population by the middle of the century (Freire and Stren, 2001).

■ This process of urbanization is concentrated disproportionately in metropolitan areas of a new kind: urban constellations scattered throughout huge territorial expanses, functionally integrated and socially differentiated, around a multi-centered structure. I call these new spatial forms metropolitan regions (Garreau, 1991; Hall, 2001; Nel.Lo, 2001; Dunham-Jones, 2000).

■ Advanced telecommunications, Internet, and fast, computerized transportation systems allow for simultaneous spatial concentration and decentralization, ushering in a new geography of networks and urban nodes throughout the world, throughout countries, between and within metropolitan areas (Wheeler et al., 2000).

■ Social relationships are characterized simultaneously by individuation and communalism, both processes using, at the same time, spatial patterning and online communication. Virtual communities and physical communities develop in close interaction, and both processes of aggregation are challenged by increasing individualization of work, social relationships, and residential habits (Russell, 2000; Wellman, 1999; Putnam, 2000).

■ The crisis of the patriarchal family, with different manifestations depending on cultures and levels of economic development, gradually shifts sociability from family units to networks of individualized units (most often, women and their children, but also individualized co-habiting partnerships), with considerable consequences in the uses and forms of housing, neighborhoods, public space, and transportation systems.

■ The emergence of the network enterprise as a new form of economic activity, with its highly decentralized, yet coordinated, form of work and management, tends to blur the functional distinction between spaces of work and spaces of residence. The work–living arrangements characteristic of the early periods of industrial craft work are back, often taking over the old industrial spaces, and transforming them into informational production spaces. This is not just New York's Silicon Alley or San Francisco's Multimedia Gulch, but a phenomenon that also characterizes London, Tokyo, Beijing, Taipei, Paris, or Barcelona, among many other cities. Transformation of productive uses becomes more important than residential succession to explain the new dynamics of urban space (Mitchell, 1999; Horan, 2000).

- Urban areas around the world are increasingly multi-ethnic, and multicultural. An old theme of the Chicago School, now amplified in terms of its extremely diverse racial composition (Waldinger, 2001).

- The global criminal economy is solidly rooted in the urban fabric, providing jobs, income, and social organisation to a criminal culture, which deeply affects the lives of low-income communities, and of the city at large. It follows rising violence and/or widespread paranoia of urban violence, with the corollary of defensive residential patterns.

- Breakdowns of communication patterns between individuals and between cultures, and the emergence of defensive spaces, leads to the formation of sharply segregated areas: gated communities for the rich, territorial turfs for the poor (Blakely and Snyder, 1997; Massey, 1996).

- In a reaction against trends of suburban sprawl and the individualization of residential patterns, urban centers and public space become critical expressions of local life, benchmarking the vitality of any given city (Hall, 1998; Borja and Zaida, 2001). Yet, commercial pressures and artificial attempts at mimicking urban life often transform public spaces into theme parks where symbols rather than experience create a life-size, urban virtual reality, ultimately destined to mimic the real virtuality projected in the media. It follows increasing individualization, as urban places become consumption items to be individually appropriated (Fernandez-Galiano, 2000).

- Overall, the new urban world seems to be dominated by the double movement of inclusion into transterritorial networks, and exclusion by the spatial separation of places. The higher the value of people and places, the more they are connected into interactive networks. The lower their value, the lower their connection. In the limit, some places are switched off, and bypassed by the new geography of networks, as is the case of depressed rural areas and urban shanty towns around the world. Splintering urbanism operates on the basis of segregated networks of infrastructure, as empirically demonstrated by Graham and Marvin (2001).

- The constitution of mega-metropolitan regions, without a name, without a culture, and without institutions, weakens the mechanism of political accountability, of citizen participation, and of effective administration (Sassen, 2001). On the other hand, in the age of globalization, local governments emerge as flexible institutional actors, able to relate at the same time to local citizens and to global flows of power and money (Borja and Castells, 1997). Not because they are powerful, but because most levels of government, including the nation states, are equally weakened in their capacity of command and control if they operate in isolation. Thus, a new form of state emerges, the network state, integrating supra-national institutions made up of national governments, nation-states, regional governments, local governments, and even non-governmental organizations. Local governments become a node of the chain of institutional representation and management, able to input the overall process, yet with added value in terms of their capacity to represent citizens at a closer range. Indeed in most countries, opinion polls show the higher degree of trust people have in their local governments, relative to other levels of government. However, institutions of metropolitan governance are rare and when they exist they are highly centralized, with little citizen participation. There is an increasing gap between the actual unit of work and living, the metropolitan region, and the mechanisms of political representation and public administration. Local governments compensate for this lack by cooperating and competing. Yet, by defining their interests as specific subsets of the metropolitan region, they (often unwillingly) contribute to further fragmentation of the spatial framing of social life.

- Urban social movements have not disappeared, by any means. But they have mutated. In an extremely schematic representation they develop along two main lines. The first is the defense of the local community, affirming the right to live in a particular place, and to benefit from adequate housing and urban services in their place. The second is the environmental movement, acting on the quality of cities within the broader goal of achieving quality of life: not only a better life but a different life. Often, the broader goals of environmental mobilizations become translated into defensive reactions to protect one specific community, thus merging the two trends. Yet, it is only by reaching out to the cultural transformation of urban life as proposed by ecological thinkers and activists that urban social movements can transcend their limits of localism. Indeed, enclosing themselves in their communities, urban social movements may contribute to further spatial fragmentation, ultimately leading to the breakdown of society.

It is against the background of these major trends of urban social change that we can understand new spatial forms and processes, thus re-thinking architecture, urban design and planning in the twenty-first century.

A THEORETICAL APPROACH TO SPATIAL TRANSFORMATION

To make the transition from the observation of urban trends to the new theorization of cities, we need to grasp, at a more analytical level, the key elements of socio-spatial change. I think the transformation of cities in the information age can be organized around three bipolar axes. The first relates to function, the second to meaning, the third to form.

Function

Functionally speaking the network society is organized around the opposition between the global and the local. Dominant processes in the economy, technology, media, institutionalized authority are organized in global networks. But day-to-day work, private life, cultural identity, political participation, are essentially local. Cities, as communication systems, are supposed to link up the local and the global, but this is exactly where the problems start since these are two conflicting logics that tear cities from the inside when try to respond to both, simultaneously.

Meaning

In terms of meaning, our society is characterized by the opposing development of individuation and communalism. By individuation I understand the enclosure of meaning in the projects, interests, and representations of the individual, that is, a biologically embodied personality system (or, if you want, translating from French structuralism, a person). By communalism I refer to the enclosure of meaning in a shared identity, based on a system of values and beliefs to which all other sources of identity are subordinated. Society, of course, exists only in between, in the inter-face

between individuals and identities mediated by institutions, at the source of the constitution of civil society which, as Gramsci argued, does not exist against the state but in articulation with the state, forming a shared public sphere, à la Habermas.

Trends I observe in the formative stage of the network society indicate the increasing tension and distance between personality and culture, between individuals and communes. Because cities are large aggregates of individuals, forced to coexist, and communes are located in the metropolitan space, the split between personality and commonality brings extraordinary stress upon the social system of cities as communicative and institutionalizing devices. The problematique of social integration becomes again paramount, albeit under new circumstances and in terms radically different from those of early industrial cities. This is mainly because of the role played in urban transformation by a third, major, axis of opposing trends, this one concerning spatial forms.

Forms

There is a growing tension and articulation between the space of flows and the space of places.

The space of flows links up electronically separate locations in an interactive network that connects activities and people in distinct geographical contexts. The space of places organizes experience and activity around the confines of locality. Cities are structured, and destructured simultaneously by the competing logics of the space of flows and the space of places. Cities do not disappear in the virtual networks. But they are transformed by the interface between electronic communication and physical interaction, by the combination of networks and places. As William Mitchell (1999), from an urbanist perspective, and Barry Wellman (1999), from a sociological perspective, have argued, the informational city is built around this double system of communication. Our cities are made up, at the same time, of flows and places, and of their relationships. Two examples will help to make sense of this statement, one from the point of view of the urban structure, another in terms of the urban experience.

Turning to urban structure, the notion of global cities was popularized in the 1990s. Although most people assimilate the term to some dominant urban centers, such as London, New York and Tokyo, the concept of global city does not refer to any particular city, but to the global articulation of segments of many cities into an electronically linked network of functional domination throughout the planet. The global city is a spatial form rather than a title of distinction for certain cities, although some cities have a greater share of these global networks than others. In a sense, most areas in all cities, including New York and London, are local, not global. And many cities are sites of areas, small and large, which are included in these global networks, at different levels. This conception of global city as a spatial form resulting from the process of globalization is closer to the pioneering analysis by Saskia Sassen (1991) than to its popularized version by city marketing agencies. Thus, from the structural point of view, the role of cities in the global economy depends on their connectivity in transportation and telecommunication networks, and on the ability of cities to mobilize effectively human resources in this process of global competition. As a consequence of this trend, nodal areas of the city, connecting to the global economy, will receive the highest priority in terms of investment and management, as they are the sources of value creation from which an urban node and its surrounding area will make their livelihood. Thus, the fate of metropolitan economies depends on their ability to subordinate urban functions and forms to the dynamics of certain places that ensure their competitive articulation in the global space of flows.

From the point of view of the urban experience, we are entering a built environment that is increasingly incorporating electronic communication devices everywhere. Our urban life fabric, as Mitchell (1999) has pointed out, becomes an *e-topia*, a new urban form in which we constantly interact, deliberately or automatically, with online information systems, increasingly in the wireless mode. Materially speaking, the space of flows is folded into the space of places. Yet, their logics are distinct: online experience and face-to-face experience remain specific, and the key question then is to assure their articulation in compatible terms.

These remarks may help in the re-configuration of the theory of urbanism in response to the challenges of the network society, and in accordance to the emergence of new spatial forms and processes.

THE URBAN THEMES OF THE INFORMATION AGE

The issue of social integration comes again at the forefront of the theory of urbanism, as was the case during the process of urbanization in the industrial era. Indeed, it is the very existence of cities as communication artefacts that is called into question, in spite of the fact that we live in a predominantly urban world. But what is at stake is a very different kind of integration. In the early twentieth century the quest was for assimilation of urban sub-cultures into the urban culture. In the early twenty-first century the challenge is the sharing of the city by irreversibly distinct cultures and identities. There is no more dominant culture, because only global media have the power to send dominant messages, and the media have in fact adapted to their market, constructing a kaleidoscope of variable content depending on demand, thus reproducing cultural and personal diversity rather than overimposing a common set of values. The spread of horizontal communication via the Internet accelerates the process of fragmentation and individualization of symbolic interaction. Thus, the fragmented metropolis and the individualization of communication reinforce each other to produce an endless constellation of cultural subsets. The nostalgia of the public domain will not be able to countervail the structural trends towards diversity, specification, and individualization of life, work, space and communication, both face to face, and electronic (Russell, 2000; Putnam, 2000). On the other hand, communalism adds collective fragmentation to individual segmentation. Thus, in the absence of a unifying culture, and therefore of a unifying code the key question is not the sharing of a dominant culture but the communicability of multiple codes.

The notion of communication protocols is central here. Protocols may be physical, social, and electronic, with additional protocols being necessary to relate these three different planes of our multidimensional experience.

Physically, the establishment of meaning in these nameless urban constellations relates to the emergence of new forms of symbolic nodality which will identify places, even through conflictive appropriation of their meaning by different groups and individuals (Dunham-Jones, 2000).

The second level of urban interaction refers to social communication patterns. Here, the diversity of expressions of local life, and their relationship to media culture, must be integrated into the theory of communication by doing rather than by saying. In other words, how messages are transmitted from one social group to another, from one meaning to another in the metropolitan region requires a redefinition of the notion of public sphere moving from institutions to the public place, away from Habermas and towards Kevin Lynch. Public places, as sites of spontaneous social interaction, are the communicative devices of our society, while formal, political institutions have become a specialized domain that hardly affects the private lives of people, that is, what most people value most. Thus, it is not that politics, or local politics, does not matter. It is that its relevance is confined to the world of instrumentality, while expressiveness, and thus communication, refers to social practice, outside institutional boundaries. Therefore, in the practice of the city, its public spaces, including the social exchangers (or communication nodes) of its transportation networks become the communicative devices of city life (Borja and Zaida, 2001; Mitchell, 1999). How people are, or are not, able to express themselves, and communicate with each other, outside their homes and off their electronic circuits, that is, in public places, is an essential area of study for urbanism. I call it the sociability of public places in the individualized metropolis.

The third level of communication refers to the prevalence of electronic communication as a new form of sociability. Studies by Wellman, by Jones, and by a growing legion of social researchers have shown the density and intensity of electronic networks of communication, providing evidence to sustain the notion that virtual communities are often communities, albeit of a different kind than face to face communities (Wellman and Haythornthwaite, 2002; Jones, 1998). Here again, the critical matter is the understanding of the communication codes between various electronic networks, built around specific interests or values, and between these networks and physical interaction. There is no established theory yet on these communication processes, as the Internet as a widespread social practice is still in its infancy. But we do know that online sociability is specified, not downgraded, and that physical location does contribute, often in unsuspected ways, to the configuration of electronic communication networks. Virtual communities as networks of individuals are transforming the patterns of sociability in the new metropolitan life, without escaping into the world of electronic fantasy (Castells, 2001).

Fourth, the analysis of code sharing in the new urban world requires also the study of the interface between physical layouts, social organization, and electronic networks. It is this interface that Mitchell considers to be at the heart of the new urban form, what he calls e-topia. In a similar vein, but from a different perspective, Graham and Marvin's (2001) analysis of urban infrastructure as splintered networks, reconfigured by the new electronic pipes of urban civilization, opens up the perspective of understanding cities not only as communication systems, but as machines of deliberate segmentation. In other words, we must understand at the same time the process of communication and that of in-communication.

The contradictory and/or complementary relationships between new metropolitan centrality, the practice of public space, and new communication patterns emerging from virtual communities, could lay the foundations for a new theory of urbanism – the theory of cyborg cities or hybrid cities made up by the intertwining of flows and places (see Part 3).

Let us go farther in this exploration of the new themes for urban theory. We know that telecommuting – meaning people working full time online from their home – is another myth of futurology (Gillespie and Richardson, 2000; see Andrew Gillespie and Ranald Richardson, p. 212). Many people, including you and me, work online from home part of the time, but we continue to go to work in places, as well as moving around (the city or the world) while we keep working, with mobile connectivity to our network of professional partners, suppliers and clients. The latter is the truly new spatial dimension of work. This is a new work experience, and indeed a new life experience. Moving physically while keeping the networking connection to everything we do is a new realm of the human adventure, on which we know little (Kopomaa, 2000; see Zac Carey, p. 133; Timo Kopomaa, p. 267). The analysis of networked spatial mobility is another frontier for the new theory of urbanism. To explore it in terms that would not be solely descriptive we need new concepts. The connection between networks and places has to be understood in a variable geometry of these connections. The places of the space of flows, that is, the corridors and halls that connect places around the world, will have to be understood as exchangers and social refuges, as homes on the run, as much as offices on the run. The personal and cultural identification with these places, their functionality, their symbolism, are essential matters that do not concern only the cosmopolitan elite. Worldwide mass tourism, international migration, transient work, are experiences that relate to the new huddled masses of the world. How we relate to airports, to train and bus stations, to freeways, to customs buildings, are part of the new urban experience of hundreds of millions. We can build on an ethnographic tradition that addressed these issues in the mature industrial society. But here again, the speed, complexity and planetary reach of the transportation system have changed the scale and meaning of the issues. Furthermore, the key reminder is that we move physically while staying put in our electronic connection. We carry flows and move across places.

Urban life in the twenty-first century is also being transformed by the crisis of patriarchalism. This is not a consequence of technological change, but I have argued in my book *The Power of Identity* (Castells, 1997) that it is an essential feature of the information age. To be sure, patriarchalism is not historically dead. Yet, it is contested enough, and overcome enough so that everyday life for a large segment of city dwellers has already been redefined *vis-à-vis* the traditional pattern of an industrial society based on a relatively stable patriarchal nuclear family. Under conditions of gender equality, and under the stress suffered by traditional arrangements of household formation, the forms and rhythms of urban life are dramatically altered. Patterns of residence, transportation, shopping, education, and recreation evolve to adjust to the multidirectionality of individual needs that have to share household needs. This transformation is mediated by variable configurations of state policies. For instance, how child care is handled by government, by firms, by the market, or by individual networking largely conditions the time and space of daily lives, particularly for children.

We have documented how women are discriminated against in the patriarchal city. We can empirically argue that women's work makes possible the functioning of cities – an obvious fact rarely acknowledged in the urban studies literature (Borja and Castells, 1997: Susser, 1996). Yet, we need to move forward, from denunciation to the analysis of specific urban contradictions resulting from the growing dissonance between the de-gendering of society and historical crystallization of patriarchalism in the patterns of home and urban structure. How do these contradictions manifest themselves as people develop strategies to overcome the constraints of a gendered built environ-ment ? How do women, in particular, re-invent urban life, and contribute to re-design the city of women, in contrast to the millennial heritage of the city of men (Castells and Servon, 1996)? These are the questions to be researched, rather than stated, by a truly post-patriarchal urban theory.

Grassroots movements continue to shape cities, as well as societies at large. They come in all kind of formats and ideologies, and one should keep an open mind on this matter, not deciding in advance which ones are progressive, and which ones are regressive, but taking all of them as symptoms of society in the making. We should also keep in mind the most fundamental rule in the study of social movements. They are what they say they are. They are their own consciousness. We can study their origins, establish their rules of engagement, explore the reasons for their victories and defeats, link their outcomes to overall social transformation, but not to interpret them, not to explain to them what they really mean by what they say. Because, after all, social movements are nothing else than their own symbols and stated goals, which ultimately means their words.

Based on the observation of social movements in the early stage of the network society, two kinds of issues appear to require privileged attention from urban social scientists. The first one is what I called some time ago the grassrooting of the space of flows, that is the use of Internet for networking in social mobilization and social challenges (Castells, 2000). This is not simply a technological issue, because it concerns the organization, reach, and process of formation of social movements. Most often these online social movements connect to locally based movements, and they converge, physically, in a given place at a given time. A good example was the mobilization against the World Trade Organization meeting in Seattle in December 1999, and against subsequent meetings of globalizing institutions, which, arguably, set a new trend of grass-roots opposition to uncontrolled globalization, and redefined the terms of the debate on the goals and procedures of the new economy. The other major issue in the area of social movements is the exploration of the environmental movement, and of an ecological view of social organisation, as urban areas become the connecting point between the global issues posed by environmentalism and the local experience through which people at large assess their quality of life. To redefine cities as eco-systems, and to explore the connection between local eco-systems and the global eco-system lays the ground for the overcoming of localism by grassroots movements.

On the other hand, the connection cannot be operated only in terms of ecological knowledge. Implicit in the environmental movement, and clearly articulated in the deep ecology theory, as reformulated by Fritjof Capra (1996), is the notion of cultural transformation. A new civilization, and not simply a new technological paradigm, requires a new culture. This culture in the making is being fought over by various sets of interests and cultural projects. Environmentalism is the code word for this cultural battle, and ecological issues in the urban areas constitute the critical battleground for such struggle.

Besides tackling new issues, we still have to reckon in the twenty-first century with the lingering questions of urban poverty, racial and social discrimination, and social exclusion. In fact, recent studies show an increase of urban marginality and inequality in the network society (HDR, 2001). Furthermore, old issues in a new context, become in fact new. Thus, Ida Susser (1996) has shown the networking logic underlying the spread of AIDS among the New York's poor along networks of destitution, stigma, and discrimination. Eric Klinenberg (2000), in his social anatomy of the devastating effects of the 1995 heat wave in Chicago, shows why dying alone in the city, the fate of hundreds of seniors in a few days, was rooted in the new forms of social isolation emerging from people's exclusion from networks of work, family, information, and sociability. The dialectics between inclusion and exclusion in the network society redefines the field of study of urban poverty, and forces us to consider alternative forms of inclusion (e.g. social solidarity, or else, the criminal economy), as well as new mechanisms of exclusion and technological apartheid in the era of Internet.

The final frontier for a new theory of urbanism, indeed for social sciences in general, is the study of new relationships between time and space in the information age. In my analysis of the new relationships of time and space I proposed the hypothesis that in the network society, space structures time, in contrast to the time-dominated constitution of the industrial society, in which urbanization, and industrialization were considered to be part of the march of universal progress, erasing place-rooted traditions and cultures. In our society, the network society, where you live determines your time frame of reference. If you are an inhabitant of the space of flows, or if you live in a locality that is in the dominant networks, timeless time (epitomized by the frantic race to beat the clock) will be your time as in Wall Street or Silicon Valley. If you are in a Pearl River Delta factory town, chronological time will be imposed upon you as in the best days of Taylorism in Detroit. And if you live in a village in Mamiraua, in Amazonia, biological time, usually a much shorter lifespan, will still rule your life. Against this spatial determination of time, environmental movements assert the notion of slow-motion time, the time of the long now, in the words of Stewart Brand, by broadening the spatial dimension to its planetary scale in the whole complexity of its interactions, thus including our great-grand children in our temporal frame of reference (Brand, 1999).

Now, what is the meaning of this multidimensional transformation for planning, architecture, and urban design?

PLANNING, ARCHITECTURE, AND URBAN DESIGN IN THE RECONSTRUCTION OF THE CITY

The great urban paradox of the twenty-first century is that we could be living in a predominantly urban world without cities – that is without spatially based systems of cultural communication and sharing of meaning, even conflictive sharing. Signs of the social, symbolic, and functional disintegration of the urban fabric multiply around the world. So do the warnings from analysts and observers from a variety of perspectives (Kuntsler, 1993; Ascher, 1995; Davis, 1992; Sorkin, 1997; Russell, 2000).

But societies are produced, and spaces are built, by conscious human action. There is no structural determinism. So, together with the emphasis on the economic competitiveness of cities, on metropolitan mobility, on privatization of space, on surveillance and security, there is also a growing valuation of urbanity, street life, civic culture, and meaningful spatial forms in the metropolitan areas around the world The process of reconstruction of the city is under way. And the emphasis of the most advanced urban projects in the world is on communication, in its multidimensional sense: restoring functional communication by metropolitan planning; providing spatial meaning by a new symbolic nodality created by innovative architectural projects; and reinstating the city in its urban form by the practice of urban design focused on the preservation, restoration, and construction of public space as the epitome of urban life.

However, the defining factor in the preservation of cities as cultural forms in the new spatial context will be the capacity of integration between planning, architecture, and urban design. This integration can only proceed through urban policy influenced by urban politics. Ultimately, the management of metropolitan regions is a political process, made of interests, values, conflicts, debates, and options that shape the interaction between space and society. Cities are made by citizens, and governed on their behalf. Only when democracy is lost can technology and the economy determine the way we live. Only when the market overwhelms culture and when bureaucracies ignore citizens can spatial conurbations supersede cities as living systems of multidimensional communication.

Planning

The key endeavor of planning in the metropolitan regions of the information age is to ensure their connectivity, both intra-metropolitan and inter-metropolitan. Planning has to deal with the ability of the region to operate within the space of flows. The prosperity of the region and of its dwellers will greatly depend on their ability to compete and cooperate in the global networks of generation/appropriation of knowledge, wealth, and power. At the same time planning must ensure the connectivity of these metropolitan nodes to the space of places contained in the metropolitan region. In other words, in a world of spatial networks, the proper connection between these different networks is essential to link up the global and the local without opposing the two planes of operation.

This means that planning should be able to act on a metropolitan scale, ensuring effective transportation, accepting multinodality, fighting spatial segregation by acting against exclusionary zoning, providing affordable housing, and desegregated schooling. Ethnic and social diversity is a feature of the metropolitan region, and ought to be protected. Planning should seek the integration of open space and natural areas in the metropolitan space, going beyond the traditional scheme of the greenbelt. The new metropolitan region embraces a vast territorial expanse, where large areas of agricultural land and natural land should be preserved as a key component of a balanced metropolitan territory. The new metropolitan space is characterized by its multifunctionality, and this is a richness that supersedes the functional specialization and segregation of modernist urbanism. New planning practice induces a simultaneous process of decentering and recentering of population and activities, leading to the creation of multiple subcenters in the region.

The social and functional diversity of the metropolitan region requires a multimodal approach to transportation, by mixing the private automobile/highway system with public metropolitan transportation (railways, subways, buses, taxis), and with local transportation (bicycles, pedestrian paths, specialized shuttle services). Furthermore, in a post-patriarchal world, childcare becomes a critical urban service, and therefore must be integrated in the schemes of metropolitan planning. In the same way that some cities require additional housing and transportation investment per each new job created in certain areas, childcare provision should be included in these planning standards.

Overall, most metropolitan planning nowadays is geared towards the adaptation of the space of places of the metropolitan region to the space of flows that conditions the economic competitiveness of the region. The challenge would be to use planning, instead, to structure the space of places as a living space, and to ensure the connection and complementarity between the economy of the metropolitan region and the quality of life of its dwellers.

Architecture

Restoring symbolic meaning is a most fundamental task in a metropolitan world in crisis of communication. This is the role that architecture has traditionally assumed. It is more important than ever. Architecture, of all kinds, must be called to the rescue in order to recreate symbolic meaning in the metropolitan region, marking places in the space of flows. In recent years, we have observed a substantial revival of architectural meaningfulness that in some cases has had a direct impact in revitalizing cities and regions, not only culturally but economically as well. To be sure, architecture *per se* cannot change the function, or even the meaning, of a whole metropolitan area. Symbolic meaning has to be inserted in the whole fabric of the city, and this is, as I will argue below, the key role of urban design. But we still need meaningful forms, resulting from architectural intervention, to stir a cultural debate that makes space a living form. Recent trends in architecture signal its transformation from an intervention on the space of places to an intervention on the space of flows, the dominant space of the information age by acting on spaces dedicated to museums, convention centers, and transportation nodes. These are spaces of cultural archives, and of functional communication that become transformed into forms of cultural expression and meaningful exchange by the act of architecture.

The most spectacular example is Frank Gehry´s Guggenheim Museum in Bilbao, that symbolized the will of life of a city immersed in a serious economic crisis and a dramatic political conflict. Calatrava´s bridges (Seville, Bilbao), telecommunication towers (Barcelona), airports (Bilbao) or Convention Centers (Valencia) mark the space of flows with sculpted engineering. Bofill´s Barcelona airport, Moneo's AVE railway station in Madrid and Kursaal Convention Center in San Sebastian, Meier´s Modern Art Museum in Barcelona, or Koolhaas´s Lille Grand Palais, are all examples of these new cathedrals of the information age, where the pilgrims gather to search for the meaning of their wandering. Critics point at the disconnection between many of these symbolic buildings and the city at large. The lack of integration of this architecture of the space of flows into the public space would be tantamount to juxtaposing symbolic punctuation and spatial meaninglessness. This is why it is essential to link up architecture with urban design, and with planning. Yet, architectural creation has its own language, its own project that cannot be reduced to function or to form. Spatial meaning is still culturally created. But their final meaning will depend on its interaction with the practice of the city organized around public space.

Urban design

The major challenge for urbanism in the information age is to restore the culture of cities. This requires a socio-spatial treatment of urban forms, a process that we know as urban design. But it must be an urban design capable of connecting local life, individuals, communes, and instrumental global flows through the sharing of public places. Public space is the key connector of experience, opposed to private shopping centers as the spaces of sociability.

Borja and Zaida (2001), in a remarkable book supported with case studies of several countries, have shown the essential role of public space in the city. Indeed it is public space that makes cities as creators of culture, organizers of sociability, systems of communication, and seeds of democracy, by the practice of citizenship. This is in opposition to the urban crisis characterized by the dissolution, fragmentation, and privatization of cities. Borja and Zaida document, on a comparative basis, the projects of reconstruction of cities and of the culture of cities around the (re)construction of public space: the synthesis between places and flows is realized in the public space, the place of social cohesion and social exchanges (Borja and Zaida, 2001, 35).

This is in fact a long tradition in urban design, associated with the thinking and practice of Kevin Lynch, and best represented nowadays by Allan Jacobs. Jacobs' work on streets, and, with Elizabeth McDonald, on boulevards as urban forms able to integrate transportation mobility and social meaning in the city, shows that there is an alternative to the edge city, beyond the defensive battles of suburbanism with a human face (Jacobs, 1993) The success of the Barcelona model of urban design is based on the ability to plan public squares, even mini-squares in the old city, that bring together social life, meaningful architectural forms (not always of the best taste, but it does not matter), and the provision of open space for people's use. That is, not just open space, but marked open space, and street life induced by activities, such as the tolerance of informal trade, street musicians etc.

The reconquest of public space operates throughout the entire metropolitan region, highlighting particularly the working-class peripheries, those that need the most attention to socio-spatial reconstruction. Sometimes the public space is a square, sometimes a park, sometimes a boulevard, sometimes a few square meters around a fountain or in front of a library or a museum. Or an outdoor café colonizing the sidewalk. In all instances what matters is the spontaneity of uses, the density of the interaction, the freedom of expression, the multifunctionality of space, and the multiculturalism of the street life. This is not the nostalgic reproduction of the medieval town. In fact, examples of public space (old, new, and renewed) dot the whole planet, as Borja has illustrated in his book. It is the dissolution of public space under the combined pressures of privatization of the city and the rise of the space of flows that is a historical oddity. Thus, it is not the past versus the future, but two forms of present that fight each other in the battleground of the emerging metropolitan regions. And the fight, and its outcome, is of course, political, in the etymological sense: it is the struggle of the polis to create the city as a meaningful place.

THE GOVERNMENT OF CITIES IN THE INFORMATION AGE

The dynamic articulation between metropolitan planning, architecture, and urban design is the domain of urban policy. Urban policy starts with a strategic vision of the desirable evolution of the metropolitan space in its double relationship to the global space of flows and to the local space of places. This vision, to be a guiding tool, must result from the dynamic compromise between the contradictory expression of values and interests from the plurality of urban actors. Effective urban policy is always a synthesis between the interests of these actors and their specific projects. But this synthesis must be given technical coherence and formal expression, so that the city evolves in its form without submitting the local society to the imperatives of economic constraints or technological determinism.

The constant adjustment between various structural factors and conflictive social processes is implemented by the government of cities. This is why good planning or innovative architecture cannot do much to save the culture of cities unless there are effective city governments, based on citizen participation and the practice of local democracy. Too much to ask for? Well, in fact, the planet is dotted with examples of good city government that make cities livable by harnessing market forces and taming interest groups on behalf of the public good. Portland, Toronto, Barcelona, Birmingham, Bologna, Tampere, Curitiba, among many other cities, are instances of the efforts of innovative urban policy to manage the current metropolitan transformation (Borja and Castells, 1997; Verwijnen and Lehtovuori, 1999; Scott, 2001). However, innovative urban policy does not result from great urbanists (although they are indeed needed), but from courageous urban politics able to mobilize citizens around the meaning of their environment.

CONCLUSION

The new culture of cities is not the culture of the end of history. Restoring communication may open the way to restoring meaningful conflict. Currently, social injustice and personal isolation combine to induce alienated violence. So, the new culture of urban integration is not the culture of assimilation into the values of a single dominant culture, but the culture of communication between an irreversibly diverse local society connected/disconnected to global flows of wealth, power, and information.

Architecture and urban design are sources of spatio-cultural meaning in an urban world in dramatic need of communication protocols and artefacts of sharing. It is commendable that architects and urban designers find inspiration in social theory, and feel as concerned citizens of their society. But first of all, they must do their job as providers of meaning by the cultural shaping of spatial forms. Their traditional function in society is more critical than ever in the information age, an age marked by the growing gap between splintering networks of instrumentality and segregated places of singular meaning. Architecture and design may bridge technology and culture by creating shared symbolic meaning and reconstructing public space in the new metropolitan context. But they will only be able to do so with the help of innovative urban policy supported by democratic urban politics.

REFERENCES FROM THE READING

Ascher, F. (1995), *La Metapolis. Ou L´Avenir de la Ville*, Paris: Odile Jacob.

Blakely, E. and Snyder, M. (1997), *Fortress America: Gated Communities in the United States*, Washington DC: The Brookings Institution.

Borja, J. and Castells, M. (1997), *Local and Global: The Management of Cities in the Information Age*, London: Earthscan.

Borja, J. with Zaida, M. (2001), *L'Espai Public: Ciutat I Ciutadania*, Barcelona: Diputacio de Barcelona.

Brand, S. (1999), *The Clock of the Long Now*, New York: Basic Books.

Capra, F. (1996), *The Web of Life*, New York: Doubleday.

Castells, M. (1989), *The Informational City*, Oxford: Blackwell.

Castells, M. (1996), *The Rise of The Network Society, Volume I: The Information Age*, Oxford: Blackwell (revised edition, 2000).

Castells, M. (1997), *The Power of Identity, Volume II: The Information Age*, Oxford: Blackwell.

Castells, M. (1998), *The End of Millennium, Volume III: The Information Age*, Oxford: Blackwell.

Castells, M. (2000), 'Grassrooting the space of flows'. In J. Wheeler, Y. Aoyama and B. Warf (eds), *Cities in the Telecommunications Age: The Fracturing of Geographies*, London: Routledge, 18–30.

Castells, M. (2001), *The Internet Galaxy*, Oxford: Oxford University Press.

Castells, M. and Servon, L. (1996), 'The feminist city: a plural blueprint', Berkeley: University of California, Department of City Planning, unpublished.

Davis, M. (1992), *City Of Quartz*, New York: Vintage Books.

Dunham-Jones, E. (2000), 'Seventy-five per cent', *Harvard Design Magazine*, fall, 5–12.

Fernandez-Galiano, L. (2000), 'Spectacle and its discontents', *Harvard Design Magazine*, fall, 35–38.

Freire, M. and Stren, R. (eds) (2001), *The Challenge of Urban Government: Policies and Practices*, Washington, DC: The World Bank Institute.

Garreau, J. (1991), *Edge City: Life on the New Frontier*, New York: Doubleday.

Gillespie, A. and Richardson, R. (2000), 'Teleworking and the city: myths of workplace transcendence and travel reduction'. In J. Wheeler, Y. Aoyama and B. Warf (eds), *Cities in the Telecommunications Age: The Fracturing of Geographies*, London: Routledge, 228–48.

Graham, S. and Marvin, S. (2001), *Splintering Urbanism: Networked Infrastructures, Technological Mobilities, and the Urban Condition*, London: Routledge.

Hall, P. (1998), *Cities In Civilization*, New York: Pantheon.

Hall, P. (2001), 'Global city-regions in the 21st century'. In A. Scott (ed.), *Global City-Regions. Trends, Theory, Policy*, New York: Oxford University Press, 59–77.

Horan, T. (2000), *Digital Place: Building Our City of Bits*, Washington, DC: The Urban Land Institute.

Human Development Report, United Nations Development Programme (2001), *Technology and Human Development*, New York: Oxford University Press.

Jacobs, A. (1993), *Great Streets*, Cambridge: MIT.

Jones, S. (ed.) (1998), *Cybersociety 2.0*, London: Sage.

Klinenberg, E. (2000), 'The Social Anatomy of a Natural Disaster: The Chicago Heat Wave of 1995', Berkeley: University Of California, Dept of Sociology, Ph.D. Dissertation (unpublished).

Kopomaa, T. (2000), *The City In Your Pocket: Birth of the Mobile Information Society*, Helsinki: Gaudemus.

Kotkin, J. (2000), *The New Geography: How the Digital Revolution is Reshaping the American Landscape*, New York: Random House.

Kuntsler, G. (1993), *The Geography of Nowhere*, New York: Simon and Schuster.

Massey, D. (1996), 'The age of extremes: inequality and spatial segregation in the 20th century', Presidential Address, Population Association Of America.

Mitchell, W. (1999), *E-Topia*, Cambridge, MA: MIT Press.

Nel.Lo, O. (2001), *Ciutat De Ciutats*, Barcelona: Editorial Empuries.

Putnam, R. (2000), *Bowling Alone: The Collapse and Revival of American Community*, New York: Simon and Schuster.

Russell, J. (2000), 'Privatized lives', *Harvard Design Magazine*, fall, 20–9.

Sassen, S. (1991), *The Global City: London, Tokyo, New York*, Princeton, NJ: Princeton University Press.

Sassen, S. (2001), 'Global cities and global city-regions: a comparison'. In A. Scott (ed.), *Global City-Regions: Trends, Theory, Policy*, New York: Oxford University Press, 78–95.

Scott, A. (ed.) (2001), *Global City-Regions: Trends, Theory, Policy*, New York: Oxford University Press.

Sorkin, Michael (1997), *Variations on a Theme Park: The New American City and the End of Public Space*, New York: Hill and Wang.

Susser, I. (1996), 'The construction of poverty and homelessness In US Cities', *Annual Reviews of Anthropology*, 25, 411–35.

Verwijnen, J. and Lehtovuori, P. (eds) (1999), *Creative Cities*, Helsinki: University of Art and Design.

Waldinger, R. (ed.) (2001), *Strangers at the Gate: New Immigrants in Urban America*, Berkeley: University of California Press.

Wellman, B. (ed.) (1999), *Networks in the Global Village*, Boulder, CO: Westview Press.

Wellman, B. and Haythornthwaite, C. (eds) (2002), *The Internet in Everyday Life*, Oxford: Blackwell.

Wheeler, J., Aoyama, Y. and Warf, B. (2000), *Cities in the Telecommunications Age: The Fracturing of Geographies*, London: Routledge.

ONE

'The Capsule and the Network: Notes Toward a General Theory'

Lieven de Cauter

Editor's Introduction

With our fourth theoretical reading we start to move back into the domains of recombinant theorisations of cities and ICTs. In this reading Lieven de Cauter, a Dutch philosopher and art historian, takes some of the analyses and ideas developed by Castells and adds a further element which draws on the analysis of media theorist Marshall McLuhan (1964): capsularisation. By analysing the ways in which a proliferating range of carefully customised 'capsules' – smart homes, gated communities, cars, aircraft, mobile phones, screens and virtualised environments – relate to the speeds, times, flows and structures of the 'network society', de Cauter's work usefully complements that of Castells. De Cauter argues that these capsules can only exist and function when they are connected to the networks that functionally link them across geographic space and time to other capsules. Crucially, if the networks collapse the capsules are virtually useless and, when they are used as housing, uninhabitable.

The capsular idea is especially useful because it allows for the multiple spaces and scales of recombination in cybercities – from the body to the globe – to be encompassed within the analysis. Capsules – as the socio-technical constructions surrounding, and extending, the human body and its senses – provide classic articulation points through which widely stretched transport and ICTs flows are produced to have meaning, in practice, in real places.

De Cauter argues that the intensification of capsularisation in contemporary cities sustains a parallel process of 'hyper individualisation'. Here, the personal organisation of the capsular experience starts to transcend the collective social and cultural politics of place. He believes that this process deepens spatial and social segmentation. This is because wealthier and globally connected citizens increasingly relate through their capsular architectures and experiences to far-off places. At the same time they tend to physically and technically fortress themselves off from the less well off communities and places that may geographically surround them in the urban regions in which they live. Finally, de Cauter suggests that capsularisation is inextricably intertwined with suburbanisation, urban sprawl, and the emergence of multicentred urban regions as the dominant spatial form of contemporary urbanisation.

All media, as extensions of man (*sic*), enhance speed in some sort of way: the speed of travelling, commerce, communication, information, transmission, etc. The speed of movement and the increase of flows and information brought about by the constant increase of technological media throughout history, means that the human species, with its fragile body and sensitive central nervous system, has to build in protections. Apart from being extensions of man (*sic*), most, if not all media, are capsules or have a capsular counterpart.

For instance, the wheel as transportation medium requires, beyond a certain speed, capsular devices: carriages, cars, coach-work. If you want to put it in biological, anatomical terms, one could say that the speed of media makes exoskeletons a necessity to the survival of man, one of the unfittest creatures since the invention of evolution.

Capsules are, both Freud and McLuhan would agree, vital for protection against shock. This leads to the structure of modern experience. Many authors, from Freud and Simmel and Benjamin to McLuhan, have argued that modern man is under constant attack of an overload of stimuli (shocks), which induce a sort of defense mechanism. Therefore *the more physical and informational speed increases, the more man (sic) will need capsules.*

This protection has a paradoxical result: the speed and the defense against the hostile environment (the air in the case of airplanes, the water in the case of boats etc.) are transferred to the tool and make the passenger immobile, passive. In primitive (cold) machines like bikes, skis, roller skates, and the like, man is moving by becoming at one with his tool (*sic*). But in sophisticated (hot) machines we transfer speed, activity and active defense against shocks to the device itself. That is capsularization. Students have pointed out to me that the motorbike contradicts my scheme. It does in a sense: it is a high-speed device without a capsule. But it is precisely this that makes the motorbike so damned dangerous and therefore adventurous if not mythical. So in the end it is the exception that proves my point. Conclusion: the more mobile we become, the more capsular our behaviour: *we are sedentary nomads* (in the literal sense of sitting travellers).

But beside real capsules there are more and more virtual capsules. Much of recent technology could be described in terms of virtual capsules: the mobile phone, the walkman or the diskman, etc. And of course all screens (film screens, television screens, computer screens) are mental capsules. *A world of screens is a capsular world.* When looking at a screen you are in a closed mental, virtual space that is far away from the actual space where you sit. *Our media are getting more capsular as technology moves on from extensions of the body to extensions of the mind* (information technology). Both high speed transportation and microelectronics obey a sort of deeply rooted capsular logic. This we would call *the technological logic of capsularization.*

THE DUAL SOCIETY: THE SOCIO-ECONOMIC LOGIC BEHIND CAPSULARIZATION

Capitalism has become transcendental: it is without counterpart, it is the most intensive and extensive concept of the world today; it is ruled by 'the holy trinity of the Divine market: liberalization, deregulation and privatization' (Petrella, 1997). Manuel Castells has, in his trilogy *The Information Age* (1996–8) proved that the restructuring of capital in the network society, corresponding to the shift from industrial to 'informational capitalism,' has brought a gigantic social exclusion, a polarization of society in the global economy (see Manuel Castells, p. 82). He pays special attention to the rise of the fourth world: disconnected groups of the population, given up zones, be it ghettos or even almost entire continents like Africa. This switching off 'from the Network of parts of the world and its population' is one of the main arguments of his trilogy. To him the rise of the network society and the formation of ghettos are intimately linked.

The de-industrialization and unemployment of unschooled workers, the individualization of labor (under the magic word flexibility) and the disintegration of the patriarchal family are the main processes of this rise of the fourth world. The infernal logic of the global economy with the destruction of the welfare state combined with demographic growth and migration waves and a new, network, 'decentralized,' scattered version of the old center–periphery logic, might change the global territory into plugged-in and unplugged areas and parts of the population. This economic logic can affect the use and control of space and territories. This is in fact the case already. The inside ordered and connected, the outside out of order and disconnected. It is what Castells calls 'the black holes of informational capitalism.'

In short, globalization has engendered a dual society and dual cities. This now well-known (alas just as often readily dismissed) constellation one could call *the economic logic of capsularization.* The capsular civilization might be a sort of return to older phases in history, where 'public space,' the world outside the fortress, was by definition unsafe and uncontrolled territory. In his balance sheet of five centuries of capitalism Wallerstein (1993) sees the rise of civil war as inevitable. The nation state, he writes, can no longer take care of its citizens. 'The scramble for protection has begun.'

In the words of a friend of mine, John King, 'privilege engenders fear' (personal communication). When fear and the mechanism of defensible space takes over then we might see that this becomes another sort of iron law: *fear leads to capsularization and capsularization enhances fear.* (Everyone who has experienced a few burglaries will know that this is not just an academic word game, but a very down-to-earth process which is hard to escape.) The one vicious circle is repeated by the other: *exclusion leads to crime and crime leads to exclusion.* This might end up changing *the world* into *an archipelago of insular entities,* fortresses, gated communities, enclosed complexes (like hotels and malls), enclaves, envelopes, cocoons, in short *capsules in a sea of chaos.*

HYPERINDIVIDUALISM AND THE SUBURBANIZATION OF DAILY LIFE

We have seen a massive disinterest in the concept of society as based on sociability and solidarity. Individualization is a theme in modernity since Weber and Simmel, but the transformation of individualism into the official ideology of neoliberalism is fairly recent. It was summarized by Margaret Thatcher: 'There is no such thing as society, there are only individuals, individuals and families.' Not much more can be added to that. It is the all too well-known syndrome enhanced by the restructuring of capital as described by Manuel Castells (notably as the individualization of labor) and the logic of temporary contracts. It is a syndrome that is both imposed upon people (in the age of ubiquitous management every thing and every body is a corporation) and it is at the same time what people seem to want. The consumer is always an isolated, atomized individual, a loner, at best in a couple or at most in a nuclear family. This last expression is telling. The nuclear family is a capsular institution. That is how we are projected in publicity, that is more and more how we live. We might call it the *logic of hyperindividualism.* It is enhanced by our technological tools and by our daily politics of space.

We are all, in a sense, suburbanites; even the convinced city dwellers have to fight the mechanism of the suburbanization of daily life: cars, telephones (mobile or fixed), televisions and computers (with Internet connection) are the basic tools (and causes) of this process. Cocooning (an activity for capsular institutions like the nuclear family) is just the sweet glossy magazine word for the hard fact of capsularization of dwelling and living. Our daily life can be exactly described as a movement via transportation capsules from one enclave or capsule (home for instance) to another (campus, office, airport, all-in hotel, mall, and so on). Continuing my rather metaphorical-didactical, tongue-in-cheek attempt to distil basic mechanisms from my observations, we could say: *neoliberal individualism plus suburbanization of daily life equals capsularization.*

THE CAPSULE AND THE NETWORK

The economic, sociological and technological logic have come together in what Castells has termed the 'Network Society' (organized under 'informational capitalism'). He claims quite rightly that the new dominant spatial structure is the 'space of flows,' and that the logic of the 'space of place' is more and more disconnected from it. I think the capsule might supply the missing link between the two. In the hype about the rise of the network society people tend to overemphasize the flow, the blurring, the sampling, the crossover, the integration, the smoothness of thresholds, etc. But maybe all that is only one side of the picture. *The Network obscures, so to speak, the capsule. We don't live in the Network but in capsules.*

All networks, from the train network, the motorcar network and the air transport network to the telephone network and the world wide web – all networks function with capsules. So from the perspective of the network I can refine or even redefine the concept of capsule: I propose capsule here as the most general concept for every closed-off and plugged-in entity, which as a sum makes the networks what they are. *No Network without capsules. The more networking, the more capsules. In other words: the degree of capsularization is directly proportional to the growth of Network.* (I would insist on this by proposing to call this not the seventh concept but the most fundamental one: the extended first law of capsularization, as it is easy to see that it is extending the first formulation: *the more physical and informational speed increases the more man will need capsules*).

For transportation networks this is obvious: cars, trains, airplanes are capsules and are becoming more so every day (the shift from train to high speed train and from car to mono-volume car is enhancing the capsular character of these vehicles). But all networks need capsules, enclaves, envelopes: both as nodes and hubs, as well as terminals (each computer terminal being a virtual capsule for our voyage through cyberspace). The house as 'machine à habiter' or 'machine for living in' (in Le Corbusier's famous terms) can be specified as an immobile capsule. Nobody can deny that today's house is only functioning when plugged in to all sorts of networks: water, electricity, gas, fax/telephone, cable television and Internet, and of course the motorway system. These networks define the house today. Without them it is like a satellite lost in space. A house unplugged is barely thinkable nowadays (apart from the fact that it is the daily reality of the majority of the world population, but we would not claim they live in houses, we should call them huts, shacks, and the like). From the perspective of the network the house is, in short, a plugged-in terminal, a capsule.

REFERENCES FROM THE READING

Castells, M. (1996–1997), *The Information Age. Economy, Society and Culture* (3 volumes), Oxford: Blackwell.

Petrella, R. (1997) *Het Allgemeen Belan: Lof van de Solidariteit*, Brussels: VUB Press.

Editor's reference

McLuhan, M. (1964), *Understanding Media*, London: Routledge.

O
N
E

'Cities Without Modernity, Cities With Magic'

Scottish Geographical Magazine (1997)

Nigel Thrift

Editor's Introduction

In our fifth theoretical reading we return to the work of Bristol University geographer Nigel Thrift (p. 39). A long-time critic of the substitutionist approach of the relations between ICTs and cities, Thrift here takes issue with the theorisations of Paul Virilio (p. 78), Manuel Castells (p. 82), and the influential neo-Marxist geographer, David Harvey (whose concept of 'time–space compression' we encountered in the Introduction to this part – see Harvey, 1989).

To Thrift, all three of these writers are guilty of perpetuating myths. At the very least, he argues, they are guilty of suggesting conceptualisations which are 'overdrawn and overdone'. In particular, Thrift criticises both Virilio's concept of a 'generalised arrival' based on absolute ICT-based mobility, and Castells's concept that cities are caught up in some globe-spanning 'space of flows'.

To Thrift, such conceptualisations, in their preoccupation with illuminating contemporary societies, tend to neglect the long traditions of ICT-based mobility and flow. They also imply some relatively simple and generalised reading off of the effects of ICT technologies from their intrinsic abilities to transmit flows (almost) instantly across space. Thus they are, to Thrift, vulnerable to accusations of technological determinism. In particular, Thrift suggests that the analyses of Virilio, Castells and Harvey neglect the ways in which culture, context, and content, mediate the contingent social constructions of ICTs, and their resulting effects, in a wide variety of different ways in different times and places.

Thrift also argues that these analyses, in concentrating on the glamorous, leading edge technologies like the Internet, ignore or neglect the importance of more banal and taken-for-granted technologies which may be of equal significance in everyday urban life.

Finally, Thrift suggests that the notions that cities are losing their meaning as centres of human communication through the growth of ICT-based communication (a particularly strong emphasis in Virilio's work) is a myth. Arguing strongly for a richly social and contingent view of the recombinations of face-to-face and mediated communications, Thrift stresses that the social practices of people in places are merely complemented, not replaced, by ICTs.

It is claimed we live in a globalising world and one of the key elements of this world is the existence of a space of flows brought into existence by modern information technologies. This space of flows is: 'the space of information. This proliferating and multi-dimensional space is virtual, densely webbed and uniformly complex; a vast and sublime realm accessed through the mediations of our imaginative and technical representations,' (Davis, 1994:86). And, in turn, the existence of the space of flows is changing our apprehension of space, time, and subjectivity. Places move closer together in time, time becomes instantaneous, and the subject becomes decentred, strung out on the wire.

The two main contemporary tellers of this myth are Harvey (1989) and Virilio (1993). For Harvey, the world is in the grip of another round of 'time-space compression':

> the processes that so revolutionise the objective qualities of space and time that we are forced to alter . . . how we represent the world to ourselves . . . space appears to shrink to a 'global village' of telecommunications and a spaceship of earth economic and ecological interdependencies . . . and as time horizons shorten to the point where the present is all there is . . . so we have to learn how to cope with our overwhelming sense of *compression* of our spatial and temporal worlds (Harvey, 1989, 240).

Thus time-space compression is simultaneously a story of the marked increase in the pace of life brought about by modern transport and telecommunications *and* the upheaval in our experience of space and time that this speed-up brings about as people, images, capital, and information all speed more and more rapidly around the world. The result is that 'time-space compression . . . exerts its toll on our capacity to grapple with the realities unfolding around us' (Harvey, 1989, 306), most especially by challenging our sense of identity and our ability to preserve tradition.

Virilio (1993) is willing to go further. For him, Harvey only describes the first effects of speed-up in which physical displacement still presupposes a journey: the individual makes a departure, moves from one location to another, and so arrives. But now, with the 'instant' transmission made possible by electronic technology, a new 'generalised arrival' has occurred, in which the element of a journey across space is lost. The individual can be in two places as once, acting as both transmitter and receiver. Thus we have arrived in a historical period in which there is 'a crisis of the temporal dimension of the present':

> one by one, the perceptive faculties of an individual's body are transferred to machines, or instruments that record images, and sound; more recently the transfer is made to receivers, to sensors and to detectors that can replace absence of tactility over distance. A general use of telecommunications is on the verge of achieving permanent telesurveillance. What is becoming critical here is no longer the concept of three spatial dimensions, but a fourth, temporal dimension – in other words that of the present itself (Virilio, 1993, 4).

In this new order, cities become interruptions in the space of flows, transient moments in the circulation of capital. Their future is to act as waystations for dominant organisational forces making their wishes known 'through the powerful medium of information technologies' (Castells, 1989, 6).

Here is a myth, which is overdrawn and overdone, and for at least four reasons. First, these revelations, which are presented as of the hour, are certainly of considerable antiquity. Through history and around the world, innovations in transport and communications have been heralded as proof positive that the world was speeding up, that places were moving closer together in time, that the world was shrinking. For example, in the eighteenth century a number of nervous disorders were thought to be the result of the faster pace of life (Porter, 1993). In the nineteenth century, as the stage coach was replaced by the train, the telegraph and the telephone, so the theme of the 'annihilation of space and time' became a favourite meditation of many writers (Thrift, 1995). In the twentieth century, the idea surfaced yet again; one author even went so far as to calculate the rate (in minutes per year) at which places were converging on one another, a phenomenon which he christened '*time-space convergence*' (Janelle, 1969). These different variants on the theme of a speeded-up world were nearly always associated with a generalised crisis of identity. From the nervous disorders of the eighteenth century onwards, the acceleration of everyday life has been thought likely to lead to volatile, fragmented, and spread-out people whose identity is in question because of the shallowness of their lives.

Second, this myth is based in a technological determinism which unproblematically reads off the characteristics of the technologies involved onto society. For example, there is little or no sense of technologies as culturally mediated. Thus, whilst it cannot be denied that 'a dynamic of thought is not separable from a physics of traces . . . the medium . . . is but the ground floor. One cannot test there' (Debray, 1996, 11). Therefore, 'when these technologies are mastered . . . their functions and cultural resonances change fundamentally' (Collins, 1995, 16). In the case of the 'new' information technologies, there is manifold evidence that this process is already taking place;

> the bloom of unlimited possibility is already passing from them. The Internet begins to shrink back from its ecstatic characterisation, (a web for democracy, a veritable set of data, an instrument of expanded perception) and settles down as a very large data-base – cum – techie salon. The smart idolatry is moving on (Spufford, 1996, 271).

Again, too often the medium is assumed to be one when it is multiple. Thus work on communications networks too often ignores the *content* of communication – from bank statements to love letters – which illustrates that the medium is in fact a set of different but intersecting elements of a whole range of actor-networks.

Third, this myth places too much of a premium on technologies which conform to dominant cultural stereotypes of what is 'new'. But, as Mark Edwards points out:

> in a survey last year, Americans were asked which piece of recent technology had most altered their lives. The most common reply was not the Internet, not even the personal computer. First, – by some way – came the microwave, followed by the video recorder (Edwards, 1996, 10).

Then, fourth, cities cannot be seen as places which are leaking away into a space of flows. This is to fundamentally misunderstand the way in which new information technologies have normally acted as a supplement to human communication rather than a replacement. Innovations like the telephone, the fax, and the computer are used to extend the range of human communication, rather than act as a substitute. It is not a case of either/or but both/and. For example, in my work on the City of London (Thrift, 1996), I have

shown the way in which the growth of information from new communication technologies has presented fundamental problems of interpretation for workers in the City which have forced greater rather than less face-to-face communication: the City has become, even more than formerly, a key storytelling node for the world as a whole, and, as a result, its importance as a spatially fixed centre has, if anything, been boosted.

REFERENCES FROM THE READING

Castells, M. (1989), *The Informational City: Economic Restructuring and Urban Development*, Oxford: Blackwell.

Davis, E. (1994), 'Techgnosis, magic, memory and the angels of information', *South Atlantic Quarterly*, 92, 518–617.

Debray, R. (1996), *Media Manifestation: On the Technological Transmission of Cultural Forms*, London: Verso.

Edwards, M. (1996), Books Section, *Sunday Times*, 10/3, 10.

Harvey, D. (1989), *The Condition of Postmodernity*, Oxford: Blackwell.

Janelle, D. G. (1969), 'Spatial reorganisation: a model and a concept', *Annals of the Association of American Geographers*, 59, 348–64.

Porter, R. (1993), 'Baudrillard: hystory, hysteria and consumption'. In C. Rojeck and B. Turner (eds), *Forget Baudrillard*, London: Routledge, 1–21.

Spufford, F. (1996), 'The difference engine and the difference engine'. In F. Spufford and J. Uglow (eds), *Cultural Babbage. Technology, Time and Invention*, London: Faber and Faber, 266–90.

Thrift, N. J. (1996), *Spatial Formations*, London: Sage.

Virilio, P. (1993) 'The third interval: a critical transition'. In V. A. Conley (ed.), *Rethinking Techologies*, Minneapolis: University of Minnesota Press, 3–12.

Editor's references and suggestions for further reading

Amin, A. and Thrift, N. (2001), *Cities: Reimagining the Urban*, Cambridge: Polity.

Harvey, D. (1989), *The Condition of Postmodernity*, Oxford: Blackwell.

Thrift, N. (1995), 'A hyperactive world'. In R. Johnston, P. Taylor and M. Watts (eds), *Geographies of Global Change*, Oxford: Blackwell, 18–35.

Thrift, N. (1996a), *Spatial Formations*, London: Sage.

Thrift, N. (1996b), 'New urban eras and old technological fears: reconfiguring the goodwill of electronic things', *Urban Studies*, 33(8), 1463–93.

Thrift, N. (1997), 'Cities without modernity, cites with magic', *Scottish Geographical Magazine*, 113(3), 138–49.

'Cyberspace Meets the Compulsion of Proximity'

Deirdre Boden and Harvey Molotch

Editor's Introduction

Our next theoretical reading is also fully immersed in the recombinant view of city–ICT relations. In it, the late Deirdre Boden, and New York University sociologist Harvey Molotch, present a highly influential riposte to the idea that the meaning and value of face-to-face contact in cities automatically, and necessarily, withers in the age of ever-expanding telecommunications bandwidth. Looking in depth at the ethnographic, cultural, and economic value of what they call the 'compulsion of proximity' (i.e. face-to-face, co-present, human interaction in places), they argue that all mediated forms of communications present, at the very best, a very pale imitation of its power.

As they put it, 'co-presence is thick with information'. All sorts of non-verbal, tacit, tactile, and sensory dimensions exist to face-to-face interaction that are all too easily ignored or underestimated within instrumental assumptions that sending more information down a wire will simply substitute for it. Moreover, the places of the city, within which the huge variety of co-present meetings take place, are richly coded with endless landscapes of cultural, social, political and sexual meaning. These help to shape the meanings and powers attributed to people meeting in those places.

Following the recombination tradition, Boden and Molotch therefore argue that interactions and communications between people are therefore radically contingent and depend utterly in their meaning and power on the particular context within which they take place. In all sorts of transactions and communications, then, the face-to-face encounter presents a much fuller opportunity to develop deep commitment and trust than do relations mediated purely by electronic and ICT-based technologies.

In their assessment of the implications of this perspective for cities, Boden and Molotch suggest that spiralling flows of mediated information may make face-to-face contact *more* important, simply to make sense of it all. Reflecting a dynamic that we will encounter again in Section V, they observe the tendency of all industries that require continuous innovation and adaptation to volatile external conditions to be particularly powerful in driving the sort of economic clustering in cities that facilitates complex and ongoing face-to-face and mediated encounters.

Finally, in an interesting twist which prefigures our discussions in Section VII, Boden and Molotch observe that there are also class-based dimensions to the separation of routinised, mediated information flows from rich, co-present interaction in place. Boden and Molotch observe that it tends to be lower-order, routinised, highly formulaic, and relatively poorly paid jobs that can be organised purely at a distance through ICTs (as in call centres, for example). Many such jobs are therefore alienating because they lack the richness of co-presence and attempt to completely substitute for it by routinised and highly structured, and highly surveilled and disciplined, phone and computer transactions.

At the same time, however, Boden and Molotch argue that the professionals of the media and informational economy – the information creators who require easy access to rich combinations of networked and co-present communication – benefit not only from better skills, better jobs and better salaries. They also benefit from their privileged access to the full range of advantages that derive from rich co-present interaction of human beings in real places.

Some view contemporary advances in communication technologies as so profound that they bring deep transformations in social relations. Invoking a well worn intellectual tradition, commentators portray the new technologies as furthering the 'impersonality' of modern life, as old-fashioned face-to-face interaction yields to the new methods of linking individuals and events, whether near or far. In the same way that previous observers credited the steam engine and atom bomb with ushering in new modes of civilization, so the communications revolution putatively brings new forms of social life.

In contrast, we think this exaggerates. Those who see revolution under-appreciate the robust nature and enduring necessity of traditional human communication procedures. Modernity, including late modernity, is achieved not just through computer circuit boards and stock exchanges, but through the daily routines – intensely social – of thinking, cooperating, and talking face to face to humans. Rather than being antithetical to one another, intimacy is the basis for advanced societies, just as surely as it was the bedrock of earlier tribal life. Co-present interaction remains *the* fundamental mode of human intercourse and socialization, a 'primordial site for sociality' (Schegloff, 1986:112). We notice no decline in this simplest and most basic of human tendencies: people need to get together. It is our purpose in this essay to indicate precisely why.

By understanding the concrete features that make co-presence so useful, we will better understand both what makes modern institutions possible as well as certain limits of their effects on human communication. By distinguishing what is 'lost' in the translation from co-present interaction to other modes, we can learn what people are trying to get back to when they say, as they often do, 'we've got to get together and talk soon' or what they appreciate when they say 'it's good to hear the sound of your voice' or 'it's really great to see you again.' This exercise will indicate some of the limitations of the new technologies' capacity to act as a substitute for hearing and seeing – for physical proximity. This in turn has implications for the future spatial and temporal patterning of organizations as well as for their internal structure. We will now unpack some of the ingredients of co-present interaction that make it so rich a form of communication.

THE THICKNESS OF CO-PRESENT INTERACTION

Co-presence is thick with information. Under any circumstance, words derive their meanings only from the context that surrounds them; co-presence gives more context. Context includes not just other words, but facial gestures, body language, voice intonation, and a thousand other particulars. Not only does this information loading consist of details of the moment, but also of past moments and future prospects that shape and reshape the flow of turns between parties. Interaction is a sequential and radically local affair, with the meaning of a particular utterance understood by reference to prior utterances in the same, previous, and anticipated conversational streams. The words, 'I'll kill you' have completely different meanings when screamed by an adult in a back alley, when yelled by a child aiming his cowboy pistol, or when squealed by a loving parent tickling a four-year-old. The statuses of the individuals, their past history together, prior statements in the talk, the intonation of voices and the physical body attitude, are all simultaneously 'read' to constitute the very different, even opposite, senses of the same set of words.

The intensity of 'focused attention' (Goffman, 1961) entailed by co-presence means that communication can occur with no words whatever, as in a pregnant silence or the gleam in a romantic eye across a crowded room. Indeed, any detail – including a word – requires that actors construct meaning by allowing each particular to inform, or 'index' every other particular. Co-presence is so information rich that we feel a need to have it to know what is really going on, including the degree that others are providing us with reliable, reasonable and trustworthy accounts. At the same time, co-presence puts us 'on the spot' by making our own dissembling, deviance or other sort of disfavored action easier to detect.

The very richness of co-presence makes it a liability when tasks are best accomplished by losing rather than gaining information. The McDonald's employee taking an order communicates directly to the cook through the touch of a hamburger pictured on the cash register keyboard. It works well for the fast-food purpose. There are numbers of reasons, some obvious and some not, why co-presence is not the universal form of communication in modern society. But co-presence does contain the kind of unique attributes that make it, in a world of technical innovation and communication bulk, the ordering device of all the other forms of communication.

In one-on-one situations, people can use their bodies to touch one another. Touching is itself a full vocabulary of 'deep significance' (Hall, 1966:57) in which different meanings are provided by the degree of touch intensity, precise location of the body used (e.g. a physical brush with the shoulder vs. the extended finger) and exact spot where the touch is placed. The importance of touch has been carefully demonstrated. Subjects receiving touch gestures in psychiatric interviews were more likely to engage in verbal interaction, rapport and approach behaviors than those in control conditions. Touch increases levels of self-disclosure from encounter group participants. In one experiment, investigators left a quarter in a public phone booth and then asked people leaving the booth if they had found a missing coin. Subjects were more likely to say 'yes' if the experimenter lightly touched them on the arm while asking. In another study a 'librarian' who provided a half-second touch to subjects checking out books caused more reported satisfaction with library services, at least among women borrowers, than levels reported by untouched control individuals. Slide projectionists who casually rest a hand on a viewer's shoulder while adjusting the equipment raise viewers' reported evaluation of the slides being shown.

COMMITMENT

Co-present interaction requires participants to set aside not only a specific time but also a shared or shareable space, as well as generally constraining other activities at the same moment or location. When we are in co-presence, we have some evidence that the other party has indeed made a commitment, if nothing else than by being there. Just getting together, either by making an appointment, or stopping one's routine business to stand and talk awhile, provides evidence of commitment. Being kept waiting is an insult because it unilaterally delays the kind of 'quality encounter' that one has mentally and physically organized one's life around. We are irritated at those who take outside calls, gaze out the window, or read newspapers during our interaction with them. Such inattentiveness not only implies that the other person is 'busy' but spills over and indicates something about the substance of the conversation: that it is not taken as very important or that the answer to a request will be 'no.' Through variation in attentiveness, co-presence maximizes the opportunity to display commitment, to detect a lack of it in others, and hence adds substantive information.

Letters, memos, fax and electronic mail are a poor substitute for gauging commitment; correspondents can do many things before answering one's pressing concerns or even *while* answering them. Neither prompt response nor level of involvement is guaranteed; they can even delegate the task of responding to a secretary or, in the modern case, a word processor with all but the opening salutation pre-programmed. The much-maligned postal service can be blamed for complete inattention ('must have been lost in the mail') or the computer network can be described as 'down'.

Telephones are better. Small silences in a telephone exchange cause us to check both the talk and the technology. Even putting someone on 'hold' or providing the sounds that begin the closing of a conversation ('well, uh I've got to . . .') requires split-second timing and collaborative effort. Still, compared to co-presence, it is easier to fake a conversation over the phone by injecting 'speech particle' utterances like 'uh huh' and other 'monitoring responses' at technically appropriate turn-transition points than when face to face. Nails can be filed and mail read with impunity, sudden exits can be explained away by pies in ovens and someone at the door.

Nor has the development of electronic mail, which is indeed far more open and interactive than the traditional memo form, obviated the advantages of co-presence which remains the far thicker medium. Judging sincerity over the phone or through electronic mail is made difficult by the fact it is easier to fake words than to disguise both words and the multiple gestures and stance that go with realtime/realspace interaction. Co-presence also affords access to the body part that 'never lies', the eyes – the 'windows on the soul' as the old sayings go. Eye contact itself signals a degree of intimacy and trust; co-present interactants continuously monitor the subtle movements of this most subtle body part. That we demand people 'look us in the eye' during a crucial conversation implies we believe we have a truth detection mechanism. This belief, regardless of its accuracy, translates into a felt need for co-presence.

IMPLICATIONS FOR CITIES

Consistent with the 'footloose' school, there has been an obvious dispersion of low-skilled back office jobs to locales remote from the headquarter control centers. Places like Omaha become the '800' phone number centers for purchasers of products made by firms located at far distant points; *New Yorker* magazine readers write to a Colorado address if they have a subscription problem.

But while it may be the case that the advent of new communication technologies may loosen some types of business concentration, there may be intensified attraction among others. For instance, despite the boom in suburban conference centers, according to Whyte (1989), 'the city is still the prime place . . . because of the great likelihood of *un*planned, informal encounters or the staging of them' (cited in Berman, 1989, original emphasis). The scattering made possible by the new technologies not only fails to eliminate the need for co-present communication, it may intensify this need among those whose job it is to coordinate dispersed activities and the plethora of information pouring in from diverse settings. In other words, the only way to effectively deal with low quality communication is with communication of the highest quality – co-presence.

Urban scholars have noted that rather than metropolitan areas becoming dispersed, concentration occurs in new forms. Business headquarters require a wide array of support services in the immediate vicinity (e.g. financial institutions, advertising agencies). While the exact reason for this need for proximity is left largely undeveloped in the urban literature, the need for face-to-face conversation clearly encourages these businesses into propinquity (rather than, for example, quick delivery time of plans, services or access to maintenance personnel). Besides such headquarter activity, innovation work like research and design development in innovation centers like Silicon Valley – seem also to require proximity among actors. Even in the case of pure science, the most crucial information that makes possible experimental replication and advancement does not rely on the journals, but mobilizes the interpersonal networks among scientists.

In those instances where geographic dispersion does separate people who need to be in touch, business and organizational actors do not rely only on teleconferencing and the other new technologies (or even the good old phone). The conference industry grows to maximize intense co-presence. The frequency of such meetings has 'skyrocketed' (White, 1988:5). A spatial consequence of this is not only sustenance for the major downtown and airport hotels that often host such meetings, but of more remote conference centers (e.g. the posh Chaminade in Santa Cruz County, CA) where interaction can be intensified without competition from big city lights.

UNEVEN ACCESS TO ORDINARY TALK

However much people seem to prefer co-present interaction, access to such ordinarily talk, at least in occupational settings, is unequally distributed. The debate over the new technologies has often been one of who will have the advantage of access to them; job programs, for example, fight poverty by bringing slum dwellers 'up' to the exciting world of computers. The issue should also arise of who will be able to have jobs that make use of the full resources of the mind and body in communicating with others.

While executives consult computers on an as-needed basis, and higher technicians gain job satisfaction from creative challenges of programming and diverse manipulations, lower-rung workers, particularly women office personnel, are routinized. Rather than working in the home or in service jobs that are interpersonally based (waitressing, retail sales), they face communication deskilling analogous to the mechanization of the human body accomplished by the assembly line. It may be that we are in an awkward transition in which humans must enter data into machines which will one day be recorded as part of transactions themselves (as now occurs universally at bank ATM machines, for example). But the fact that US job expansion has been most rapid in the lower reaches of the business service economy, in part because of the new technologies, implies that these low-grade high-tech jobs will grow in number.

In the rush to automate, some organizations are unwisely designing jobs with an overly mechanistic view of how the new technologies can be effective. Stripping away thick communication inhibits workers' capacity to do their jobs well, even jobs that are centered on repetitive data entry tasks. Workplace research makes it clear that organizational success happens not by workers' unthinking following of rules, filling out forms or inputting their computers. Instead, jobs can get done when workers mobilize their capacity for ad hoc solutions, sometimes in opposition to organizational rules.

CONCLUSION

We have insisted in this essay that specific features of co-present interaction make it fundamental to social order, both local and global. The immediacy and inherent indexicality of all human existence means that the fine, fleeting, yet essentially social moments of everyday life anchor and articulate the modern macro-order. Through the trust, commitment and detailed understandings made possible in situations of co-presence the essential time–space distanciation of modern society is achieved.

Our proposal, in this reading, is that there is a hierarchy among forms of human intercourse such that:

- ▪ People still prefer co-presence, compared to other forms of communication, particularly in those activities that set the conditions under which more routine communication events occur;
- ▪ Co-present interaction is just as frequent as it ever was and among some kinds of actors, perhaps more frequent;
- ▪ Co-present interaction dominates over other forms of communication in that other forms take their shape through recall or anticipation of co-present talk, rather than the other way round;
- ▪ Co-present interaction is crucial in that the most important communication among the most consequential actors is done with co-presence and this will continue to be the case;
- ▪ This need for co-presence puts limits on the degree and kind of organizational, temporal, and spatial reshaping that the new technologies can induce.

The so-called communications revolution adds to, rather than replaces, co-presence, indeed requiring large numbers of people to engage in intense face-to-face interaction as means of coordinating, interpreting and designing the information and messages that the advanced technologies transmit. At the same time, some kinds of actors, those lacking the social power to insist on routine access to ordinary talk in their workplace and whose role is not judged to necessitate it, are subject to spatial dispersal, organizational isolation, as well as interactional deskilling.

This world requires a delicate balance of co-presence and absence. This balance is a struggle to achieve because it is often not apparent just which tasks require what kinds of communication media or their appropriate ratio. The balance is also a struggle because there may be opposition from those called upon to play the least interesting roles in this production of modernity. Problems of social justice thus remain because even though the available technologies may change, the interactive skills and social yearnings of humans have not.

REFERENCES FROM THE READING

Berman, M. (1989), 'Two thousand years of street smarts', *Village Voice: Literary Supplement*, Nov. (80), 9.

Goffman, E. (1961), *Encounters*, Indianapolis: Bobbs-Merrill Educational.

Hall, E. (1966), *The Hidden Dimension*, New York: Doubleday.

Schegloff, E. (1986), 'The routine as achievement', *Human Studies*, 9(2–3). Special issue on Interaction and Language Use, edited by G. Button, P. Drew and J. Heritage.

White, G. (1988), 'Face to face: preference for round table spawns need for conference centers, planners', *Los Angeles Times*, March 28, IV–5.

Whyte, W. (1989), *City: Rediscovering the Center*, New York: Doubleday.

Editor's suggestion for further reading

Boden, D. and Molotch, H. (1994), 'The compulsion of proximity'. In R. Friedland and D. Boden (eds), *Now/Here: Space, Time and Modernity*, Berkeley: University of California Press, 257–86.

'The Co-Existence of Cyborgs, Humachines and Environments in Postmodernity: Getting Over the End of Nature'

Timothy Luke

Editor's Introduction

Our final theoretical reading is taken from the work of US political scientist Timothy Luke, who is based at the Virginia Polytechnic Institute and State University. In it Luke develops a sophisticated, recombination-based analysis of the cybercity condition through the lens of what is known as 'cyborg' theorisation.

Luke starts by rejecting the very idea that the social and the technical, the natural and the cultural, the place-based and the ICT-mediated, and the 'cyber' and the city, can or should be separated into different conceptual domains. Instead, he draws on the work of anthropologists of technology such as Donna Haraway (1991) and Bruno Latour (1993) to develop a variety of 'hybrid' and 'socio-technical' concepts. These concepts – such as the blended human-machine or 'humachine' – are so recombinant that they make it impossible, and indeed pointless, to try to distinguish where the social stops and the technological starts. They thus usefully point to ways of imagining cybercities that transcend the traditional binary separations of city and culture, city and countryside, subject and object, technology and society, place and flow, and male and female, that have dominated social sciences, the humanities, and politics in the west since the Enlightenment.

To Luke, the saturating of our world with complex ranges of ICTs (and ICT-based biotechnologies) means that *all* environments are now socio-technical constructions which blend organisms and machines, objects and subjects, into inseparable, dynamic, and relational, hybrids. These processes of 'cyborganization' require us to analyse the world from the point of view of the agency of technologies as well as people. They mean that we have to confront the rise of 'denature' – the technological construction of all environments, whether city, countryside, or human body. And they force us to change the way we think about cities, as they become utterly constructed through whole varieties of machinic and flow systems within which cyborganised relations become social relations (and vice versa).

To Luke, the utter reliance of contemporary urban dwellers on vast ranges of superimposed cyborganised systems – ICTs, water, logistics, airlines, highways, commodity chains, energy systems, financial systems, food systems, medical systems, and so on – means that our societies are especially fascinated by what happens when such systems collapse (whether through technical failure, war, earthquakes, social unrest, or terrorist attack). 'Cyborgs, like us', writes Luke, 'are endlessly fascinated by machinic breakdowns, which would cause disruptions in, or denials of access to, their megatechnical sources of being.'

Luke's arguments are especially useful as we move from conceptualising cybercities to imagining how recombinant forms, linking ICTs and urban landscapes into inseparable wholes, are being constructed in a range of contexts around the world. This task we turn to in the next part of this book.

NEW LIFE FORMS? CYBORGS AND HUMACHINES

Anthropologist Donna Haraway's (1991) celebration of cyborg consciousness and agency provides a new heading for our investigation of the connections between nature and society. If, as she maintains, 'a cyborg is a cybernetic organism, a hybrid of machine and organism, a creature of social reality' in which social realities must be seen as 'lived social relations, our most important political construction, a world changing fiction' (Haraway, 1991: 149), then one must carefully re-examine where and how cyborgs – 'creatures simultaneously animal and machine, who populate worlds ambiguously natural and crafted' – actually live (Haraway, 1991: 149). Using a world changing fiction signals the necessity of searching for facts of world change. The science fiction of cyborgs illuminates the science fact of cyborganizing environmental sites and structures, giving both forms to cyborgs for living and ranges for them to roam.

Cyborgs are not creatures of pristine Nature; they are the planned and unplanned offspring of manufactured environments, fusing into new organic compounds of naturalized matter and artificialized anti-matter. Haraway sights them at breaches in the borders dividing humans and animals, organisms and machines, or physical and non-physical realms. In the swirl of postmodernity, little now convincingly separates human beings from other animal beings, especially now as all lifeforms are being reinterpreted as diverse iterations of a shared genetic code. In practice, 'language, tool use, social behavior, mental events' (Haraway, 1991: 151–2) all fail as definitive distinctions.

Similarly, contemporary machines confuse the differences between organisms and machines. In fact, 'our machines are disturbingly lively, and we ourselves frighteningly inert' (Haraway, 1991: 152), because, at this juncture, 'the boundary between the physical and non-physical is very imprecise for us . . . modern machines are quintessentially microelectronic devices; they are everywhere and they are invisible . . . they are about consciousness . . . or its simulation' (Haraway, 1991: 153).

These ambiguities result, as social theorist Bruno Latour (1993) asserts, in the subjectification of objects, and the objectification of subjects. They transform much of the living and dead world into 'quasi-objects' that are neither entirely alive nor merely dead. Haraway's deployment of this science fiction is aimed at illuminating these science facts. That is, 'a cyborg world might be about lived social and bodily realities in which people are not afraid of their joint kinship with animals and machines, not afraid of permanently partial identities and contradictory standpoints' (Haraway, 1991: 154).

Cyborganizing processes are integrating humans and machines so completely into new syncretic totalities of synergistic behavior that the once obvious borders between human/machine, human/animal, physical/non-physical blur away. Out of the blur, Haraway draws the figure of the cyborg, but does she miss other life forms also out there in these spaces? This may happen because maybe her conceptual arrays, strangely enough, privilege the organismic over the mechanistic in detecting these new life forms.

For Haraway, 'the cyborg is a creature in a post-gender world; it has no truck with bisexuality, pre-oedipal symbiosis, unalienated labor, or other seductions to organic wholeness through a final appropriation of the powers of the parts into a higher unity' (1991: 150). Although cyborgs do not identify with nature and organic wholeness, they are still, in Haraway's imagination of them, organismic figures – individual subjectivities, discrete bundles of being, or personal agents.

Here, however, we must ask whether this is a biocentric reading of cyborg life forms that might marginalize the equally significant machinic lineage in the cyborg family. For example, in cinematic terms, could cyborgs be much more machinic, like a *Robocop* subjectivity or a *Terminator* civilization? Do fusions of animal and apparatus, organism and mechanism, physical and non-physical also transform machinic realms, which are the other pole and the hemisphere in the lifeworlds of cyborg becoming?

With biocentric images, Haraway's depiction of the cyborg – as man-machine fusions – discloses a new world. Her cyborgs mostly inhabit the organic hemisphere near the pole of naturalistic organism, but the new world of cyborg life arguably also has a machinic hemisphere around the pole of technic mechanism. For nearly a century now, advanced industrial societies have set themselves and their human inhabitants apart from Nature's organic economies in new urbanized habitats, which are artificially powered, mechanically maintained, and scientifically constructed. In resisting anthropocentric readings of man and nature with her discovery of cyborgs, has Haraway downplayed other possibly new life forms in advancing her analysis of quasi-objects in the present by biocentering cyborg life on the human, animal, physical lineage in cyborg descent? If cyborgs are the individualized subjects formed out of blurred borders, then what are the collectivized subjects fabricated from the fusion of machine/human, animal/human, non-physical/physical regions? Without being machinocentric, why not trace the machinic, humane, and non-physical quasi-object lineage in cyborg descent?

AT THE END OF NATURE: 'DENATURE'

Haraway's cyborg manifesto struggles to express awe and amazement over 'the end of Nature.' The actual eclipse of pure wildness in the natural environment or the apparent end of sublime depth in nature's vastness are scientific facts in an era that inventories every square meter of ground with satellite mapping technologies, catalogues the molecular codes of the human genome, or zones land masses for bureaucratically defined 'wilderness use.' After two centuries of industrial revolution and three decades of informational revolution, nature no longer can be assumed to be God-created (theogenic) or self-creating (autogenic). What is taken to be 'nature' now is largely human-created (anthropogenic), not only in theory but also in practice. One need not wait for the science fiction of advanced space travel technologies to contact other 'extra-terrestrial life forms'; the science facts of altered atmospheric chemistry, rampant genetic engineering, and unchecked species extinctions suggest that urban industrial humanity is a race of extra-terrestrial intelligent beings already intent upon imperializing the earth in cyborg colonies with humachinic technologies.

In *Star Trek*, for example, two of the *Trek* classic feature films center their plot around how entire planets might be 'terraformed' with genetically encoding, meteorologically programmable, and geophysically stabilizing extra-terrestrial probes to simulate earth ecologies on other worlds in biotronic genesis events, rendering new planetary surfaces hospitable to terranean life forms of the M Class. Seeing Haraway's cyborgs shifts our register, allowing us to recognize how the earth itself already has submitted to the terraforming of multinational capital formations. Latour's 'constitution' has already been amended by new agencies. Are not humachinic ensembles an extra-terrestrial intelligence working at the level of loosely coupled collectives of machinic production, consumption, and administration, but also reconfiguring everything terrestrial as quasi-objective/quasi-subjective hybrids to fit their basic codes of operation?

Nature is no longer the vast realm of autonomous, unmanageable, or nonhuman wild activity; in being enmeshed in networks of cyborg scientific rationalization and commercial commodification, nature becomes denature(d). Nature is being terraformed, or denatured, to embed humachines into cycles of natural reproduction as well as to establish artificialized ranges for cyborg life forms, who never can be wholly natural. The entire planet now is increasingly a 'built environment' or 'planned habitat' as pollution modifies atmospheric chemistry, urbanization restructures weather events, biochemistry redesigns the genetics of existing biomass, and architecture accretes new biotic habitats inside of sprawling megacities.

The informational revolution with its denaturing administration of production on a global scale and cyborganizing management of consumption at a local level is a rolling constitutional convention, amending all of our existing covenants between the organic and inorganic. Informationalization also might be regarded as the biotronic genesis of transnational capitalism's terraforming of the earth (Luke, 1994). This anthropogenic (de)nature, then, proliferates spaces and networks in extra-terrestrialized regions where most cyborgs can live only through and within humachinic environments. As the collapse of Kobe, the bombing of Grozny, or the siege of Sarajevo show, cyborg life amidst broken-down humachinic operations also can be very inhumane – or, to follow Hobbes, solitary, nasty, brutish, and short.

Denature is nature, deformed and reformed simultaneously by anthropogenic transformations. Yet, for all intents and purposes, large expanses of denature continue to be (mis)taken for pristine natural corners at the periphery of manufactured denatural cores. These naturalist fantasies mystify denaturalist realities, which are not so easily marketed. Once one escapes the intimidation of such sierra clubbing, however, denatured realities, like the Aral Sea, Chernobyl, or the Boswash Corridor, can exert their communicative authenticity. Whatever nature once was cannot be regained, because it existed as a set of forces, settings, or conditions when the human influences upon planetary ecologies were very low impact. Its original condition, therefore, cannot be gauged apart from these minimal human-and-environment interactions.

Here is one of the deepest contradictions in cyborg life today. The formation of states around territorial sovereignty has distorted, and still disrupts, the rational operation of megatechnical cyborg living. Following the rules of territorial sovereignty, the historical organization of capital, labor, and technology typically created machinic ensembles for service to a few (and disservice for the many) only within and for territorialized sovereign space. Powerful nation-states to a significant extent also had to become efficient 'machinations,' articulating their own complex humachinic infrastructures to service cyborg populations occupying their sovereign territories inside of elaborate urban-industrial environments.

STRANGE NEW WORLDS AND NEW CIVILIZATIONS

The old constitution does not apply here: denature usually is now known as 'the environment' in deliberations by 'the parliament of things' (Latour, 1993: 142–5). Environments are not nature as such, but the denatured planetary condition constructed by machinic ensembles. After nature is reconstituted by the terraforming probes of fishing machines, farming machines, mining machines, timbering machines, ranching machines, its matter and energy flow through transport machines and communication machines to manufacturing machines, managing machines, military machines as well as living machines, leisure machines, labor machines. Their growth, reproduction, excretion, respiration, constitute 'the environmental' spaces of transnational capitalist exchange where cyborg beings are cyborganizing/ed in the ecological carrying capacities of these machinic ensembles' systems. These beings, as the *Star Trek* classic foretells, populate the strange new worlds or new civilizations where no one has gone before.

The cyborg myth engages the present in 'transgressed boundaries, potent fusions, and dangerous possibilities' in which 'cyborg unities are monstrous and illegitimate' (Haraway, 1991, 154). And, in both accepting monstrosity and embracing illegitimacy, cyborg mythologies provide tremendously potent possibilities for both resisting and recoupling with the systems of power. Only by resorting to science fictions can the full dimension of the world's extra-terrestrialization by science fact be brought to our conscious awareness. Corporate capital has been

conscious of its terraforming transformative powers since at least the 1930s. After all, a major theme of the 1939 New York World's Fair was 'It's a Man Made World' as its Trialons and Perispheres reconfigured Long Island. But very few social theorists have been willing to face this evidence as forthrightly as Donna Haraway and Lewis Mumford. Indeed, as this preliminary exploration indicates, these science facts even now might only be glimpsed through refraction of science fictions.

Once science fictions bracket the terraforming impact of science facts, however, the extra-terrestrial origins of denatured second and third nature appear more clearly. And, while some small groups of human beings maybe still can live pre-machinically, like the Kung of the Kalahari, Haitian fishermen on Hispaniola, Mongol herdsmen in Siberia, or even the House Amish of Pennsylvania, many more human beings live highly cyborganized lives, totally dependent upon the denature of machinic ensembles with their elaborate extra-terrestrial ecologies of megatechnical economics. This is as true for the Rwandans in the refugee camps of Zaire as it is for the Manhattanites in the luxury coops of New York City. Without the agriculture machine, the housing machine, the oil machine, the electrical machine, the media machine, or the fashion machine, almost all cyborganized human beings cannot survive or thrive, because these concretions of machinic ensembles generate their basic environment.

Cyborgs, like us, are endlessly fascinated by machinic breakdowns, which would cause disruptions in, or denials of access to, their megatechnical sources of being. Beirut in the fifteen year civil war, Sarajevo in its two year long siege, or the Roadwarrior's travels in post-megamachinic Australia (depicted in the *Mad Max* films) are all dark revelations of what once were highly evolved cyborg beings, struggling to survive decyborganized societies without all of the life-support systems of megatechnics – oil machines, light machines, water machines, food machines, clothing machines, transport machines, or medical machines. These societies or spaces reveal something far worse than 'the end of nature.' They display the 'death of denature' without whose bounty cyborgs lose both their identities with humachines and differences from ordinary terrestrial life forms.

REFERENCES FROM THE READING

Haraway, Donna (1991), *Simians, Cyborgs and Women: The Reinvention of Nature*, New York: Routledge.

Latour, Bruno (1993), *We Never Have Been Modern*, London: Harvester Wheatsheaf.

Luke, Timothy W. (1994), 'Worldwatching at the limits to growth', *Capitalism, Nature, Socialism*, 18 (June), 43–63.

Editor's references and suggestions for further reading

Amin, A. and Thrift, N. (2001), *Cities: Reimagining the Urban*, Cambridge: Polity.

Bingham, N. (1996), 'Object-ions: from technological determinism towards geographies of relations', *Environment and Planning D: Society and Space*, 14, 635–57.

Gandy, M. (2002), *Concrete and Clay: Reworking Nature in New York City*, Cambridge, MA: MIT Press.

Gray, C. (ed.) (1995), *The Cyborg Handbook*, New York: Routledge.

Gray, C. H. (2001), *Cyborg Citizen*, New York: Routledge.

Haraway, D. (1991), 'A manifesto for cyborgs: science, technology, and socialist-feminism in the late twentieth century'. In D. Haraway (ed.), *Simians, Cyborgs and Women: The Reinvention of Nature*, New York: Routledge, 149–81.

Kaika, M. and Swyngedouw, E. (2000), 'Fetishising the modern city: the phantasmagoria of urban technological networks', *International Journal of Urban and Regional Research*, 24(1), 122–48.

Latour, B. (1993), *We Have Never Been Modern*, London: Harvester Wheatsheaf.

Latour, B. and Hermant, E. (1998), *Paris ville invisible*, Paris: La Decouverte.

Luke, T. (1994), 'Placing power/siting space: the politics of global and local in the New World Order', *Environment and Planning D: Society and Space*, 12, 613–28.

Luke, T. (1997), 'At the end of nature: cyborgs, "humachines", and environments in postmodernity', *Environment and Planning A*, 29, 1367–80.

Luke, T. (1999), 'Simulated sovereignty, telematic territoriality: the political economy of cyberspace'. In M. Featherstone and S. Lash (eds), *Spaces of Culture: City-Nation-World*, London: Sage, 27–48.

Luke, T. and Ó Tuathail, G. (1998), 'Global flowmations, local fundamentals and fast geopolitics'. In A. Herod, G. Ó Tuathail and S. Roberts (eds), *Unruly World? Globalization, Governance and Geography*, London: Routledge, 72–94.

Picon, A. (1998), *La ville territoire de cyborgs*, Paris: L'Imprimeur.

Swyngedouw, E. (1997), 'The city as hybrid: On nature, society and cyborg urbanisation', *Capitalism, Nature, Socialism*, 65, 93–106.

Thrift, N. (1996), 'New urban eras and old technological fears: reconfiguring the goodwill of electronic things', *Urban Studies*, 33(8), 1463–93.

Zimmerman, R. (2001), 'Social implications of infrastructure network interactions', *Journal of Urban Technology*, 8(3), 97–119.

III CYBERCITIES

Hybrid forms and recombinant spaces

Plate 9 Travellers using an Internet and information kiosk in London's Heathrow airport, 2002.
(Photograph by Stephen Graham.)

INTRODUCTION

This is an informational world where increasingly our self is linked to the world (or divided from it) through the screen – the glass pane of a car windscreen, the computer terminal or the television set

(Crang and Thrift, 2000, 9)

The dominant dynamic in cybercity development involves the intimate recombination of urban places, the corporeal presence of people's bodies, physical mobilities, and complex, multi-scaled mediations by all sorts of ICT and mobility systems. As urban regions drive the growth of ICT infrastructures and applications, they themselves are becoming saturated by a myriad of computerised interactions. These are now seamlessly co-produced with the processes of urban life (in fact, in most cases they *are* now the processes of urban life). Urban actions thus have their electronic shadow. Every physical movement has its computerised trace. Every urban landscape crosscuts, and interweaves with, multiple and extended sets of electronic sites and spaces. Most of these remain invisible. Many are simply unknowable.

Increasingly, then, cities and urban life can be seen, at least partly, to be computerised constructs as microchips and computerisation become increasingly miniaturised, invisible, and ubiquitous – especially in the richer and technologically advanced cities and parts of cities. Separating the city from the cyber is thus becoming increasingly untenable, and, indeed, pointless. They are one and the same. They are two sides of the same process. And they are mutually constructed and utterly fused. Understanding one without the other will produce a blinkered analytical poverty.

The challenge, thus, is to start off any analysis by focusing on the hybrid forms and 'recombinant' encounters and spaces that continually emerge in the dynamic and ongoing (re)creation of cybercities. It is to this challenge that we now turn. To successfully meet it we need to draw on the recombinant thinking about city–ICT relations analysed in Part II. We need to focus at the outset at the ways in which complexes of ICT applications and urban structures, spaces and domains are constructed, maintained, and experienced, in parallel. And we need to understand the ways in which the fast-changing physical landscapes and spaces of contemporary cities are actually being shaped and reshaped in intimate combination with the social shaping of a growing universe of systems which sustain electronic interaction at a (more or less significant) distance.

This last point is especially important. For, as most cities spread into sprawling urban regions, it is less and less clear what a 'city' actually is. The traditional and classical notions of the city as a definable, separate cultural, physical, and political space is becoming untenable in a world of multiplying mobilities and 100-mile, multicentred city regions. Notions of accessibility and proximity, centre and periphery, city and countryside, the social and technical, gathering and public interaction, presence and absence, and public and private space are being remade. This is being done in a complex variety of ways in different cities through the spread and re-partitioning of urban landscapes, the urbanisation of whole societies, and the folding of urban landscapes into a wide range of electronic and physical systems of flow, mobility, and action-at-a-distance.

Rather than being preoccupied with purely global and transnational flows, however – as is too often the case in the debates about cities and 'globalisation' – we must be conscious of how cybercity recombinations involve multiple geographical scales simultaneously. Electronic and physical flows, and social relations, continuously telescope between multiple scales – from bodies to rooms, buildings, streets, neighbourhoods, districts, highways, cities, urban regions, and nations to the blossoming scales of transnational, and even global, interaction.

Such are the complexities of the recombined and hybrid spaces where experiences of cybercities are being remade, that some authors have argued that there is a palpable sense of crisis in the ways in which cities are represented in our age of ceaseless electronic mobility (Boyer, 1996, 139; see Paul Virilio, p. 78). To the British urbanist Richard Skeates, the growth of 'megacities' and transnational urban corridors means that the very distinctiveness of a place called the 'city' is now threatened. 'We can no longer use the term "city"', he writes:

> in the way it has been used to describe an entity which, however big and bloated, is still recognisable as a limited and bounded structure which occupies a specific space. In its place we are left with the urban: neither city in the classical sense of the word, nor country, but an all-devouring monster that is engulfing both city and country and in so doing effectively collapsing the old distinction (1997, 6).

The spaces of contemporary cities thus tend to resist easy categorisations of function and interaction. Time patterns in cities are being stretched and reconfigured beyond the more rigid routines of work, commuting, and home time that tended to characterise the classic industrial metropolis. It is even possible to regard the contemporary city as 'a gearbox full of speeds' (Wark, 1988, 3). The contemporary urban fabric, as both the hybrid product and the site of multiplying and diverse relational networks, offers stark contradictions and huge tensions. Flows sometimes connect in place to be more than the sum of their parts; often they do not. Time and space, within the hybrid forms and recombinant spaces of contemporary cities, thus present what the British geographers Mike Crang and Nigel Thrift have called a 'multiple foldable diversity' (2000, 21).

References and suggestions for further reading

Ascher, F. (1995), *Métapolis ou l'avenir des villes*, Paris: O. Jacob.

Adams, P. (1995), 'A reconsideration of personal boundaries in space-time', *Annals of the Association of American Geographers*, 85(2), 267–85.

Adams, P. (1997), 'Cyberspace and virtual places', *Geographical Review*, 87(2), 155–71.

Batty, M. and Miller, H. (2000), 'Representing and visualising physical, virtual and hybrid information spaces'. In D. Janelle and D. Hodge (eds), *Information, Place and Cyberspace*, Berlin: Springer, 133–47.

Beckman, J. (ed.) (1998), *The Virtual Dimension: Architecture, Representation, and Crash Culture*, New York: Princeton Architectural Press.

Boyer, C. (1996), *Cybercities: Visual Perception in the Age of Electronic Communication*, New York: Princeton University Press.

Crang, M. and Thrift, N. (eds) (2000), *Thinking Space*, London: Routledge.

De Meyer, D., Versluys, K., Borret, K., Eeckhout, B., Jacobs, S. and Keunen, B. (1999), *The Urban Condition: Space, Community and Identity in the Contemporary Metropolis*, Rotterdam: 010 Publishers.

Engeli, M. (ed.) (2001), *Bits and Spaces: Architecture and Computing for Physical, Virtual and Hybrid Realms*, Basel: Birkhäuser.

Fathy, T. (1991), *Telecity: Information Technology and its Impact on City Form*, London: Praeger.

Forer, P. and Huisman, O. (2000), 'Space, time and sequencing: substitution at the virtual/physical interface'. In D. Janelle and D. Hodge (eds), *Information, Place and Cyberspace*, Berlin: Springer, 73–90.

Graham, S. and Marvin, S. (2001), *Splintering Urbanism: Networked Infrastructure, Technological Mobilities and the Urban Condition*, London: Routledge.

Koolhaas, R. and Mau, B. (1994), *S, M, L, XL*, Rotterdam: 010 Publishers.

Koolhaas, R., Boeri, S. and Kwinter, S. (2000), *Mutations*, Bordeaux: ACTAR.

Kural, R. (2000), *Architecture of the Information Society*, Copenhagen: Danish Academy of Fine Arts.

Latour, B. and Hermant, E. (1998), *Paris ville invisible*, Paris: La Decouverte.

Massey, D. (1992), 'Politics and space/time', *New Left Review*, 196, 65–84.

Mitchell, W. (1995), *City of Bits: Space, Place and the Infobahn*, Cambridge, MA: MIT Press.

Mitchell, W. (1999), *E-Topia: Urban Life Jim, But Not as We Know It*, Cambridge, MA: MIT Press.

Moore, R. (ed.) (1999), *Vertigo: The Strange New World of the Contemporary City*, London: Laurence King.

Mulder, A. (ed.) (2002), *Transurbanism*, Rotterdam: NAi Publishers.

Office of Technological Assessment (1995), *The Technological Reshaping of Metropolitan America*, Washington, DC: Congress of the United States.

Riewoldt, O. (1997), *Intelligent Spaces: Architecture for the Information Age*, London: Lawrence King.

Skeates, R. (1997), 'The infinite city', *City*, 8, 6–20.

Wark, M. (1988), 'On technological time: Virilio's overexposed city', *Arena*, 83, 82–100.

Woodroffe, J., Papa, D. and MacBurnie, I. (eds) (1994) *The Periphery*, London: Wiley/Architectural Design.

'Eclectic Atlases'

Stefano Boeri

Editor's Introduction

Our first analysis of the hybrid forms and recombinant spaces of cybercities comes from the Italian urbanist, and Director of the Urban States of Europe (USE) Project, Stefano Boeri (see Boeri, 2002). Boeri looks at the question of how we represent our sprawling cities in the satellite age. He addresses a central paradox: just as new communications technologies reveal more about the state and form of our cities from space, so the pictures they reveal seem less and less to resemble the forms that we conventionally think of as 'cities'. Instead of clear, distinct and centralised settlements, with definable boundaries, the spread of urbanisation means that we have urban regions instead of cities. These sweep amorphously over huge distances. They coalesce along corridors and around highway interchanges, malls, 'Edge Cities' and airports. And they blur seamlessly into urbanising countryside, rendering the classical urban–rural distinction less and less useful.

Boeri contends that, when we are confronted by the transfigured image of the cities in which most of us live, the technical questions of aerial imaging and remote sensing are transformed into an epistemological trauma as we are forced to confront the question: what *is* a contemporary city? Given that the 'city' has always had as much power as an idea as it has as a form, this debate is of profound significance (see Donald, 1999).

To Boeri such a trauma underlines a growing impotency of urban planning, architecture and design to shape the city in any meaningful, purposeful way. It forces urbanists to invent a new language to try and deal with, and describe, the amorphous, decentralised, generic and all-encompassing urban forms that they now confront. And, paradoxically, it forces urbanists to move away from the representational tool that has been central to the invention of modern urban planning: looking downwards on cities from the air (see Vidler, 2000; Campanella, 2001).

In response to this representational and definitional crisis, Boeri urges analysts and urbanists alike to construct what he calls 'eclectic atlases'. These are multidimensional and multi-positional views of the complex hybrid and recombinant forms, spaces, and experiences of (cyber)cities. Mixing texts, photos, interviews and maps, these representations of cities take on board recombinant positions in denying that there is now any privileged vantage point from which to see 'the city'. Such eclectic atlases help us to assert the social and technical recombinations that help to forge new urban worlds, new urban mobilities, and new urban identities. They are, to Boeri at least, a vital stepping stone if we are to realign our idea of what 'the city' actually is with the lived urban worlds of contemporary reality. 'In the scepticism of an impersonal and synoptic view', suggests Boeri, 'these atlases more often use several angles simultaneously to look at the territory'.

DISTANCE AND CHAOS

Satellites have undermined a deep conviction in both architecture and town planning: in order to *understand* the territory better, it is necessary to *see* more of it.

Thanks to the proliferation of satellite surveys we can see vast areas of space reduced to a few synthetic images. We can also see more time: 'live' and sequenced images, seasonal cycles or nocturnal shots. With infrared photography, we are now able to capture some vital phenomena that cannot be seen topographically: traffic flow, large crowd concentrations, seasonal migration. We can view immense sections of territory in synthetic images, and at the same time examine minute details with the aid of digital enlargements.

Our eyes have conquered a viewpoint that aerial photography and thematic cartography were never able to offer us, other than in an artificial manner through montage.

In this way, we have finally managed to be able to see the state of European cities frozen in an instantaneous representation. Be that as it may, it is a state that suddenly seemed unrecognisable: many of the large urbanised areas of Europe (the sprawling city of Milan and the Italian north-east, the conurbation of Amsterdam–The Hague–Rotterdam, the urban region of Zurich and Basil, the city-territories of Madrid and Athens and the Parisian metropolis . . .) appeared to us as strange and shapeless figures, lacking in any clear divisions from the countryside, with no obvious centre and without a sharp distinction between their parts. They are no longer the large urban areas with a clear profile that we learnt to recognise in our geography books, but rather indefinite and borderless entities dispersed across the territory.

Confronted by the transfigured image of the cities in which we live, this technical conquest was suddenly transformed into epistemological trauma.

Seeing the disappearance of what were once the borders of peripheral districts reduced to a pulp of built objects spread over what was once countryside; watching outlying urban centres swallowed up by borderless sprawl; observing the open spaces that once separated the central metropolitan areas being fragmented by the growth of an irregular network of built-up bands, we suddenly realised that the images with which we continued to represent the geography of our territory had become useless, along with the rigid, binary terminology used to describe them: centre/suburbs, interior/exterior, city/country, public space/private space.

The democratisation of a powerful technology for territorial observation has had the paradoxical effect of spreading a sense of impotency among the disciplines that study inhabited space. This is to the point where more conscientious researchers, when trying to explain contemporary urban phenomena, have frequently had to call upon the chaotic nature of territory and the impossibility of constructing any kind of overall representation.

The rhetoric of chaos is in fact a product of the vigorous innovation that has taken place in the field of visual technology. It is an innovation that has brought us face to face with stupefying and disquieting images, because they cannot be deciphered using either the concepts of our encyclopaedia, or our vocabulary. 'Megalopolis', 'urban constellation', 'diffused city', 'city of dispersal' and 'low density habitat', are but some of the neologisms with which we have attempted to define the chaotic entities that we have finally been able to see, but not to explain.

In this way, satellites have both realised and ridiculed the dream of a global vision: we have sent them up there to see something that unexpectedly appeared to be incomprehensible. This is perhaps because the implicit code in the point of observation that we instructed them to take has nothing to do with the codes of that which we finally saw.

ZENITH ARROGANCE

It would be sufficient to drive along any large road that enters or exits one of our cities to realise that the European territory has, over the last twenty or so years, radically changed both quantitatively and above all qualitatively.

We would see that what has changed our territory has not been new districts, large buildings and infrastructures (roads, flyovers, rail tracks, tunnels), but rather a multitude of solitary and amassed buildings: detached houses, hangars, shopping centres, apartment blocks, garages and office complexes. A reduced range of manufactured objects incongruously thrown together, one against the other. Although they are modest constructions, they are at the same time concerned with distinguishing themselves from their surroundings. Scattered and heterogeneous groups of buildings that are expressions of small fragments of our society (the family, small industry, corporations, the shop, the club . . .), intentionally isolated from public space and disinterested in its rules.

Within just a few years, amid the indifference of politicians and the architectural elite, a scattering of isolated objects has literally disrupted our territory, strewn along the roads and borders of the compact city, unifying distant urban centres, climbing up mountainsides, and rolling down to lap the shores of the sea and riversides.

The delay with which European town planning and architecture have come to realise the aesthetic chaos produced by this impetuous wave of individual and uncoordinated tremors, has encouraged many experts and researchers to try and describe the minimal units, origins, evolutionary patterns and hidden common aspects of this new city composed of a multitude of solitary objects.

This breathless concern with description (which in many cases has produced some fairly miserable examples of contemporary urban kitsch) has often remained the prisoner of an old rule of discourse: zenith morphology, where meaning is attributed only to those figures that express themselves in a complete form and with a visible, two-dimensional surface. It also requires considerable 'distancing' between observer and territory, as if this 'stepping back' were a necessary condition for the knowledge of territorial phenomena, and thereby reducing the observer's subjectivity to an impersonal entity, which is outside his/her field of observation.

The code of zenith vision, which has formed the basis of modern town planning vocabulary, is a resistant and overbearing paradigm that tends to cancel out the others and interprets irreducible local codes as chaos.

Moreover, the zenith paradigm is misleading because it prompts the observer to distance himself from the territory, and gives the illusion of having the same impersonal and powerful point of view as the techniques of representation he is using. This is hypocritical because it allows the observer to avoid responsibility; it shows the surface of the territory from a distance, yet continues to insist that the laws and rules of the phenomena before us are actually situated somewhere else; behind or 'under' the visible space: in economy, society, in deep underlying structures (see Boeri, 1989).

Yet, it is by no means certain that we can escape the rhetoric of chaos by attaching some inclusive descriptions of society's economic and institutional relations to the zenith representation of the new 'diffused city'. It is not by placing one of these territorial representations on top of the other that we will ever be able to grasp the essence of contemporary inhabited space. This would be to look at the territory as if it were composed of flat specialised layers, which would not provide insight into the vertical, mobile energies, and the physical and psychological land-scapes that extend their roots between them. We would produce thematic 'maps' crammed with useless and highly ordered information, but be incapable of appreciating the multifaceted and dynamic nature of urban phenomena.

By busily seeking invisible, underlying structures able to make sense of the visible phenomena on the surface of the territory, we would neglect to consider the fact that contemporary territory is shaped by tensions resting between space and society, which are not expressed through the code of zenith morphology. Neither are these tensions exposed in configurations that are often precisely that which appear superficially, and make no attempt to be anything other than that. In conclusion, we can say that a paradigm that is strong in its visual tools but weak in its interpretative ability is insufficient when attempting to explain this chaos. We cannot ask it to resolve the problem it has created.

LATERAL THOUGHT

Fortunately, a sceptical minority attitude has been gathering momentum in the wake of the great visual power of the structuralist 'zenith' paradigm. It is convinced that the city is not only composed of stratified 'levels of reality', but is also a collective way of conceiving space. Furthermore, it is persuaded that every stage in a city's evolution also implies and requires a jump in its forms of representation. This minority attitude attempts to infiltrate the ranks of the enemy paradigm with small acts of sabotage. Sitting on the shoulders of the giant, it continually throws pebbles in his eyes.

In some parts of Europe this attitude is producing 'eclectic atlases' (Boeri, 1997a), that propose new ways of examining the correspondences between space and society. These are heterogeneous texts (reports, photo-graphic surveys, geographic and literary descriptions, classifications, research reports, qualitative investiga-tions, essays and articles, anthologies and monographs, collections of plans or projects . . .) but similar to the visual approach. They tend to be 'atlases' because they seek new logical relationships between special elements, the words we use to identify them, and the mental images we project upon them. They tend to be 'eclectic' because these correspondences are based on criteria that are often multidimensional, spurious and experimental.

This variegated family of studies and inquiries does not believe that chaos could be a reflection of external phenomena, but rather the effect of fatigue in the conceptions of territory. For this reason the eclectic atlases attempt to construct representations with multiple entrances, and they provide a counterpoint to the dominant paradigm. They attack it laterally, moving simultaneously towards both physical and mental space, because they believe in the existence of profound connections between forms of vision and the forms of things we see. They look on the inhabited territories of Europe in search of individual, local and multiple codes, which link the observer in each single case to the phenomena observed: the physical city, its inhabitants and the internal city of the observer.

With their dissatisfaction in representations that reduce the changes in space to a linear succession of historical documents, the eclectic atlases prefer to use forms to represent the flow of time in territory. They produce provisional and unusual maps, in which the territory is not represented as a continuous mineral substrate or the layering of stable states of things. Instead, territory is expressed as an intertwining of sinuous and multidimensional configurations that are reversible and never coeval.

In the scepticism of an impersonal and synoptic view, these atlases more often use several angles simultaneously to look at the territory. They look not only from above, but also through the eyes of those who live in the space, or experimenting with new unprejudiced perspectives. The utmost concern is placed on not confusing the subjectivity of the real observer (we who look at the landscape or observe its representations) with the mere pretence of subjectivity incorporated in the technologies of representation. It demonstrates a conviction that if our identity is realised through the fundamental act of seeing – within the scenario of the eye – this does not mean mistaking the slow construction of a conceptual paradigm – which revolves around the act of vision – with the continuous mutation of optical technology (Chevrier and Catherine, 1996).

In their skilful ability to intertwine various viewpoints, the eclectic atlases propose plural visual thinking, which has abandoned the utopia of a global vision from an optimal point of observation. The most interesting characteristic of these eclectic atlases is that they seem to be sympathetic towards their field of observation. They use an eclectic eye to look at an eclectic territory.

The eclectic atlases experiment unsystematically with 'lateral' ways of seeing and representing the territory of European cities (De Bono, 2000). These approaches seem to converge on some principal movements in the 'themes of the eye' regarding territory.

INSIDE THE SPACE, WINKING: AN INVESTIGATIVE EYE

The work entitled *La Disparition* by Paola Di Bello (1998), represents the map of the Paris underground in a montage of points where the users of all the stations place their fingers to indicate their position in the network thinking, I am here. This is a hybrid zenith vision made from the traces left by infinite individual eyes and producing an eclectic map where daily life is represented as physically crossing out the most intensely used locations. The map shows a network of obligatory paths and, simultaneously, the traces of erratic and changing mental maps of its users. It holds many different functional and interpretative codes of the city together, leaving us to decide which to activate, or use, and how to literally enter the representation.

This work demonstrates the first shift suggested by the eclectic atlases, which implies the observer being considerably closer to his field of observation in physical terms. It is the invitation to enter the inhabited space with our bodies to seek out the traces left by new lifestyles, and to look at the 'small' in order to see more (Geertz, 1999). For example, hidden behind the apparent standardisation of our habitual territories (wherever we go we find the same new buildings: shopping centres, motorway grills, apartment blocks, single-family homes etc.) there are many clues to ways of life that are protective of their local identity, and tend to privatise the space by nesting within its folds. They leave brief, changing, and temporary traces that can only be collated by the sensitive, 'rhizomatic' eye of a detective (Boeri, 1997b). It is a way of looking that produces local maps, a kind of interspersed or punctuated sampling and biographies of places. The result is a method that tells the narrative of an individual route through space, and does not use the representation as a means of distancing itself from the territory.

Let us take as an example a group of houses scattered across a suburban area: a typical situation in the European territory. Its representation on a topographical map only reveals some indecipherable, irregular geometric figures. However, when we observe this situation from within, searching for clues of the relationships that link the inhabitants to their place of residence, we are told much more. The inclusion of unusual spaces and objects within this enclosed site (such as the tavern, the office, the vegetable garden, the amusements) tells us, for example, that the living spaces of grandparents, which are more complex than others, frequently become central places for the other nuclei of an extended family, accustomed to living separated, but nearby in the diffused city. To a significant degree, this explains the distribution of single family homes in clusters.

We must become accustomed to seeking the laws and regulations that govern extensive parts of the territory with regular, locally targeted surveys. The use of an investigative paradigm (see Ginzburg, 1979) and the selection and interpretation by abduction (Ecco and Seboek, 1983) of the physical traces of new lifestyles, do not only extend from the need to put into play the researcher's sensorial abilities, and thereby increasing his investigative capacity as has often been suggested. Nor is it merely a question of research subjectivity, given that an investigative approach would necessarily imply a significant critical distance between the observer and the phenomena being observed.

What is more consequential is the relative slowness with which physical space reflects the (much faster) changes in lifestyle, the inertia and the friction between them, that makes the territory a treasure chest of clues and traces of the new. This is precisely because new behavioural patterns do not immediately succeed in shaping a space and leaving enduring traces, but rather leave more fleeting, discontinuous and shifting signs.

Each new behavioural pattern, every social and economic energy, every new interaction, no matter how immaterial, must all touch, run and organise themselves along the mineral sphere of the territorial crust. They must pass through the eye of the needle of the physical territory.

The Urban States of Europe (USE) Project (Boeri, 2002) utilised this investigative approach in both the selection of cases for research and in the exploration

itself. From observations of the refuse that systematically invades Belgrade's public spaces every evening (rubbish, boxes, damaged or broken goods), from the extent of the unordered street vending phenomenon, or from the number of vans that travel night and day in the residential areas of Elche, they were able to make deductions as to the extent of domestic manufacture (see Dzokic *et al.*, 2000). Following some privileged witnesses around the circuit of their daily lives collected further precious clues.

An investigative approach is today perhaps the most appropriate way in which we can evaluate the proximity that territorial disciplines maintain with contemporary spaces. It could be a means of recuperating the huge interpretative delays accumulated.

REFERENCES FROM THE READING

Boeri, S. (1989), 'Le Città Scritte. Riflessioni su tre Libri di Carlo Aymonino, Vittorio Gregotti, Aldo Rossi', doctorate thesis, Venice.

Boeri, S. (1997a), 'Eclectic atlases', *Documents* 3, Kassel.

Boeri, S. (1997b), 'I detective dello spazio', *Il Sole 24 Ore*, Sunday cultural supplement, 16 March.

Chevrier, J.-F. and Catherine, D. (1996), 'Actuality of the image, between the fine arts and the media', *Documents* 2, Kassel.

De Bono, E. (2000), *Il Pensiero Laterale*, Milan: Rizzoli.

Di Bello, P. (1998), *La Disparition*, Paris: l'Editions.

Dzokic, A., Topalovic, M., Kucina,I., Djura, M. and Leelen, M. (2000) 'USE 01. Belgrade Gray Realm'. In R. Koolhaas, S. Boeri and S. Kwinter, *Mutations*, Bordeaux: ACTAR.

Ecco, U. and Seboek, T. (1983), *Il Sego dei Tre*, Milan: Bompiani.

Geertz, C. (1999), *Mondo Globale, Mondi Locali*, Bologna: Il Mulino.

Ginzburg, C. (1979), *La Crisi Della Ragione*, Turin: Einaudi.

Editor's references and suggestions for further reading

Basilico, G. and Boeri, S. (1998), *Italy: Cross Sections of a Country*, Berlin: Scalo Zurich.

Boeri, S. (ed.) (2002), *Uncertain States of Europe*, Milan: Skira Editore.

Campanella, T. (2001), *Cities From the Sky*, New York: Princeton Architectural Press.

Donald, J. (1999), *Imagining the Modern City*, London: Athlone Press.

Graham, S. and Healey, P. (1999), 'Relational concepts of space and place: implications for planning theory and practice', *European Planning Studies*, 7(5), 623–46.

Koolhaas, R. (1995), 'Whatever happened to urbanism?'. In R. Koolhaas and B. Mau, *S, M, L, XL*, New York: Monacelli Press.

Koolhaas, R., Boeri, S. and Kwinter, S. (2000), *Mutations*, Bordeaux: ACTAR.

Lerup, L. (2000), *After the City*, Cambridge, MA: MIT Press.

Lewis, P. (1983), 'The galactic metropolis'. In R. Platt and G. Macuriko (eds), *Beyond the Urban Fringe*, Minneapolis: University of Minnesota Press, 23–49.

Polo, A. (1994), 'Order out of chaos'. In J. Woodroffe, D. Papa and D. and I. MacBurnie (eds), *The Periphery*, Architectural Design, London: Wiley, 108, 24–29.

Skeates, R. (1997), 'The infinite city', *City*, 8, 6–20.

Vidler, A. (2000), 'Photourbanism: planning the city from above and below'. In G. Bridge and S. Watson (eds), *A Companion to the City*, Oxford: Blackwell, 35–45.

Wall, A. (1994), 'The dispersed city'. In J. Woodroffe, D. Papa and D. and I. MacBurnie (eds), *The Periphery*, Architectural Design, London: Wiley, 108, 68–11.

'The City of Bits Hypothesis'

from *High Technology and Low Income Communities* (2000)

William Mitchell

Editor's Introduction

Our second analysis of the hybrid forms and recombinant spaces of cybercities draws on the influential work of the architect William Mitchell, who is Dean of the School of Architecture and Planning at MIT in Cambridge, Massachusetts. Mitchell's work, best appreciated in his two recent books *The City of Bits* (MIT Press, 1995) and *E-Topia* (MIT Press, 1999), analyses the complex and recombinatory interplays between urban forms and spaces and ICT-mediation with unusual subtlety and sophistication.

In this piece, Mitchell summarises what he calls the 'city of bits' hypothesis which was initially developed in the first of the above books. This hypothesis posits that contemporary cybercities are shaped by a dynamic, ongoing, and recombinant interplay between urban spaces, physical movements, and ICT mediation. Using this starting point Mitchell analyses a set of broad alternatives for making communication in the cybercity. These rely either on (some form of physical) presence or on (ICT mediated) telepresence. For each, there are, in turn, options to communicate synchronously (that is, in real time, like a telephone call) or asynchronously (that is, with a delay between the sending and the receiving of the message, as in an email or postal exchange).

By analysing how these communications options are selected and combine together within the hybrid forms of contemporary cities, Mitchell's work usefully illuminates the subtle geographies and complex trade-offs and combinations which shape cybercity development. This allows him to discuss a range of key recombinant urban policy and planning challenges: maintaining 'cohesion' in the relations between virtual and face-to-face community; addressing the links between virtual and physical 'real estate'; and the possibility for developing innovative urban–ICT policies, strategies and designs that attempt to shape city–ICT recombinations to meet wider policy goals (see Part VIII).

The implication of Mitchell's work is that urban researchers and policy-makers need to be extremely sensitive to the complex pluralities of communication flows within and between contemporary cities. He suggests that focus must fall on how the various combinations of presence and telepresence, and synchronicity and asynchronicity, become subtly combined in different ways in different places. The challenge which follows for research is to address the subtle linkages between such urban communicational ecologies and the geographies, economies and landscapes and social worlds of cities. Mitchell's work demonstrates why urban policy, planning, design and governance must develop policy paradigms which accommodate, and harness, the subtle interplays of physical presence and telepresence, as they help to shape the combination of physical transportation, ICT use and the use of urban places. As we shall see in Section VIII, such policy approaches, whilst extremely challenging and difficult, are now emerging rapidly in a wide variety of disciplinary contexts and urban environments across the world.

THE NEW ECONOMY OF PRESENCE

To gain a concrete idea of the implications of the digital revolution, let us begin with a simple thought experiment. Imagine that you want to convey some information to a coworker. You can walk to her office and discuss the matter face-to-face; the two of you are physically present in the same place at the same time. If she is not there, you can leave a written note on her desk, and she will find it and read it at some later point; this is using a low-tech form of information storage technology to accomplish a transaction through asynchronous physical presence. A third alternative is to call her telephone extension; if she answers, you accomplish the transaction through synchronous telepresence. Finally, you can exchange e-mail or voicemail; now, you accomplish your goal through asynchronous telepresence – without the need ever to be in the same place or to be available at the same time. It is convenient for our purposes to array these four possibilities in a two-by-two table, as illustrated in Table 2.

If we had to rely on our unaided human capabilities, and no technology intervened, all human transactions would fall in the upper left, synchronous presence quadrant of this table. That is how it was in the past, and still is in some traditional societies. But the development of recording technologies created the possibility of asynchronous communication; writing and associated media for recording and storing text came first, followed by mechanical recording devices such as the camera and the Edison phonograph, and now digital storage of numbers, text, images, sound, and video in RAM and on disks and tape in computer systems. Similarly, the emergence of transmission technologies yielded telepresence; drums, bonfires, and smoke signals, and semaphores were early means of communication at a distance, then came the telegraph, the

Table 2 Communications alternatives

	Synchronous	*Asynchronous*
Presence	Talk face-to-face	Leave note on desk
Telepresence	Talk by telephone	Send email or leave voice mail

Source: Mitchell, 1999, 110

wireless telegraph, modern radio and television, and now worldwide digital telecommunication networks – such as the Internet – carrying packets of bits.

The combination of recording and transportation technologies produced mail services – a limited capacity, snail's-pace first approximation to asynchronous telepresence. Finally, since the 1960s, digital recording, processing, and telecommunication capabilities have converged to create true asynchronous telepresence – the most familiar forms of which are e-mail, voicemail, World Wide Web (WWW) sites, and video-on-demand servers. So, by the mid-1990s, human transactions were being distributed over all quadrants of the table.

These various modes of communication and conduct of transactions are partially equivalent to each other but not completely interchangeable; they vary considerably in their respective qualities, costs, and most appropriate uses. Face-to-face provides intensity, multimodality, and immediate feedback. It requires putting yourself on the line, sometimes even in physical danger. And when you consider space, transportation, and opportunity costs, it is often very expensive to achieve. Telepresence greatly reduces transportation costs while retaining the possibility of immediate feedback, but the quality of communication is limited by available bandwidth and interface technologies, it still requires coordination of availability, and – as the telephone vividly demonstrates – it can generate unexpected and unwanted interruptions. It also allows some division of attention; you can talk on a cellular telephone while supervising a child's play, for example. Asynchronous communication eliminates the need for coordination of availability, but it also eliminates the possibility of immediate feedback; it does not reduce transportation costs, and it is limited in quality by available recording, storage, and playback capabilities. The most recently emergent possibility – asynchronous telepresence – effectively eliminates transportation costs and the need to coordinate availability, and it allows multiple transactions to take place in parallel through use of software agents and the like, so it is extremely inexpensive and convenient, but it is limited by available storage and bandwidth, and it certainly does not have the intensity and immediacy of a face-to-face encounter. These commonalities and differences are summarized in Table 3.

Given this extended range of possible means for conducting our affairs, a number of questions present themselves. What are the best uses of our limited opportunities for face-to-face interaction? When does

Table 3 Advantages and disadvantages

	Synchronous	Asynchronous
Presence	Intense, multimodal Immediate feedback High transportation costs High space costs Requires coordination of availability Requires full attention	Limited by storage and playback capacities No immediate feedback No reduction in transport costs No reduction in space costs No need to coordinate availability May allow some division of attention
Telepresence	Limited by bandwidth and interface capabilities Retains immediate feedback Reduces transportation costs Requires coordination of availability May allow some division of attention	Limited by storage, bandwidth and interface capabilities No immediate feedback Reduces space costs No need to coordinate availability Allows multiple activities and transactions in parallel

the value added by the immediacy and high quality of face-to-face interaction justify the high transportation, space, and coordination costs? Conversely, when do the savings on transportation costs outweigh the disadvantages of telepresence? When is immediate feedback essential, and when can we dispense with it? When can we take advantage of the convenience of asynchronous communication without losing something essential? What are the best uses of the low-cost, potentially very high-volume communications capabilities provided by asynchronous telepresence? How can it broaden our opportunities and contacts? How can it free up time and reduce transportation and space demands by eliminating unnecessary face-to-face interactions? More subtly, how can we most effectively use the various available modes in combination – for example, by using many low-cost, convenient e-mail messages to arrange and coordinate much rarer, higher-cost, face-to-face meetings?

Often, the ways that people respond to these sorts of questions are constrained by particular contexts and circumstances. If they are in remote locations that impose particularly high travel costs for face-to-face interaction, for example, then they are likely to rely more heavily than they otherwise would on telepresence. Similarly, if they need to communicate across widely separated time zones, they may prefer asynchronous to synchronous communication because

it eliminates the need to coordinate availability. If they have plenty of time on their hands and a narrow range of daily duties they may enjoy doing everything face to face, but if they have numerous responsibilities and are very busy, they may have to rely much more on e-mail and the telephone.

In addition, technological progress influences choices by reducing the costs of the electronically mediated alternatives and increasing their quality – while the cost and quality of face-to-face interactions remain much the same, or even become more expensive as cities become more crowded and jammed with traffic. Processing power, storage capacity, switched connections to remote locations, and high-bandwidth communications channels were all once extremely scarce and expensive commodities but are now becoming very inexpensive; in many practical contexts, we can begin to treat them as almost free resources. Increases in bandwidth have allowed progress from the telegraph's Morse code to high quality voice communication, to increasingly effective teleconferencing, and now (in research contexts at least) to sophisticated shared virtual environments. In the early days of the Internet, remote asynchronous communication was limited to clumsily downloading files from one machine to another and exchanging short, entirely textual e-mail messages; now, the World Wide Web allows smooth navigation of multimedia material at scattered sites simply by pointing and clicking, e-mail is evolving into multimedia mail, and software agents can roam the Net to perform tasks for you while you sleep. The magnitude of these recent changes in cost and quality is so great that the digital world seems qualitatively very different from the world of the early days of electronic recording and tele-communications.

One effect of these increasingly attractive and inexpensive electronic alternatives is to create new kinds of competition for our attention, our business, and our loyalties. You may, for example, have bought this book in the old-fashioned way – by going to an actual bookstore, handing over your cash, and carrying away your purchase – or you may have surfed into an online "virtual" bookstore, browsed through the offerings displayed there, electronically provided your credit card number and address, and had the product delivered from the warehouse directly to your door. In other words, you may have carried out the transaction face to face, during business hours, in a traditional sort of architectural setting, or you may have conducted it remotely and asynchronously through use of electronic infrastructure and appropriate software.

By the mid-1990s growing competition from electronic sites and services was already producing some substitution of connectivity for contiguity, of electronic interaction for travel, and of virtual venues for real estate, and there is likely to be much more of this as high-bandwidth digital telecommunication systems with capabilities far beyond those provided by the Internet and the World Wide Web in the late 1990s become widely accessible. But even more important, we can anticipate that new opportunities for interaction will create new and higher demands, and that both traditional and electronic means will play parts in satisfying these – just as photography did not simply substitute for the sketchbook and the painting but created whole new roles and demands for images, and as the telephone not only substituted for movement, but also created new motivations for travel and an efficient means of arranging and coordinating it.

In sum, we now have a new economy of presence – within which we continually choose among the possibilities of synchronous and asynchronous com-munication, and presence and telepresence. Given that all the different modes have their respective advantages and disadvantages, we should not (as techno-romantics gleefully anticipate and traditionalists fear) expect wholesale substitution of electronically mediated com-munication for face-to-face interactions. We are not all going to end up alone in darkened rooms typing e-mail messages to each other. If people are left to themselves, they will make more-or-less informed and rational choices among the available alternatives as appropriate to their particular needs and circumstances. This will result in varying distributions of activity over the quadrants of our table at different historical moments and in different geographic and cultural contexts . . .

MAINTAINING COMMUNITY COHESION

Urban theorists from Plato and Aristotle onward have argued that successful towns and cities require an appropriate balance of residential space, work areas, and essential services [. . .] tied together into a coherent unit by circulation and utility networks and containing generally accessible places for public life and political and religious activities. From this, they have derived ideal populations and dimensions for settlements. Aristotle's *Politics*, for example, recommends a city that is small enough to be held together by pedestrian circulation, for the citizens to know each other, and for public life to take place in the central agora. Today's New Urbanists propose neighborhoods along very similar lines. And planners of modern cities, generally, have been concerned with creating and maintaining the right balance of land use and transportation – with making the best use of available land, keeping travel times to work and services within acceptable bounds, and with minimizing transportation congestion. Now, however, we need to consider virtual places as well as physical ones, telecommunication links in addition to transportation connections, and the interrelationships among all of these, as illustrated in Figure 1.

A traditional city – as it appears on a map – is a structure of physical places and transportation links. Analogously, the World Wide Web is a structure of sites (virtual places) tied together electronically by hyperlinks. But the physical world and the elec-tronic virtual world are not separate, as much current discussion might lead one to believe; in fact, as the light-colored links in Figure 1 suggest, they are intricately intertwined. Consider, for example, the connection of physical places by telecommunication links; this creates the potential for telecommuting, and for the establishment of geographically distributed virtual communities. Conversely, the combination of

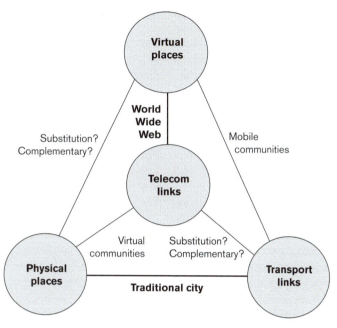

Figure 1 Physical and virtual structures and their interrelationships

Source: Mitchell, 1999, 126

virtual places with transportation supports such practices as taking your office on the road – for example, by carrying a laptop with appropriate software and dialing in to a server to gain access to e-mail, files, and other resources. Furthermore, we can imagine that telecommunication and transportation links may either substitute for one another or complement each other. And the same goes for virtual and physical places.

The burden of sustaining a community interaction and cohesion may now be carried by any or all of these means. We should not expect uniformity. In different contexts, very different patterns and mixes may make sense; architects, urban designers, and planners will need to consider the tradeoffs among the many emerging possibilities. Services may be delivered in traditional sorts of physical settings, by electronic means, or through some mix of the two. Community members may travel to work to conduct transactions and to socialize, or they may accomplish many of the necessary connections electronically. Efficient transfer of people and goods may be accomplished by locating origins and destinations as near to each other as possible, or by building high-speed transportation linkages between them. Thus the old yardsticks of accessibility and overall cohesion – walking distance and travel time – no longer apply in straightforward ways.

Within these hybrid structures of real estate and virtual places, held together by both transportation and telecommunication linkages, public space may appear in a variety of guises – some ancient and some very new. There may be traditional, physical venues for face-to-face encounters – agoras, forums, squares, piazzas, streets, public parks and gardens, and so on. Public monuments, museums, galleries, and libraries may be important repositories of community memory, and allow asynchronous communication across generations. Radio and television talk shows, online chat rooms, and MUDs, MOOs, and other kinds of shared virtual environments may employ telecommunications capabilities to create placeless, electronic agoras. And online newsgroups, bulletin boards, and Web sites function as remotely accessed, asynchronously visited public places.

With these observations in mind, we can now develop a new kind of urban typology. At one end of the spectrum are completely traditional, place-based communities – composed entirely of physical spaces, organized around traditional types of public spaces, and held together by physical circulation through streets and transportation networks. At the other end of the spectrum are fully virtual communities. These consist of widely and randomly scattered physical places, together with many virtual places – including virtual public places – held together primarily by electronic connection rather than physical transportation. In between are numerous possible hybrids, in which both physical and virtual places play significant roles, and in which linkages are formed both by transportation and electronic connections. These hybrids include established cities and university campuses on which community networks have been overlaid, scattered rural communities tied together by networks, small communities that extend themselves virtually, and so on.

Communities like these are not necessarily discrete and clearly bounded, as more traditional ones typically have been. They may overlap and intersect in complex ways. And, given that electronic connections are much more easily changed than physical adjacencies and transportation linkages, they may be a lot less stable in their sizes and configurations. The Deleuzian terminology is useful here; we might say that the global urban system is becoming less hierarchical and more rhyzomic in character [p. 73].

CONCLUSIONS

The digital revolution, like the agricultural and industrial revolutions before it, opens up new possibilities for urban form and organization and creates powerful pressure for change. Under the emerging new conditions, established concepts and methods of urban analysis and design may no longer suffice, and familiar nostrums for urban problems may no longer work. We need to consider the roles of virtual places as well as physical ones, of electronic connections as well as transportation linkages, and of asynchronous encounters and transactions in addition to synchronous ones. If we want to understand the plight of low-income communities in the twenty-first century, and find policies and design strategies to alleviate it, we will have to set the problem firmly in the context of the unfolding digital revolution and its urban consequences.

Editor's references and suggestions for further reading

Batty, M. and Miller, H. (2000), 'Representing and visualising physical, virtual and hybrid information spaces'. In D. Janelle and D. Hodge (eds), *Information, Place and Cyberspace*, Berlin: Springer, 133–47.

Engeli, M. (ed.) (2001), *Bits and Spaces: Architecture and Computing for Physical, Virtual and Hybrid Realms*, Basel: Birkhäuser.

Forer, P. and Huisman, O. (2000), 'Space, time and sequencing: substitution at the virtual/physical interface'. In D. Janelle and D. Hodge (eds), *Information, Place and Cyberspace*, Berlin: Springer, 73–90.

Janelle, D. and Hodge, D. (eds), *Information, Place and Cyberspace*, Berlin: Springer.

Mitchell, W. (1995), *City of Bits: Space, Place and the Infobahn*, Cambridge, MA: MIT Press.

Mitchell, W. (1999), *E-Topia: Urban Life Jim, But Not as We Know It*, Cambridge, MA: MIT Press.

'Urban Morphology and the Shaping of the Transmissible City'

City (2000)

Mike Crang

Editor's Introduction

A growing range of architects are now exploiting the powers of advanced computing and digital media to design spaces and forms which help to illustrate the complex, hybrid and recombinant interactions within and between cybercities. In this reading Mike Crang, a geographer at Durham University in the UK, analyses the work of one of the most influential of such architects: Marcos Novak.

To Crang, the work of architects such as Novak challenges the ways in which we represent cybercities as they blur into complex and dynamic informational spaces. Novak's designs of multidimensional 'transarchitectures' – electronic forms constructed using advanced computers – challenge our notion of what architecture and design are in contemporary cities. Finally, Crang argues that Novak's work forces us to rethink the nature of presence in the city.

Architecture, for Novak, is not just about heavy and settled formations erected in physical space. Rather, it is a complex, dynamic and continuous combination of many points, spaces and dimensions. These link locations at many geographical scales through simultaneous physical and virtual means. The challenge, to Novak, is to develop aesthetic styles which capture the free and flexible meshing of physical and electronic, presence and absence, and mobility and settledness, within the hybrid realities of cybercities. This, in essence, is the purpose of his *Transarchitectures*. They are, as Crang puts it, 'transmissible cities which are designed both to represent and shape the extended cyborg mobilities as complex, socio-technical constructions, rather than as separated domains of physical and virtual space'.

DIRECTIONS

In this essay I want to rethink notions of urban substance and form. Recently there has been a trend away from accounts that focus on the stuff of cities, their built fabric, the classical territory of urban morphology, to pay welcome attention to the city of signs, the city as text, as representation and mediated environment. Studies in these fields have shown up the

problems in the mapping, typologizing, categorizing and segmentation of urban space implied in the study of urban morphology. Increasingly how the city is imagined and represented has come to be seen as both contestable and malleable. I want to think about how to imagine the shape of the city in an unstable informational space in a way that accommodates difference and change. Rather than a morphological definition that sees the city as stable and fixed, I am seeking one that can cope with the vision of the city expressed in Musil's *Man Without Qualities*, a city of:

> irregularity, change, sliding forward, not keeping in step, collisions of things and affairs, and fathomless points of silence in between, of paved ways and wilderness, of one great rhythmic throb and the perpetual discord and dislocation of all opposing rhythms, and as a whole resembl[ing] a seething, bubbling fluid in a vessel consisting of the solid material of buildings, laws, regulations, and historical traditions (4, cited in Robins, 1999, 53) . . .

UN-ANCHORING ARCHITECTURE: FRAGMENTATION AND MORPHOSIS

The architect Marcos Novak offers one way of thinking through these issues in his projects to create, first, a liquid architecture of cyberspace, which he suggests offers an 'augmented space' (1997a), that is thinking through what worlds of information might be shaped like, and, second, a 'transarchitecture' of their intersection with the material world. In the latter, we have to think of ourselves including a delegated agency, so that an actor also includes a variety of machinery able to create effects at a distance. Novak suggests that both mean redefining the urban field by challenging three deeply embedded assumptions of urban studies. First, that space is three-dimensional and shared between actors. Second, space is either solid or void. And third, you can only be in one place at one time. Novak suggests that, with mediated presence, distantiated contact and delegated agency, these conditions no longer hold. For instance, the ability to be in one room or building, linked to a computer accessing data elsewhere, while phoning someone who is at another point, begins to unravel any simple time-geography based on physical presence. An actor is working at multiple levels in different dimensions. The assumption of autonomous objects as having localized presence with enduring substance, as enshrined in an architectural canon dominated by metaphysics of presence is past, and, as Wes Jones (1998, 41) argues, the challenge of the moment is 'that which Architecture most takes for granted – presence – is least secure today'. The development of temporally simultaneous activities in spatially discontinuous locations calls for a different imagination of presence and the shape of urban time-space [. . .]

Novak is an enthusiast for these possibilities, but he is also clear that this is not a world without constraints. He outlines a world where information, imagination and the self can be externalized in a new playful arena (Novak, nd). But it is not a freeform to simply conjure up. He suggests we need to think rigorously about the architecture of these worlds – with principles of access, storage and movement – that defined the use of buildings – remain crucial but their functioning is transfigured (Novak, 1995, 4/4) when, say, proximity is replaced with bandwidth; it may now be bits instead of people moving but their routes remain canalized and determinate. Moreover, electronic communication is often not a direct substitute but a complement for other forms of communication, and different media get used for different purposes. To think through these multiple topographies we need to move beyond three-dimensional space. The idea of creating and acting in warped, multi-dimensional space offers something beyond most accounts of the unrepresentability and fragmentation of urban life [. . .]

What Novak offers is a vision of creating new aesthetic forms that enable new forms of action. In its strongest version, it is about information made concrete in the way depicted in cyberpunk writing [p. 395]. But he offers a new aesthetic idiom, suggesting that just as 'chaos and complexity have switched polarities from negative to positive, so too are all the expressions of disjunction and discontinuity being revisited as forms of a higher order. Unlike the disjunction of collage that has characterized much of this century, the new disjunction is one of morphing' (Novak, 1995) and to make the distinction clearer the latter involves 'warpage, not mechanics, not even alchemy, but the curving of the underlying spatial matrix itself' (Novak, 1997a). That is not just a space of representation but a space of action – and one that might be experienced as tactile rather visible. If we return to the idea of the city on a map where the connectivity of different places creates a crumpled and folded surface then we can pursue the idea further. If we move from conventional two- and three-dimensional space, we can go beyond the sequential folding of origami and think of a map on a mobius band or as part of tora or other complex shapes where the relationship between places relate to each other in multiple and part dimensions, rather than north and south or top and bottom (Plate 10). The complex dimensions allow the multiple connectivities of different places to occur and tie in to notions of qualified and partial presence as fields and degrees

transarchitectures

Plate 10 Transarchitectures: Marcus Novak's outline of complex spaces. On the right is a series of progressively more deformed 'toras' with warped spaces and complex dimensions. In the main picture is a representation of an architectural space that has been subject to that warpage.

of informational linkage. Space ceases to be the ground for the juxtaposition or distribution of given elements, geography ceases to be a context for building or a question of scale but becomes a rupture of scale, and buildings cease being discrete elements but are topological operators crossed by different dimensions (Caché, 1995, 40, 71). Space does not separate structures, but is a field linking sites together. Since these connections and degrees of linkage are variable we have a warping and folding field, and as this occurs

in different patterns for different people the space is differentially shaped [. . .]

As Robins (1999) has pointed out, too often futuristic accounts of the electronic city are driven by a resurrected modernism that, like the Saint Simonian's of the 19th century, looks to technology to offer transparency, efficiency and thus social harmony. Indeed visions of the city in accelerated time-space very often assume the desirability of instantaneity and speed, in the 'real-time city'. Robins (1997) has called for a revaluation of

Byzantine complexity, and social complexity rather than transparency in thinking through the city – rather than the often unthinking celebration of ordered, purified digital space that is somehow friction-free and the assumption that the absence of dirt and disorder is a good thing. The visions outlined above offer a version of transphysical, electronic and real space that is about complex folding, disorder and multiplicity rather than corporate homogeneity. Thus Novak sees it as an attempt to address a fragmented reality:

> Body, cyborg, avatar: To see through a multiplexed kaleidoscope; to read and speak thoughts in parallel, bifurcating sentences to the limits of chaos and still to understand and be understood. When I am multiple, an ensemble of particle selves, who are we? Where are we? How are we embodied? How are we empowered to act in the cities of the spaces of consciousness? (Novak, 1997a)

This is a sense of cyber-cities 'in-here'. Not a separate virtual realm that resolves difficulties but a world where physical space is thoroughly contaminated by the virtual (1997a). In interviews he has stressed that although his 'worlds' are electronic, he does not see this as opposed or liberating from the body in a simple sense: 'there is no actual state of disembodiment, but that there are only alternative states of embodiment in media that are more or less solid' (in Mork, 1995). What these architectures are feeling towards are modes of construction and behaviour that respond to the possibilities of new media in the light of failures learnt all too painfully in other arenas of urban design. They force us to rethink the possible shapes of electronic spaces in light of architecture, and the architectural shape of cities in light of possible spaces.

REFERENCES FROM THE READING

Caché, B. (1995), *Earth Moves: The Furnishing of Territories*, Cambridge: MIT Press.

Jones, W. (1998), *Instrumental Form: Woods, Buildings: Machines*, New York: Princeton Architectural Press.

Mork, K. (1995), *Interview with Marcos Novak*. Alt-X. http://www.altx.com/interviews/marcos.novak.html (accessed 29 August 2000).

Novak, M. (nd), Morphing media interview with Clay Graham at *Cyber23: Virtual Architecture: Liquid Architectures*, http://www.best.com/~cyber23/virarch/novak.htm (accessed 3 September 2000).

Novak, M. (1995), 'Transmitting architecture: the transphysical city', *Ctheory: Theory, Technology and Culture*, 19(1/2), article 34.

Novak, M. (1997a), 'transArchitectures, avatararchitectures: the architecture of augmentation', http://www.archi.org/transarchitectures/02_catalogue/ctap30.htm (accessed 3 September 2000).

Robins, K. (1997), 'Global cities: real time or . . . byzantine?', *City* 7, 40–6.

Robins, K. (1999), 'Foreclosing on the city? the bad idea of virtual urbanism'. In J. Downey and J. McGuigan (eds), *Technocities*, London: Sage, 34–60.

Editor's suggestions for further reading

Beckman, J. (ed.) (1998), *The Virtual Dimension: Architecture, Representation, and Crash Culture*, New York: Princeton Architectural Press.

Crang, M. (2000a), 'Public space, urban space and electronic space: would the real city please stand up?', *Urban Studies*, 37(2), 301–17.

Crang, M. (2000b), 'Urban morphology and the shaping of the transmissable city', *City*, 4(3), 304–15.

Crang, M., Crang, P. and May, J. (eds) (1999), *Virtual Geographies: Bodies, Space and Relations*, London: Routledge.

Engeli, M. (ed.) (2001), *Bits and Spaces: Architecture and Computing for Physical, Virtual and Hybrid Realms*, Basel: Birkhäuser.

Mulder, A. (ed.) (2002), *Transurbanism*, Rotterdam: NAi Publishers.

Novak, M. (1991), 'Liquid architecture in cyberspace'. In M. Benedikt (ed.), *Cyberspace: First Steps*, Cambridge, MA: MIT Press.

Zellner, P. (1999), *Hybrid Space: New Forms in Digital Architecture*, New York: Rizzoli.

'Generation Txt:
The Telephone Hits the Street'

Zac Carey

Editor's Introduction

One of the most visible and important sites of recombination within cybercities is that which occurs continually between the mobile phone and that long-standing conduit of communication and information exchange: the city street. Within an astonishingly short time telephones have been untethered from the ties of fixed installations and static wires in specific locations. Now they roam the spaces within and between cities in their billions. In fact, by mid-2003, it was expected that 1,467 million mobile phones would be in use worldwide. For the first time this figure was expected to be higher than the corresponding figure for conventional landline phones.

This process has led to a dramatic transformation in the nature of city street life as the public spaces of streets and transport spaces are continuously recombined with millions of private electronic interchanges. As mobile phones are reshaped to offer third-generation functionality – text and email messaging, file transfer, picture and video transmission, gaming, e-commerce, digital cameras and online information, global positioning information and georeferenced or 'location based' services – they are becoming, in a sense, mobile digital appliances. Mobiles and Personal Digital Assistants (PDAs) are thus key interfaces through which many urban residents shape and experience city life. They herald a radical, but poorly understood, transformation in the ways in which city spaces are lived, constructed and consumed.

In the following Reading the London architecture student Zac Carey reflects on the complex interplays between mobile phones and city streets. Carey analyses the free and flexible meshing of cultures of information exchange that subtly combine street encounters and mobile phone encounters in cybercities. He considers what these developments mean for the ways in which we think about public space in cities (as flexible and indeterminate coordinations of the personal and collective use of cities start to replace more formal and rigid approaches). Finally, Carey illustrates the dramatic physical reconstruction of many streets as a huge range of service providers and small retailers have started to exploit the fast-growing mobile (and non-mobile) telephonic economy of the cybercity.

To Carey, the saturation of the city with mobile phones and other personal, mobile, ICT technologies heralds a reconstruction of the way city spaces are used, appropriated, and mediated. This transformation brings with it a subtle combination of the need for physical proximity. It involves a very public choreography of physical movement in the city (which demands that mobile phone conversation invariably starts with the sentence 'I'm at the . . .'). And it is facilitated by an increasingly powerful set of electronic tools of mobile communication, transaction, information exchange and surveillance.

Through these combinations, Carey argues that the complex ecologies of city street life are subtly remade as recombinations of physical and electronic mobility support new development and social dynamics (see Timo Kopomaa, p. 267).

TEXT REVOLUTION

In July, 2000, the *New York Times* reported on the growing culture of text messaging in the war in the Philippines between Muslim insurgents and government troops. Six months later the President, Joseph Estrada, was deposed by a military-backed coup. Several days of unrest culminated in a mass demonstration, in the country's capital, Manila. This was orchestrated by mobile phones and, specifically, the exchange, within the space of twenty-four hours, of some 18 million text messages.

Within an urban context, there is something appealing about the idea of uncontrolled, unsanctioned and unpredictable occupation of the city's ambiguous, public and semi-public spaces. With the advent of the mobile phone, this becomes a real possibility – as rapid, remote and clandestine forms of information exchange begin to undermine the status quo and urban spaces become no longer defined by landmarks, but by coverage. The widespread dissemination of the Internet and mobile phone into mainstream society has prompted a new apprehension of the city – by both its users and its designers.

THE TELEPHONE HITS THE STREET

It's Wednesday. 11:30 a.m. A ten-minute walk down a high street typical of an inner London borough reveals roughly one in forty people engaged in conversation on mobile phones. Some 2.5 per cent might not sound like much, but it is peak time on the mobile networks, and if these people are here now, then they are, for the most part, pensioners, the unemployed or individuals engaged in only part-time employment (see Plate 11). Further, of the seventy-one shops, stalls and kiosks between my starting point and the bus-stand, twelve are actively selling paraphernalia relating to mobile telecommunications – including the handsets

Plate 11   Mobile phone users on a central London Street. (Photograph by Zac Carey.)

Plate 12 The reinvention of street stalls as communication stalls in central London. (Photograph by Zac Carey.)

state-of-the-art systems can precisely locate any handset (within 10 to 100 metres in most British cities, and 100 metres to 10 km in the countryside) and even, because of the usual proximity of the device to the chest cavity, the carrier's heart-rate.

The traditional stumbling blocks – bandwidth, battery power, coverage and portability – have largely been overcome. Mobile phone modems are now as fast as, or faster than, those most commonly used by home-users for connection to the Internet. By 2005, Nokia predict, you will be able to view true video on your mobile. Wireless technology looks set to radically alter the design of the modern office as communication becomes a function of mere proximity. The technologists go further. We will bank from our phones – more securely than with current internet protocols – and, with specially designated accounts, be billed for train journeys without the need for tickets, simply by stepping through sensors affixed to the train doors. If we return the same day, the system will amend the transaction accordingly. The prediction is that traditional language barriers, too, will be overcome – by means of real-time, automatic translation – provided, of course, we communicate by phone.

Today, the American Telephone & Telegraph Company claims its stock to be the most widely held in the US. The big telecommunications companies are amongst the very largest corporations in the world. In Britain, 10 per cent of the companies listed on the FTSE 100 cite telecoms manufacture or supply as their principal business. The mobile telephone, specifically, has invaded every aspect of our daily lives: from the boardroom, where its presence is expected, to the classroom, where children cheat at exams by texting answers to one another, to the bedroom, or so the Taiwanese media would have us believe, where its vibrating operation has come into play as a surrogate sex toy.

themselves, a myriad range of accessories, and the top-up cards of pay-per-use subscriptions (see Plate 12). That's an impressive 17 per cent, or roughly one-sixth of all the business taking place on the street. It would be difficult, in a street that boasts electronics stores, butchers, travel agents, newsagents, video-rental, a health food shop and clothing stores, to identify any other single retail activity that comes close to this figure.

Research published by the UK telecoms regulator Oftel in February 2001 revealed that between 1998 and 2001 the number of mobile subscribers in Britain rose nearly fourfold from 9 million to 34 million. By February 2001 penetration reached about 60 per cent of the UK population. By 2001, 6 per cent of homes had a mobile phone instead of a fixed phone while 42 per cent of households had Internet access.

While in the UK revenues from text messaging still only accounted for roughly 7.5 per cent of the total by 2001, the trend here, as elsewhere, is upward. The technology is becoming, of course, ever more dazzling and the handsets or user-agents, as the industry prefers to call them, ever more affordable. Legalities aside, the

THE TEXT GENERATION

The public has always dictated the use of mobile phone technology, and hence the shape and pace of its development. At first, the industry was slow to grasp this. They would include premium-rate services or features within their handsets which were never in fact used, or conversely, discounted services, gimmicks, that later became the focus of their business. The most famous of these is SMS or text-messaging. In the

Philippines, Globe Telecommunications introduced text-messaging as a low-interest extra feature of its GSM digital cell phone technology, aimed to entice the middle-class and affluent youth. But the low cost – free at first, then 1 peso per message compared with the 8 pesos (20 cents) a minute cost of a phone call, and similar expense for mainstream Internet access – soon seduced a much broader customer base. Within months, more messages were being sent per day in the Philippines than in Germany, France, Italy and Britain combined. Dubbed by the press 'generation text', the majority of texters were teenagers happy to invent a mobile subculture, complete with its own netiquette, dialect and the tactics necessary to negotiate the inherent difficulties of trying to exchange messages rapidly on a 12-key keypad. Partly because of its tolerance of contraction and concision, English quickly became the preferred texting language. Phrases such as CU L8R, WRU and I LUV U are all now part of the Philippines lexicon. As with email, emoticons are common, too. These are the emotional icons of email which, when viewed sideways, supposedly convey expression more briefly than words can muster:

:-)   :-(: * -

Thus a form of communication previously confined to the desktop computer came out into the throng of the city, to the malls, cafés and streets. Meanwhile, the cell phones of soldiers fighting in the southern island of Mindanao were frequently winding up in the hands of the guerrillas, complete with their memory full of phone numbers. Soon, along with grenades, the rebels were lobbing text messages at soldiers. Evidently, the bombardment was not ineffective.

THE MOBILE AND THE INDETERMINATE CITY

By virtue of the cellular phone, meeting places have become indeterminate; fluid territories rather than precise spots. Discreet locations outside the shops, on the corner have given way to strips of territory – walking past the museum, getting on a bus, behind the red van stopped at the lights. We're no longer required to make prior arrangements – activists can keep their chosen sites for demonstration (physical and virtual) secret until the last minute, thus stealing a march on the authorities. These facts are not lost on the

underworld. Although the technology has traditionally emerged from within the secretive practices of the police and intelligence bureaux – it was, after all, the Detroit Police department who first utilised a prototype of the car telephone – it has always been used in ever more inventive ways by those who seek to undermine the forces of law and order.

Presently, Geographic Information Systems (GISs) map crime-prone locations. Surveillance cameras record drug-deals, muggings, hold-ups and street vandalism. Gunshots are precisely located using an array of directional microphones connected by phone to a central computer, and police in the United States now perform live scanning of fingerprints from their laptop-equipped patrol cars. The dealers and prostitutes, meanwhile, use ever more intricate combinations of beepers, web sites and prepay mobile phones to arrange their covert transactions, obviating the necessity to frequent their streets and stoops for extended periods of time, and so evade detection. The telephone kiosks of central London are now unceasingly wallpapered in flyers touting the wares, and mobile numbers, of local prostitutes, testifying to this condition.

If the industry was at first slow to grasp the potential power of the technology, then they have been quick to make amends. In 1998, a research project conducted by the Royal College of Art (RCA), at the industry's behest, concluded that the network-enhanced city enjoyed a geography of hotspots defined by waiting points and population flows, and that it was in these spaces that the industry could best engage its market. Information about local activities, similar to a community noticeboard, or the message cards pinned up in newsagents might be automatically retrieved as a consequence of proximity, as well as information about local real-estate and crime statistics. Temporary, amorphous and anonymous communities would be created, like those that form at music festivals, football matches, concerts and street demonstrations, not by virtue of shared experience so much as shared access to information. In this city, journeys, themselves, become shared and curated spaces, as content providers determine what information we should or should not have access to, and when. The RCA group's own investigations were conducted, of course, in the interests of the mobile telecommunications industry. But inevitably, with the technology in place, other organisations – the dealers, pimps and anarchists – will find the means to appropriate bandwidth for the purposes of peddling their own brand of (mis-)information via unlicensed wireless broadcast, exactly as the Filipino guerrillas did with SMS technology.

TELEMATIC COMPRESSION

In London's Soho, the installation of the high-speed communications network known as 'Sohonet' has reinforced the dominance of the area as a productivity centre for the media industry. The ability to move huge quantities of data at high speeds between London and Hollywood has actually served to condense media industry activity within the territory rather than disperse it. Despite having offices less than five minutes from Dean Street in the heart of Soho, an agent for one such media company recently complained to me of being based too far from the rabbit-warren within which his clients burrow. Their reasons for courting proximity are manifold; the requirement for rapid production combined with the need to share and exchange hardware and specialist knowledge, and the sheer density of support infrastructure – 24-hour photo and reprographics facilities, as well as courier services – all serve to reinforce the enclave community. To date, the telephone, far from alleviating this condition, has simply facilitated yet faster rates of production and, by allowing back-office activities to take place elsewhere, ever-denser occupation of these sorts of territories.

The mobile telephone was, like its precursors, initially associated with powerful men – the stereotypical 1980s high-powered business executive in the west, or the gunrunning yakuza gangster in the east.

The 1990s saw the industry deliberately target a new customer base, the young affluent professional, male and female. But the real success of the phone is its newest market: highly sociable schoolchildren and teenage boys and girls. It is this group, the text generation, who are inventing its language and exploiting its potential. Not yet old enough to vote, their consumer decisions have, as recently as the last five to ten years, come to dominate the fortunes of the home-entertainment industries. One only has to look towards the popular music industry to see the importance of their influence. For the most part they, the texters, are docile, preoccupied with the inanities of their conversation but, as recent history reveals, when moved to strike, this community of mobile phone users can overthrow the regime.

The phone's substantial benefit to business, covert or otherwise, does not preclude its social role. Instead, the phone augments both global and local businesses, and animates social relationships at both scales. The tendency is towards a global network of patches of communities; physical and virtual sites of fervent activity, and both fortified by telecommunications technology. A middle ground emerges, a gradation of community from the local to the global; one which traditional fixed telecommunications previously diminished. This ground is not defined by location, or at least not solely by location, but by the activities that take place within it. It is the street, the inner-city hamlet, the park, coffee-shop and shopping-mall, our privatized public spaces. It is no coincidence that my Wednesday morning stroll revealed the observations that it did, for these public and semi-public spaces are precisely the domain of the mobile phone.

Editor's references and suggestions for further reading

Brown, B., Green, N. and Harper , R. (eds) (2001), *Wireless World: Social and Interactive Aspects of the Mobile Age*, Cologne: Springer-Verlag.

Katz, J. and Aakhus, M. (eds) (2002), *Perpetual Contact: Mobile Communication, Private Talk, Performance*, Cambridge: Cambridge University Press.

Kopomaa, T. (2000), *The City in Your Pocket: Birth of the Mobile Information Society*, Helsinki: Gaudemus. (See Timo Kopomaa, p. 267 of this volume)

Sussex Technology Group (2001), 'In the company of strangers: mobile phones and the conception of space'. In S. Munt (ed.), *Technospaces: Inside the New Media*, London: Continuum, 205–23.

Townsend, A. (2000), 'Life in the real-time city: mobile telephones and urban metabolism', *Journal of Urban Technology*, 7, 85–109.

'Excavating the Material Geographies of Cybercities'

Stephen Graham

Editor's Introduction

In our penultimate analysis of the recombinations which occur between city space and ICTs the Editor analyses some of the material geographies of telecommunications hardware and equipment that were constructed within and between major cities during the ICT boom of the late 1990s. Graham argues that the 'Death of Distance' is not some 'virtual' process. Whilst it may be largely invisible, it actually involves the messy, complex and expensive construction of real wires, servers, and installations within, between and under the fabric of cities. Whilst these spaces and infrastructures are generally ignored, they are critical to the functioning of cybercities.

First, Graham looks at the ways in which transoceanic optic fibre networks are being linked seamlessly to high capacity local fibre networks which cover the expensive 'last mile' in and under the streets and buildings of major cities. This is being done in order to offer excellent bandwidth capabilities direct to the desks of stock-market traders, digital media companies, and all manner of other information-rich businesses. In increasingly liberalised global telecommunications markets such local-global (or 'glocal') connections between the globe and the desk drive market access for infrastructure providers in key 'hot spot' cities. Graham analyses the particular case of MCI-WorldCom, an innovative telecommunications company that constructed a global archipelago of local and global networks linking major cities in the 1990s – before it went bust through massive debts and accountancy failures as part of the 'dot.com' meltdown in 2001.

In his other two examples, Graham explores further instances of the recombinant infrastructural spaces that invisibly underpin cybercities. First, he analyses the growth of 'telecom hotels' – highly secure, anonymous buildings and bunkers which house the equipment for telecommunication switching and web server hosting. Despite the distance-transcending capabilities of telecommunications, these need to be located as closely to major urban markets as possible in order to offer the best quality of service to users there.

Finally, Graham looks at the ways in which, even before September 11th, 2001, rising demands for completely secure data storage facilities were leading to the refurbishment of disused Intercontinental Ballistic Missile silos and Cold War bunkers in the USA and the UK. With firms and government organisations now reliant on sets of precious digital data records, the Internet increasingly prone to viruses and attacks, and major urban cores at risk from catastrophic terrorism, a growing range of hybrid and recombinant spaces of data storage are being constructed which are deliberately not connected to the Internet. These combine intense fortification and physical remoteness as a deliberate strategy to manage and reduce risks.

These days, telecommunications and digital media industries endlessly proclaim the 'Death of Distance' and the 'ubiquity of bandwidth'. Paradoxically, however, they actually remain driven by the old-fashioned geographic imperatives of putting physical networks (optic fibres, mobile antennas and the like) in trenches, conduits and emplacements to drive market access. The greatest challenge of the multiplying telecommunications firms in large cities is what is termed the problem of the 'last mile': getting satellite installations, optic fibres, and whole networks through the expensive 'local loop'. In other words, the challenge is to thread networks under the congested roads and pavements of the urban fabric, to the smart buildings, dealer floors, headquarters, media complexes and stock exchanges that are the most lucrative target users (Plate 13). Because high-bandwidth networks have to be end-to-end it is not enough to construct networks to main exchanges and city cores; fibre must be threaded

through the curbs of users and beyond the actual computers inside buildings. Consequently, fully 80 per cent of the costs of a network are associated with this traditional, messy business of getting it into the ground in highly congested, and contested, urban areas (see Pile, 2001). This hard material basis for the 'digital revolution' is neglected but crucial. Focussing on it allows analysis to begin to reveal the complex social and technological practices that surround and support the explosion of digitally mediated economic and cultural flows (Zimmerman, 2001).

Such an approach also begins to reveal the subtle and powerful reconfigurations of urban space that are the result of such changing technological practices. Take an example. In a frenzied process of competition to build or refurbish buildings in the right locations for booming new media, telecommunications, and e-commerce companies, a New York agent reported in 2000 that 'if you're on top of an optic fibre line, the property is worth double what it might have been' (Bernet, 2000, 17). With this, and many other examples of the reconfigurations of urban space, we see that the 'information age', or the 'network society', is not some immaterial or anti-geographical stampede online. Rather, it encompasses a complex and multifaceted range of restructuring processes that become highly materialized in real places, as efforts are made to equip buildings, institutions, and urban spaces with the kinds of premium electronic and physical connectivity necessary to allow them to assert nodal status within the dynamic flows, and changing divisions of labour, of digital capitalism (Castells, 1996; Schiller, 1999).

These restructuring processes are intrinsically bound up with changing governance and power relations and patterns of uneven development at all spatial scales, from the transnational to the body (see Smith, 2001). In general, they tend to support a complex fracturing of urban space as premium and privileged financial, media, corporate and tele-communications nodes extend their connectivity to distant elsewheres whilst stronger efforts are made to control or filter their relationships with the streets and metropolitan spaces in which they locate (through defensive urban design, closed circuit surveillance, the privatization of space, intensive security practices, and even road closures).

To demonstrate these processes, in what follows I will explore a few examples of how urban spaces of premium connectivity are being constructed in global financial and media centres. I will also look at the

Plate 13 Telecommunications workers install a new fibre grid in downtown Manhattan, 1999. (Photograph by Stephen Graham.)

parallel development of data havens for storage and co-locating web servers in a variety of lower-cost, peripheral zones – literally the geographies of back-up.

DIGITAL ARCHIPELAGOS: GLOBAL CITY OPTIC FIBRE NETWORKS

First, it is clear that the pull of so-called global city cores is strongly shaping the global geography of telecommunications investment (see Graham, 1999). One of the world's fastest growing telecommunications firms, up until the crises in the industry in 2001, for example, was MCI-WorldCom. It emerged as a global player by constructing dedicated fibre networks for global city cores and few other places. This completely 'unbundled' solution avoided the costs of building networks to serve less lucrative spaces. In the 1990s MCI-WorldCom built over 60 city fibre optic infrastructures in major city centres across the world, in carefully targeted, financially strong city centres (45 of which were in the USA). A total of 130 MCI-WorldCom city grids were eventually planned – 85 in the USA, 40 in Europe, with the rest in Asia, Latin America and the Pacific. Each was to be carefully targeted onto 'information-rich' global cities and parts of global cities which have sufficient concentrations of large corporate or government offices to ensure high levels of international revenue relative to miles of network constructed.

MCI-WorldCom's network in London, built since 1994, was a particularly revealing example of how global connectivity was combined with subtle intensifications of uneven development within global cities. With just 180 kilometres of fibre constructed to serve only extremely communications-intensive users in the financial and corporate districts of London, the London WorldCom network had, by 1999, secured fully 20 per cent of the whole of the UK's international telecommunications traffic, which was, in turn, a good proportion of Europe's (Finnie, 1998).

Before its collapse in 2001, MCI-WorldCom also worked hard to build the transoceanic and transcontinental fibre networks to tie the urban grids together into global archipelagos – a global market which absorbed $22 billion between 1988 and 1998 – largely on direct city-to-city global links (*Communications International*, July 1999, p. 47). As well as constructing a transatlantic fibre network known as Gemini between the centres of New York and London, WorldCom built

their own pan-European Ulysses network linking its city grids in Paris, London, Amsterdam, Brussels, and major UK business cities beyond London. The strengthened importance of direct city-to-city connection was not lost on telecommunications commentators. As Finnie (1998) argued:

> it should be no surprise . . . that when London-based Cable and Wireless plc . . . and WorldCom laid the Gemini transatlantic cable – which came into service in March 1998 – they ran the cable directly into London and New York, implicitly taking into account the fact that a high proportion of international traffic originates in cities. All previous cables terminated at the shoreline (Finnie, 1998, 20).

The operations of global cities simultaneously reach out, extending their influence further across the globe via dedicated global fibre networks, whilst withdrawing into their ever-larger, mixed-use corporate plazas. These electronic superbanks are not skyscrapers but groundscrapers: 'huge nine-to-eleven-storey buildings with immense floor plates' to accommodate the remarkable IT needs of global financial institutions today (Pawley, 1997, 59).

'LOCATION, BANDWIDTH, LOCATION': URBAN TELECOM HOTELS AS THE MATERIAL GEOGRAPHY OF THE WEB

In our second example, secure developments for the mushrooming telecommunications industry are proliferating, clustered around the invisible terminals to super-high capacity interurban optic fibre trunk lines. These, in turn, tend to be laid along highways or railway tracks to minimise construction costs. In a privatised and competitive telecoms environment development is tending to concentrate around such optic fibre 'points of presence' to reduce the costs of laying high capacity lines, ensure maximum quality of connection, and allow multiple fibre operators to be accessed. As a result the edges of major global city cores are now being equipped with portfolios of 'telecom hotels' (also known as co-location facilities).

Akamai, an international provider of web server systems to major firms, is typical of the companies constructing telecom hotels. They now operate 'the largest, most global network of servers in the industry, deployed across multiple carriers'. Akamai's state-of-the-art server or 'co-location' farms are housed in highly secure building complexes located in the major global cities of the world. This offers the 'closest proximity to users possible', a factor of continuing importance in the location of heavily trafficked web sites because of Internet congestion, bandwidth bottlenecks and the dominance of global telecoms capacity by major metropolitan regions (see http://www.akamai.com).

'What's critical to these companies is access to business centers, access to fiber routes, and access to physical transportation', writes the *New York Times*, (21 March, 2000, 4). Complex and highly uneven geographies are emerging here. These 'defy predictions that the Internet and new technologies will undo cities. On the contrary, the reliance of contemporary communications on optic fiber creates a new centrality, a concentration of strategic resources in giant metropolitan areas' (Varnelis, 2002, 11). Within such cities, however, spaces and communities even short physical distances from main market nodes are often bypassed by fibre networks. 'In telecom terms a fiber-bereft desert can easily lie just a mile from [a telecom hub]. Uneven development will be the rule as the invisible city below determines construction above' (Varnelis, 2002, 11).

Telecom hotels are necessary because much of the world's most popular web content needs to be continuously copied (or 'cached') so that it is locally accessible for the major metropolitan markets across the world. As Kazys Varnelis (2002, 9) explains:

> The expense and low speed of interurban [ICT] connections leads corporations with big web presences to upload data to servers maintained by [major web service companies like] Akamai in major metropolitan markets. Thus, when you access a web site such as CNN.com or Apple.com you don't pull data from Atlanta or Cupertino but rather from a server near your metropolitan area's telecommunications exchange.

Telecom hotels are anonymous, windowless buildings and massive, highly fortified spaces which house the computer and telecommunications equipment for the blossoming commercial Internet, mobile and telecommunication industries. Their growth has done much to fill city cores as their office markets have decentralised in a kind of 'stealth occupation of the city' (Varnelis, 2002, 10). By 2001 New York, for example,

had 64 of these, each with an average size of over 50,000 square feet (Evans-Cowley *et al.*, 2002). To occupying companies, the physical qualities of the chosen buildings (high ceiling height, high-power and back-up electricity supplies) need to be combined with nodal positions on fibre networks. 'Whose fibre (and what type of fibre for that matter) will be a major consideration in the site selection process. A perfectly built building in the wrong part of town will be a disaster' (Bernet, 2000, 17). Emergency power back up, 24-hour security and advanced fire suppression systems are also provided. A full suite of high-power electrical systems are especially important as 'most buildings today are equipped with only 10 per cent of the necessary power requirements' of a telecom hotel (Bernet, 2000).

THE GEOGRAPHIES OF BACK-UP: THE EMERGENCE OF ULTRA-SECURE ENCLOSURES

In our final example, the imperative of security for data storage amongst many e-commerce and corporate firms is such that a wide range of peripheral, isolated and ultra-secure spaces are currently being configured as spaces for the remote housing of computer and data storage operations. There are several elements of this process. In the first element, a variety of offshore small island states – Anguilla and Bermuda to name two – are currently packaging themselves as free Internet zones – secure locations for web server platforms which conveniently minimise corporate taxation liabilities, vulnerabilities to Internet regulation, and operating costs.

In the second part of the process, old disused sea forts and oil rigs are now being actively reconfigured by e-commerce entrepreneurs, in attempts to secede from the jurisdictions of nation states altogether. For example, the self-styled Principality of Sealand – a disused World War II anti-aircraft fort 6 miles off the coast of Essex, England – is being used as an ultra-secure space for corporate web servers and e-commerce platform. The platform is beyond interference from any crackers, hackers or terrorist attacks, whilst maintaining high-capacity 20 millisecond links with all the world's data capitals. Since September 11th many of London's financial and corporate headquarters have installed servers in the platform's massive concrete legs to improve their resilience against catastrophic terrorism in the City of London. These are linked to the Internet only through encrypted links.

But perhaps even more bizarre is the third part of the process: the reconstruction of old Cold War missile launch sites and bunkers to offer the ultimate in security against risks of both electronic and physical incursion (D'Antonio, 2000, 26; Bright, 2003). A representative of AI Bunker, which will open a data storage bunker in the old US Cruise Missile base at Greenham Common in the UK in 2003, captured this post 9/11 logic simply. 'Do you put your computers in a great big glass building that's vulnerable to attack?' he asked. 'Or do you put them in a bunker that's completely protected?'

Conversions of old Cold War bunkers and silos are increasingly common in the US and UK. Developers of an old Titan facility at Moses Lake in Washington State are exploiting the old ICBM launching and control bunkers to offer 166,000 square feet of the most dependable and secure data storage spaces on the planet. The buildings are 'tremor proof, fireproof and impervious to even the most powerful tornado. Their three-foot thick concrete walls, reinforced with steel and lead, could withstand a truck bomb the size of the one that brought down the Murrah building in Oklahoma City or a 10-megaton atomic explosion just one quarter-mile away' (*ibid.*). All infrastructures are backed-up for guaranteed uninterrupted power supplies. The computers are separated from the public Internet to deter hackers; the service and manufacturing firms that use the space are required to have private intranets that offer the best electronic firewalls available.

REFERENCES FROM THE READING

Bernet, B. (2000), 'Understanding the needs of telecommunications tenants', *Development Magazine*, spring, 16–18.

Bright, M. (2003), 'Cold war bunkers back in business', *The Observer*, 23 February, 5.

Castells, M. (1996), *The Rise of the Network Society*, Volume 1 of *The Information Age: Economy, Society and Culture*, Oxford: Blackwell.

D'Antonio, M. (2000), 'Bunker mentality', *New York Times Magazine*, 26 March, 26.

Evans, M. (1999), 'It's a wired world', *Journal of Property Management*, November/December, 42–7.

Evans-Cowley, J., Malecki, E. and McIntee, A. (2002), 'Planning responses to telecom hotels: what accounts for increased regulation of co-location facilities?', *Journal of Urban Technology*, 9(3), 1–18.

Finnie, G. (1998), 'Wired cities', *Communications Week International*, 18 May, 19–22.

Graham, S. (1999), 'Global grids of glass: on global cities, telecommunications and planetary urban networks', *Urban Studies*, May, 36(5–6), 929–49.

Pawley, M. (1997), *Terminal Architecture*, London: Reaktion Books.

Pile, S. (2001), 'The un(known) city . . . or, an urban geography of what lies beneath the surface'. In I. Borden, J. Kerr, J. Rendell and Pivaro, A. (eds), *Unknown City: Contesting Architecture and Social Space*, Cambridge, MA: MIT Press, 263–78.

Schiller, D. (1999), *Digital Capitalism: Networking the Global Market System*, Cambridge, MA: MIT Press.

Smith, M.-P. (2001), *Transnational Urbanism: Locating Globalization*, Oxford: Blackwell.

Varnelis, K. (2002), 'Towers of concentration, lines of growth'. Available at http://www.varnelis.net/projects/onewilshire/index.html (Feb. 2003).

Zimmerman, R. (2001), 'Social implications of infrastructure network interactions', *Journal of Urban Technology*, 8(3), 97–119.

'Learning from September 11th: ICT Infrastructure Collapses in a "Global" Cybercity'

Anthony Townsend

Editor's Introduction

The complex and multifaceted infrastructures that sustain cybercities are often invisible or taken for granted – at least when they work well. Paradoxically, it is often the sudden absence of the material and technical infrastructures of cybercities that reveals their importance. This process fuels our fascination with 'decyborganised' societies, as discussed by Timothy Luke (p. 106).

In an economic context where dominant firms demand 24-hour ICT-access to the four corners of the world – and the unbreakable electricity supplies necessary to power this – the withdrawal of these services through technical malfunction, war, electricity outages, hacking, saturated demand or terrorist attack, becomes all the more catastrophic. Such instances of collapse are episodes which powerfully reveal the critical importance of the recombinant and hybrid forms, spaces and infrastructures that make cybercities work. Such collapses also hammer home their extreme vulnerabilities.

When ICT systems collapse this process is a social as much as a technical process. For, just as an enormous amount of social work by people needs to be done to make the complex ICT systems of cities function in the first place, the process of diagnosis, repair and restoration that is necessary to make them function again, following collapse, is a fundamentally social process, too.

The collapse of ICT systems is most spectacular when they are under strain in cities where ICT infrastructures, services and applications are most concentrated. In wealthy, advanced cities such collapses can be devastating simply because reliance on electrically powered and telecom-connected computers is becoming so complete across all walks of life. The explosive recent growth of electronic commerce, consumption, and distribution and production systems – infrastructures that are mediated at every level by electrically powered computer and telecommunications – means that these days we are all, in a sense, 'hostages to electricity' (Leslie, 1999, 175). The economic consequences of collapse and outages can be extremely expensive and even catastrophic. 'The always-on economy, by definition, depends upon continuous energy. For a large business online, the cost of a power interruption can exceed $1 million per minute' (Platt, 2000, 116–28). For stock markets and electronic financial service firms, the costs can be much greater still.

This point is not lost on the infrastructure firms themselves in their recent advertising, or in their increasing investment in duplicate and back-up power systems to protect online service providers, Internet backbones, cable TV and phone companies. Nor is it missed by leading IT and software entrepreneurs. For example, an article in *Wired* (2000) by the co-founder of Sun Microsystems, Bill Joy, caused a major furore. Joy suggested that the mediation of human societies by astonishingly complex computerised infrastructure systems will soon reach the stage when 'people won't be able to just turn the machines off, because they will be so dependent on them that turning them off would amount to suicide' (2000, 239).

Because of these factors there is no more revealing case study of the collapse of ICT systems in cybercities than the ways in which phone, mobile, electricity and Internet systems responded to the now iconic terrorist attacks on Manhattan – heart of one of the world's most computerised cities – on September 11th, 2001. In this specially commissioned analysis of the events which followed the attacks, Anthony Townsend, an urban researcher at New York University, details how the various mobility and ICT systems that sustained Manhattan's 'global city' economy responded to the catastrophic damage caused by the collapse of the World Trade Center towers. He shows how the crisis was further intensified by unprecedented peaks of demand as virtually the entire population of the city immediately tried to contact friends and family all over the world (and vice versa); as the global media sought to find channel capacity to sustain intensive coverage of the events; and as emergency services went into action.

The attacks, as Townsend suggests, sparked the largest telecommunications event in human history. Detailing the ways in which telephone operators and electricity and landline, mobile and Internet providers responded to the attacks, and to the unprecedented demand peaks that followed, Townsend's analysis hammers home the ways in which new ICT infrastructures are not somehow separate, technical, and automatic. Rather, they are actually cyborganised socio-technical systems which are utterly embedded into the social and professional contexts of the many human beings that are still necessary to actually make them work.

The resilience of urban ICT systems to risks of collapse, thus, depends on the combined resilience of people, hardware and software working in subtle synchrony. Urban ICT systems are also physically embedded, as material systems of switches, services, wires, tubes and ducts, within the murky, dirty and difficult spaces of the city's often ignored underground realm (see Pile, 2001). They share this space with highways, bridges, sewers, power networks, electricity and gas networks, water tunnels, civil defence installations, and subways. Finally, urban ICT systems are utterly reliant on less glamorous, but still vital, electricity systems, which may actually be the most vulnerable part of the whole system (as was demonstrated by the US power outages in August 2003).

Since September 11th, 2001 the world's major financial institutions have worked hard to reduce the vulnerabilities of their systems, and their data, to 'catastrophic terrorism'. Many have already moved their systems to decentralised webs of hardened offices in the suburbs of major finance capitals. This, indeed, has fuelled the latest version of the prediction that the days of concentrated city cores are over, especially in the USA. For example, Mike Davis (2002, 11–12) argues that:

> It is conceivable that bin Laden et al. have put a silver stake in the heart of the 'downtown revival' in New York and elsewhere . . . The current globalization of fear will accelerate the high-tech dispersal of centralized organizations . . . into multisite networks. Terror, in effect, has become the business partner of technology providers like Sun Microsystems and Cisco systems, which have long argued that distributed processing . . . mandates a 'distributed workplace'.

Major firms will certainly work to reduce the vulnerabilities of their operations by decentralising their ICT systems. However, it seems likely that the central business cores of global financial cities will continue to operate in the face of the risks of catastrophic terrorism, albeit in modified form.

▓ ▓ ▓ ■ ▓ ▓

Since the invention of the telegraph over 150 yeas ago, lower Manhattan has been the world's leading telecommunications hub. Telecommunications technologies evolved symbiotically with the financial services industry, providing the means to quickly and cheaply communicate with colleagues across the city or across the globe. As a result, there is more fiber optic cable strung beneath the streets of Manhattan than on the entire continent of Africa. The two main telephone switches in the financial district each house more lines than many European nations. The ether above is saturated through an endless variety of wireless transmissions, such as those that emanated from the more than 1,500 antenna structures mounted atop the World Trade Center's north tower.

The September 11, 2001 terrorist attacks sparked the largest telecommunications event in human history and in the process focused attention on the pervasiveness of digital network infrastructure in today's cities. While transportation, water, and power networks are all critical to the proper functioning of a modern metropolis, during crisis and times of uncertainty communications networks play a crucial role in urban survival.

This reading examines the events of September 11, 2001 and the following months in New York City by focusing on the role of digital communications networks.

RESPONSES

Almost as soon as the first plane struck 2 WTC at 8:48 a.m., telecommunications systems throughout the world began experiencing what would soon become the biggest surge of use in history. As many have pointed out, New York's role as a media center was undoubtedly one reason it was chosen as a target. It was almost certain that video cameras would inadvertently capture the attacks live.

Even before emergency workers had begun to respond to the raging high-rise fire caused by the impact of the first plane, television networks and news web sites were pumping images of urban catastrophe across the globe. While most people choose the richness and immediacy of broadcast media such as radio and television over the Internet, popular news web sites were swamped with requests for images and streaming video. Traffic to the CNN web site doubled every seven minutes between 9 and 10 a.m. that morning until its servers were no longer able to keep up with demand and crashed. Even after stripping it down to the so-called 'end of the world' page, CNN could not keep up with an estimated 2 million page requests per minute.

The telephone network under strain

Tens of millions of Americans instinctively reached for their telephones as news of the plane crashes began to spread, and historic call volumes were sustained throughout much of the week. On September 11th alone, AT&T connected some 431 million calls, 20 per cent more than normal. Based on AT&T's long-

distance market share, it is likely that over 700 million long-distance calls were connected nationwide that day. Approximately double this number of calls were attempted, but AT&T's experience with previous localized disasters such as earthquakes and hurricanes had prepared them. Giving priority to the needs of the affected people, AT&T blocked incoming calls to keep outgoing lines from New York and Washington open. Local telephone systems were swamped as well, with Verizon reporting over 250 million local phone calls in the New York region in the week of September 11–18. Phone traffic around in New York City was double the normal volume – an unprecedented level.

The government response

While millions watched helplessly and tried to reach loved ones, the men and women of the Police (NYPD) and Fire (NYFD) departments, along with state and federal authorities, were rushing into the burning buildings. They evacuated lower Manhattan for the second time in a decade, and cordoned off what had quickly been transformed from disaster area to crime scene after the impact of the second plane at 9:03 a.m. The failure of mobile wireless emergency communications that these brave men and women experienced is perhaps the most tragic part of this story. The breakdowns fell into two main categories. First, the repeater systems for fire and police radios (installed inside the World Trade Center towers after the 1993 terrorist bombing) failed when power was cut after the planes' impact. Few can forget the voices of firefighters captured on documentary footage as they desperately tried to understand the weak voices coming through their radios. A report prepared by McKinsey found that problems with the radio system put many of the fire companies out of touch as they ascended the stairs into the twin towers. As a result, at least 120 firefighters did not receive the order to evacuate and were killed in the second collapse.

Even when radio equipment worked perfectly, first responders were hindered due to the lack of interoperable radios. Supporting firefighters arriving from New Jersey were unable to communicate by radio with their New York City counterparts, who used a different communications frequency. Other problems, caused by psychological factors rather than technological ones, also strained the capacity of emergency radio channels. The system used by New York

City's Emergency Medical Service was swamped with unnecessary transmissions from panicked operators, causing congestion that interfered with important messages.

The sheer variety of first responders further contributed to the breakdown of communications. Dozens of local, state, and federal authorities responded to the attacks. Yet, since few law enforcement agencies share compatible communications technologies, many rely upon public networks like the cellular telephone system for interagency communications. Throughout the day, congestion plagued the public cellular systems as historic call volumes combined with oversubscribed networks. During the morning of September 11th, fewer than one in twenty mobile calls were connected in New York City. Since no system was in place to prioritize calls for emergency workers, response was confused and uncoordinated throughout much of the day.

The financial industry response

While law enforcement and local government responded haphazardly due to a lack of effective communications, the financial services industry responded rapidly and decisively to the attacks and resulting destruction. This quick and capable response was the result of preparations dating back to the 1993 World Trade Center bombing and beyond.

One of the most severely affected firms, Lehman Brothers, responded almost instantly. Lehman CTO Bob Schwartz was able to activate the company's disaster recovery plan just after 9 a.m. by sending a message from his portable Blackberry messenger while descending the stairwell of 1 WTC. While Lehman's Manhattan data center was destroyed in the collapse of the towers, operations were seamlessly transferred to a twin backup center in New Jersey. The firm was ready for trading at backup facilities in New Jersey the next day, five days before the stock markets were ready to reopen.

While Lehman Brothers was one of the most severely affected firms, it was by no means unique. By noon on the 11th, every major bank in New York City had activated some form of disaster recovery plan. By the next morning, nearly fifty firms in and around the World Trade Center had declared disasters and called upon their disaster recovery contractors firms such as SunGard and Comdisco to retrieve sensitive data from offsite backups. Many firms (like Lehman Brothers) were able to switch over instantaneously to backup systems in other locations mainly in midtown Manhattan, New Jersey, and downtown Brooklyn. Cantor Fitzgerald, which lost some 700 employees in the WTC, was up and running when the bond market reopened on Thursday morning, 47 hours later. This was possible because the firm had duplicate data centers in both New Jersey and London.

NATIONAL TRENDS

Outside New York City, September 11th was marked by abrupt shifts in telecommunications use. As the afternoon wore on, telephone traffic began to subside and many turned to the Internet for more detailed information. Around 4 p.m., a surge of users rushed to America On-Line (AOL) and other online services, flooding the Internet with short mails and instant messages. Presumably, once closest contacts had been reached by telephone, people were broadcasting their whereabouts and status through blanket mailings to their entire address book. AOL reported an astonishing 1.2 billion instant messages per day throughout September, far higher than before the 11th.

Across the Internet, there were similar patterns. At Columbia University in New York City, email use surged 40 per cent to nearly one million messages per day. The Associated Press photo archive on AOL, which had only received 30,000 requests daily before the attacks, received over one million daily afterwards. Content distribution networks like Akamai observed a three to fourfold increase in streaming media demand (see Stephen Graham, p. 138). Ironically, however, the insatiable demand for Internet pornography could not be deterred. World Trade Center still ranked just seventh among keyword searches on AOL, behind #1 Nostradamus and five pornographic terms!

As quickly as Internet use had surged, it faded away as President Bush took the stage late in the evening to calm a frightened nation. Much like during major television events such as the Super Bowl, AOL and other Internet Service Providers (ISPs) recorded record low usage during the President's speech. Millions logged off the Internet and gathered around television sets to hear how their leaders would react.

THE RESILIENCE OF THE INTERNET

Not surprisingly, the Internet was one of the few telecommunications networks that was not seriously disrupted by the terrorist attacks. Unlike the voice network, destroyed infrastructure did not prevent traffic getting through. Designed around Cold War ideas for attack-proofing communications networks and built by a wide variety of competing firms, the Internet's present architecture is a dense web of overlapping, redundant, and interconnected networks. According to Matrix.net, the leading analysts of traffic patterns on the Internet, network disruptions caused by the September 11th attacks were few and highly localized in the New York region.

The ability of the Internet to withstand damage stems from its reliance on two protocols for routing traffic around damaged areas. The first is TCP/IP. TCP/IP's 'packetized' nature means that data doesn't 'care' how it gets to its destination – data is separated into packets which can take multiple routes to the destination. The second, less well-known cousin of TCP/IP is BGP, or the Border Gateway Protocol. Because it is fairly common for a packet to transit several networks between its origin and destination, a system needs to be in place to determine the shortest path along each multi-network journey. This is complicated by the fact that each network has multiple entry and exit points in multiple cities. BGP provides a way for each network to advertise its shortest routes, while revealing nothing about its internal structure (which is a competitive secret). The advantage of BGP is that it provides for the dynamic reassignment of routes, which comes in handy during a disaster like September 11th. While it does take time (and often some human intervention) for networks to adjust, BGP is remarkably effective at keeping localized disruptions from propagating throughout the network. This is in contrast to other tightly coupled networks like the air transport system, in which localized problems tend to propagate quickly throughout the network. (i.e. 'Your flight from Los Angeles to New York is delayed because of bad weather in the Midwest').

These protocols helped the US Internet remain remarkably resilient while facing the challenges posed by September 11th, which fell into two categories. The first was the surge in traffic cause by a major media event. Globally, this surge was barely noticed and most major ISPs reported a 5 per cent increase in data traffic across the board, which was easily handled by existing capacity. However, in the New York region late-night Internet activity caused significant congestion from around 9 p.m. on September 11th until early the next morning. The congestion on the New York area Internet lasted several days, according to John Quarterman of Matrix.net.

The second category of challenges had to do with physical damage and outages of Internet routing equipment. Intermittent power problems at two major Internet switching stations in New York did produce some routing problems for US–European Internet connections. However, BGP was able to quickly and effectively route around these problems. Many carriers such as Sprint and Qwest offered each other capacity on transatlantic lines to deal with these outages.

THE INTERNET NEEDS POWER: THE WEB OF URBAN NETWORKS

For their resilience, digital networks remain highly dependent upon older, less flexible infrastructure systems. Much of lower Manhattan's telecommunications would have remained online if Con Edison had been able to maintain the electricity grid. The lack of physical access to Manhattan also means that many telecommunications networks must share the same physical conduits into the Manhattan central business district: the Lincoln, Holland, and Battery Tunnels. Should these transportation arteries be compromised by a terrorist plot (such as the one that had been uncovered by federal investigators in the mid-1990s), Manhattan might be cut off from Europe or lose much of its ability to talk to the rest of the United States.

Editor's references and suggestions for further reading

Davis, M. (2002), *Dead Cities*, New York: New Press, Chapter 18.
Leigh-Star, S. (1999), 'The ethnography of infrastructure', *American Behavioural Scientist*, 43(3), 377–91.
Leslie, J. (1999), 'Powerless', *Wired*, April, 119–83.

Joy, W. (2000), 'Why the future doesn't need us', *Wired*, April, 238–60.

Pile, S. (2001), 'The un(known) city . . . or, an urban geography of what lies beneath the surface'. In I. Borden, J. Kerr, J. Rendell and A. Pivaro (eds), *Unknown City: Contesting Architecture and Social Space*, Cambridge, MA: MIT Press, 263–78.

Platt, C. (2000), 'Re-energizer', *Wired*, May, 114–30.

Rochlin, G. (1997), *Trapped in the Net: The Unanticipated Consequences of Computerization*, New Jersey: Princeton.

Zimmerman, R. (2001), 'Social implications of infrastructure network interactions', *Journal of Urban Technology*, 8(3), 97–119.

PART TWO

Cybercity Dimensions

Plate 14 Physical and virtual: global port and aiport infrastructures in Tokyo Bay, June 2003. They rely on seamless logistics systems based on ICTs in order to function. (Photograph by Stephen Graham.)

IV CYBERCITY MOBILITIES

Plate 15 Reaction of the London *Evening Standard* cartoonist Patrick Blower to the news that the London Underground was being wired to allow travellers to use mobile phones (24 November 1999). (Source: Patrick Blower.)

INTRODUCTION

The town is the correlate of the road. The town exists only as a function of circulation and of circuits; it is a singular point on the circuits which create it and which it creates. It is defined by entries and exits: something must enter it and exit from it

(Deleuze and Guattari, 1997, 186)

A focus on the infrastructures that sustain the various mobilities and flows which move within and between cities – transportation, telecommunications, energy, water, and streets – offers a powerful and dynamic way of seeing contemporary cities and urban regions. When our analytical focus centres on how the wires, ducts, tunnels, conduits, streets, highways and technical networks that interlace and infuse cities are constructed and used, modern urbanism emerges as an extraordinarily complex and dynamic socio-technical *process*. Contemporary urban life is revealed as a ceaseless and mobile interplay between many different scales, from the body to the globe. In fact, mobile interactions across distances and between scales, mediated by telecommunications, transport, energy and water networks, are the driving connective forces of much-debated processes of 'globalisation' (see Urry, 2000a,b). Drawing on the work of Deleuze and Guattari, the architectural theorists Laurent Gutierrez and Valérie Portefaix (2000, 20) argue that advanced and high-tech cities like Hong Kong, in particular, can now be viewed as 'fluid machines'. In these:

> there is a continual fluctuation of people, goods, data, and services as moving entities, together forming a society where the whole structure is in movement. This dynamic is supported by thousands of signs indicating both movement and intensity of urban flows . . . Each flow individually forms its complex horizontal network, further linked vertically through different transportation systems . . . Both new infrastructures and the hyper-concentration of facilities [in such cities] create a strategic terrain for a network of international corporate cultures (international finance, telecommunications, information technology). These networks are open systems capable of absorbing new centres without causing instability.

In this perspective, cities and urban regions become, in a sense, staging posts in the perpetual flux of infrastructurally mediated flow, movement and exchange. They emerge as processes in the distant sourcing, movement and disposal of water reserves and the remote dumping of sewerage, pollution, and waste. They are the hotbeds of demand and exchange within international flows of power and energy resources. They are the dominant sites of global circulation and production within a burgeoning universe of electronic signals and digital signs. They remain the primary centres of transnational exchange and distribution of products and commodities. And they are overwhelmingly important in articulating the corporeal movements of people and their bodies (workers, migrants, refugees, tourists etc.) via complex and multiple systems of physical transportation.

The constant flux of this urban process is constituted through many superimposed, contested, and interconnecting infrastructural 'landscapes'. These provide the mediators between nature, culture, and the production of the 'city'. There is the 'electropolis' of energy and power. There is the 'hydropolis' of water and waste. There is the 'informational' or 'cyber' city of electronic communication. There is the 'autocity' of motorised roadscapes and associated technologies. And so on. Importantly, however, these infrastructural 'scapes' are not separated and autonomous; they rely on each other and co-evolve closely in their interrelationships with urban development and with urban space.

Infrastructure networks are thus the key physical and technological assets of contemporary cities. As a 'bundle' of materially networked, mediating infrastructures, transport, street, communications, energy and water systems constitute the largest and most sophisticated technological artefacts ever devised by humans. In fact, the fundamentally *networked* character of modern urbanism as Gabriel Dupuy (1991), reminds us, is perhaps its single dominant characteristic. Much of the history of modern urbanism can be understood, at least in part, as a series of attempts to 'roll out' extending and multiplying road, rail, airline, water, energy, and telecommunications grids, both within and between cities and metropolitan regions. These vast lattices of technological and material connections have been necessary to sustain the ever-expanding demands of contemporary societies for increasing levels of exchange, movement, and transaction across distance. Such a perspective leads us to highlight four critical connections between mobilities, infrastructure networks and contemporary urbanism that together form the starting points of this part of the book.

Complementarities between electronic and physical mobilities

The very first words uttered by Alexander Graham Bell, when he made the first ever telephone call to his assistant in Boston in 1876, were 'Watson – come here I want to see you!' From the outset, electronic communications was inseparable from physical movement. All the evidence today suggests that, whilst there may be some substitution of routine face-to-face exchanges by ICT-mediated flow today, at every geographical scale the explosion of digital mobility is going hand in hand with a parallel explosion of physical mobility. Overall, transport and telecommunications mobilities tend to grow in parallel, within and between cities. This means that demands for physical mobility and movement are unlikely to reduce even whilst telecommunications use continues to burgeon. Overall, transport and telecommunications actually feed off and fuel, more than simply substitute, each other.

On the one hand, ICTs undoubtedly do sometimes substitute for some journeys and interactions – as with phone and PC banking, travel reservations, or the use of home DVDs to replace cinema trips, for example (although DVDs are still physical products that need to be made, transported and purchased). As we saw (Paul Virilio, p. 78; Lieven de Cauter, p. 94), there are major fears here that such uses of ICTs will support an intensely capsularised urban life, as people fortify their homes, neighbourhoods, and cars and interact more and more through telemediated exchanges which are themselves capsularised by software. In some cases this seems to be happening. For example, Rebecca Solnit writes of an advert for a CD-ROM encyclopaedia that she found in the *Los Angeles Times*. 'You used to walk across town in the pouring rain to use our encyclopaedias', this read. 'We're pretty confident that we can get your kid to click and drag' (2001, 10).

But the relationship between transport and telecommunications goes beyond such simple substitution. Overall, ICTs have highly complementary relationships with physical travel, and with other mobilities (for example electricity flows). Whilst substitution does occur, these broader complementarities are resulting in the growth of physical transportation flows. This growth takes four forms.

First, ICTs play a central role in improving the efficiency and effectiveness of transport networks. This reduces the real cost of travelling (at least for those who can afford access to the networks). As major physical transport developments become harder to finance and legitimise within many cities, so more attention is focusing on how existing spaces and infrastructures can be remade and reconfigured through ICTs. Complex computer ticketing, transaction systems and air-traffic control systems help reduce the costs of air travel. This makes it more attractive as people's perceptual awareness of distant people, places and opportunities increases. Electronic information exchange also plays a major role in the organisation and management of transport networks. Such innovations help to extend the reach, reliability and usefulness of transport flows through air travel, auto-route guidance, fax, mobiles, email, real time information, and electronic data interchange-based 'just-in-time' logistics systems. A single flight of a 747, for example, has been estimated to generate 50,000 electronic exchanges in booking, maintenance, refuelling, airport management, catering, staffing, air traffic control, logistics etc.

Second, ICT infrastructures are overwhelmingly being built to parallel the locations and geographies of streets, utilities and transport systems. As we saw (Stephen Graham, p. 138), investments like optic fibres are being channelled within, through and underneath the main corridors of physical movement within and between cities. The economic logic of minimising risk in today's highly competitive and liberalised infrastructure markets therefore compounds the already strong complementarities between the physical geographies of transport, utilities and ICT infrastructure. 'New telecom infrastructure like fiber optic is run between the same information hubs that already dominate via copper connections, often on rights-of-way of even older infrastructure, such as railroads and highways. This reinforces those networks' traditional, city-centric, organisation of space' (Cohen, cited in Meyer, 2001, 4).

Third, access to cheaper and more powerful telecommunications and ICTs tends to increase an individual's, group's or organisation's 'perceptual space'. This tends to create more incentive, and desire, for physical travel for those who can afford to gain access to transport networks. The extending transnational personal, business, social and organisational webs of contact that are facilitated by ICTs, whether driven by leisure, recreation, migration, or business, in turn fuel demands for physical travel. This is necessary to experience directly the quality of interaction that can only be achieved through physical contact – usually in cities. Here we recall the 'compulsion of proximity' analysed by Boden and Molotch (p. 101).

Finally, congested transport systems create new demands for telecommunications. Mobile phones, Personal Digital Assistants (PDAs) and Internet-equipped cars may actually help to sustain *larger* traffic jams because they allow 'dead time' to be converted to 'live' working or contact time. It is no accident that some car manufactures now sell cars with car phones, faxes, Internet access or mobile computers – the ideal way of staying in touch with work and home once a driver is stuck in a gridlock or slow moving traffic. In this way telecommunication links help to overcome people's resistance to travelling by cars on roads which are congested, as real time information and contact overcomes many of the uncertainties and difficulties of congested, physical travel (p. 167).

To sum up, then, what appears to be happening is a major expansion in *all forms* of communications to sustain an (extremely uneven, unequal, and partial) speeding-up of mobility interchange at all scales. Although some substitution may undoubtedly be taking place, overall growth of electronic and real mobility tends to overwhelm the contribution of substitution. Such processes drive a remaking of the spaces and times of everyday urban life. They sustain increasingly fluid reworking of the rhythms and patterns of the city, as activities like work, shopping, tourism, leisure and social interaction become increasingly uncoupled from specific sites and places and knitted together into new and more extended geographies and fluid temporal arrangements. Finally, the new hybrid interchanges of mobility and flow, as ICTs fuse with, and reconfigure, the other mobility spaces and systems of urban life, become critical and strategic sites at which the very political organisation of space and society becomes continually remade.

The 'power geometries' of physical and electronic mobility: immobility as a form of marginalisation and humiliation

Any discussion of the growth in physical and electronic mobilities, however, must be extremely conscious that it tends to be dominated by a wealthy minority of our planet. As we shall see in Section VII of the book, social inequalities tend to be reinforced by unevenness in access to both electronic and physical forms of mobility (Kenyon *et al.*, 2002). Doreen Massey speaks of a highly complex and uneven *power geometry* of time–space compression within which different social and cultural groups have extremely uneven levels of control and access to means of both electronic and physical mobility. She argues that 'different social groups and different individuals are placed in very distinct ways in relation to these flows and interconnections' (Massey, 1993, 61).

Massey identifies three main groups within this broad and uneven picture: those who are 'hypermobile' who tend to be in control of the 'time space compression' (dominant business leaders, media companies, affluent western tourists); those who communicate and travel a great deal but aren't in control (for example, call centre workers); and those who are on the receiving end or are excluded from these processes (people on very low incomes who cannot access ICT, phone and transport systems). 'Immobility', suggests Jennifer Wolch (2000, 3), 'is punishment for the weak and for those unable to compete in modern society. Immobility is a source of humiliation and degradation for those in society who do not fit into the "traveller" category.'

Our concern in this part of the book is with the subtle recombinations and hybridisations of mobility that occur within and between cybercities, as ICTs and physical movement systems and spaces work together and interrelate. Focusing on the ways in which the flows of people, goods and information are sustained by combinations of physical transport systems, ICT systems, and other utilities, this part examines some of the key mobility sites of the contemporary city – the street, the airport, the highway, the car, the transport network, and the logistics chain – to see how ICTs are shaped together with other mobility systems. In so doing, this part of the book attempts to develop a socio-technical perspective of the multi-scale interplays which link flows of atoms and bits, people, goods, services, messages and meaning, through the hybrid ICT-transport systems and spaces that increasingly characterise cybercities.

References and suggestions for further reading

Adams, P. (1995), 'A reconsideration of personal boundaries in space-time', *Annals of the Association of American Geographers*, 85(2), 267–85.

Adams, P. (1997), 'Cyberspace and virtual places', *Geographical Review*, 87(2), 155–71.

Barley, N. (2000), *Breathing Cities: The Architecture of Movement*, Basel: Birkhäuser.

Bell, J. (ed.) (2001), *Carchitecture: When the Car and the City Collide*, Basel: Birkhäuser.

Branscomb, L. and Keller, J. (1996), *Converging Infrastructures: Intelligent Transportation and the National Information Infrastructure*, Cambridge, MA: MIT Press.

Couclelis, H. (2000), 'From sustainable transport to sustainable accessibility: can we avoid a new *tragedy of the commons*?'. In D. Janelle and D. Hodge (eds), *Information, Place and Cyberspace*, Berlin: Springer, 341–57.

Deleuze, G. and Guattari, F. (1997), 'City/state', *Zone*, 1–2, 195–9.

Dupuy, G. (1991), *L'urbanisme des réseaux: théories et méthodes*, Paris: Armand Colin.

Easterling, K. (1999), *Organization Space*, Cambridge, MA: MIT Press.

Forer, P. and Huisman, O. (2000), 'Space, time and sequencing: substitution at the virtual/physical interface'. In D. Janelle and D. Hodge (eds), *Information, Place and Cyberspace*, Berlin: Springer, 73–90.

Giannopoulos, G. and Gillespie, A. (eds) (1993), *Transport and Communications in the New Europe*, London: Belhaven.

Gottdeiner, M. (2001), *Life in the Air: Surviving the New Culture of Air Travel*, New York: Rowan and Littlefield.

Graham, S. (1997), 'Telecommunications and the future of cities: debunking the myths', *Cities*, 14(1), 21–9.

Graham, S. (2000), 'Constructing premium networked spaces: reflections on infrastructure networks and contemporary urban development', *International Journal of Urban and Regional Research*, 24(1), 183–200.

Graham, S. and Marvin, S. (2001), *Splintering Urbanism: Networked Infrastructures, Technological Mobilities, and the Urban Condition*, London: Routledge, Chapter 7.

Gutierrez, L. and Portefaix, V. (2000), *Mapping HK*, Hong Kong: Map Books.

Hamilton, K. and Hoyle, S. (1999), 'Moving cities: transport connections'. In J. Allen, D. Massey and M. Prycke (eds), *Unsettling Cities*, London: Routledge, 49–94.

Hanson, S. (1993), *The Geography of Urban Transportation*, New York: Guilford.

Hepworth, M. and Ducatel, K. (1992), *Transport in the Information Age: Wheels And Wires*, London: Belhaven.

Hodge, D. (1990), 'Geography and the political economy of urban transportation', *Urban Geography*, 11(1), 87–100.

Janelle, D. and Hodge, D. (2000), *Information, Place and Cyberspace: Issues in Accessibility*, Heidelberg: Springer.

Kenyon, S., Lyons, G. and Rafferty, J. (2002), 'Transport and social exclusion: investigating the possibility of promoting inclusion through virtual mobility', *Journal of Transport Geography*, 10, 207–19.

Massey, D. (1993), 'Power-geometry and a progressive sense of place'. In J. Bird, B. Curtis, T. Putnam, G. Robertson and L. Tickner (eds), *Mapping The Futures: Local Cultures, Global Change*, London: Routledge, 59–69.

Meyer, M. (2001), 'Delivering the future: E-freight'. Available at http://www.intermodal.org/FIRE/meyerpaper.html.

Moss, M. and Townsend, A. (2000), 'The role of the real city in cyberspace: understanding regional variations in Internet accessibility'. In D. Janelle and D. Hodge (eds), *Information, Place and Cyberspace*, Berlin: Springer, 171–86.

Nilles, J., Carlson, F., Gray, P. and Hanneman, G. (1976), *The Telecommunications–Transport Trade Off*, Chichester: Wiley.

Pascoe, D. (2001), *Airspaces*, London: Reaktion Books.

Rosler, M. (1998), *In the Place of the Public: Observations of a Frequent Flyer*, Osfildern: Cant Verlag.

Solnit, R. (2001), *Wanderlust: A History of Walking*, New York: Viking.

Swyngedouw, E. (1992), 'Communication, mobility and the struggle for power over space'. In G. Giannopoulos and A. Gillespie (eds), *Transport and Communications in the New Europe*, London: Belhaven, 305–25.

Urry, J. (2000a), *Sociology Beyond Societies: Mobilities for the Twenty-First Century*, London: Routledge.

Urry, J. (2000b), 'Mobile sociology', *British Journal of Sociology*, 51(1), 185–203.

Williams, H. (1991), *Autogeddon*, London: Jonathan Cape.

Wolch, J. (2000), 'Implications of time space compression'. Mimeo.

Wollen, P. and Kerr, J. (eds) (2002), *Autopia: Cars and Culture*, London: Reaktion Books.

TWO

'People'

from *Breathing Cities: The Architecture of Movement* (2000)

Nick Barley

Editor's Introduction

In our first reading on cybercity mobilities the British architectural writer Nick Barley reflects on the ways in which movement and mobility come together to shape the identification, experience and conceptualisation of contemporary London.

As Europe's premier 'global' city, London is experiencing a saturation of mobilities. Its systems for moving people, especially – pavements, roads and the 'tube' subway – are stalling. This is occurring through extreme congestion and the results of long-term physical and financial neglect. But the chaos has been exacerbated by the massive rise in streetworks caused by the hundreds of utility firms in the city (who now operate in one of the world's most liberalised utility markets).

This situation sustains paradoxical extremes of juxtaposition. London is the site of the world's most connected international airport (Heathrow). It has some of the world's best global ICT linkages and is amongst the world's four most important concentrations of Internet activity (see Stephen Graham, p. 138). And yet it has an increasingly strangled streetspace and underground subway network. Average automobile speeds are barely above those for horse drawn traffic at the start of the twentieth century. Thus, extremes of mobility and ICT-based, seamless flow, jostle, cheek-by-jowl, with the stresses and strains caused by extremes of immobility and the physical, social, environmental and economic costs of frustrated flow and congestion. This crisis led the mayor, Ken Livingstone, to introduce the world's largest central area road tolling initiative in February 2003, replete with a huge CCTV cordon and massive ICT support systems.

As the complex choreographies of travel within, around and through the London urban region are continuously acted out and constituted through diverse mobilities, so the meanings of 'London' as a city are perpetually renegotiated and remade. In this context Nick Barley reflects on where the 'centre' and the 'edge' of London might now be, given that the city is an immense sprawl of geographical and technological spaces which are infused with countless mobility systems and a myriad of connections to more or less distant elsewheres. Thus, Barley concludes, the very spiralling mobilities that life in urban regions now sustains means that cities need to be thought of not as stable, static or bounded entities, but, rather, as shifting sets of conceptual possibilities.

Every city-dweller is familiar with the experience of inching forwards in an interminable traffic jam, or waiting three-quarters of an hour for a bus – of all the things that infuriate people about living in a European city, among the highest-ranking must be congestion. The majority of people who work in cities spend at least an hour travelling to and from their jobs, and some more long-distance commuters spend over two hours every day in transit. But although individual journey times may be longer than we would like, the extraordinary ability of European urban transport systems to support the ceaseless daily expansion and contraction of the city's population borders on the fantastic.

The sheer volume of the city's elastic population is difficult to comprehend. In London each weekday morning, in the three-hour period between 07.00am and 10.00am, the number of people in the central area of the city swells by 1.3 million. The majority of these people – 85% – are carried on public transport: 41% arrive on trains, 35% use the Underground, and 8% travel by bus; only 13% drive private cars, and 2% ride bicycles. It's possible to imagine then, that commuter flow in London is hidden away below street level. But think about those figures another way: even if cyclists represent just one fiftieth of London's 1.3 million morning arrivals, they nonetheless amount to around 26,000 additional bicycles moving around the city centre during this part of the morning. Similarly, although the proportion of people using private cars makes up just 13% of commuters, this amounts to 130,000 additional cars in the city centre by 10.00am.

And this movement of people around London doesn't stop at 10.00. The 'rush hour' is only the beginning of a steady increase in communal activity that peaks in the mid-afternoon, much of it involving short journeys which ensure pressure on the city's transport network is fairly constant. The cumulative statistics are breathtaking: during an average day, from 12,000 bus stops around the city, travellers make four million journeys on 5,000 buses; while below street level, with its 274 stations, 392km of track, 500 trains and a staff of 16,000, London Underground facilitates some 2.5 million journeys.

If this accumulation of statistical data begins to fade to a blur, it's worth remembering that all this movement, as well as being an essential factor in the way the cities are planned for the future, is equally importantly one of the key reasons why our cities have developed their unique history. London looks like it does, and is organised in the way that it is, precisely because people have always moved around it. It may be the landmark buildings which characterise individual cities in our minds, but while Big Ben and Nelson's Column let us know we are in London, the way we move around these immobile monuments has been equally important in the definition of the city. London is London because of flux.

Still more people descend out of the sky onto London every day. The airports of Heathrow, Gatwick, Stansted, City and Luton welcome between them more than 100 million passengers a year. All these people, arriving in something calling itself 'London', have nevertheless – if they actually want to be in London – to make their way into the centre of the city. Ironically, almost the only places in London which announce you are actually in London are the airports – and most of them fall outside the administrative boundaries of the city. Heathrow and Gatwick are 80km apart, while Stansted is 85km from Heathrow by the shortest route. And the identity, and location, of 'Central London' shifts according to who is looking for it: arrive in City airport a banker due urgently at a meeting at Lloyds, and you are almost there; touch down at Heathrow a student looking to learn English, and it's a long trek up the Piccadilly Line to Leicester Square; a tourist shuttled in from Stansted, and you have to find a route out of Liverpool Street station before you can even begin to negotiate your way through the crowds and 'bright lights' of the West End.

Although all of these locations embody a different 'centre' of London, none of them declare themselves as such. And London has many other undeclared centres – maybe as many as it has residents. For every person living in the city, daily life can support the sense that the idea of 'London' can be defined by reference to more than one geographical sector. The patch of streets in which someone works every day, or the area in which they go out drinking are more central to what they recognise as London than the notional 'heart of the city' peddled by tourist brochures. And in addition to these everyday centres, each of us carries in our head a shifting picture of an ideal, assimilated London, an abstraction we sense only sporadically in normal life: through the rain-smeared window of a late-night bar, or driving over the Thames at dusk. But these personal versions of essential London are undiluted by the recognition that The Mall, Oxford Street and Soho are all parts of the 'centre', the area we collectively agree to call the most 'London-like' of our city.

The 100 million airborne arrivals who descend on London each year are equal to almost twice the population of Britain. Travel on this scale now makes it impossible to characterise cities as stable entities. They're no longer simply geographical locations but urban contexts adapting themselves to constant flux. As much as it is a collection of buildings, a city is a shifting set of conceptual possibilities, robust enough to expand and contract on demand without losing its essential identity. But predominantly, cities still seem to be pushing outwards. Even back in 1845, Friedrich Engels had trouble finding 'the beginning of the end' of London, feeling that it might be difficult for him to walk out of the city. Today, we know we haven't a hope of doing so. When one of London's airports is in fact in Cambridgeshire, with kilometres of rolling countryside in between, the city has become more a territory for the imagination than one with a measurable physicality.

Editor's suggestions for further reading

Allen, J., Massey, D. and Cochrane, A. (1998), *Rethinking the Region*, London: Routledge.
Sinclair, I. (2002), *London Orbital: A Walk Around the M25*, London: Granta.

TWO

'Do Telecommunications Make Transportation Obsolete?'

Pnina Ohana Plaut

Editor's Introduction

Our second reading on cybercity mobilities considers an age-old question: does the growing use of electronic interactions help to reduce, or even make obsolete, propensities for physical movement and the transport flows that derive from them? Examining the evidence to support such a substitution hypothesis, the Israeli transport researcher Pnina Ohana Plaut deploys an economic perspective. She concludes that, whilst some substitution does occur, this happens within a situation of rising overall mobility flows (both physical and electronic).

Ohana Plaut identifies, ultimately, a complementary relationship between transport and telecommunications. This means that ICTs are just as likely to enhance the capabilities of transport systems as they are to replace physical trips. ICTs, she notes, are being widely used to improve the efficiency and attractiveness of physical movement, as well as for reducing trips. And the fact that transport and ICT interactions tend to be shaped together, to recursively interact, means that each is closely dependent on the other.

Mobilities of people and physical goods tend to be closely linked with electronic and digital flows, within complex, hybrid formations. In order to understand mobilities within and between cities, therefore, Plaut concludes that the exploitation of the capabilities of ICTs within industry and logistics, in particular, is leading to a reconfiguration of cities as industrial locations. To her, cybercities are systems which support the simultaneous growth of physical and electronic mobilities.

Transportation and communication are rarely used or provided for their own sakes. Rather, they are generally regarded as 'derived demand', that is, as means for achieving other objectives of everyday life, such as getting to and from work, shopping, delivering products to markets, maintaining social and business links and relations, etc. Transportation and communication act as sorts of 'inputs' into these other economic and social processes, which cannot take place in their absence.

Transportation and communication systems perform functions that are partially overlapping. Both serve to bridge the gap, as it were, between separated physical locations, although they do so through very different technologies. The transport system enables the physical movement of people and goods, while communication systems engender information flows exclusively. The areas in which the two industries 'cohabit', however, are growing, as the information-based and other service industries grow. Many of the service industries that have been growing so rapidly may themselves be regarded, to a large extent, as forms of information production (e.g. medicine, law, financial services, etc.).

Transportation costs play a dominant role in all theories of location and land use, as well as in city planning and urban structure. Locational advantage (or disadvantage) is based on the costs (including time costs) of gaining access to various destinations and services. Over the past two decades, the costs of communication have plummeted relative to transportation costs, reflecting changing technology and the changing prices of other commodities (such as energy). Moreover, new technologies have made it possible to exchange or transmit materials such as written pages, data, drawings, etc. through communications lines, whereas only a few years back these would all be necessarily transferred through physical transport. Thus, for those markets in which communication and transportation coexist or compete, the 'comparative advantage' of communication has been rising.

There is a widely accepted presumption that the spread of advanced information technologies means that location no longer matters, and that the uses of transportation will decline as these technologies become more widespread. These expectations stem from the fact that the modern economy is more concerned with the flows of information, rather than the flows of things or people.

These conjectures are equivalent to presuming that modern technologies, such as telematics and information transfer, may serve as alternatives to travel. If true, they can act to redress and resolve problems of the transportation system, including congestion, energy consumption, air pollution and safety costs. In addition, any reduced utilization of the transportation system would imply significant changes in land use patterns and constraints, for example, by freeing up land uses for parking, roads and office space for other uses. If these conjectures are correct, they carry important implications for sustainable transportation and development and for urban structure and land uses. There are also equity implications, as low-income people can presumably afford a telephone, cell phone, or computer more easily than an automobile.

THEORIES OF THE RELATIONSHIPS BETWEEN TRANSPORTATION AND COMMUNICATIONS

Two basic hypotheses have shaped the thinking about the relationships between transportation and communication: substitution and complementarity. Substitution, the more common hypothesis, is usually defined as the elimination of travel altogether, as when a physical trip or shipment is entirely replaced by a transmitted message. This hypothesis predicts that, as communication technology becomes more advanced and cheaper, communication will replace some travel. Underlying the substitution hypothesis is

an assumption that the total volume of interactions, whether by travel or communications, is roughly constant.

Although the substitution hypothesis has a wide following among both academics and the general public, its scientific basis is not clear. Casual observation suggests that persons who work at home can effectively substitute electronic communication for commuting trips. Similarly, a telephoned inquiry to determine whether a needed commodity is available can save a fruitless shopping trip that finds it out of stock. And, surely, in the absence of an effective telephone system, a lot of cars would be circulating, each carrying a single piece of paper for delivery to an office or store. Sadly, this may be what lies behind much of the traffic congestion in underdeveloped countries. In any case, in all these examples, substitution between communications and transportation does occur. The common speculation among transportation professionals in recent times predicts substantial displacement of trips by telephone and Internet contacts.

Anecdotal evidence also suggests that the total volume of interaction is increasing, in part as a reflection of the increased ease of both movement and discourse. A phone call may stimulate an agreement to meet for lunch, adding a new trip. Or it may produce an agreement to meet to draft a contract, which may necessitate additional meetings for further negotiation. Following the contract, physical products may be shipped in response to electronically transmitted purchase orders. In these circumstances, more communication stimulates more transportation of people and goods, so that the transport flows stimulated lead to more use of communications, and so on. Certainly at the macro level, the volumes of both transportation and communication services have been growing rapidly. Hence, the alternative to the substitution hypothesis is that the two sets of services may be mutually stimulating, and so may exhibit complementarity.

Complementarity encompasses two distinct phenomena: *enhancement*, where the use of telecommunications causes additional travel that would not have occurred in the absence of the augmented use of communications; and *efficiency*, in which one of the two services contributes to reduce the resource requirement, and hence the costs, of the other. For example, it may be possible to increase the efficiency of transportation systems through more intensive use of telecommunications.

INDUSTRIAL USES OF TRANSPORTATION AND COMMUNICATION

To date, most of the attempts, both theoretical and empirical, to assess the relationships between transportation and communications have focused on individual commuters or households. Concentrating on a single communication mode (for example, tele-working, teleconferencing, teleshopping, telecommuting, domestic telephone habits and so on), they have attempted to discover how it affects travel behavior and commuting. While these studies have derived important results, their conclusions are limited because the analysis was confined to the household sector. In reality, the bulk of the use of both transportation and communications services is not by households but rather by industry. For example, in the European Union, about two-thirds of all transportation and communications services are used by industry. The proportion of transportation and communications services consumed by industry in the United States is of roughly the same order of magnitude.

Since the consumption of transportation and communication services by households, including commuters, represents less than half of the entire output of the transportation and communications service sectors, a comprehensive understanding of the relationships between transportation and communications must include analysis of their industrial uses. These industrial uses have been largely overlooked in the analytic literature. A second limitation of previous research has been its focus primarily on single countries, with little, if any, cross-national comparison.

In my research (Plaut, 1997), these issues regarding the relationships between transportation and communications uses by industry were addressed for a number of regions, including the European Union, the United States, Canada and Israel. Among the important questions raised was whether or not the common presumption of substitutability between the two services could be detected for industrial uses. Understanding of these roles can shed light on the process and consequences of economic unification and integration in Europe and elsewhere.

The industrial uses of transportation and communications services can be examined through the analysis of economic input–output data. These are numbers that measure the flow of products and services from one industry to another, throughout the national economy. An input–output 'observation' is the level of inter-industry use of some product or service, such as the use of transportation services by the farming or the steel sectors. Analyzing input–output flows can reveal whether transportation and communications tend to be used together or as substitutes for one another by other industries throughout the economy. Since different industries are of different sizes with different levels of output, the best way to analyze the question is by analyzing $1 representative units of output for each sector. In other words, while comparing dollar outputs across different industries or countries, the basic questions are: do industries tend to use transportation and communications together as complements or as substitutes for one another, where using more of one is associated with less of the other? Are there differences in these patterns across countries?

The relationships, of course, vary from one industrial sector to another. Overall, however, national patterns can be evaluated by treating dollar outputs from all industrial users of transportation and communications as comprising a 'sample' and then applying specially adapted sample statistical analytic tools. Accordingly, one can ask not only whether the two services tend to be used by industry as substitutes or together as complements, but how *significant* are these differences, in the statistical sense of 'significance'. Can we reject the hypothesis that a complementarity or substitution pattern was produced by chance? How sure can we be that the evident pattern 'really exists"?

COMPLEMENTARITIES PREVAIL

In the above cited work, the evidence is overwhelmingly in favour of the complementarity relationship between transportation and communications, and *against* the substitution hypothesis. In every country or region examined the pattern was the same: industrial users tend to use transportation and communications *together* and not as substitutes for one another. Moreover, the pattern carries over even when one disaggregates the transportation sector into smaller transportation subsectors, for example, by treating road transport, air transportation, marine transportation, and other transportation separately. For almost every transportation subsector, the pattern of complementarity with communications emerges with 'statistical significance', meaning that communications appears to be a complement for all forms of transportation, not just for 'transportation' in aggregate.

Complementarity prevails in nearly all cases for both direct purchases of transportation and communications and 'total' purchases. The latter takes into account the indirect use of transportation and communications through other intermediate purchases. In other words, if the farm sector purchases transportation directly, but also indirectly through the machinery, energy and equipment used on the farm whose production *also* used transportation in their own production process, then all these 'remote' indirect uses of transportation enter the 'total use' measurement.

While it appears that communications do not serve as substitutes for transportation, the next question is: why not? There are a number of possible explanations, and probably additional explanations will be developed by researchers in the future. Since the mid-1980s, there has begun to develop a literature on the possible 'travel-enhancement' effects of communications in some capacities. Some have argued that telecommunications are as likely to induce additional travel as to reduce travel, where the net impact is unclear. These claims rely upon trip generation models and the functional approach to explain trip-making behavior. Underlying this approach is the question of how travel time saved due to telecommunications is then to be used.

For industry, especially for the freight sector, transportation-communications complementarity may derive from the more intensive use of communications as logistics support for transportation. In the freight transport system the decision to travel is determined by the need to move a physical item from one location to another.

A possible explanation for complementarity in industrial uses of transportation and communications services may be due to enhancement and efficiency effects. These may capture the role of communications in improved logistics support for transportation and freight movement. Expressions such as 'just in time' and 'total quality control,' or references to the five 'Olympic zeroes' – no stock, no time lag, no fault, no breakdown, no paper – point to a new form of logistics based on tighter flow management and the desire to enhance productivity and meet the challenges of the competitive economy. For example, there is evidence that the introduction of new information management technology, including data sharing and electronic data interchange networks, produced a sharp upturn in railroad productivity in the United States in the early 1990s.

Transportation costs may be reduced with improved telecommunications by eliminating delays or improving the level of service through the timely coordination of pickups. Improvements in load factors, routing and back-haul planning raise the efficiency of transport systems. Warehousing operations and utilization are also influenced by telecommunications technologies, shifting toward a smaller number of larger depots serving larger areas. Real-time scheduling of shipping containers from ship, to truck, and rail for just-in-time delivery at destinations is yet another example. All these are examples of how communications make the transport system more efficient, flexible and responsive.

Enhanced efficiency may create greater demand for transportation, for example, in cases where a smaller fleet of electronically scheduled vehicles can make frequent deliveries to meet no-stock and just-in-time production. Efficient scheduling of freight movement will rely increasingly on the precision permitted by real-time communication.

The success of Federal Express and other courier services is partly based on their capacity to keep tabs on the location and status of every parcel at every moment – to use the electronic media for tracking physically transported objects.

In sum, news of the demise of location and transportation appears to be premature. Those who believe that advanced telecommunications and information technologies have made geography and distance irrelevant appear to be mistaken. If anything, more communications appear to be producing an expanded use of the transportation system, and vice versa.

REFERENCE FROM THE READING

Plaut, P. O. (1997), 'Transportation-communications relationships in industry,' *Transportation Research A* 31(6), 419–29.

Editor's references and suggestions for further reading

Couclelis, H. (2000), 'From sustainable transport to sustainable accessibility: can we avoid a new *tragedy of the commons*?'. In D. Janelle and D. Hodge (eds), *Information, Place and Cyberspace*, Berlin: Springer, 341–57.

Forer, P. and Huisman, O. (2000), 'Space, time and sequencing: substitution at the virtual/physical interface'. In D. Janelle and D. Hodge (eds), *Information, Place and Cyberspace*, Berlin: Springer, 73–90.

Giannopoulos, G. and Gillespie, A. (eds) (1993), *Transport and Communications in the New Europe*, London: Belhaven.

Hepworth, M. and Ducatel, K. (1992), *Transport in the Information Age: Wheels And Wires*, London: Belhaven.

Janelle, D. and Hodge, D. (2000), *Information, Place and Cyberspace: Issues in Accessibility*, Heidelberg: Springer.

Mokhtarian, P. L. (1990) 'A typology of relationships between telecommunications and transportation', *Transportation Research*, 24A (3), 231–42.

Nilles, J. (1988), 'Traffic reduction by telecommuting: A status review and selected bibliography', *Transportation Review* 22A, 301–17.

Salomon, I. (1985), 'Telecommunications and travel: Substitution or modified mobility?', *Journal of Transport Economics and Policy*, September, 219–35.

Salomon, I. (1986), 'Telecommunications and travel relationships: a review,' *Transportation Research*, 20A(3), 223–38.

'The City and the Cybercar'

Mimi Sheller and John Urry

Editor's Introduction

Our fourth reading on cybercity mobilities develops a sophisticated sociological perspective of how the growing interpenetration of cars and ICT systems helps to bring with it a reconstitution of time, space and sociality within and between cybercities. In this specially commissioned piece, Lancaster University sociologists Mimi Sheller and John Urry draw on their work on the sociologies of mobility to reflect on what happens when cars, ICT systems and cities are shaped in parallel.

Sheller and Urry reflect on the paradoxical freedoms and coercions that emerge when cars and automobiles become dominant transport systems in cities. They consider what automobilism does to those who are excluded from it. They reflect on its implications for other forms of transport (walking and public transport). And they analyse the complex socio-technical constructions of road and automobility systems as they shape the bodies and subjectivities of car dwellers in the city.

Sheller and Urry use this background to analyse the implications of the wiring of cars into complex computer and ICT systems: what they call the city of cybermobility. Arguing that the increasingly intense articulation between cars and ICTs represents the reconnection of physical transportation and information transmission, Sheller and Urry suggest that current efforts to equip cars with ICT systems are of profound significance.

Rather than replacing car use, Sheller and Urry thus argue that the infusion of ICTs into cars and car use represents a convergence between physical mobility and cybermobility. If sensitively handled, the authors are hopeful that the design of the recombined spaces of the transport-ICT hybrid may offer potential for developing cybercities where the social and environmental costs of automobility are ameliorated; where the public face of city spaces is strengthened; and where the balance between private and public transportation in cities is creatively blurred.

Achieving these things, however, will require major shifts in the paradigms that shape transport and urban planning. Moreover, it will be essential to overcome many of the vested interests that benefit from the status quo situation: the dominance of dumb, internal combustion-driven automobiles in cities.

INTRODUCTION: TIME, SPACE IN THE CITY

Automobility is a complex amalgam of interlocking machines, social practices and ways of dwelling – not in a stationary home, but in a mobile, semiprivatised and hugely dangerous iron cage. It permits multiple socialities, of family life, community, leisure, the pleasures of movement and so on, which are interwoven through complex jugglings of time and space that car journeys both facilitate but also necessitate. These jugglings result from two interdependent features of automobility: that the car is immensely flexible *and* wholly coercive.

Automobility is a source of freedom, the freedom of the road. Its flexibility enables the car-driver to travel at speed, at any time in any direction along the complex road systems of western societies that link together most houses, workplaces and leisure sites. Cars extend where people can go to and hence what as humans they are able to do. Much social life could not be undertaken without the flexibilities of the car and its availability 24 hours a day. It is possible to leave late by car, to miss connections, to travel in a relatively timeless fashion. People find pleasure in travelling when they want to, along routes that they choose, finding new places unexpectedly, stopping for relatively open-ended periods of time, and moving on when they desire. Cars are one of the convenience devices of contemporary society, devices that make complex, stretched patterns of social life, especially in the city, just about possible – at least for those who own them.

But at the same time this flexibility and these rights are themselves necessitated by automobility. The structure of auto space (Freund, 1993) forces people to orchestrate in complex and heterogeneous ways their mobilities and socialities across very significant distances. The urban environment built for the convenience of the car has unbundled those territorialities of home, work, business, and leisure that were historically closely integrated. Automobility has fragmented social practices that occurred in shared public spaces within each city. In particular, automobility divides workplaces from homes, so producing lengthy commutes into and across the city. It splits homes and business districts, undermining local retail outlets to which one might have walked or cycled, thereby eroding town-centres, non-car pathways, and public spaces. It also separates homes and various kinds of leisure sites often only available by car. People are trapped in congestion, jams, temporal uncertainties and health-threatening city environments. At the same time automobility disables those who are not car-drivers (older people and children, the sight impaired, those without cars) by making their habitats dangerously non-navigable.

Automobility thus coerces people into an intense flexibility. It involves an individualistic timetabling, forcing people to juggle tiny fragments of time so as to deal with the temporal and spatial constraints that it itself generates. The car-driver operates in instantaneous time that contrasts with the official timetabling of mobility that accompanied the growth of the railways . This was modernist clock-time based upon the public timetable. But, as a car-driver wrote in 1902, 'travelling means utmost free activity, the train however condemns you to passivity . . . the railway

squeezes you into a timetable' (cited in Morse, 1998, 117). Objective clock-time is replaced by personalised, subjective temporalities, as people live their lives in and through their car(s). Automobility develops instantaneous time to be juggled and managed in complex and heterogeneous ways.

As personal times are desynchronised, spatial movements are synchronised to the rhythm of the road. The loose interactions and mobilities of pedestrians are forced to give way to the tightly controlled mobility of machines. Automobility dominates how both car-users and non-car-users organise their lives through time–space. Car-travel rudely interrupts the taskscapes of others (pedestrians, children going to school, postmen, garbage collectors, farmers, animals and so on), whose daily routines are merely obstacles to the high-speed traffic that cuts mercilessly through slower-moving pathways and dwellings. Junctions, roundabouts, and ramps present moments of carefully scripted inter-car-action during which non-car users of the road present hazards or obstacles to the drivers intent on returning to their normal cruising speed.

Automobility also generates new scapes that structure the flows of people and goods along particular routes, especially motorways or interstate highways (see Urry, 2000, on scapes and flows). There is a rewarping of time and space by advanced transportation structures, as scapes pass by some towns and cities while connecting other areas along transport-rich tunnels. Such tunnels also shape urban geographies of social exclusion and ghettoisation. Public housing estates in the UK, or so-called 'projects' in US cities, are often cut off both from bus, rail or subway links to employment-rich business districts within the city and from automobile roadways linked to more desirable (middle-class) residential and leisure areas outside the city. At the same time tolls and parking fees can deter rural and suburban dwellers from entering the city too easily. Thus the inequalities among multiple publics are entrenched in urban spaces of unevenly distributed access and exclusion, what has been termed a splintering urbanism (Graham and Marvin, 2001).

DWELLING IN MOBILITY

The shortage of time resulting from the extensive distances that increasingly have to be travelled means that the car is the only viable means of highly flexibilised mobility. Also other forms of mobility in the city are relatively inflexible and inconvenient, judged by criteria that automobility itself generates and generalises. In particular, the car enables *seamless* journeys: home–away–home. It does away with the stationary pauses necessitated by stations, apart from the occasional stop at the gas station. And this is what the contemporary traveller has come to expect.

The system of automobility thus undermines other forms of mobility. The predominance of the car in government policy and planning afforded seamless car journeys while breaking down those linkages that once made other forms of transport possible. Thus, the seamlessness of the car journey makes other modes of travel inflexible and fragmented. So-called public transport rarely provides that kind of seamlessness (except for first-class air travellers with a limousine service to and from the airport). More directly, car manufacturers in the USA, such as General Motors, bought and dismantled electric tramway systems in order to make suburbs car-dependent. Auto-intensive middle-class suburbanisation resulted in the auto sprawl syndrome in which cars make urban suburbanisation/sprawl possible, and in so doing they make those living in such areas dependent upon the use of cars. Freund argues that 'Modernist urban landscapes were built to facilitate automobility and to discourage other forms of human movement . . . [Movement between] private worlds is through dead public spaces by car' (1993, 119). Indeed, large areas of the globe now consist of car-only environments – the quintessential non-places of super-modernity (Augé, 1995). About one-quarter of the land in London and nearly one-half of that in LA is devoted to car-only environments. And they then exert an awesome spatial and temporal dominance over surrounding environments, transforming what can be seen, heard, smelt and even tasted.

Thus car-environments or non-places are neither urban nor rural, local nor cosmopolitan. They are sites of pure mobility within which car-drivers are insulated as they dwell-within-the-car. Such non-places represent the victory of liquidity and placelessness over urbanity.

Protected by seatbelts, airbags, crumple zones, roll bars and bull bars, car-dwellers boost their own safety and leave others in the city to fend for themselves in a nasty, brutish and short world of such mobility. In each car the driver is strapped into a comfortable armchair and surrounded by microelectronic informational sources, controls and sources

of pleasure, what Raymond Williams calls mobile privatisation. Car-drivers are located within a place of dwelling that insulates them from the risky and dangerous urban environment that they pass through. The sights, sounds, tastes, temperatures and smells of the city are reduced to the two-dimensional view through the car windscreen, with the environment beyond an alien other, to be kept at bay through the diverse privatising technologies incorporated within the contemporary car. These technologies ensure a consistent temperature (with the standardisation of air-conditioning), large supplies of information, a relatively protected environment, high-quality sounds and sophisticated systems of monitoring and navigation.

The driver's body is itself fragmented and disciplined to the machine, with eyes, ears, hands, and feet, all trained to respond instantaneously, while the desire to stretch, to change position, or to look around must be suppressed. The car becomes an extension of the driver's body, creating new urban subjectivities. A Californian city planner declared even in 1930 that it might be said that 'Southern Californians have added wheels to their anatomy' (cited Flink, 1988, 143). Features such as automatic gearboxes, cruise control, GPS navigation, and CD-changers free drivers from direct manipulation of the machinery, while embedding them more deeply in its sociality.

Given these powerful restructurings of time, space, and self within the car-driver matrix, the potential impacts of new information and communication technologies upon urban life have to be rethought in terms that go beyond current debates about the potential replacement of physical travel by virtual travel. Here we consider the potential convergence of these two mobilities within cities of cybermobility.

THE CITY OF CYBERMOBILITY: A VISION OF THE FUTURE

For too long transport has been theorised and planned as if it were a free-standing system disconnected from other technologies and socialities. True, for most of the twentieth century the revolution in communication technology was driven by the separation of information transmission from the physical means of transportation.

However, the current trend involves re-embedding information and communication technologies (ICT) into moving objects: through satellites, mobile phones, palmtop computers and the general notion of mobile computing. At the same time that information has been digitised and released from location, so cars, roads, and buildings have been rewired to send and receive digital information. Mostly the effects here have been thought of interest to urban planners and car manufacturers – a question of traffic control or car and road safety, for example through computer-assisted operation control systems, dynamic route guidance, and traffic information systems.

But if we consider civil societies as involving modes of mobility as well as communication, then there may be crucial processes of contention and democratisation at work in the reflexive social adaptation of these new technologies, which will contribute to the further (post)modernisation of cities. Emerging technologies are not only creating new human-machine hybrids such as the car-driver, but are also grafting together existing machines to create new intermodal scapes and hybrid flows: PCs with telephones; televisions with satellites; mobile phones with cars. Highly complex information, communication, and simulation systems (designed originally for military vehicles) are already making the jump to civilian applications.

Given the environmental pressures upon the system of automobility, including political contention around the control of pollution, traffic reduction, management of risk, and addressing social exclusion, then the politics of urban sustainability will play a major part in designing new built environments. Thus far environmentally friendly policy initiatives have focused on reducing car use, stemming the flow of traffic, and shifting people onto public transport and non-motorised mobility through penalties for car-use and incentives for biking, walking, or riding the bus, tram, or train. These will still be crucial tools in transforming the existing car-driver matrix.

However, such policies do not address the continuing production of a system of automobility of cars, road, and drivers which dominates both the built environment of cities and their hinterlands and the scapes of time–space through which people organise and literally mobilise their lives. Until these fundamental social aspects of mobility are addressed, we continue in a deadlock between the profit-making imperatives of car manufacturers and the decaying semi-public infrastructure of transportation starved of investment and of the necessary time-horizons and physical space in which to expand. More encompassing approaches to changing cultures of

automobility and of cities include the creation of auto-free zones in city centres, fostering access by proximity through denser living patterns and integrated land use patterns, and promoting greater coordination between both motorised and unmotorised transport systems.

Some futurologists have proclaimed the 'electronic cottage' to be the way out of the traffic, as well as out of the city and its urban problems (Toffler, 1980). But this would be disastrous for the public life of cities, and for those marooned in them. Rather than the replacement of physical travel by virtual or weightless travel (see Urry, 2000), what may occur is a convergence between them. Through an interlocking of smart transportation systems and the urban info-structure a new mode of cybermobilisation could be created that would integrate private and public transport, motorised and non-motorised transport, and information transmission and human mobility. Crucial to this de-traditionalisation of urban transportation will be a re-design of both public mass (and mini) transportation systems and of private or semiprivate vehicles. Telecommuting will not be *the* key to transforming urban life because people do *like* to be physically mobile, to see the world, to meet others and to be bodily proximate, and to engage in locomotion (Boden and Molotch, p. 101). Current developments such as mobile telephones and, especially, short text messaging suggest that many people will engage in communication simultaneously with locomotion (Zac Carey, p. 133). Mobile ICT is also increasingly central to work-practices and information gathering in contexts of unavoidable time–space distanciation and fragmentation. The introduction of flexitime in order to smooth out and redistribute rush-hour peaks of transportation demand, for example, would be enhanced if communication, banking, or shopping occurred in transit on the move.

Thus the question is how these technological developments can be used to redesign urban public spaces so as to address the negative constraints, risks, and impacts of automobility. This will require the intermodality of multiple transport and communication flows to allow various mobile publics to switch across a range of spaces and zones easily. Creating smaller, lighter, cars not fuelled by petrol is a major start. However, such micro-cars which break with steel-and-petroleum technology would have to be integrated into a mixed system allowing more room for bikes, trams, hybrid vehicles (such as the ULTra automated taxis), pedestrians, public transportation and modes yet to be imagined. This would require the redeployment of existing urban zoning laws to exclude or severely delimit steel-and-petroleum cars (as has already begun to occur in various cities such as London) and to place lower speed limits on them (as in many Swedish urban areas). The aim would be not only to free up space for new kinds of intermodal flows but also *time* for new socialities that would juggle the complex timing of schedules in more varied ways.

There would also have to be incentives to both car manufacturers and consumers to produce a new culture of automobility (through extensions of already existing legislation, taxation, and penalties). Various objectives could be met: reduced energy consumption and polluting emissions, use of fewer private vehicles, and curtailment of traffic proliferation and road building; redirection of investment to new and better modes of public transportation, bike and soft vehicle lanes, and more diversified multifunctional stations and public spaces; reduced risks to human safety inside and outside the slower and lighter car; and minimising social exclusion through better planning of networks and intermodality.

But we think the key to such a system would be the use of a multifunction smart-card that would transfer information from home, to car, to bus, to train, to workplace, to web site, to shop-till, to bank (a system already under development). Cars for cities could then be partially deprivatised by making them available for public hire through the use of a smart-card which would also be used to pay fares on buses, trains, or more flexibly routed collective mini-vans. All vehicles would be *more than* technologies of movement; they would be hybridised with the rapidly converging technologies of the mobile telephone, the personal entertainment system, and the computer.

The incentive for car manufacturers is that small cars would no longer be at the bottom of the profit scale; the innovation of new ICT applications would provide an endless source of novelty, desirability and profitability. The 'hook' for car-drivers is that the micro-cars and all other forms of transport would be personalised with their own communication links (email addresses, phone numbers, world wide web addresses, etc.) and entertainment applications (digitally stored music, programmed radio stations, etc.), but only when initiated by inserting the smart-card. Thus, any public vehicle could instantly become a home away from home: a link to the reflexive narratives of the private self in motion through public time–space scapes. The streetscapes of global cities could thus be transformed through a more mixed flow of slow-moving semi-public micro-cars, bike lanes, pedestrians and improved mass transport.

Public-friendly cars would allow people to travel lighter, if not weightlessly and would restore some civility to urban public space destroyed by current traffic flows and by the spatial patterns of segregation and fragmentation generated by automobility. The urban smart-car could just be the best way to lure twentieth-century speed-obsessed car-drivers to give up their dependence on dinosaur cars and fossil fuels, a system unsustainable on every measure. Urban planning that recognises the need for a radical transformation of transport can use existing legislation and regulation in new ways, to build integrated and intermodal public transport systems. We have suggested that a truly intermodal transport system will depend on the integration not only of private and public transport, but also of intelligent transport systems and the emerging ICTs that enable new ways of dwelling while on the move.

REFERENCES FROM THE READING

Augé, M. (1995), *Non-Places*, London: Verso.

Flink, J. (1988), *The Automobile Age*, Cambridge, MA: MIT Press.

Freund, P. (1993), *The Ecology of the Automobile*, Montreal and New York: Black Rose Books.

Graham, S. and Marvin, S. (2001), *Splintering Urbanism*, London: Routledge.

Morse, M. (1998), *Virtualities: Television, Media Art and Cyberculture*, Indiana: Indiana University Press.

Toffler, A. (1980), *The Third Wave*, New York: Morrow.

Urry, J. (2000), *Sociology Beyond Societies*, London: Routledge.

Editor's references and suggestions for further reading

Bell, J. (ed.) (2001), *Carchitecture: When the Car and the City Collide*, Basel: Birkhäuser.

Sheller, M. and Urry, J. (2000), 'The city and the car', *International Journal of Urban and Regional Research*, 24(4), 737–57.

Swyngedouw, E. (1992), 'Communication, mobility and the struggle for power over space'. In G. Giannopoulos and A. Gillespie (eds), *Transport and Communications in the New Europe*, London: Belhaven, 305–25.

Wollen, P. and Kerr, J. (eds) (2002), *Autopia: Cars and Culture*, London: Reaktion Books.

'Cybercommuting on an Information Superhighway: The Case of Melbourne's CityLink'

David Holmes

Editor's Introduction

ICTs are often discussed using the metaphor 'information superhighway'. This depiction is a way of making sense of their complex effects by drawing an analogy with a massive, visible, and physical system of transportation. Ironically, however, attention rarely focuses on the opportunities that ICTs offer to reshape real, motorised, 'superhighways'. In a context where many cities face gridlocked highways, a public sector starved of resources to invest in new or improved road spaces, and a growing environmental opposition to new road building, ICTs allow major private companies and public agencies to offer new, automated, ways of organising the access to road space in cities through the price mechanism. Since the mid-1990s cities as diverse as Toronto, Los Angeles, San Diego, Singapore, London and Manila have all involved private companies in various ways in the electronic charging of either city centre road space or highway road space (Graham and Marvin, 2001, Chapter 6).

None, however, have gone so far as Melbourne, Australia – a story examined by the Australian sociologist David Holmes in this specially commissioned piece. Holmes argues that Melbourne's CityLink project – the largest electronically tolled highway network in the world – is leading to a dramatic reconfiguration of the mobility and urban spaces of the city. The new and reconfigured 'premium' highway spaces generated by the project have supported a rapid acceleration of car traffic flow for those who can afford access to the charged network. It has generated a further round of spatial sprawl in Melbourne. And it has led dominant flows of car traffic to bypass many traditional shopping streets in the city whilst bringing large malls and corporate sites and airport spaces closer together in travel time terms.

The result, Holmes argues, is a profoundly ambivalent change in the mental geographies and social experiences of Melbourne, as notions of mobility, flow and movement become reconfigured and starkly commodified. Drawing parallels with the wider experience of cyberspace, Holmes argues that, whilst users of the CityLink experience greater logistical freedom, this is a programmed freedom which accentuates some possibilities whilst diminishing others. It is also a closely surveilled freedom which, in the longer term, is likely to trigger the greater and greater spatial separation between land uses within a city which, as a result, will sprawl even further. Thus the improved ICT-supported physical mobility brought by CityLink may, for those who can afford it, be ultimately self-defeating, as the time spent to articulate with the functions of the city will grow further as a result of a new round of urban sprawl.

Holmes's discussion of the CityLink project demonstrates clearly that the use of ICTs to prioritise the mobility of certain city residents on city-wide highways comes at a cost. The commodification process, as 'dead', free highway space becomes 'smart', charged highway space, benefits the rich at the expense of the poor. This is because the standard costs of using CityLink are relatively much smaller to upper income than to lower income groups. The use of ICTs in this way tends to support major reconfigurations in city form which may make more sustainable and socially equitable urban transport systems even less possible in the future (as cities spread and become less dense and more car oriented). Finally, such initiatives makes mobility more reliant on complex ICT systems which are, as he demonstrates, inherently prone to the crashing and failure analysed by Timothy Luke, p. 106.

> The car has come to show the world that the heart is just a poetic relic, that a human being contains two standard gauges, one indicates miles, the other minutes.
> (Illy Ehrenberg (1928) *The Life of the Automobile*)

Melbourne's massive electronic toll road project, CityLink, which fully opened in the first months of the millennium, has changed both the traffic dynamics of Melbourne's sprawling suburbs and the way many Melburnians feel about their city.

As the then largest private construction project in the world, Melbourne's CityLink project was envisaged to provide motorists with a new mobility which eases the conventional gridlock–freeway impasse. Based on an electronic toll system it also represents an interesting case study of how by-pass freeway construction design combined with information technology can facilitate improved urban transit systems. At the same time, where such a development radically replaces traditional transportation infrastructure it can, as in the Melbourne experience, have an adverse impact on the aesthetic, environmental and community welfare of large cities.

Figure 2 Map of the Melbourne CityLink network on completion

Source: CityLink Melbourne Limited

THE CITYLINK

CityLink, built in a joint venture by Transfield-Obayashi, is 22 km of privately controlled link road which joined together three of Melbourne's freeways between Melbourne Airport in the north west, major industrial centres in the south east and the port area (see Figure 2). Servicing a population of 3.1 million and a potential registration of over one million motorists, CityLink features two extensive tunnel projects which pass under the Yarra River, as well as a bridge over the river and two elevated roadways linking two freeways, and an official gateway to the city centre.

With the opening of the project, the social and technical acceptance of electronic tolling by Melbourne motorists has been tested with contradictory results.

The limits of the commodification of car travel have been trialed in the context of a transport infrastructure, which, in times past, has been considered basic facilities provided for in government spending. Whilst there have been all kinds of resistances to using CityLink, the absence of alternatives for those dependent on cross-city automobility has effectively normalised and legitimated new tolerances for what the broad base of the population may not be prepared to see being commodified.

There is, for many, a necessary ambivalence that is thrown up by CityLink which goes right to the core of what it means to live in a large metropolis. Whilst there may be substantial opposition to the concept of being sold key gateways in and out of the city, at the same time, such connections, when they are defined by their efficiency, can be seen to overcome the familiar distance and alienation which less planned suburbanisation brings. This latter view, which relies on appreciating the short-term immediacy of CityLink, is one which the marketing of the project relied upon. Transurban lists as the positive features of the project that it is faster, cheaper, safer and greener, all of which are based around the advantage of the electronic management of tolls and vehicles not having to stop to pay them. Average speeds across all Melbourne roads are estimated to be 22 kilometres per hour, a state of congestion which the advertising claims 'costs Victorians $5 million (Australian) per day' as well as precious time. It is claimed that the expressways of the project will help to ease this situation.

Second, because CityLink overcomes stop-start kinds of driving, it promises to reduce fuel consumption by 30 per cent for the same distance of road travelled and save Victorian businesses over 20,000 tonnes of fuel each year. Third, because of proposed smoother traffic flow, there is allegedly less risk entailed in being on a CityLink road. And finally, there is also less pollution because vehicles will not be slowing to inefficient speeds, where the distance-to-emission ratio becomes higher.

Of course, short of pointing out that several of these arguments by Transurban are implicitly anti-motor vehicle, it is certainly true that information technology can help mitigate the negative aspects of modern motoring. Stephen Graham argues that:

> more effective methods of managing travel networks can help increase the efficiency of transport – road, rail, air travel – so lowering costs and increasing the attractiveness of travel as an option. It has become increasingly clear that the new technologies of road transport informatics provide ways of overcoming the problems of congested road networks and increasing the effective capacity of these networks at a fraction of the cost of constructing entirely new transport infrastructure (Graham, 1996).

The use of informatics as a solution to the technical problems of road-based transport can only be encouraged insofar as it is roads that provide the dominant mode of commuting in any given city. What is difficult to sustain however, is Transurban's assertion that such a technical solution is 'bringing our [Melbourne's] community' together. Whilst CityLink may help overcome the physical fragmentation of urban dispersal, what is

forgotten, in the midst of the high speed possibilities which CityLink brings, is that it creates the conditions for even greater cycles of dispersal, requiring continuous upgrading of highway systems which can somehow overcome the suburban sprawl that it itself is helping to create.

Where CityLink differs from tollway freeways and bridges which have existed for many years, is the fact that the new toll system is so extensively integrated with the urban and suburban arterial network (see Figure 2). In this respect CityLink is unique. It is the first electronically regulated *radial* link toll in the world, one which will set an interesting precedent for how information technology is able to regulate as well as re-constitute human geography.

In doing so CityLink may well prove to be decisive in short-circuiting the role that urban and transport planners could have in preserving less car-centred forms of urban development. CityLink is a capital-intensive project which, first and foremost, promotes a long-term dependence on the motor car as a privileged mode of transport. Developments like CityLink call attention to the need for much greater research into the cultural, geographic and economic conditions governing how more people could live closer to where they work and shop, thereby taking pressure off the gridlock–freeway approach to urban form. Similarly, questions surrounding the increasing exclusion of public transport from the highway could be examined.

COMPUTERISED CAR-SPOTTING

On the toll links an investment of 40 million dollars was made to ensure stringent compliance to tolls including the deployment of military technology. The software which is used in the information operating system is supplied by Vector International (a program that is also used by Shell Corporation at a gas plant in Nigeria and in a contactless smart-card fare collection system in Hong Kong). This not only records the toll charges but manages traffic flow as though it is one collective nervous system. The hardware for reading the electronic tags utilises military missile imaging systems for drawing a box around moving vehicles so that the differences between charging for cars, vans or trucks can be determined.

Traffic flow is monitored by CCTV. This approach to vehicle identification has in turn brought up further anxieties from the public about invasions of privacy and unnecessary control of detailed information by a private company. In March 1999, the morning presenter for Melbourne's Radio 3LO was able to speak directly with the manager of CityLink's Information Technology department, who proudly explained the ability of the database to track individual vehicle movements as well as make available e-tag credit for paying speeding and parking fines. The integrated nature of the database was hailed as furnishing information that could be sold to commercial interests. Perhaps confirming the extent of this integration was the case of the e-tag registrant who received a bank statement containing seventeen pages of accidental deductions authorised by Transurban. In February 2002, almost 8,000 credit card numbers were discovered to have been stolen from Transurban's database, raising further fears about the integrity and confidentiality of customer information.

SYSTEM CRASH

Consistent with the nature of highly programmed environments, everything about CityLink depends on continuous flow, from the registration system to the automated call centre which handles billing, to the cars on the expressway itself. Like programmed systems, which are designed for throughput, raising productivity and efficiency, when CityLink is working it really can deliver. But when it crashes its built-for-efficiency design realises havoc in place of planned order. Paradoxically, this introduces a random form of uncertainty for motorists which runs counter to planned time-saving attraction that CityLink has been sold with.

CityLink uses 80 CCTV cameras across its 22 kilometres to monitor and regulate all traffic flow. Before it opened, Vector International, the company that installs the surveillance system explained the precautions:

> In the event of a traffic incident (such as a breakdown or traffic snarl), the operator will be immediately notified and presented with a menu of possible response plans that the system can perform. Response plans can bring up messages on tollway signs, close or open lanes, call emergency services and actuate tunnel safety fire and ventilation systems. An integrated traffic forecasting module will assist the operator to compare actual traffic loads with expected peaks to assist with the early identification of potential problems. A feature of the system is the complete integration of video displays. Video signals selected from 80 CCTV cameras will be superimposed on graphical maps of roads and junctions. In many cases, an incident will automatically select the most appropriate CCTV camera for immediate operator attention.

However, on February 7, 2001, a series of accidents during morning peak hour created standstills throughout the day for 5 to 10 kilometres. A truck which overturned just after midnight on the Bolte Bridge, spilling oil on the dockland areas of the expressway was not cleared for 9 hours. Before this cleared another truck lost its load on the south-east link followed by a truck with a ruptured fuel tank in the Burnley tunnel. On this occasion there was no choice for motorists but to sit it out. With few exits, and no turning back the fully paid-for time saved on hundreds of days of guaranteed flow, was rapidly eaten up during a single system-crash.

A special act of parliament, the Melbourne CityLink Act, was also passed to provide CityLink with special powers. This gives Transurban extensive control over the enforcement of its tolls. These powers have proven to be quite sweeping as in the case of one non-English speaking motorist who used the road 223 times in 4 months of 2001 without an electronic tag and accumulated A$31,000 in fines. CityLink threatened that these would escalate to over A$110,000 if its rigid time-penalty system was not observed.

THE TOLL ON URBAN COMMUNITY

The effect of accenting private over public transport infrastructure is one which may act as a catalyst for new trends in urbanisation, in Melbourne and other Australian cities. In the case of CityLink, such infrastructure has been promoted on the basis of 'world first' innovation in the scale and sophistication of the engineering and electronic dimensions of the project. However, the technical problems encountered in the electronic tolling and construction may discourage such innovations elsewhere. But certainly the opportunity to promote toll roads, in urban centres characterised by sparse population density spread, is very attractive to engineering and development consortiums.

CityLink has also marketed itself as a redemptive agent for rebuilding community and as the traffic breakthrough Melbourne has needed. By the end of 1999, its operators promised that it will be:

bringing our community together, faster, safer, and with less wear and tear on your vehicle . . . With CityLink there's virtually no stopping. Using advanced engineering and electronic toll technology, it creates a continuous expressway across Melbourne. It means you can spend more time where it counts most, at home and with friends.

CityLink contributes towards a double-edged privatisation process, the increasingly solipsistic retreat into the private home, and the breaking up of the openness and fluidity of the city by highly controlled, high-speed traffic vectors which preclude spontaneous or unplanned forms of interaction. Instead, the one-way no-turning-back culture of freeways further elevates the motor car as a factor in all of our social interactions, because we are no longer predisposed to stopping and negotiating non-freeway kinds of spatial dimensions in our lives. The more this logic is accentuated, and the more commuters come to depend on the motor car, the less momentum there is to justify public transport development and forms of built environment that are premised on the more lateral, open and negotiable spaces that pre-existed the freeway.

The no-turning-back culture that is fostered by CityLink is self-reproductive. It promotes ever-greater scales of sub-urbanisation and in doing so produces the demand for ever faster, ever more extensive freeway provision to service the abstract car-defined urban world that is made in its own image.

Here too, an intensification of the cellular aspects of the suburbanisation process follows. This contributes to the further ghettoisation of distinct settings of workplace and home, as well as concentrating shopping and consumption into fewer and fewer zones – the mega shopping malls. This aspect, the cocooning of a city's geography, is evident wherever the efficiency, speed and scale of freeway technologies have come to dominate a city. The motor car and dependence on radiated media are, arguably, two sides of the same coin: where our social world becomes geographically fragmented we increasingly rely on an electronic assembly of some kind to overcome urban atomisation.

At the end of the freeway, where the middle classes find themselves in quarantined suburbs, cycles of psychological disconnection are also possible. In Australia this relative isolation brings fear of incursion from 'undesirables', leading to the development of so-called security suburbs such as Sanctuary Cove on the Gold Coast, with their own electronic surveillance and personnel to exclude those who are unable to attain such a lifestyle of architectural seclusion. As the cycle of fear created by such isolation grows towards strangers the motor car becomes an agent not simply of transport, but of protection. Meanwhile, those strangers who are also protected in their vehicles become the objects of road rage . . . and so these cycles continue.

Projects like CityLink can become pivotal in determining the evolution of a city's form, because of the fact that they are structural and tend to set the agenda of what sort of urban space is being created for future generations. At issue is the future of public space itself in its social, technical and aesthetic forms. This is true from the point of view of the by-passing of traditional agora like markets and parking-based streetscapes, to the further privileging of the super-regulated private spaces of shopping complexes, another cocoon for which the freeway is the link.

But it is also true from the point of view of the extra time that is spent on freeways because of the greater distances that become acceptable to travel as the scale of urbanisation increases. With CityLink, whilst each individual trip might be made faster by the tollway, it also redefines lifestyle possibilities around drive-time. Choices about what to do in one's leisure time are opened up by the reduction of drive-time. But, for these choices to be meaningful, in a lifestyle sense, requires repetitive use of the same technical means of bringing these spaces within the daily routine. Driving long distances for work, leisure of maintaining family ties becomes normalised through this repetition. Such normalisation may even lead to active acceptance of toll-roads which can accelerate such cycles of repetition.

As a technology which is designed to shrink distances and re-establish connections, CityLink is not dissimilar in many of its features to the function which the new information highway perform in our lives. Technologies like the Internet paradoxically bring individuals together at the same time as they separate and fragment them.

In the use of these technologies, be they information superhighways or vehicular superhighways, the individual usually experiences a greatly enhanced autonomy by way of speed and control as long as s/he stays within that environment. However, the 'programmed' nature of the technology actually prohibits the forming of mutual relations of reciprocity outside the operating design of the technological environment. At the same time, individuals are typically removed from having control over the structure of the technology and increasingly lack the means to form relations independent of that structure. Moreover, we have little control over the fact that such environments, which allow us to overcome suburban isolation whilst at the same time contributing to this isolation, can be sold to us as a commodity.

REFERENCES FROM THE READING

Ehrenberg, I. (1928), *The Life of the Automobile*, London: Pluto.

Graham, S. (1996), 'Telecommunications and the future of cities: debunking the myths', *Cities*, 141, 21–9.

Editor's references and suggestions for further reading

Graham, S. (2000), 'Constructing premium networked spaces: reflections on infrastructure networks and contemporary urban development', *International Journal of Urban and Regional Research*, 24(1), 183–200.

Graham, S. and Marvin, S. (2001), *Splintering Urbanism: Networked Infrastructures, Technological Mobilities and the Urban Condition*, London: Routledge, Chapter 6.

Hodge, D. (1990), 'Geography and the political economy of urban transportation', *Urban Geography*, 11(1), 87–100.

Swyngedouw, E. (1992), 'Communication, mobility and the struggle for power over space'. In G. Giannopoulos and A. Gillespie (eds), *Transport and Communications in the New Europe*, London: Belhaven, 305–25.

'The New Orgman: Logistics as an Organising Principle of Contemporary Cities'

Keller Easterling

Editor's Introduction

Our penultimate reading on cybercity mobilities is a specially commissioned piece from Yale architectural theorist and historian Keller Easterling. Her work analyses the architectures of mobility and logistic systems. Here she reflects on the growing role of tightly organised systems of logistical flow as a dominant structure which shapes cities and urbanism across the world. With the growing orchestration of cities, transport systems, and consumption and investment patterns by large transnational corporations, and their affiliates, Easterling suggests that the dominant shapers of cities – what she calls, following Harold Rosenberg, the 'new orgmen' – actually sell various types of logistical connectivity. For, as Amin and Thrift (2002, 69) argue, 'the imperative of rapid and flawless provision has stimulated the rise of powerful independent logistics corporations with their own transport fleets, IT systems and distribution centres, to further complicate the geography of supply and demand'.

Contemporary logistics systems allow for the precise organisation of time and space through extended, and minutely coordinated, technologised flow. They are therefore of critical importance in helping to configure the strung-out interurban geographies of globalisation. Logistics systems organise movements of people, goods, information, money and capital with unprecedented precision. They form critical supports to the extending dominance of global capitalist neoliberalism.

Merging globe-spanning seaport, airport, rail and highway systems, and organised through sophisticated ICTs and financial systems, Easterling argues that it is thus through logistical configurations that contemporary urban mobilities arise to have meaning. It is within such systems that the cultural and social segmentations of the global city system are accommodated and organised. It is here that the social practices of managing and organising flow around the world blend into the softwares, hardwares, and 'orgwares' (i.e. organisational protocols) of ICT, transport, finance and construction systems. Finally, it is in the selling and construction of logistics systems that specific urban spaces are being (re)constructed as what Easterling terms 'piratical transhipment sites' within the wider corporate and neoliberal divisions of labour of global, urbanising, capitalism.

Easterling's perspective is notable for the breadth of her analysis. Her research helps to makes sense of the ways in which globally stretched sites and buildings are now linked, through logistics, into the building blocks of a hypermobile and global capitalist system. Easterling reveals how the orchestration of logistics systems is a critical realm of contemporary politics, as they privilege the mobility and power of some things, people and ideas whilst compromising and eroding the mobility of power and others. Above all, Easterling's analysis reveals how all sorts of practices and sites – containers, franchises, software, urban real estate investments, ports, zones and transport networks – are constituted together into what she calls logistics cities. This is done, she argues, in ways that make the world the shape it is. It is thus to the world of the new orgman, Easterling suggests, that critical analyses of cybercity mobilities, and, indeed, politics, need to turn.

There is a new orgman. Orgman was Harold Rosenberg's nickname for America's 'organization man,' so named by sociologist William Whyte in the book of the same title (1959). The mid-twentieth-century orgman was a docile individual inhabiting a sea of suits and tract houses whose only mission was to ascend through a hierarchical corporate structure. The new orgman, descended from this fabled character, inherited that field of houses, an architectural field measured not by aesthetics, but by the almost agricultural logistics for producing a series of identical building operations in succession. The inheritance has attracted a new set of logistical organizations and a new field of commercial production within which organization is content and procedure is product.

The new orgman sells logistics. He sells management styles and networking protocols. Similarly, logistical specifications and rules direct the arrangement of 'real estate products' like resorts, franchises, logistics parks, IT campuses, airport cities or intermodal ports. Each set of instructions outlines a very specific spatial protocol, literally the software for new fields of production, games played with the same ingredients and rules whether they are in Texas or Taiwan. This software generates the most valuable property in the world: spin and logistics. But it also generates distinct forms of urbanism, even distinct species of global city.

TERRA INCOGNITA

While there may be no physical terrain left to discover on the planet, there are territories of desire, timeframes and under-marketed populations. Orgmen have precise tools for identifying these unconventional sites. They have a frontier enthusiasm for abstract territory, a pioneering sense of creation in conglomerating a labor cost, time zone and desire into a new spatial product. These products are global familiars, made by abstraction and repetition no less hyperbolic, volatile and extravagant. They gain entry into any situation, freighted with desires, sporting their global currency, and their duty-free legalities that can slither through any jurisdictional shallows. Their north arrows spinning, these formats index the world not by locality, but by annual days of sunshine, ocean temperatures, flight distances, runway noise restrictions, the time needed for a round of golf, time needed for a shopping spree, SKU's ('Stock Keeping Units' i.e. Bar Codes), TEU's

('Twenty Foot Equivalent Units' or Containers), layovers, numbers of passengers, bandwidth, time-zones and labor cost.

The orgman's argot of acronyms and figures reinforces a belief in logics, in the possibility of optimization and error elimination. With either evangelism or subterfuge, he longs to control more and more territory by simply incorporating and reformatting contradictory information according to the protocols of his organization. He must expand into the world and script its behavior according to his repertoire, formatting bytes, containers, ships, and infrastructures for compatibility. The territory devours information but often only increases its size without increasing its intelligence. These most logical and rational scenarios, when pursued outside of interference and contradiction, may produce a kind of violence necessary to remain intact. But they may also produce elaborate, and hilarious fictions that surprisingly have enormous political power and consequence whether by accident or design.

The orgman occasionally merges with all of the other pirates and confidence men riding the many seas of a global capitalist empire, seas found in the ocean, land and air. Whether sea-faring or satellite-faring, pirates fly many flags, leverage advantages in the differential values of labor and currency, brandish national identity one moment and launder it the next, using disguises to neutralize difference. Yet, while many of these pirates are the engines of translation and cross-pollination between formats, the means of producing and harvesting error, the orgman's resistance to inevitable error and contradiction presents, to all the other pirates of various stripes, a huge volatile surface area for manipulation and plunder. These are the fault lines between segregated fictions or fields of expertise, the areas of intolerance, logical exception and organizational fallout. Our latest development cocktails, the new mixtures of persuasion and data that are constantly traveling around the world, create in each new location something ever more particular, producing in these fault lines a kind of perpetual wilderness or terra incognita.

We had been warned that, culturally, the Japanese refuse to carry to-go food or beverages on the street. Yet many customers were walking out the door proudly carrying their Starbucks cups with the logo showing. I stood there watching with Howard Behar, architect of our international expansion. He turned to me with tears in his eyes (Schultz and Young, 1997).

Korea Container Terminal Authority will make every effort possible to achieve the goal of Korea becoming the most wealthy logistics nation in the new millennium maritime era. Looking forward to readers' boundless love, concern and unsparing lashing and encouragements to us (Kwang-Soo, Kim, President, Korea Container Terminal Authority http://www.kca.or.kr/eng/KCTA/index.html – accessed 2001).

SOFT

The orgman as evangelical wants to be identified with a very specific profile that is globally recognizable. He must saturate every strata and species of territory, absorb any desire for consumption and proselytize about his spatial product. The inevitable biography of the orgman Chief Executive Officer (CEO), together with the attributes of the product function as a kind of self-help manual that weaves together insider information, emotional messages and logistical jargon. New age CEOs have replaced the word 'monopoly' with 'family.' They are 'sharing' intelligence when they enter into massive affiliations that numb the competition. Many of the companies want their spatial brand to mean everything, to encompass an approach to living, often, by affiliation with other spatial products. More products come into the constellation, and more potential competitors are converts to the fold. Associations with infrastructure, advertising, entertainment, or celebrities provide flattering accouterments to serve a company that no longer wants to look omnifunctional and self-sufficient. It often means, however, that the constellation of spatial products is simply going through another cycle of expansion, incorporation and risk spreading.

Subterfuge, disguise and camouflage also accompany the orgman's desire for control and territorial expansion. Far from wishing to be identified to potential converts, some orgmen, wishing to remain undetected, shelter offshore in any one of a number of enclave formations. These enclaves, often called 'parks' in the orgman patois, signal streamlined logistics, customs waivers and loosened legalities. The orgman drops into a Free Economic Zone (FEZ) or a paper sovereignty long enough to avoid taxes, engage

inexpensive labor, or re-label a product. Bounded by, for instance, an economic imperative, a natural resource or a strain of expertise that keeps it segregated, his organization is sometimes more strongly linked to similar organizations or networks of similar enclaves than to a local condition. Occasionally the organization appears to surface for discreet public relations efforts, but it is largely reclusive. Their segregation from other worlds sometimes temporarily garners power and helps to shape these worlds into distended and dominating territories.

Attending both evangelism and subterfuge is the masquerade of soft, a special intelligence, even an ethos, about organizations. Portrayed as a means to grow and learn, soft is often an apology for maintaining a domain and avoiding contradiction. It references the late twentieth-century revivals of cybernetic thinking, fascinations with recursive organizations, automatic biological systems and loops. Soft adopts swarms, tsunamis, fractals and other mascots of chaos and favors organizations easily translated into a mathematical or logical system or a software. For soft, 'cybernetic' implies a generic organic, holistic or natural constitution. Hybridized into corporate names, it may indicate that a company is pursuing a new organizational paradigm, not a corporate hierarchy but a flat, fluid, feminized, connected structure. Orgmen discover on the front lines the organizational paradigms that theorists discover in circuits, neurons and microorganisms. In the US they have a literature in magazines like *Wired* or *Forbes ASAP* or best-selling books by writers like George Gilders, Nicholas Negroponte or Kevin Kelly that are characterized as the 'required reading for all executives.' Soft executives are team players. They are enlightened, go-playing oligarchs. They speak of synergy and feedback, often pretending that this internal regulation of information represents a kind of differentiation. Recursivity produces an organization with its own steady state and even its own catastrophes. Soft is smooth, but smooth and soft are often hard.

We've been on forty cruises, one every two months. We're addicted. We have 40,000 miles on Holland America, they gave us a special emblem. We like other ships, too. The Costa Riviera was fun. They had a man with a monkey going around the ship and served pizza all day long. We also love Carnival's Jubilee. It's a beautiful ship (cited in Dickinson and Vladimir, 1997, 217).

we have now moved beyond the simplistic arguments between, on the one hand, a Silicon Valley replica and on the other, a financial centre. Instead there has now emerged, we believe, a consensus that Hong Kong should position itself as the world-class city of Asia by, inter alia, making better and more sophisticated use of technology. The latter should not be developed for its own sake but as an important ingredient in the integral development of Hong Kong as a world city. If all the above developments are put together, we believe a strong case can be made of developing Hong Kong into the world's leading logistic hub of a global scale ('Hong Kong as the World's Leading Port,' A Concept Paper on e-commerce and logistics, December 1999 http://www.hkcsi.org.hk/papers/submit/9912ecomm.htm).

ORGMAN URBANISM

The urbanism of the evangelical orgman is a scripted and synchronized component of most cities. Franchises and brands affiliate and time their offerings to deliver and disgorge audience members from meals to movies to shopping. The frequency with which the real estate product is repeated is a volumetric version of the repetition of advertisements in broadcast media. The nation of tourism maintains a Vatican-like state within a state. Under its own sovereignty, it offers political asylum to all those converts who recognize and endorse its images. The colors, accouterments, costumes and even smells, whether the out-gassing of plastic or the aroma of coffee, are part of the property of ritual. These are the cloying new urbanists of Empire who believe themselves to be creating public space, a 'front porch,' a 'third place' for interaction in the community. Evangelical spatial products are occasionally effective in even the most obdurate political situations, and it is often their capacity for fiction and segregation that make them even more plausible as political pawns.

The orgman of sheltering enclaves and 'parks' has developed a new species of global city. From New York to São Paulo to Tokyo, vertical circulation and climate control generated the skyscraper urbanism of the global city as financial city. From Singapore to Hong Kong to Alliance, Texas, a twenty-four hour conveyance urbanism of infrastructures, containers

and specialized vehicles generates its alter-ego, the global city as logistics city. Hong Kong, for instance, characterizes itself as a logistics city rather than a financial capital or Silicon-Valley-style global city. The logistics oligarch takes over from the financial oligarch, offering the can-do of a military logistician. Urbanism is a live three-dimensional enactment of the orgman's data-tracking software. It is an urbanism of the work addict, of stacking and sorting behaviors necessary to move containers from ships and pallets from warehouses with increasingly high expectations of efficiency. Obdurate physical material ideally behaves more like information, sorting itself and thus further enticing the addict to his obsession. In fact, automation devices, in a bid to replace or enhance labor, are sometimes the featured inhabitants in these logistical fields, not belts or cogs in the machinery, but chips in an information *landschaft*.

The logistics city attracts not only warehouse space but intelligent office space, export processing centers, IT campuses, calling centers, conference/exhibition centers and other programs that thrive in the slippery space between national jurisdictions. It is not sited in its locality but rather positioned within a global network of similar enclaves serviced by a specialized infrastructure. Conurbations of ports and logistics enclaves surround Singapore, Hong Kong, Tokyo, Kaohsiung and Schiphol's Y-town, and they spread over hundreds of square kilometres in China and Russia. Logistics parks also locate inland and their developers may build parks all around the world from North Carolina to Thailand, the Philippines and Germany. Alliance, just north of Fort Worth, Texas, is a logistics park spreading over 65 square miles. These global logistics or port management companies develop similar installations all around the world with a peculiar form of sovereignty that brings to mind the mercantile companies of another time.

Although a space of exemption, as they become pawns in regional rivalries, these Special Economic Zones (SEZs) ironically land in the cross-hairs of political and territorial conflicts. Moreover, the sea, now carrying 95 per cent of the world's trade, is now no longer the peripheral territory of the state. Nowhere is this more clear than along the Asian coast and South China Sea where transnational 'growth triangles' like the 'SIJORI' triangle that links Singapore to southern Malaysia and north-east Indonesia are part of a complex political game. Many of these new logistics conurbations develop in archipelagic formations that are already fraught with legal and territorial disputes over claims to the ocean's oil riches. Piracy, terrorism, tourism, refugees, tax sheltering, labor migrations and labor exploitation also haunt such formations.

> Error is a straying from the truth, something believed wrongly; it is a mistake, a misplay, a deviation from a standard of expectation or judgment; it is the result of ignorance, inadvertence, or the inability to achieve what is right (Benveniste, 1994, 60).

> The truth of the matter is that every circuit of causation in the whole of biology, in our physiology, in our thinking, our neural processes, in our homeostasis, and in the ecological and cultural systems of which we are parts, every such circuit conceals or proposes those paradoxes and confusions that accompany errors and distortions in logical typing (Bateson, 1979).

ERROR

Whatever the claims of dominant logics or optimized procedures, error is inevitable and sometimes even more catastrophic relative to attempts to banish it. Error specialists even attempt to categorize these errors, chief among them articulation and control errors. An articulation error is 'the total lack of fit between what is wanted and what is available.' It is not isolated and circumstantial but produced in large quantities, even occasionally mass-produced. The maker may not be able to respond to fashion or to force the changing trends. At this juncture the company can learn and attempt to alter the product or it can retreat and attempt to control its customers. Control errors are those in which the company manages to sell or force the acceptance of unwanted goods. The company denies, or hides the inappropriateness of the product, keeping secrets from the public as well as secrets from itself. It will deny the addictive properties of tobacco or an enforceable global compact about the rights of labor, assuming that the problems are merely articulation errors or isolated problems. The company will attempt to diffuse responsibility in the hopes that the error will not be detected. Someday, the consumers they long to control will finally come around to seeing things their way. When the product is finally rejected, it becomes an articulation error, and the company can

decide whether or not to recognize the failure, learn and adapt.

Error always escapes its cage. Claims of optimization often quickly become impossible within the comedies of the marketplace, a marketplace based not on inherent values or logic, but rather reliant on the largely unpredictable desires of consumers. Orgmen are the perfect comedians, delivering power-point presentations, boasting of optimization that will inevitably implode as a result of its attempts to control the unpredictable. Comedy somehow suits the broad antics and characters of the marketplace. Perhaps comedy is more constantly and abundantly present in a world that attempts to foreground stability, unity and recursivity that it does not have. The orgmen that succeed are able to dodge consequence by being responsive to their contradictions and by normalizing products that approach that error threshold.

Beyond systemic error, the orgman's segregated worlds, even when masquerading as soft, are rarely so elegant or flexible that they can avoid political conflict. Their fluidity and mobility, the craftiness and complexity of an essentially hermetic organization, may even make the confrontation with error more explosive. A legal reckoning, a better product or any kind of extrinsic information may release overcharged frictions between segregated worlds that have fattened for too long in the safety of territorial enclosure.

Error is not the shameful, unwanted, catastrophic event, but the cultivated crop, perhaps one of the chief by-products of the orgman's real estate formats. Exceptions in organizations like networks, for instance, increase pathways, as new material that skews and grows the network. Error is the antigen, the extrinsic intelligence that reseeds a network, a wild card that, matching many formats, has unintended effects. Cities themselves amass an excess of circumstance, error, and contradictory evidence, multiplying meaning, enemies, and friends. This terra incognita, this territory of translation and piracy can be selectively compatible with the orgman's argot and his special species of site, capable of disassembling or invigorating his hermetic worlds and extreme fictions.

REFERENCES FROM THE READING

Bateson, G. (1979), *Mind and Nature: A Necessary Unity*, New York: Dutton.

Benveniste, G. (1994), *The Twenty-First Century Organization*, San Francisco: Jossey Bass.

Dickinson, R. and Vladimir, A. (1997), *Selling the Sea: An Inside Look at the Cruise Industry*, New York: Wiley.

Schultz, H. and Jones Young, D. (1997), *Pour Your Heart Into It*, New York: Hyperion.

Whyte, N. (1959), *The Organization Man*, New York: Doubleday.

Editor's references and suggestions for further reading

Amin, A. and Thrift, N. (2002), *Cities: Reimagining the Urban*, Cambridge: Polity, Chapter 3.

Easterling, K. (1996), 'Switch'. In P. Phillips (ed.), *City Speculations*, New York: Princeton Architectural Press, 31–5.

Easterling, K. (1999a), *Organization Space*, Cambridge, MA: MIT Press.

Easterling, K. (1999b), 'Interchange and container', *Perspecta*, 30, 112–21.

Graham, S. and Marvin, S. (2001), *Splintering Urbanism: Networked Infrastructures, Technological Mobilities, and the Urban Condition*, London: Routledge, Chapter 7.

Gutierrez, L. and Portefaix, V. (2000), *Mapping HK*, Hong Kong: Map Books.

Klein, N. (1999), *No Logo*, New York: Picador.

'Deterritorialisation and the Airport'

from *Life in the Air: Surviving the New Culture of Air Travel* (2001)

Mark Gottdeiner

Editor's Introduction

Our final reflection on cybercity mobilities concentrates on the most important urban logistical and mobility site of all – the airport. In it, Mark Gottdeiner, an influential American urban sociologist from the University of Buffalo, reflects on the astonishing importance of airports as portals linking cities to the globe.

As mass air travel and the normalisation of aerial mobility interacts closely with extending ICT-based systems of exchange, so the airport itself burgeons as a deterritorialised connection point between the worlds of the urban region and the strung-out ultra-rationalised flow systems of global airline, tourism, migration and cargo networks. Saturated with global retailing and franchising, and utterly infused with all manner of information and communications systems and financial networks, in many cases, airport spaces, and the huge economic concentrations around them, are starting to develop as 'cities' in their own right. Sometimes, these 'airport cities' are as important in economic and jobs terms as the old cores of the cities they are designed to serve.

Gottdeiner focuses, in particular, on the sociology of airport space. Drawing on the work of French anthropologist Mark Augé, he identifies the airport as a 'nonplace' where encounters tend to be routinised, silent or transient. To Gottdeiner airports are largely impersonal realms where the local and the global interpenetrate, utterly and continuously, through an infinite stream of personal mobilities (supported, of course, by vast electronic ones).

Gottdeiner's discussion recalls Lieven De Cauter's notion of capsularization (p. 94). Gottdeiner notes that, as people leave the capsule of the car and wait in the airport to enter into the ultimate capsule – the airline interior – they actually tend to capsularise themselves further – through the use of personal music players, mobile phones, mobile computers, TVs, and a voluntary silence – as they wait to be processed through passport and ticket controls.

Thus, in bringing together constantly shifting worlds of myriads of strangers from all over the world, airports represent highly rationalised, instrumentalised and deterritorialised realms. To Gottdeiner, this ambivalent sociology of encounter – both alienating and liberating at the same time – makes the airport the quintessential sociological space of mobility in the contemporary city. 'We may pause to wonder', he suggests, 'if these increasingly visited terminals are not the new generators of our future civilization.'

New social relations based on air travel are challenging our concepts of home and community. They are summed up by the concept of de-territorialization, or a general de-emphasis of the importance of significant places, like home and community, in contemporary daily life. According to one observer, "Today's executives live everywhere at once, inhabiting a growing community of the air that is responsible to no nation or governing body but only to the imperatives of commerce" (Iyer 1998, 37).

Current changes caused by frequent air travel and de-territorialization are quite profound. Until quite recently in human history, space and time were considered absolutes. The here and now were graphically and fundamentally distinguished from both the past and the future. Dreams of time travel were inspired precisely by this unconquerable barrier of absolute time that hemmed in our lives by birth at one end and death at the other. The very same was true for absolute space, which constituted a barrier of its own. For thousands of years, the concept of travel meant something done with arduous labor that involved considerable time and energy. Travel was the ultimate and most distinct form of displacement.

In an opposite sense, our domicile or domestic location represented a formidable anchoring point. The home, in particular, was a place with a specific ambiance, surroundings, and personalized touches that represented the accumulation of our daily life over many years.

Now transportation, information, and communication technologies have erased the once insurmountable barriers of space and time that hemmed in our daily lives. By disentangling the idea of the home as a personalized place from any reference to a given space, and by allowing us instantaneous access to places across the globe, transport, information, and communication technologies are deeply reshaping the experience of both home place and travel. According to Salvino Salvaggio:

> For several million people a telephone number, a house, a local bank, a post office, and a local newspaper, have already been replaced by an entirely deterritorialized cellular phone and answering service, an e-mail address, a set of credit cards, as well as media of world communication (Salvaggio, 1999).

For many business travelers, the home as a unique place condensing family relations across the generations has been de-emphasized in favor of the serial use of temporary locations, such as hotels, that advertise themselves as "a home away from home." Many jobs are less dependent now on any one location than the use of many locations woven together as spaces for work using the laptop, the cellular phone, the Internet, or the fax. This mode of de-territorialization involves a progressively greater population and results in a redefinition of home, place, space, and local community.

THE AIRPORT AND DE-TERRITORIALIZATION: THE CREATION OF A NEW CHARACTER

Decades ago, the academic field of urban sociology argued that a person's character was produced by the particular place where he or she lived. City people, for example, were said to be distinctively, qualitatively different from country people. The cause of such difference was claimed by these early researchers to be based on the separate molding patterns of the urban and rural material realms. The nature of space, in turn, affected social interaction. Different environments produced different realities which, it was believed, produced different kinds of character, different personality types.

We still believe in these stereotypes and differences today, although we may not attribute their generative cause to the quality of the environment alone. Character is more rightly believed to be produced by countless causes, especially the cultural modes of behavior we inherit from our families and intimate social contacts. Today we know that people are not the products of cities or suburbs, they are simply passing through and dancing to the tune of other, more powerful social forces. Most Americans move more than once in their lifetimes, mainly for employment purposes. Less and less can we say that some single individual is a product of a particular place but – and this is the point – more and more people are products of transient lives and transient spaces.

The airport is the ultimate symbol of the new American reality since we are all just passing through from one location to another. Our friends are more for the moment than for a lifetime; the places we stay and the people we interact with come to us because of situations. We are less the stalwarts of stable communities than temporary residents in a transitory social communion of fellow travelers. Whenever we are forced to spend long periods of time at an airport, we may become part of a larger community, the communion of fellow stranded travelers. We share some laughs, some strong negative opinions about air travel, some personal confidences, and sometimes even a night in the terminal lounge, but when the boarding announcement for our flight finally comes, we walk onto the plane without so much as a goodbye to our new friends, never to see them again.

Transitory social communion and the foregrounding of the airport as a character in personal drama was depicted in the recent film *In the Company of Men*. Two young men who work for a generic company as executives are seen in an equally generic airport as the movie opens. According to the Blockbuster cover, "Two executives on assignment at a branch of the anonymous company they work for decide to get back at all the women they believe have hurt them professionally and personally." They hatch a bizarre plot of aggression aimed at a generalized object – all women – while lounging in the placelessness of the airport terminal on the way to their out-of-town assignment. This opening sequence recalls another classic film by Alfred Hitchcock, *Strangers on a Train*. Here the premise is derived from the anonymity and class segregation of train travel in the 1940s. Two upper middle-class young men meet each other in the lounge car of a transnational train and hatch a murder plot that one of them does not take seriously until it is much too late.

Against the backdrop of the airport, the two protagonists in the newer film conspire but with a similar difference in motives. Yet they are not murderers. The damage they intend to commit is purely emotional. They decide to find a "plain Jane" who has obviously not been dating and to shower her with affection so that, for the first time, she has not only one suitor but two. Their object is to make her fall in love with at least one of them and then to expose their ruse, thereby destroying her with the realization of false love and the open manipulation of her emotions for sport. As in the earlier classic, however, there is a catch: only one of these men is playing and he is actually manipulating the other toward his doom.

In the Company of Men brings out in sharp relief the inhumanity of lives spent in transition, of men without community, without serious attachment to others, who wander the nation by air on company business and who have so little left of true character that they derive entertainment from creating and feeding off of human misery.

Could the airport, as a distinct milieu that is increasingly dwelled in as we adjust to lives spent in air travel, be helping to create or amplify a new social character – the uncaring, detached, self-contained individual armed with laptop, Walkman, credit cards, cellular phone, Palm Pilot, and business agenda? Observers of first-class travel, for example, note that passengers in that privileged section are too busy doing work while traveling to communicate with each other. In fact, being able to do work while having almost every other comfort catered to is the very premise of first class. The same observations have been made about the premier membership clubs in the terminal. They are set up with first-class business services – fax, copy machines, comfortable chairs, and tables for laptops. Hours and hours of watching ordinary people waiting for planes confirms that they rarely talk to each other. Walkmen with personal earphones are ubiquitous. Sitting with novel or newspaper in hand is most common. Airport lounges have their own TV broadcasts courtesy of the airport satellite channel, a division of the old Turner network. Departure lounges are not commonly a place for social communion.

At the turn of the century, Georg Simmel, a brilliant urban sociologist living in Berlin, observed a version of this social isolation and considered it a defining characteristic of the city. The "urban" is precisely that place where we encounter each other as *strangers*. Urban life, Simmel argued, consists of the most focused instrumental relations with others. We interact with people who sell us things for the purpose of buying, who process things for us in bureaucracies, and who share the same space but do not seek to be our friends. Simmel demonstrated that this characteristic urban life contrasted greatly with rural life. The latter involved people who knew each other intimately, who may have relied on each other for services but often did not demand cash payment, and who had a more multi-dimensional connection with each other through long-term family and community friendships. For many people in the city, especially newcomers, the relative coldness of urban life was difficult to take.

The airport has taken on the characteristics of Simmel's city to an extreme. It has all the trappings of a thoroughly instrumental space with even less of a need for people to interact. In fact, the airport norm is one of *non-interaction*. People are *expected* to keep to themselves in airports. We celebrate the anticipated transition to the social status of traveler by collapsing into our shell with the typical props of magazine, laptop, or blasé gaze that means for us all, "I'm just killing time." We do not bother others and they are not expected to bother us. When they do, the situation thus created becomes quite uncomfortable. One of the pet peeves of frequent fliers is the encounter with a chatty neighbor.

Simmel, however, also mentioned that urban life has a positive side. Freed from the bounds of traditional society that dictates how we engaged in life, the city enables us to invent a new one. In the role of stranger, we are no longer bound by the social attributions of others. Simply put, in the large, urban environment, I not only can be anyone I want, I can also find the services and social infrastructure to pursue my personal goal of becoming who I want to be. For a person fleeing the farm to pursue a career in the theater many years ago, for example, there was no other place to go but the large city. Generations upon generations of city people have discovered and invented new modes of living by exploiting the freedom of anonymity that urban life offered over the course of history. The city for Simmel was nothing less than the veritable site of civilization. As the airport takes on more of the complex shadings of the city as well as its multifunctional public space, we may pause to wonder if these increasingly visited terminals are not the new generators of our future civilization.

REFERENCES FROM THE READING

Iyer, Pico (1998), 'The new business class', *New York Times Magazine*, 8 March, 37–8.

Salvaggio, Salvino (1999), Letter to author.

Editor's suggestions for further reading

Augé, M. (1995), *Non-Places: Introduction to the Anthropology of Supermodernity*, London: Verso.

Gottdeiner, M. (2001), *Life in the Air: Surviving the New Culture of Air Travel*, New York: Rowan and Littlefield.

Pascoe, D. (2001), *Airspaces*, London: Reaktion Books.

Rosler, M. (1998), *In the Place of the Public: Observations of a Frequent Flyer*, Osfildern: Cant Verlag.

V CYBERCITY ECONOMIES

Plate 16 The Eurotéléport Roubaix: a major hub of telecommunications and media infrastructures and service companies in north-east France which was developed with support from local municipal authorities. (Photograph by Stephen Graham.)

INTRODUCTION

The economic dynamics of cybercities are dominated by complex combinations of geographical centralisation and decentralisation. These are being forged as the economic processes of capitalist cities adjust to form new relations between telepresence and face-to-face presence, and fixity and mobility, based on harnessing the capabilities of both ICTs and transport and logistics systems. Thus we must not see city economies as fixed, contained or bounded entities. Rather, they must be thought of as 'assemblages of more or less distanciated economic relations which will have different intensities in different locations' (Amin and Thrift, 2002, 52). The broad geographical patterns of these distanciated intensities reveal at least three key dynamics.

'Sticky spaces': the global metropolitan heartlands

First, so-called 'global' and second-tier cities are tending to keep, and sometimes strengthen, their roles as key anchor points within the internationalising, and increasingly knowledge-based, economy. Very broadly, those global and second-tier cities, parts of cities, and the socioeconomic groups involved in producing high value-added goods, services and knowledge outputs, are tending to become intensively interconnected internationally (and sometimes even globally) (see Sassen, 2001a,b; Markusen et al., 1999). Using the capabilities of high quality communications and transport infrastructures, zones of intense international articulation – business spaces, new industrial spaces, corporate zones, airports, new cultural or entertainment zones, logistics areas – are emerging in such cities, albeit to highly varying degrees.

Such 'sticky spaces' (Markusen, 1999; Markusen et al., 1999) include global financial capitals like Manhattan or the City of London; high-tech industrial districts like Silicon Valley, Cambridge, Seattle, or Bangalore; government complexes like Washington DC; cultural production centres like Hollywood (Scott, 2000); or the emerging digital innovation clusters like New York's Silicon Alley. Within them a tight degree of interaction on the industrial district model may survive and prosper. In such places, flexible, continuous, and high value-added innovation continues to require intense face-to-face learning and co-location in (the right) place, over extended periods of time.

These critical strategic sites of global capitalism are the real drivers of global finance, technological innovation, international telecommunications traffic, logistics and transportation flows, governance, and global property markets. Such cities often tend also to drive international cultural trends in film, new media, TV, radio, fashion, the visual and performing arts, literature, music, fashion, design, architecture, food, and publishing. In many cases the gross economic products of such metropolitan areas now surpass those of sizeable nations. New York's Gross Product ($829 billion) is greater than that of Brazil ($778 billion). That of Paris ($374 billion) exceeds that of Australia ($361 billion). And London's ($239 billion) in greater than that of Sweden ($226 billion) (all figures dated 2000 – source Koolhaas et al., 2001, 51).

Current markets for international telecommunications, and ICT services, are overwhelmingly driven by the economic growth of these dynamic, internationally oriented, global and second-tier metropolitan regions. Such city-regions have important assets in an internationalising economy based more and more on flows of information, services and symbolic products like media, advertising, and electronic entertainment. They support *relational proximity* in a world of fast flows. They support concentrated and intense interaction in place whilst also having the high-quality physical, service and telecommunications infrastructures to extend access to distant, often dominated, places. Today's volatile and globalising economies make a prerequisite of trust, constant innovation, and 'reciprocity', which, as we saw with Deirdre Boden and Harvey Molotch's work (p. 101), can best be fully forged through the kinds of face-to-face contact that are most usually practised within and between major business capitals.

Stressing the extraordinarily social nature of modern economies, geographers Nigel Thrift and Kris Olds (1996; 314–16) write that 'it is clear that face-to-face interaction has not died out. Indeed, in some sense it has become more important as reflexivity (including an enhanced ability to see oneself as others see us) has become built into economic conduct.' Recalling Deirdre Boden and Harvey Molotch's work, Tony Fitzpatrick, the director of the engineering consultancy firm Ove Arup, argues that:

> cities reflect the economic realities of the 21st century. Remote working from self-sufficient farm-steads via the Internet cannot replace the powerhouses of personal interaction which drive teamwork and creativity. These are the cornerstones of how professional people add value to their work. Besides, you cannot look into someone's eyes and see that they are trustworthy over the Internet (1997, 9).

Cities on the economic periphery: cost-based competition and the distant delivery of routinised products

Most cities, however, are never going to be international powerhouses of financial, media, biotech or research and development industries. Here, a second dynamic of selective decentralisation and reconfiguration is underway through which a range of more peripheral cities and urban spaces in developed, developing, postcommunist or newly industrialising nations are also able to attain some sort of economic position within circuits of internationally integrated industrial and economic development. Such places are able to capture more decentralised flows of investment driven by the liberalisation of trade and investment, as key decision-makers attempt to reduce costs and exploit untapped labour resources whilst delivering goods and services to external markets.

These decentralising logics – which drive routinised or land and labour-hungry uses away from dominant global and second tier cities – might involve global nodes for the production of high volume manufacturing goods and services; places that can deliver routine services online or via telephone or Internet links to the core city regions; or sources for the extraction and production of various types of raw materials. Here we see the familiar scramble of entrepreneurial and increasingly internationally oriented localities for foreign direct investment (FDI) in routine manufacturing, mobile services and resource extraction.

Importantly, when new media do support decentralisation from large cities like New York and London – as is the case with many 'back-offices' delivering routine services online – their destinations tend not to be rural spaces but smaller, more peripheral, cities. In sectors like online computer programming and software support, such shifts can link cities like Bangalore in India via ICTs to distant markets in the main metropolitan areas of the global north (see Shirin Madon, p. 309). Because of their overwhelming external orientation, economic and technological connections elsewhere in many of these spaces – for example the burgeoning clusters of call centres and back offices in North American, Caribbean and European cities – now tend to far outweigh connections with the local 'host' space.

E-commerce: ICTs and the remediation of urban consumption

A third key economic dynamic of cybercities is emerging. This entails the use of ICTs to reorganise the internal and consumption-driven economic dynamics of cities and systems of cities. It is important, given the hyperbole about globalisation, not to underestimate the degree to which telemediated flows still predominantly work to sustain the reorganisation of very *local* economic relations within major cities. After all, the majority of jobs in all cities are generated by local demands for goods and services and not by high-level flows of finance capital or research and development. In addition, the vast bulk of electronic exchanges are very local indeed: 60 per cent of all phone calls and emails are estimated to move within single buildings. Even when stories are apparently about decentralisation and substitution, essentially local, urban dynamics are often at play. Most teleworking, for instance, is done for part of the week in the zones within and around the large cities that allow people to go to the office on one or a few days a week for face-to-face meetings.

Of particular importance here is the growth of e-commerce at various scales within and between cities. This involves the use of ICTs to organise consumption and distribution patterns both between businesses (B2B) and between businesses and consumers (B2C). As increasing numbers of consumers use personal or public computers and mobile phones to order goods and services, so new geographical dynamics are emerging. Whilst they have been substantially affected by the dot.com collapse since 2001, these services are starting to have important implications for the balance between 'clicks' and 'bricks' and retail and 'e-tail' within cities. New logistical spaces and systems are necessary to sustain the delivery of groceries, books and consumer durables from major warehouses and distribution hubs direct to people's homes via post or courier. Demand for certain types of shop within traditional malls and high streets may dwindle. And a range of new hybrid and recombinatory spaces are emerging, from local e-commerce equipped neighbourhood stores, to new consumption spaces in city cores which are owned and run by entertainment and media transnationals. In these e-commerce and face-to-face commerce are sustained in parallel.

References and suggestions for further reading

Alaedini, P. and Marcotullio, P. (2002), 'Urban implications of information technology and new electronics for developing countries', *Journal of Urban Technology*, 9(3), 89–108.

Amin, A. and Thrift, N. (2002), *Cities: Reimagining the Urban*, Cambridge: Polity, Chapter 3.

Braczyk, H.-J., Fuchs, G. and Wolf, H.-G. (eds) (1999), *Multimedia and Regional Economic Restructuring*, London: Routledge.

Castells, M. (ed.) (1995), *High Technology, Space and Society*, London: Sage.

Castells, M. (1989), *The Informational City: Information Technology, Economic Restructuring and the Urban-Regional Process*, Oxford: Blackwell.

Castells, M. and Hall, P. (1994), *Technopoles of the World: The Making of 21st Century Industrial Complexes*, London: Routledge.

Clarke, S. and Gaile, G. (1998), *The Work of Cities*, Minneapolis. University of Minnesota Press.

Fitzpatrick, A. (1997), 'A tale of tall cities', *Guardian Online*, 6 February, 9.

Gibson, D., Kozmetsky, G., and Smilor, R. (1993), *The Technopolis Phenomenon: Smart Cities, Fast Systems, Global Networks*, Lanham, MA: Rowman and Littlefield.

Gottmann, J. (1990), *Since Megalopolis: The Urban Writings of Jean Gottmann*, Baltimore: Johns Hopkins University Press.

Graham, S. and Marvin, S. (2001), *Splintering Urbanism: Networked Infrastructures, Technological Mobilities, and the Urban Condition*, London: Routledge, Chapter 7.

Hepworth, M. (1989), *The Geography of the Information Economy*, London: Belhaven.

Knox, P. and Taylor, P. (eds) (1995), *World Cities in a World System*, Cambridge: Cambridge University Press.

Komninos, N. (2002), *Intelligent Cities: Innovation, Knowledge Systems, and Digital Spaces*, London: Spon.

Koolhaas, R., Boeri, S., Kwinter, S. (2001), 'Gross product of cities'. In R. Koolhaas, S. Boeri and S. Kwinter, (eds) *Mutations*, Bordeaux: ACTAR, 51.

Kotkin, J. (2000), *The New Geography: How the Digital Revolution is Reshaping the American Landscape*, New York: Random House.

Leinbach, T. and Brunn, S. (2001), *Worlds of E-Commerce: Economic, Geographical and Social Dimensions*, London: Wiley.

Markusen, A. (1999), 'Sticky places in slippery space: A typology of industrial districts'. In T. Barnes and M. Gertler (eds), *The New Industrial Geography*, London: Routledge: 98–123.

Markusen, A., Yong-Sook, L. and DiGiovanna, S. (eds) (1999), *Second Tier Cities: Rapid Growth Beyond the Metropolis*, Minneapolis, MN: University of Minnesota Press.

Moss, M. (1987), 'Telecommunications, world cities and urban policy', *Urban Studies*, 24, 534–46.

Mulgan, G. (1991), *Communication and Control: Networks and the New Economies of Communication*, Oxford: Polity.

Sassen, S. (1994), *Cities in a World Economy*, London: Pine Forge.

Sassen, S. (2001a), *The Global City: New York, London, Tokyo*, Princeton: Princeton University Press, 2nd edn.

Sassen, S. (ed.) (2001b), *Global Networks, Linked Cities*, New York: Routledge.

Schiller, D. (1999), *Digital Capitalism: Networking the Global Market System*, Cambridge, MA: MIT Press.

Scott, A. (2000), *The Cultural Economy of Cities*, London: Sage.

Scott, A. (2001), *Global City-Regions: Trends, Theory, Policy*, Oxford: Oxford University Press.

Sussman, G. and Lent, J. (eds) (1998), *Global Productions: Labor in the Making of the Information Society*, Cresskill, NJ: Hampton Press.

Storper, M (1997), *The Regional World*, New York: Guilford.

Thrift, N. and Olds, K. (1996), 'Refiguring the economic in economic geography', *Progress in Human Geography*, 20(3), 311–17.

Veltz, P. (1996), *Mondialisation: Villes et territoires*, Paris: Presses Universitaires de France.

Zook, M. (2003), *The Geography of the Internet*, Oxford: Blackwell.

'Agglomeration in the Digital Era?'

from *Cities in a World Economy* (2000)

Saskia Sassen

Editor's Introduction

In our first reading on cybercity economies Saskia Sassen, author of *The Global City* (1991, 2001) and an influential sociologist who works at the University of Chicago, addresses the apparently paradoxical agglomeration of major financial and business or global cities in an age saturated with digital technologies. Why, asks Sassen, does a digitized global financial industry, whose products are now nothing but continuous streams of computerised information, continue to cluster in tiny districts within the world's most expensive, and congested, elite financial cities? Her answer stresses the intensely social nature of these high-tech currency, exchange and financial industries; their demands for an extraordinary range of infrastructures, skills, and support services; and the growth – following pervasive privatisation – of certain global cities to emerge as de facto financial capitals for huge chunks of the planet, as their social and economic elites, and financial systems, effectively denationalise.

Combined, these forces mean that the truly global financial and business cities – notably New York, Tokyo, and London – have increased their dominance and 'reach' over the global financial architectures and systems of contemporary capitalism in the past twenty or so years. In fact, the growing digitization of financial capitalism has, paradoxically, allowed these cities to extend their control over financial transactions and markets over transoceanic scales.

Thus, whilst much routinised activity – such as call centres – is being pushed away from the cores of these cities to lower cost locations in the suburbs or in nationally or globally peripheral cities, the centrality of global city cores shows a good deal of resilience. Post-September 11th there may be smaller numbers of elite financial traders and stockbrokers in the strategic sites of these city cores (as firms colonise docklands and suburbs to reduce their vulnerabilities to major terrorist attacks). But the key central business district locations of global cities continue to retain their dominant hold over global financial industries.

The continuing weight of major [financial] centers is, in a way, countersensical, as is, for that matter, the existence of an expanding network of financial centers. The rapid development of electronic exchanges, the growing digitalization of much financial activity, the fact that finance has become one of the leading sectors in a growing number of countries, and that it is a sector that produces a dematerialized, hypermobile product, all suggest that location should not matter. In fact, geographic dispersal would seem to be a good option given the high cost of operating in major financial centers, and differentiation would seem to eliminate most reasons for having a geographic base. Furthermore, the last 10 years have seen an increased geographic mobility of financial experts and financial services firms.

There has been geographic decentralization of certain types of financial activities, aimed at securing business in the growing number of countries becoming integrated into the global economy. Many of the leading investment banks have operations in more countries than they had 20 years ago. The same can be said for the leading accounting and legal services and other specialized corporate services. And it can be said for some markets: For example, in the 1980s, all basic wholesale foreign exchange operations were in London. Today, these are distributed among London and several other centers (even though their number is far smaller than the number of countries whose currency is being traded).

There are, in my view, at least three reasons that explain the trend toward consolidation in a few centers rather than massive dispersal. I developed this analysis in *The Global City* (Sassen 1991), focusing on New York, London, and Tokyo, and since then, events have made this even clearer and more pronounced.

SOCIAL CONNECTIVITY

First, while the new telecommunications technologies do indeed facilitate geographic dispersal of economic activities without losing system integration, they have also had the effect of strengthening the importance of central coordination and control functions for firms and, even, markets. (Let's remember that many financial markets have "owners," are run by firms so to speak, and hence also contain central management functions of sorts.) Indeed, for firms in any sector, operating a widely dispersed network of branches and affiliates and operating in multiple markets has made central functions far more complicated. Their execution requires access to top talent, not only inside headquarters but also, more generally, to innovative milieus – in technology, accounting, legal services, economic forecasting, and all sorts of other, many new, specialized corporate services. Major centers have massive concentrations of state-of-the-art resources that allow them to maximize the benefits of telecommunications and to govern the new conditions for operating globally. Even electronic markets such as NASDAQ and E*Trade rely on traders and banks located somewhere, with at least some in a major financial center.

One fact that has become increasingly evident is that to maximize the benefits of the new information technologies, you need not only the infrastructure but a complex mix of other resources. Most of the value-added that these technologies can produce for advanced service firms lies in the externalities. And this means the material and human resources – state-of-the-art office buildings, top talent, and the social networking infrastructure that maximizes connectivity. Any town can have the fiber optic cables. But do they have the rest?

A second fact emerging with greater clarity concerns the meaning of "information." There are, one could say, two types of information. One is the datum: At what level did Wall Street close? Did Argentina complete the public sector sale of its water utility? Has Japan declared such and such bank insolvent? But there is a far more difficult type of "information," akin to an interpretation/evaluation/judgment. It entails negotiating a series of data and a series of interpretations of a mix of data in the hope of producing a higher-order datum. Access to the first kind of information is now global and immediate thanks to the digital revolution. You can be a broker in the Colorado mountains and have access to this type of information. But the second type of information requires a complicated mixture of elements – the social infrastructure for global connectivity – that gives major financial centers a leading edge.

You can, in principle, reproduce the technical infrastructure anywhere. Singapore, for example, has technical connectivity matching Hong Kong's. But does it have Hong Kong's social connectivity? When the more complex forms of information needed to execute major international deals cannot be gotten from existing databases, no matter what one can pay, then one needs the social information loop and the associated de facto interpretations and inferences that come with bouncing off information among talented, informed people. The importance of this input has given a whole new weight to credit rating agencies, for instance. Part of the rating has to do with interpreting and inferring. When this interpreting becomes "authoritative," it becomes "information" available to all. The process of making inferences/interpretations into information takes quite a mix of talents and resources.

Risk management, for example, which has become increasingly important with globalization due to the growing complexity and uncertainty that comes with operating in multiple countries and markets, requires enormous fine-tuning of central operations. We now know that many, if not most, major trading losses over the last decade have involved human error or fraud. The quality of risk management will depend heavily on the top people in a firm rather than simply on technical conditions, such as electronic surveillance. Consolidating risk management operations in one site, usually a central one for the firm, is now seen generally as more effective. We have seen this in the case of several major banks: Chase and Morgan Stanley Dean Witter in the United States, Deutsche Bank and Credit Suisse in Europe.

In brief, financial centers provide the social connectivity that allows a firm or market to maximize the benefits of its technological connectivity.

NEED FOR ENORMOUS RESOURCES

Global players in the financial industry need enormous resources, a trend that is leading to rapid mergers and acquisitions of firms and strategic alliances between markets in different countries. These are happening on a scale and in combinations few would have foreseen just a few years ago. In 1999 alone, we saw a whole new wave of mergers, notably Citibank with Travellers Group (which few would have predicted just two years earlier), Salomon Brothers with Smith Barney, Bankers Trust with Alex Brown, and so on. This wave has been so sharp that now when firms such as Deutsche Bank and Dresdner Bank want to purchase a U.S. security firm, they complain of a lack of suitable candidates. Many analysts now think that midsize firms will find it difficult to survive in the global market, when there are global megafirms such as Merrill Lynch, Morgan Stanley Dean Witter, and Goldman, Sachs. We are also seeing mergers between accounting firms, law firms, insurance brokers – in brief, firms that need to provide a global service. Analysts foresee a system dominated by a few global investment banks and about 25 big fund managers. A similar scenario is also predicted for the global telecommunications industry, which will have to consolidate to offer a state-of-the-art, globe-spanning service to its global clients, among which are the financial firms.

Another kind of "merger" is the consolidation of electronic networks that connect a very select number of markets. One of Chicago's futures exchanges, the Board of Trade (BOT) is loosely linked to Frankfurt's futures exchange and the other to Paris's futures exchange (MATIF). The New York Stock Exchange is considering linking up with exchanges in Canada and

Latin America and has opened talks with the Paris Bourse. NASDAQ's parent is having similar talks with Frankfurt and London. Perhaps most spectacular was the announcement in July 1998 of a linkup between the London Stock Exchange and Frankfurt's Deutsche Borse; the goal is to attract the top 300 shares from all over Europe – a blue-chip European exchange. Paris reacted by proposing that some of the other major European exchanges should create an alternative alliance.

[. . .]

DENATIONALIZATION OF THE CORPORATE ELITE

Finally, national attachments and identities are becoming weaker for these global players and their customers. Thus, the major U.S. and European investment banks have set up specialized offices in London to handle various aspects of their global business. Even French banks have set up some of their global specialized operations in London, inconceivable even a few years ago and still not avowed in national rhetoric.

Deregulation and privatization have further weakened the need for *national* financial centers. The nationality question simply plays differently in these sectors than it did even a decade ago. Global financial products are accessible in national markets, and national investors can operate in global markets. It is interesting to see that investment banks used to split up their analysts team by country to cover a national market; now they are more likely to do it by industrial sector.

In my *Losing Control?* (Sassen 1996), I have described this process as the incipient denationalization of certain institutional arenas. I think such denationalization is a necessary condition for economic globalization as we know it today. The sophistication of this system lies in the fact that it needs only to involve strategic institutional areas – most national systems can be left basically unaltered. China is a good example. It adopted international accounting rules in 1993, necessary to engage in international transactions. But it did not have to go through a fundamental reorganization to do this. Japanese firms operating overseas adopted such standards long before Japan's government considered requiring them. In this regard, the wholesale side of globalization is quite different from the global mass consumer markets, in which success necessitates altering national tastes at a mass level. This process of denationalization will be facilitated by the current acquisitions of firms and property in all the Asian countries in crisis. In some ways, one might say that the Asian financial crisis has functioned as a mechanism to denationalize, at least partly, control over key sectors of economies that, while allowing the massive entry of foreign investment, never relinquished that control.

Editor's suggestions for further reading

Corbridge, S., Martin, R. and Thrift, N. (eds) (1994), *Money, Power and Space*, Oxford: Blackwell.

Graham, S. (1999), 'Global grids of glass: on global cities, telecommunications and planetary urban networks', *Urban Studies*, 36(5–6), 929–49.

James, J. (2001), 'Information technology, cumulative causation and patterns of globalisation in the Third World', *Review of International Political Economy*, 8, 147–62.

Sassen, S. (1994), *Cities in a World Economy*, London: Pine Forge.

Sassen, S. (2001a), *The Global City: New York, London, Tokyo*, Princeton: Princeton University Press, 2nd edn.

Sassen, S. (ed.) (2001b), *Global Networks, Linked Cities*, New York: Routledge.

Thrift, N. (1996), 'New urban eras and old technological fears: reconfiguring the goodwill of electronic things', *Urban Studies*, 33(8), 1463–93.

Warf, B. (1995), 'Telecommunications and the changing geographies of knowledge-transmission in the late twentieth century', *Urban Studies*, 32, 361–78.

Warf, B. (1989), 'Telecommunications and the globalization of financial services', *Professional Geographer*, 41(3), 257–71.

'Webs of Myth and Power: Connectivity and the New Computer Technopolis'

Vincent Mosco

Editor's Introduction

Our second analysis of cybercity economies also addresses dynamics of centralisation and agglomeration in some of the 'sticky places' of global capitalism. Here Vincent Mosco, a leading Canadian communications scholar, analyses the construction, or reconstruction, of privileged spaces of centrality for technological innovation in cities. Mosco focuses on two particular such 'technopoles' and dicusses how they bring together the skilled labour, key infrastructures, links to higher education, public support, and venture capital necessary to sustain ongoing innovations within contemporary capitalism.

In a frenzied wave of competitive and copycat urban planning, city development agencies around the world have sought to use the language of 'technopoles' to try to attain nodal status on the global innovations networks of major transnational firms. The glamorous prefixes 'Silicon', 'Cyber', 'Intelligent', 'e-', and 'Multimedia' have been splashed across both old urban neighbourhoods and ambitious and futuristic new cityscapes. The symbolism of ICTs has been ruthlessly exploited by planners, governments and real estate companies to try to add value and cachet to specific locations, with extremely varying results (see Rebecca Solnit and Susan Shwartenberg, p. 296; Shirin Madon, p. 309; Stephen Graham and Simon Marvin, p. 341; Tim Bunnell, p. 348; Neil Coe and Henry Wai-chung Yeung, p. 354). The technopole phenomenon hammers home the symbolic work that ICTs play in the material and social reconstruction of urban places.

Mosco's work is notable for the way it seeks to look beyond the glamorous place marketing, and the technological and economic spin of 'information age' cities, to analyse the social, distributional and political dimensions of technopole planning and development, as it occurs in practice. Drawing on the influential experiences of the growth of California's Silicon Valley since the 1950s, and the planned construction of a range of technopoles across Japan since the 1970s, Mosco compares the development dynamics of two particular technopoles between the late 1980s and late 1990s. The first case analyses the reconstruction since the mid-1990s of parts of midtown Manhattan as a major multimedia technopole labelled 'Silicon Alley'. The second, contrasting, case examines the spectacular construction of an entirely new thirty-mile urban corridor known as the Multimedia Super Corridor south of Kuala Lumpur by the Malaysian government. (This is a case we will return to: see Tim Bunnell, p. 348).

In drawing conclusions from his analyses of the two cases, Mosco sounds some cautionary notes about the ways in which spaces for democratic citizenship seem to be squeezed out of technopole spaces and planning strategies. Mosco argues that technopoles demonstrate the continued importance of critical innovation sites within cybercities – ones which can furnish the face-to-face potential, high level skills, built property, infrastructures, capital and symbolic power necessary to make that innovation work.

And yet this obsession with production-side dynamics has led to a neglect of the nature of technopole spaces as political and social arenas. To Mosco, the mythologising of technopoles actually deflects attention form their problematic exclusionary and political effects. Within them, urban space is often privatised, controlled and rationed off to technological elites, either through urban design strategies or the extreme price rises associated with gentrification (see Rebecca Solnit and Susan Shwartenberg, p. 296; Shirin Madon, p. 309). The reach of surveillance is often deepened as corporate and private security practices extend their powers to protect technopoles as key strategic sites. All the while, Mosco believes that the glossy information age ideologies of technological 'progress' and corporate globalisation are used to conveniently camouflage the disturbing social and geographical restructuring of real urban spaces and places within technopole strategies.

The situation in these two technopoles has changed radically since this piece was written by Mosco (see Indergaard, 2003). To a considerable extent Silicon Alley has unravelled since the dot.com 'crunch' of 2000–2002. 'The combined values of 46 Silicon Alley stocks that were publicly traded fell from US$1519 in January 2000 to US$112 in March 2001' (Indergaard, 2003, 390). There have been very high levels of bankruptcy, a process compounded by the economic fallout of the September 11th attacks (see Anthony Townsend, p. 143). Between the end of 2000 and April 2003 41 per cent (21,000) of computer industry posts, 25 per cent of telecommunications jobs, and 15 per cent of media and communications jobs were lost in New York City. As Michel Indergaard suggests, it remains 'unclear how the resources that were assembled as elements of Silicon Alley will be recombined' in the light of these drastic cutbacks (2003, 390). The credibility of MSC, in contrast, 'is on the rise' (Indergaard, 2003, 396). This is so because its massive public subsidies and support, and the developmentalist, rather than neoliberal state model that it has benefited from, have allowed it to survive the impacts of the Asian Financial crisis in the mid- to late 1990s (Indergaard, 2003: see Tim Bunnell, p. 348).

MAGIC PLACES: THE TECHNOPOLES

The technopole is one reason why specific local places are taking on a growing significance in spite of the end-of-geography talk. For the technopole is a specific place which brings together institutions, labor, and finance that generate the basic materials of the information economy. They result from various local, national, and, in some cases, international, planning activities by public and private sector organizations to promote systematic technological innovation. The term technopole originated in the Japanese government's effort of the 1960s to build a science-based technopole Tsukuba about forty miles outside of Tokyo and most would see Silicon Valley in California as its icon and most successful form. In their global survey, Castells and Hall (1994) refer to two dozen or so technopoles, many eager to emulate the Silicon Valley model.

The most interesting of the newer technopoles draw from the Japanese and Silicon Valley models but more importantly take them a significant step further. These earlier successes built resilient and fluid production networks that brought together diverse knowledge professionals who could manage innovation from idea to market. Newer technopoles integrate these production networks into similarly resilient and fluid consumption networks located within the technopole. The reading proceeds to take up two types of this newer computer technopolis in New York's Silicon Alley and Malaysia's Multimedia Super Corridor.

NEW YORK: FROM BROADWAY TO SILICON ALLEY

Silicon Alley is the technical hub of an agglomeration of New York's media industries connecting advertising, publishing, broadcasting, telecommunications, mass entertainment, contemporary art, and fashion, all concentrated in a collection of overlapping districts from Broad Street at the south end of Manhattan, moving north to the SoHo and Greenwich Village artistic communities, through the Times Square entertainment district and on up to the publishing houses north of Times Square and the advertising agencies along Madison Avenue (Plate 17). Filling office buildings left vacant by financial services firms that shed workers with new technologies or relocation, and giving a post-industrial economic allure to a city once bankrupt and out of manufacturing alternatives, Silicon Alley embodies a cyber version of the *phoenix* myth: in this case the city reborn from the ashes of its industrial past. Even so, it also propels a transformation of urban politics and power as corporate-controlled bodies like

Plate 17 A sign announcing the northern entrance to Silicon Alley in Manhattan's Flatiron district, New York. (Photograph by Stephen Graham.)

Business Improvement Districts (BIDs) remake public spaces into private enclaves and rewrite the rules of policing, civic activity and public spectacle. All of this takes place in the name of connectivity, in this case referring to the connections among the convergent computer, communication, and cultural sectors in Manhattan and to the market potential of a web industry built on enhancing electronic connectivity worldwide.

In a short time, Silicon Alley became a global center for multimedia design and development. According to a 1997 Coopers and Lybrand report, the district anchors a new media industry that employs 56,000 in New York City and 106,000 in the metropolitan area's 5,000 new media firms, making it one of the largest employers of computer communication workers in North America, on a par with Silicon Valley. Annual revenues climbed 56 per cent over 1996 to $2.8 billion in the city and 50 per cent in the metro area to $5.7 billion. Full-time jobs in new media now match those in the premier media industries of New York: advertising and print publishing (Coopers and Lybrand, 1997). In addition, Silicon Alley has become a model for the kind of mobile production that is increasingly common in web work. Casual workers move in and out, benefiting from physical proximity when necessary and returning to other office or home sites. In fact, Silicon Alley has pioneered in the short-lease, prebuilt, prewired office market, what it calls the Plug'n'Go system, which allows small businesses and casual workers to move and plug into physical and cyber networks. The district itself is expanding in ways that place it in closer proximity to traditional media industries. With rents rising and vacancies declining in the heart of the district, companies are moving north as far as mid-town, west to the Hudson River, and down Broadway through SoHo as far as the Harbor, where the tax rebates and refurbished buildings are attracting new firms.

Silicon Alley is unique in its integration of media (especially publishing and advertising), the arts (particularly the development of SoHo neighborhood) for attracting talent to multimedia design and production, and telecommunications (for example the regional Teleport). According to one commentator, 'a lot of the style of . . . Silicon Alley may be new, but the muscle behind it is not. Established New York industries, especially advertising, publishing, fashion and design . . . are now leading much of the expansion of the new technologies' (Johnson, 1997). A 1996 Coopers and Lybrand report highlighted the significance of close ties among businesses in these several communities. Forty-three per cent of new media companies surveyed worked principally for advertising firms and forty-two percent for print publishing and entertainment firms (Coopers and Lybrand, 1996). By the start of 1997, New York had surpassed all rivals in the number of registered commercial and nonprofit Internet domain sites, twice as many as its nearest rival, San Francisco and 4.3 per cent of the entire US (Johnson, 1997).

The advertising firm Saatchi & Saatchi demonstrates a vital way in which this system of commercial connectivity works. The S&S Silicon Alley web unit Darwin Digital, brought together Procter & Gamble with Time Warner at the address Parenttime.com. The project links Time Warner's childcare magazine division with an integrated advertising package supplied by P&G, all created and run from New York City. Darwin Digital has also helped create a new network for children's games on the Internet with the major sponsorship of General Mills, the Minneapolis-based cereal company, whose creative staff came to New York to produce the site with Darwin. Indeed, some of the most successful Silicon Alley firms are exclusively devoted to advertising. For example, the Alley firm AdOne was in 1998 the largest online classified network in the United States with over 500 local, regional and national partner publications combining a print readership of 40 million. Advertising has also attracted venture capital to the district. It is certainly the case that Silicon Alley has generally attracted considerable venture capital, an unusual occurrence for this region. Investment in the New York's new media industry grew from $167 million in 1996 to $240 million one year later, boosting the state of New York to fourth behind the high-tech states of California, New York and Texas. But the biggest deals are in advertising. Whereas investments of around $500,000 are common for Silicon Alley's typically small web design firms, in 1997 the Silicon Alley Internet advertising firm Double Click, Inc. was on the receiving end of a $40 million deal with the Silicon Valley-based venture firm Weiss, Peck & Greer (Ravo, 1998). Online advertiser 24/7 Media Inc. leaped into the competition with Double Click with a $10 million venture deal.

Finally, another online advertiser Agency.com was voted the top Silicon Alley firm in 1997 by the New York business magazine @ny. New York expects continued growth in the new media sector because of the prominence of its top advertising agencies, close ties to Wall Street finance, and, following its historic role as a world capital of commerce, as the Internet medium develops, it will concentrate on electronic commerce which is expected to become the biggest revenue maker.

CREATIO EX NIHILO IN MALAYSIA

If New York is viewed as the information age phoenix rising from the ashes of manufacturing decline, then Malaysia is the magic land where palm-oil plantations become Multimedia Super Corridors (MSC) almost overnight. The Multimedia Super Corridor enacts an alternative but related myth, *creatio ex nihilo*, as the Malaysian national government creates a completely new built environment out of what it views as the raw material of 400 square miles of rain forest and palm-oil plantation south of Kuala Lumpur. The increasingly

celebrated place is where the Malaysian government proposes to spend between $8 and $15 billion of public and private money to turn this area of rolling countryside, rain forests and palm-oil plantations into a post-industrial district where multinational corporations will develop and test new software and multimedia products.

Malaysia's 'nothing' is giving way to two new high-technology cities: Cyberjaya, what one pundit called an 'info tech omphalos,' and Putrajaya, a new cyber-ready capital including an administrative capital and a new international airport. Today their major highway is a $2 billion fiber network under construction. But the plan is that in these cities bureaucrats will serve the public in cyberspace, consumers will shop with smart cards, children will attend virtual schools, professors will lecture electronically at the planned Multimedia University, executives will manage through tele-conferencing, and patients will be treated through telemedicine. The government has struck deals with many of the world's leading computer communication companies, including Microsoft, IBM, and Nortel which will establish development sites in the region to test new products and services such as electronic commerce, telemedicine, virtual education, paperless administration, and state-of-the-art electronic surveillance and policing. The companies will enjoy substantial tax freedom, the opportunity to import their own labour and technology, and to export all capital and profit. Malaysia hopes to use this project to jump start a stagnant economy and move beyond a low wage platform as the basis for growth.

The implications of purchasing this hoped-for growth by turning its land into the locus for transnational cyber-development projects and its citizens into beta testers for electronic capitalism are profound. Like the Silicon Alley project, the Super Corridor is to be built on two conceptions of connectivity, including the idea that the creation of a dense web of multinational businesses in a new space can propel national development and the idea that social progress grows out of fully integrating citizens into the electronic web. The MSC is an effort to stem the erosion in the massive growth that Malaysia experienced based on a labor cost advantage it enjoyed in computer and telecommunication hardware production. Having lost that advantage to other Asian nations, particularly to Bangladesh, Vietnam, and China, the Malaysian government believes it can pioneer in software and product development. Malaysia proposes nothing short of making a national model out of the city-state Singapore's centrally directed, export-oriented, high-technology approach to development [p. 354]. Indeed, although the MSC is concentrated in one soon-to-be developed region, plans exist to support the MSC with a hardware corridor in the north of Malaysia, including the island of Penang, that would attract national and foreign businesses interested in higher-end production with more skilled labor than can be found in the lowest wage regions of Asia.

Malaysia marks an important test of whether the once super fast-growing regions of Asia can continue to grow in the highly competitive area of software engineering and information technology product development. It also bears close scrutiny because Malaysia proposes to retain tight censorship, strong libel laws, and a patriarchal Islamic culture, even as it welcomes foreign multinationals, inviting them to test the full range of new media products on its citizens. Recent developments in global financial and equity markets also mark this as a case to watch because massive declines in currency values, near collapse of stock prices and the withdrawal of foreign capital have created huge rifts between Malaysia (joined by Indonesia, Singapore and Thailand) and first world powers that once pointed to these so-called Asian tigers as evidence for the success of traditional modernisation schemes (see Tim Bunnell, p. 384).

CONCLUSION: SPACE FOR CITIZENSHIP?

Notwithstanding all the talk about the death of distance, research demonstrates that, whether defined in geographical or electronic terms, place matters. Rather than annihilating space with time, the application of communication and information technologies transforms space, including both the space of places and of flows. This reading has addressed one leading edge example of transformed spaces, the technopole, or high-tech centre. Drawing on recent exemplars, it examined the evolution of technopoles from a largely productivist space to one based on connectivity among producers and consumers, both collective and individual. The New York and Malaysia examples are important because they represent both an expansion and a tightening of the technopole net and because they attract widespread attention by enacting myths of

birth and rebirth which provide a particularly magical allure to the space these technopoles occupy.

These are also important cases because they raise fundamental questions, which have not received much attention in a largely technicist and economistic literature, about the nature of governance and citizenship in the technopole. Both New York and Kuala Lumpur are pioneering in redefining the control of space by privatizing and internationalizing it in new and more penetrating ways. They are providing striking examples of just what local citizenship may come to mean in high-technology places. It is indeed hard to find in the technopoles of the world any genuine source of inspiration for fresh thinking about citizenship at the local level, for ways to return to its original of meaning of citizenship in the city or the community.

This reflection on the local raises important issues about the metaphor of connectivity which is so prominent in contemporary discourse. Connectivity does not mean that distance is dying or that geography is at an end. Rather, the architecture of connectivity accentuates the importance of certain nodes in its global networks making particular spaces, such as the Silicon Alley–mid-town Manhattan nexus and the Kuala Lumpur–Cyberjaya–Putrajaya nexus in Malaysia. Additionally, it offers powerful tools to deepen and extend existing practices that tighten certain power relations. The Silicon Alley–Madison Avenue connection forms a central node in the global advertising business, serving as the springboard for the commercialization of the web.

Connectivity also justifies transformations in governance that accelerate the privatization of public services and, as the evidence from Malaysia attests, the deepening of a surveillance society. Connectivity is a powerful metaphor, indeed a time–space transcending myth. It is all the more powerful because the comforting image of a shrinking globe of connected peoples masks fundamental transformations, both globally and locally, in political, economic and cultural activity.

REFERENCES FROM THE READING

Castells, M. and Hall, P. (1994), *Technopoles of the World*, London: Routledge.

Coopers & Lybrand (1996), *The New York New Media Industry Survey*, New York: Coopers & Lybrand.

Coopers & Lybrand (1997), *The Coopers & Lybrand Money Tree*, New York: Coopers & Lybrand.

Johnson, K. (1997), 'The place for the aspiring Dot Com: Internet industry's most popular address is Manhattan', *The New York Times*, September 30, B–1.

Ravo, N. (1998), 'Silicon Alley seeing more deals and cash', *The New York Times*, February 12, p.B–9.

Editor's suggestions for further reading

Mosco, V. (1999), 'Citizenship and the technopoles'. In A. Calabrese and J.-C. Burgelman (eds), *Communication, Citizenship and Social Policy*, New York: Rowman and Littlefield, 33–48.

Indergaard, M. (2003), 'The webs they weave: Malaysia's multimedia super-corridor and New York's Silicon Alley', *Urban Studies*, 40(2), 379–401.

Pratt, A. (2000), 'New media, the new economy and new spaces', *Geoforum*, 31, 425–36.

'Cyberspace and Local Places: The Urban Dominance of Dot.Com Geography in the Late 1990s'

Matthew Zook

Editor's Introduction

Our third analysis of cybercity economies deals with the extremely uneven geography of Internet content production. Through a sophisticated empirical analysis of the geography of dot.com registrations at various scales within the United States, Matthew Zook, a geographer from the University of Kentucky, demonstrates that some cities are emerging as key sites for the 'production' of cyberspace whilst others are being bypassed by it. As the economic value-added of the Internet shifts to involve a greater emphasis on the production of content, Zook shows that only certain advantaged cities are able to bring together the venture capital, skills, background in publishing and media, attractiveness to entrepreneurs, and innovation capacity to genuinely sustain major concentrations of Internet content production.

Zook's data gives an excellent overview of the ways in which ICTs, in supporting a complex and multifaceted restructuring of a wide range of economic sectors, are changing the economic landscapes of cybercities. Using the location of dot.com registrations as a useful proxy for these dynamics, Zook discovers that the economic landscapes of Internet content production has three key features. First, the major urban corridors of the east and west coast are extremely dominant at the continental scale (although major concentrations of activity also occur in some key mid west cities like Chicago and Austin, Texas).

Second, certain major metropolitan regions – notably San Francisco, New York and Los Angeles – are the key absolute concentrations of Internet content production. However, when account is taken of the sheer size of city economies, highly concentrated specialisations in cyberspace content production also emerge in Provo-Orem (Utah), San Diego, San Francisco and Austin.

Finally, Zook shows that Internet content production is concentrated in particular districts within cities – the major business districts and gentrifying multimedia neighbourhoods like Silicon Alley (New York) and the South of Market (SoMA) district in downtown San Francisco. The social and economic conflicts that have arisen in the latter case are considered in detail by Rebecca Solnit and Susan Shwartenberg (p. 296).

The power of the Internet has provided new opportunities for interaction between spatially distant firms and individuals. This ability, however, does not end the importance of place and geography in the economy. The companies that create content for and conduct commerce via the Internet have clustered with the same unevenness that has characterized economic development throughout history. These firms, often grouped under the rubric of 'dot.coms', are the most active experimenters in Internet use but still find it desirable to locate in certain key regions. A simple fact that is often overlooked by those who see the Internet as the harbinger of the end of geography is that while electronic data and information can flow more freely, many important aspects of business such as face to face interaction, remain tied to specific locations. The result is a complicated dynamic involving the connection of specific places to global networks creating a system of production that is both place-rooted and networked at the same time.

EXPLANATION OF DATA

This reading defines the commercial Internet content business as enterprises which use the Internet to create, organize, and disseminate informational products to a global marketplace. These informational products could be the sale of physical items, (e.g. eBay) the sale of digital products, (e.g. CD-Now), the sale and use of services, (e.g. Travelocity or Thestreet.com), the use of a database search engine, (e.g. Google), or the delivery of convenience through of portals or destination sites, (e.g. Yahoo, Amazon or AOL).

This definition includes a wide array of traditional industries because the new methods for commu-nication and distribution offered by the Internet have a wider impact than any one particular sector. In a very real sense, these firms are actively engaged with a technology that is enabling the restructuring of the current organization and boundaries of their respective industries. While the exact nature of this reorganization is still undetermined, it promises to have a profound effect on the spatial organization of many portions of the economy. Because this definition of the Internet content business is based on individual firms rather than of a traditional industry as a whole, it is difficult to use standard sources of sectoral and geographically based data such as the census or county business patterns.

Given this data availability problem, it is crucial to find alternative indicators for economic activity on the Internet. Moss and Townsend (1997) use the regis-tration addresses of domain names, e.g. yahoo.com or wired.com, to determine a geographical location. This system, originally designed as a convenience for a small number of computer specialists and academics, has now become the ubiquitous means of brand identification (dot.com) for Internet companies. This arguably makes clusters of registered .com domain names the best available indicators of where the

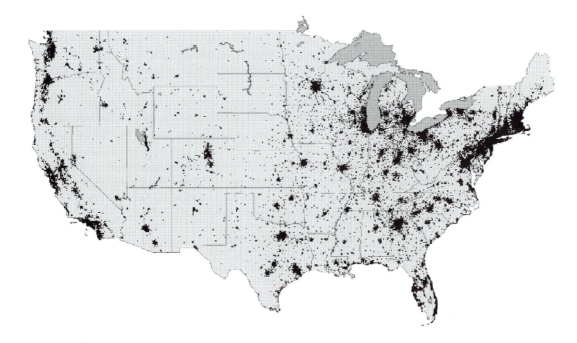

Figure 3 Distribution of commercial domain names in the United States, 1999

Source: Domain name data from author survey, July 1999

Internet content business is locating. The domain name data set for this article is based on tabulation conducted by the author during July 1999. The data was generated through the use of an Internet utility program known as 'whois' which returns contact information for a particular domain.

THE CURRENT GEOGRAPHY OF INTERNET CONTENT PRODUCTION

This section of the paper presents the empirical results of the mapping of commercial ('.com') domain names in the United States. Figure 3 shows the distribution of .com domain names around the United States which represents 75 per cent of all '.com' domains registered worldwide. One immediately sees significant concentrations in Northern and Southern California, the north west, the eastern seaboard, and scattered throughout the rest of the country. In many ways this pattern follows the distribution of population with most all cities emerging as notable sites of domain names registrations. In fact, a simple linear regression – a procedure which measures the probable causal connection between two variables – at the level of Metropolitan Statistical areas (or 'MSAs'), with number of .com domain names as the dependent variable and population as the independent variable, yields an r-squared of 0.63. This implies a fairly strong connection between the two.

However, the distribution of the Internet content business, as indicated by .com domain names, is not simply a straightforward story of correlation to population. Moving from the national level to the Consolidated Metropolitan Statistical Area (CMSA) – i.e. the city region level – it is possible to note significant differences between regions. A useful technique for comparing regions is the domain name specialization ratio. This indicates the extent to which a region is specialized in domain names compared to the United

Table 4 Top 15 concentrations of dot.com/dot.net/dot.org domain names by central metropolitan statistical area (CMSA) in the USA, 1999

CMSA name	Com/Net/Org Domains	Domains per 1,000 firms	Firm specializations ratio
1 Los Angeles-Riverside-Orange County, CA CMSA	553,325	919.41	2.04
2 New York-Northern New Jersey-Long Island, NY-NJ-CT-PA CMSA	552,750	623.14	1.38
3 San Francisco-Oakland-San Jose, CA CMSA	378,075	1,254.44	2.78
4 Washington-Baltimore, DC-MD-VA-WV CMSA	204,900	687.00	1.52
5 Boston-Worcester-Lawrence, MA-NH-ME-CT CMSA	158,850	659.91	1.46
6 Chicago-Gary-Kenosha, IL-IN-WI CMSA	154,050	496.63	1.10
7 San Diego CA	130,150	1,270.46	2.82
8 Miami-Fort Lauderdale, FL CMSA	122,400	707.42	1.57
9 Dallas-Fort Worth, TX CMSA	113,300	549.08	1.22
10 Philadelphia-Wilmington-Atlantic City, PA-NJ-DE-MD CMSA	105,850	444.25	0.99
11 Seattle-Tacoma-Bremerton, WA CMSA	100,125	633.95	1.41
12 Atlanta GA	98,225	618.38	1.37
13 Houston-Galveston-Brazoria, TX CMSA	87,475	481.99	1.07
14 Denver-Boulder-Greeley, CO CMSA	67,050	588.90	1.31
15 Phoenix-Mesa AZ	64,725	649.79	1.44
National Totals	4,849,225	450.50	1.00

Source: Domain name data from author survey, July 1999; firm data from Dun & Bradstreet MarketPlace, 1998

States as a whole. A value greater than 1.00 indicates a higher specialization than the national average and a value less than 1.00 indicates a lack of specialization.

Although most CMSAs have specialization ratios higher than the national average, the variance between these regions can be quite extreme. For example, as Table 4 illustrates, the San Francisco Bay region has almost three times the number of domain names per firm as either the Chicago, Philadelphia or Houston metropolitan regions. Also of interest is the noticeable discontinuity between the top three regions and the rest in terms of total numbers of domain names. Together, the Los Angeles, New York and San Francisco regions have more domain names than the next twelve largest metropolitan regions combined.

However, the geography of the Internet cannot simply be described in terms of total numbers of domains since in many ways this simply reflects size. As Table 5 illustrates, one obtains a very different ordering of metropolitan regions on the basis of domain name specialization ratios – the New York region drops off the table completely and Provo-Orem,

Utah moves to the top. Additionally, new regions, such as Boise, Idaho; Austin, Texas; and Las Vegas, Nevada, appear as smaller but highly specialized areas of .com domain names.

Furthermore, domain names are not evenly distributed within regions but are clustered in particular locations. In the largest three regions there are high concentrations of domain names in the city of San Francisco, Manhattan, around San Jose and Silicon Valley, and in the Santa Monica–Hollywood area of Los Angeles. In addition there are numerous other smaller concentrations such as Berkeley–Emeryville, Brooklyn, and Long Beach.

Given these sub-regional clusters, it is useful to take the analysis to the next smallest geographical category, i.e. the city. Interestingly, the top three cities mirror the experience of the metropolitan regions with New York, Los Angeles, and San Francisco containing more domain names than the next twelve cities. As one would expect, Table 6 shows even more variation in specialization ratios at this finer grain of analysis. For example, the cities of Houston, Denver, Chicago, and

Table 5 Top 15 specialised central Metropolitan statitstical areas (CMSAs) in dot .com/dot .net/dot .org domain names in the USA, 1999

CMSA name	Com/Net/Org domains	Domains per 1,000 firms	Firm specializations ratio
1 Provo-Orem UT	17,375	1,618.09	3.59
2 San Diego CA	130,150	1,270.46	2.82
3 San Francisco-Oakland-San Jose, CA CMSA	378,075	1,254.44	2.78
4 Austin-San Marcos TX	48,550	1,002.31	2.22
5 Los Angeles-Riverside-Orange County, CA CMSA	553,325	919.41	2.04
6 Las Vegas NV-AZ	34,900	835.35	1.85
7 Boise City ID	13,575	744.33	1.65
8 Roanoke VA	7,300	725.36	1.61
9 Tucson AZ	19,175	723.45	1.61
10 Gainesville FL	5,350	714.67	1.59
11 Miami-Fort Lauderdale, FL CMSA	122,400	707.42	1.57
12 Santa Barbara-Santa Maria-Lompoc CA	12,275	699.75	1.55
13 Washington-Baltimore, DC-MD-VA-WV CMSA	204,900	687.00	1.52
14 Boston-Worcester-Lawrence, MA-NH-ME-CT CMSA	158,850	659.91	1.46
15 Phoenix-Mesa AZ	64,725	649.79	1.44
National Totals	4,849,225	450.50	1.00

Source: Domain name data from author survey, July 1999; firm data from Dun & Bradstreet MarketPlace, 1998

Table 6 Top 20 concentrations of commercial domain names by US cities, 1999

City name	Commercial domains	Domains per 1,000 firms	Firm specialization ratio
1 New York NY	149,225	1,075	2.39
2 San Francisco CA	78,075	1,595	3.54
3 Los Angeles CA	71,200	783	1.74
4 San Diego CA	66,525	1,553	3.45
5 Houston TX	53,825	468	1.04
6 Miami FL	52,000	624	1.38
7 Chicago IL	43,750	493	1.10
8 Dallas TX	39,375	606	1.35
9 Atlanta GA	33,800	681	1.51
10 Austin TX	33,425	965	2.14
11 Seattle WA	31,725	700	1.55
12 San Jose CA	28,450	1,020	2.26
13 Washington DC	25,850	711	1.58
14 Minneapolis MN	24,750	566	1.26
15 Las Vegas NV	23,700	791	1.76
16 Fort Lauderdale FL	22,725	602	1.34
17 Portland OR	21,175	540	1.20
18 Phoenix AZ	20,350	488	1.08
19 Tucson AZ	16,175	639	1.42
20 Denver CO	16,025	411	0.91

Source: Domain Name data from author survey, July 1999; firm data from Dun & Bradstreet MarketPlace, 1998; Cities defined by zipcodes

Phoenix are hovering close to the national average and San Francisco stands out as the most highly specialized major center of .com domain names. Of course, in many ways these findings reflect the larger urban structure of these cities and regions. For example, although Denver, defined as a city, falls below the national average for domain names, this is probably due to the sprawling nature of recent urban development around the historic urban core.

Figures 4 and 5 illustrate how these commercial domain names are distributed in the New York and San Francisco Bay regions. There are significant concentrations of domain names in the financial district of San Francisco as well as the South of Market or Soma district also known as Multimedia Gulch [p. 296]. The New York City Map shows concentrations in domain names near the tip of Manhattan in the TriBeCa district,

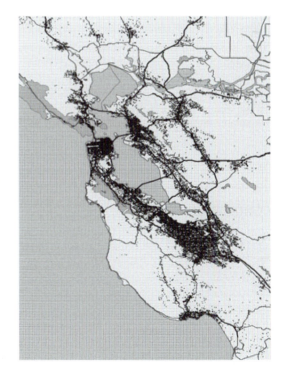

Figure 4 Location of commercial domain names in the San Francisco Bay region, 1998

Source: Domain name data from author survey, June–July 1998

Figure 5 Location of commercial domain names in New York metropolitan region, 1998

Source: Domain name data from author survey, June–July 1998

SoHo, and Greenwich Village which generally corresponds to the parts of the city referred to as Silicon Alley (Vincent Mosco, p. 199). One can also see a dense concentration of domain names on the east side just south of Central Park where many corporations and businesses are located.

CONCLUSION

The production of Internet content exhibits a remarkable degree of clustering despite its much-ballyhooed spacelessness. Despite the possibility of every region developing its own particular brand of Internet content, it appears that, in fact, distinct inter-regional differences in domain names are occurring. Given the fact that the Internet offers enormous returns to scale in which small first mover advantages can quickly translate

into meaningful competitive advantage, the geography described here has significant implications for future trajectory of the Internet content business.

However, one of the most interesting aspects about the three leading regions is that, although they all have a significant Internet presence, they are very dissimilar to one another in basic industrial makeup. This suggests that there are multiple pathways towards the development of an Internet content specialization. Industry analysts often point to the existing high-technology industries of Silicon Valley, the financial and publishing industries of New York, and the entertainment industry of southern California to explain their high concentrations of Internet related activity.

This Reading's most significant finding is that the Internet is not bringing about the wholesale elimination of place-based networks in favor of cyberspace. As Graham (1997) argues, 'time and space barriers are only selectively being overcome. Place-based and place-bound ways of living, and the social, economic institutional, and cultural dynamics that can arise where urban propinquity does matter are still critically important in shaping how cities and localities are woven into global lattices of mobility and flow.' Just as the earlier technologies of the telegraph and railroad upset the standards upon which the competitive advantage of firms and regions were based, the Internet promises a reorganization of the production systems of a wide range of industries. The challenge for future research will be understanding and analyzing the way specific regional networks and individual firms adapt to and exploit the opportunities offered by the Internet.

REFERENCES FROM THE READING

Graham, S. (1997), 'Cities in real-time age: the paradigm challenge of telecommunications to the conception and planning of urban space', *Environment and Planning A*, 29, 105–27.

Moss, M. and Townsend, A. (1997), 'Tracking the net: using domain names to measure the growth of the Internet in U.S. cities', *Journal of Urban Technology*, 4(3) 47–60.

Editor's suggestions for further reading

Dodge, M. and Kitchin, R. (2001), *Mapping Cyberspace*, London: Routledge.

Drennan, M. (2002), *The Information Economy and American Cities*, Baltimore: Johns Hopkins University Press.

Kellerman, A. (2002), *The Internet on Earth: A Geography of Information*, London: Wiley.

Moss, M. and Townsend, A. (1997), 'Tracking the net: using domain names to measure the growth of the Internet in U.S. cities', *Journal of Urban Technology*, 4(3) 47–60.

Moss, M. and Townsend, A. (2000a), 'The Internet backbone and the American metropolis', *The Information Society*, 16, 35–47.

Moss, M. and Townsend, A. (2000b), 'The role of the real city in cyberspace: understanding regional variations in Internet accessibility'. In D. Janelle and D. Hodge (eds), *Information, Place and Cyberspace*, Berlin: Springer, 171–86.

Pratt, A. (2000), 'New media, the new economy and new spaces', *Geoforum*, 31, 425–36.

Scott, A. (2000), *The Cultural Economy of Cities*, London: Sage.

Townsend, A. (2001), 'The Internet and the rise of the new network cities, 1969–1999', *Environment and Planning B: Planning and Design*, 28, 39–58.

Zook, M. (2000), 'The web of production: The economic geography of commercial Internet content in the United States', *Environment and Planning A*, 32, 411–26.

Zook, M. (2001), 'Old hierarchies or new networks of centrality? The global geography of the Internet content market', *American Behavioural Scientist*, 44(10), 1679–96.

Zook, M. (2002), 'Grounded capital: venture financing and the geography of the Internet industry, 1994–2000', *Journal of Economic Geography*, 2(2), 171–7.

Zook, M. (2003), *The Geography of the Internet*, Oxford: Blackwell.

'Teleworking and the City: Myths of Workplace Transcendence and Travel Reduction'

from *Cities in the Telecommunications Age* (2000)

Andrew Gillespie and Ranald Richardson

Editor's Introduction

Our fourth cybercity economy reading examines the changing work practices which surround teleworking, mobile working and remote working. Our interest, in particular, is centred on the complex relationships between ICT-mediated work, physical travel and the city. The authors of this extract are Andrew Gillespie and Ranald Richardson, both of whom are geographers at Newcastle University's Centre for Urban and Regional Development Studies in the UK (a major centre in the analysis of communications geography over the past twenty-five or so years).

Drawing on extensive empirical work on these relationships, Gillespie and Richardson examine whether or not the relationships between teleworking, physical commuting and the city are characterised by broad substitution tendencies (with ICT-mediation being used to reduce physical trips). They argue that, despite long-standing assumptions that people will increasingly work from isolated 'electronic cottages', true electronic home working is actually likely to occur on only a limited scale. This is because most workers will still need to physically travel to work for part of the working week. Thus, the vast majority of people remain tied to locating their homes within easy reach of urban regions so that they can physically commute to work within a reasonable time and cost.

Gillespie and Richardson next examine the phenomenon of mobile teleworking through which workers operate on the move to service an extended geography of clients, while keeping in touch through ICTs. With the growth of 'hot desking' firms in some sectors are attempting to reduce the fixed costs of maintaining offices for every staff member whilst using mobile ICTs to support nomadic working practices. This can reduce the amount of travelling done by workers. However, as service personnel may spend more time with clients, actual distance travelled by nomadic workers may also increase.

The third type of teleworking involves group or team teleworking. This involves cross-organisational teams working together on tasks through ICTs. Whilst research on the travel and urban impacts here is very limited, Gillespie and Richardson suggest that, because such practices involve extending ICT-based contacts with wider and wider groups of personnel, efforts to back these up with physical trips and face to face meetings may actually lead to greater physical, as well as virtual, mobilities.

Gillespie and Richardson's final category involves the use of ICTs to support remote offices (also called back offices or call centres). These deliver mass services to more or less distant markets. A rapidly growing phenomenon – as firms restructure face-to-face service outlets distributed across geographic space – and replace them with large, centralised call centres in low-cost locations, these offices are probably the most important, but also the most neglected, form of telework. Their urban and travel implications are ambivalent. As Gillespie and Richardson argue, the restructuring which leads to the establishment of call centres often involves the geographical decentralisation of administration centres from high-cost global and second tier cities to low-cost, peripheral urban regions (see Ewart Skinner, p. 218). It can also involves the replacement of small, customer-facing service outlets across cities and regions, and the commuting that made them function, with large scale call centre complexes, usually in new edge of city business parks of the old industrial or low-cost peripheral regions. The latter are usually almost completely reliant on car-based commuters because of their 24-hour shift patterns and locations away from good public transport facilities.

Gillespie and Richardson's work is notable for the way in which they analyse teleworking as a complex range of ICT-mediated practices, rather then just the substitution of traditional city commutes by homeworking (as is usually the case). As a result, their analysis suggests that the links between teleworking, physical transport and cities are far more complex and complementary than the usual perspective which simply emphasises teleworking as a simple threat to the traditional economic centrality of the corporate city core.

TELEWORKING AND THE NEED TO TRAVEL

Much of the rhetoric around teleworking, particularly in the United States, has concerned its potential to substitute for travel, specifically the journey to work. Although the implications for travel would therefore at first sight appear to be obvious, in reality the outcomes can be rather complicated. Mokhtarian (1990) reminds us that there are four possible interactions between telecommunications and travel:

(i) substitution (i.e., telecommunications decreases travel);
(ii) enhancement (i.e., telecommunications directly stimulates travel);
(iii) operational efficiency (i.e., telecommunications improves travel by making the transportation system more efficient);
(iv) indirect, long-term impacts (e.g., telecommunications may affect locational and land use decisions, thereby affecting travel).

Based on an extensive review of the literature, Graham and Marvin (1996, p. 269) conclude that:

the relationship between telecommunications and the urban environment is not as simple as the substitutionist perspective would imply. Instead, electronic and physical transformation proceed in parallel, producing complex and often contradictory effects on urban flows and spaces.

As a means of illustrating the validity of this conclusion, this essay will examine the travel implications of various types of teleworking, drawing upon a report undertaken for the U.K. Parliamentary Office of Science and Technology by the authors (Gillespie, Richardson, and Cornford, 1995) . . .

Electronic homeworking

From the evidence reviewed, we can surmise that the growth of electronic homeworking (EHW) will be relatively modest, and that it will in the main be part-time in nature, perhaps being undertaken for a day or two a week (this is certainly the case in the United States). Its impact upon urban form and travel will be hardly revolutionary therefore, both because its incidence will be relatively limited and because most electronic homeworkers (except those who are self-employed) will still need to commute to their office for the majority of their working days.

Because of the latter feature, and because the incidence of electronic homeworking is higher in those professional occupations and service activities that are concentrated in cities, EHW in the United Kingdom is overrepresented in metropolitan regions, particularly London. Huws's (1993) survey of employers in the United Kingdom established that while London accounted for 16 percent of the country's total sample of employers, it accounted for 24 percent of employers with teleworkers. Although we do not yet have access to data on the residential location of teleworkers, we can assume that they are geographically constrained by the need to travel to their employer's premises and, frequently, to the premises of clients as well, for home-based teleworkers working for a single employer spend on average a quarter of their time on the employer's premises and a further quarter elsewhere. Even freelance teleworkers, who might be assumed to have the greatest degree of locational freedom and who are most usually associated with rurally based lifestyles, are often constrained by the need to be close to clients, and survey evidence reveals that more than half of such teleworkers live in the centers or suburbs of cities

[. . .]

Mobile teleworking

Although there is agreement that nomadic or mobile teleworking is growing significantly, there are few reliable statistics on its incidence or rate of growth. Gray, Hodson, and Gordon (1993) estimate that there are more than 7 million nomadic desk jobs in the United States, and over 1.5 million in the United Kingdom, but the basis for their estimates is not stated. There has always been mobile work, of course, such as sales staff and field engineers; our interest is in the way some firms are starting to look strategically at how new technologies can be used to change working practices, with a view to reducing costs and improving customer service, and the locational and travel implications of such changes.

One example of a new working practice with potentially significant implications for the demand for office space, the location of office space, and the substitution/generation of travel is "hot-desking," introduced first in computer companies such as IBM and Digital, but now spreading more widely into firms with other mobile staff. In the case of IBM, the stimulus for hot-desking was cost competition, coupled with a recognition that the company was overprovided with expensive office space, given that many of the staff spent much of their working days out on the road or at customers' premises. The hot-desking scheme developed and introduced by IBM involved providing an average of one desk for two workers, with all of the desk space shared, and increasing the amount of space for meetings. The new working practices have proved successful in terms of productivity, generating the following changes in employee time use:

- travel time –13%
- time with customer + 36%
- time in office –23%
- total space saved –30%

A significant increase in nomadic working in conjunction with concepts such as hot-desking, if taken up widely, would have obvious implications for office space demand, with a significant reduction in the average space requirements of certain sectors. The increased emphasis on the mobility of the workforce is also likely to encourage firms to locate in out-of-town sites with easy access to the road network and with plentiful parking space rather than in city-center locations, thus reinforcing existing trends toward out-of-town developments.

To the authors' knowledge, no published studies have been carried out on the transport implications of this form of mobile teleworking, and there are a range of possible outcomes – some contradictory – if the practice continues to grow. Where workers have traditionally been mobile but are now being discouraged from traveling to their office base, the number of miles traveled per worker should fall, as in the IBM case reported above. However, as part of the rationale behind strategic changes in working practices such as hot-desking is to "get closer to the client," one would also expect that there would be more visits to each client than previously. If customers come to expect more site visits as a matter of course then more, rather than fewer, miles may be traveled. A further possibility is that if mobile working spreads to new areas of work, the overall travel could increase even though time traveled per worker falls (new groups of mobile workers could also have an impact on existing modal splits, with more workers having to take the car). Finally, new travel patterns may emerge as the classic commuting trip to the office declines and workers stay at home until it is time to visit the client. We can assume that these trips are likely to be less city-center-dominated than the classic commute trip, as well as being more car-dependent.

Group or team teleworking

Despite the existing technological limitations on the development of team teleworking, there can be little doubt that new forms of work organization are pulling in this direction. An increasing business focus on quality and customer service, the need for flexibility to cope with turbulent markets, and an emphasis on innovation are leading to new structures within information-based organizations in which "task-focused teams," often crossing organizational boundaries, are becoming the new paradigm of work organization. Of course, not all teamworking involves spatially distributed teams, but multiple pressures are pushing in this direction. The process of globalization, the increasing need for organizations flexibly to combine and recombine their spatially dispersed specialized human resources, and the requirement to forge strategic alliances with other organizations possessing complementary assets are all leading to the construction of task-focused teams with geographically distributed participants.

As with the other forms of telework considered above, one might assume that team telework would cut down on demand for travel. As far as we are aware, however, no detailed studies have been carried out into this aspect of team telework, so we cannot make detailed comments on the travel patterns it generates. We would, however, make the following observations. Generally speaking, computer-supported team telework not only means more telemediated contact with groups across space; it also means telemediated contact with groups with whom contact has been limited or nonexistent, as firms or networks reorganize to take advantage of distributed resources (for example, skilled labor and laboratory facilities). These new contacts also generate new travel demands as groups find that technological and organizational capacity is not (for the moment, at least) sufficiently developed to take shared tasks from inception to fruition. So, for example, it is likely that distributed R&D teams will travel for face-to-face meetings across the world, whereas previously they may have worked only locally. Even for teams working within national boundaries, more travel can be generated by teamwork. In the case of IBM, for example, distributed teams focused on business sectors, so as to get "closer to the client," mean that teamworkers may be physically further away from both their clients and their fellow teamworkers. Despite the sophisticated supporting electronic networks, face-to-face meetings are still required, both with clients and with other team members, but now instead of popping next door to meet work colleagues, or traveling a few miles to meet clients, workers have to travel up and down the motorway on a regular basis.

We would therefore anticipate that team tele-working, in expanding the geographical spread of participants in the virtual work activity space, is likely to lead to new demands for travel and to substantial increases in the distances over which business travel takes place.

Remote offices: the example of call-centers

The locational and travel implications of call-centers are particularly complex and interesting, due to two features; first, the work concerned frequently moves between cities as well as between different types of location within cities; and second, the travel implications extend beyond work travel to also encompass travel to consume.

In the case of telebanking in the United Kingdom, for example, there are two clear locational implications with respect to employment. First, the possibility of separating production and consumption is allowing the relocation of substantial parts of the production process to lower-cost parts of the country, with cities such as Leeds, Edinburgh, and Glasgow gaining appreciable numbers of telebanking jobs. To an extent, therefore, travel to work to a bank branch in say, London, or to a small town in the outer Southeast of England is being replaced by a journey to work to a call center in, say, Leeds (with an appreciable degree of job downsizing en route, due to the much higher levels of labor productivity associated with telemediated service delivery).

Second, "in contrast to most bank branches, nearly all telebanking operations are on business parks on the edge of cities, rather than in town or city centers. There is no need for an expensive city-centre location" (Marshall and Richardson, 1996, 1855). The two locational effects are usually compounded such that jobs are in effect moving from the center or suburban high street of one city to an out-of-town business park location in another city. The shift in the type of intraurban location is inducing a clear modal shift in that out-of-town call centers tend to have a much higher car mode share than the jobs they are replacing. The modal shift due to locational change is exacerbated by the greater incidence of shift-working in telebanking operations, militating against the use of public transport

[. . .]

CONCLUSION

What can we conclude about the implications of teleworking, broadly defined, for urban form and for travel patterns and travel behavior? First, it might be observed that it is remarkable that so much research effort has been expended on studying the locational and travel implications associated with a handful of electronic homeworkers, when so little has been expended on studying the locational and travel implications associated with a very much larger number of workers whose working practices are being radically changed by new ICTs [. . .]

Second, the notion that teleworking will lead to reduced travel, and hence to more environmentally sustainable cities, is, at the very least, open to question. Even with respect to EHW, where the most obvious potential for travel substitution is to be found, we have concluded that the most likely long-term effect in the United Kingdom is that the geographical extent of the London "daily urban system" will be expanded, and the nodality of the region in terms of travel patterns will be further reduced. When we consider the likely travel impacts of the growth of mobile working and of spatially dispersed teamworking, we are at once confounded by the almost complete absence of empirical research. However, both of these significant developments in working practices appear likely to expand the daily activity spaces of individual workers and to lead to significantly increased journey distances. It also seems likely that significant modal shifts in the direction of increased car dependency will be associated with these new ways of working. Finally, the location of teleservice employment in large call centers has, within the context of the particular planning regime in the United Kingdom over the last ten years, clearly been associated with a shift from city center and high-street locations to out-of-town/edge-of-town business park locations, and will have helped fuel the growth in car dependency in the journey to work.

The "reduced demand for travel" scenario, which is usually invoked with respect to teleworking may, then, be decidedly misleading in terms of its apparently positive contribution to building more sustainable cities. Not only are communications technologies expanding the "activity spaces" within which work takes place, leading to longer distances traveled, but in addition, journey patterns associated with new ways of working are becoming more diffuse and less nodal, and hence more difficult to accomplish by public

transport. This effect is exacerbated by companies adjusting their premises stock to accommodate more effectively new ways of working, leading to a reduction in demand for conventional city-center offices and an increase in demand for office space in office park environments with high levels of accessibility to the motorway system. At the same time, the substitution of telemediated for face-to-face banking and other services risks further undermining the role of city centers and high streets, as branch offices are closed and customers are served from large teleservice centers, themselves usually located on business parks. Far from contributing to more sustainable urban ways of life and travel behavior, therefore, teleworking and teleservices seem to be developing hand in hand with lower-density, less nodal urban forms and with travel behavior that is more car-dependent than before. Teleworking and tele-activities are, then, perhaps best understood not as developments that suppress the demand for mobility but, rather, as forms of what might best be described as "hypermobility."

REFERENCES FROM THE READING

Gillespie, A., Richardson, R. and Cornford, J. (1995), *Review of Telework in Britain: Implications for Public Policy*. Report prepared for the UK Parliamentary Office of Science and Technology, Newcastle upon Tyne: Centre for Urban and Regional Development Studies.

Graham, S. and Marvin, S. (1996), *Telecommunications and the City: Electronic Spaces, Urban Places*, London and New York: Routledge.

Gray, M., Hodson, N. and Gordon, G. (1993), *Teleworking Explained*, Chichester: John Wiley and Sons.

Huws, U. (1993), *Teleworking in Britain*, Employment Department Research Series No. 18, Sheffield: Employment Department.

Marshall, J. N. and Richardson, R. (1996), 'The impact of "telemediated" services on corporate structures: the example of "branchless" retail banking in Britain', *Environment and Planning A*, 28, 1843–58.

Mokhtarian, P. L. (1990), 'A typology of relationships between telecommunications and transportation', *Transportation Research*, 24A(3), 231–42.

Editor's suggestions for further reading

Forer, P. and Huisman, O. (2000), 'Space, time and sequencing: substitution at the virtual/physical interface'. In D. Janelle and D. Hodge (eds), *Information, Place and Cyberspace*, Berlin: Springer, 73–90.

Giannopoulos, G. and Gillespie, A. (eds) (1993), *Transport and Communications in the New Europe*, London: Belhaven.

Gillespie, A. (1992), 'Communications technologies and the future of the city'. In M. Breheny (ed.), *Sustainable Development and Urban Form*, London: Pion, 67–77.

Nilles, J. (1988), 'Traffic reduction by telecommuting: a status review and selected bibliography', *Transportation Review* 22A, 301–17.

Nilles, J., Carlson, F., Gray, P. and Hanneman, G. (1976), *The Telecommunications–Transport Trade Off*, Chichester: Wiley.

Salomon, I. (1985), 'Telecommunications and travel: substitution or modified mobility?', *Journal of Transport Economics and Policy*, September, 219–35.

Salomon, I. (1986), 'Telecommunications and travel relationships: a review', *Transportation Research*, 20A(3), 223–38.

T
W
O

'The Caribbean Data Processors'

from *Global Productions: Labor in the Making of the Information Society* (1998)

Ewart Skinner

Editor's Introduction

Our fifth reading on cybercity economies engages further with the decentralising logics which surround the establishment of call centres and back offices across transnational scales. These are hybrid urban ICT sites through which large numbers of routine service workers can service major metropolitan, regional, and national markets, from a distance, using telephones and ICTs. Call centre flows are now occurring at transnational and transcontinental, as well as sub-national scales. For example, workers in call centres in Delhi, India now watch British soap operas to help them understand the regional accents of British callers enquiring about their utility bills, insurance policies or computing software. They log on to web sites telling them what the weather is like in the UK so they can empathise with their often cold and rain-sodden clients in Manchester or Glasgow, even though it may be 45 degrees Celsius outside the office window.

In this piece the author, Ewart Skinner, a telecommunications scholar at Bowling Green State University in the USA, addresses the growing transnational scales of such decentralising tendencies through a detailed case study of the growth of data processing offices in Jamaica which service the USA to the north. Skinner focuses in particular on the new urban spaces that have been customised in Jamaica – as in many developing and peripheral economies – to try and tempt in the data processing and call centre activities of major transnational corporations from the global north. Skinner describes one such new urban zone – Jamaica Digiport International (JDI) – in detail. Closely linked with wider efforts to develop Jamaica as a home for freeport and tax-free trade and manufacturing, JDI is a classic example of the new urban spaces that are emerging far away from the centralising and high value-added heartlands of global, second tier and high-tech cities.

Instead of a logic of face-to-face contact and reflexive, innovative work, the logic here is one of using ICTs to exploit cheap information labour. This is done to offer routinised and standardised labour in extremely low-cost locations, whilst still using high capacity telecommunications links to connect to distant northern markets. As Skinner notes, new urban 'digiport' and remote information processing zones like JDI are heavily subsidised by public money, tax breaks and infrastructural support. They benefit from ICT services that are far superior to those available just outside their (often fortified) walls and gates. And they specialise in providing the relatively low skill information services that support the hidden, taken for granted supports to the economies of wealthy cybercities in the global north: answering 0800 calls, processing of vouchers and advertising returns, organising telemarketing, and so on.

Analyses like Skinner's suggest that, every time one rings a call centre, it is worth asking the answering person where they are geographically in the world. The answer may be both a surprise and an indicator of the often hidden, transnational flows of routinised electronic labour, that, just as importantly as the better researched flows of global financial services, elite analysts, Internet entrepreneurs and stockbrokers, do so much to tie city economies together in today's world.

Driven by a mix of utopianism and pragmatism, Caribbean governments have undertaken sophisticated public relations campaigns to attract foreign data-processing firms. Working through international organizations such as the World Bank, United States Agency for International Development (USAID), and the Economic Commission for Latin American and the Caribbean (ECLAC), and consultancy firms such as INTEX USA and The Services Group, both of Arlington, VA, Caribbean nations have developed extensive blueprints for survival in the international data- processing sector. The consensus is that employment expansion will be an immediate payoff. Recent unemployment rates of 24% in Barbados, 30% in the Dominican Republic, and 25% in Jamaica, are distressingly high. To unemployed and underemployed nationals, data entry operators receive what are perceived to be relatively good wages, and workers are easily captive thereby of the sophisticated information campaigns promoting the industry.

TRANSNATIONAL ENTERPRISES: ADVANTAGES

It was not lost on the more aggressive foreign data processing investors (FDPIs) that cultural proximity and advantages in labor conditions, limited regulation, politics, technology, culture, and the already vulnerable economic situation of Caribbean nations provided optimum opportunities for firms wishing to outsource. For transnational firms, the pluses (abundant, inexpensive labor; low staff turnover; high staff loyalty; underemployed, educated workers with English fluency in some countries; increases in use of idle computer resources, particularly night downtime, and government incentives) far outweigh the disadvantages. Concern about losing vital data and product control, privacy, security, cultural and language differences, political and economic instability, inadequate and inefficient telecommunications links, productivity shortfalls because of poor labor standards, opposition from labor groups, adverse laws and regulations, and the high cost of training and staff relocation are all disincentives for potential FDPIs. Yet given that these barriers have not slowed industry growth in the region, it appears that for U.S. firms, the growing Caribbean economy and political atmosphere are positive prospects.

FDPIs that Caribbean governments have been courting are service bureaus such as SAZTEC and foreign companies with large, daily internal data entry requirements, especially American Airlines and other corporate "end users." Other major companies that use Caribbean labor through direct outsourcing and subsidiary operations or through service bureaus include: Sears & Roebuck; Dialog (financial services); BRS (financial services); Mead Data Central (publishing); McGraw Hill Publishing Company; NPD/Nielson, Inc. (research and publishing); Texas Instruments, Inc.; Digital, Inc. (computer software company); and BPD Business Data Services Ltd. Caribbean local firms sometimes collaborate with service bureaus in joint ventures.

JAMAICA

Jamaica is the largest of the 13 English-speaking Caribbean Community (Caricom) countries. With a population of 2½ million, Jamaica has undergone one of the most ambitious data entry programs in the region. These enterprises have a choice of locating either inside or outside the publicly owned free trade zones. The government prefers the former because its strategy appears to be geared toward free trade zone-based, foreign, export-oriented, data processing enterprises. Three free trade zones exist: the Kingston Free Zone, the Garmex Free Zone, and the Montego Bay Free Zone. The Montego Bay Free Zone is the location of the country's model data processing facility, Jamaica Digiport International (JDI). Information processing inspires heady optimism about its short-term employment potential. In a document titled, "Going for the Iceberg," the Exporters of Information Services, a subgroup of the Jamaica Exporters' Association, projected that by 1994, export earnings from the sector would reach $140 million, and employment would skyrocket to 18,000 [...]

JAMAICA DIGIPORT INTERNATIONAL

Earliest of the information processing firms to develop in Jamaica was Telmar Jamaica Ltd., jointly owned by Radio Jamaica and New Jersey-based Telmar Data Systems. In 1984, the firm installed a satellite dish that transmitted processed data back to the firm's U.S. clients.

Currently, the standard in the data-processing industry and government FTZ approach is the Jamaica

Digiport International (JDI), opened in 1989 to provide low-cost satellite communications for data-processing and telemarketing companies in the Montego Bay Free Zone. It is a joint venture between Telecommunications of Jamaica (30%), AT&T (35%), and Britain's Cable and Wireless (35%). Voice, text, and data handled by JDI are beamed to a satellite off the African coast and then to clients in the United States, Canada, and the United Kingdom. (Chang, personal communication, May, 28, 1993). The JDI earth station in Montego Bay includes a 15-meter, 125 Watt, C-Band antenna. It offers its clients speeds between 0.096 and 1.544 megabits per second, international toll-free (800) numbers, credit card authorization for direct selling, and rates as low as US$0.24 per minute for calls to the United States.

The facility can be viewed as an example of electronic labor integration. Virtually every AT&T-type service is available from JDI at prices well below standard international rates. JDI provides technical and managerial support for the Jamaican data entry/data processing industry. Its license allows them to provide telecommunications services only to FTZ businesses. Operators outside the zone must use the more expensive Telecommunications of Jamaica (TOJ), the national telecommunications facility. JDI optimistically expects that 30,000 Jamaicans will eventually be employed by the industry.

Editor's suggestions for further reading

Bain, P. and Taylor, P. (2000), 'Entrapped by the electronic panopticon? Worker resistance in call centres', *New Technology, Work and Employment*, 15(1), 2–17.

Bristow, G., Munday, M. and Grippaios, P. (2000), 'Call centre growth and location: corporate strategy and spatial divisions of labour', *Environment and Planning A*, 32, 519–38.

Freeman, C. (2000), *High Tech and High Heels in the Global Economy*, Durham and London: Duke.

Mitter, S. and Bastos, M.-I. (eds) (2001), *Europe and Developing Countries in the Globalised Information Economy*, London: Routledge.

Richardson, R. (1999), 'Call centres and the prospects for export-oriented work in the developing world: evidence from western Europe'. In S. Mitter and M.-I. Bastos (eds), *Europe and Developing Countries in the Globalised Information Economy*, London: Routledge, 74–94.

Richardson, R. and Marshall, J. N. (1999), 'Teleservices, call centres and urban and regional development', *The Service Industries Journal*, 19(1), 96–116.

Skinner, E. (1998), 'The Caribbean data processors'. In G. Sussman and J. Lent, (eds), *Global Productions: Labor in the Making of the 'Information Society'*, Cresskill, NJ: Hampton Press, 57–90.

Wilson, M. (1998), 'Information networks: the global offshore labor force'. In G. Sussman and J. Lent (eds), *Global Productions: Labor in the Making of the 'Information Society'*, Cresskill, NJ: Hampton Press, 39–56.

'Geographies of E-Commerce: The Case of Amazon.com'

Martin Dodge

Editor's Introduction

Our sixth cybercity economy reading is the first of three which together address the relationships between online and electronic commerce and city economies. Fuelled by the hype of the dot.com boom in the late 1990s, a wave of rhetoric spread across the western world suggesting that the consumption side of city economies was facing a radical, and staggeringly rapid, revolution. Replacing the traditional physical trip to the high street, store or mall – with all its expense, congestion, pollution, and consumption of time – retailing was, it seemed, moving into the realms of cyberspace. 'E-tailing' challenged retailing. 'Clicks' challenged bricks. And more and more of the consumption transactions and experiences that people used to access by physically moving around urban consumption spaces were being recreated, or at least simulated, through web sites. These were constructed as elements within remote logistics and financial transaction systems and allowed people to use their credit cards to access goods, commodities and services from their homes or whilst on the move (Keller Easterling, p. 179).

In this reading Martin Dodge – a leading geographer of ICTs at University College London – analyses the urban and geographical implications of the world's shining example of global e-commerce success: Amazon.com. One of the few success stories of global e-commerce that has managed to sustain and grow despite the fallout of the bursting of the dot.com bubble in 2000–1, Amazon.com is shown as exemplifying the complex relationships between online and material flows and forms which link consumer e-commerce to city economies in complex and multidimensional ways.

Dodge emphasises four key relationships in his analysis. First, Amazon.com is a powerful global company with a priceless web site address which serves as its 'brand'. It is a centralised US transnational which uses the distributed systems of electronic finance and logistical flow to draw capital and investment away from its markets across the world to its main headquarter region (Seattle in the USA). This global system allows it to continually extend the range of goods it offers beyond books (to CDs, DVDs, software, toys, tools, clothes and so on).

Second, Amazon.com actually uses the flexible 'screen geography' of the Internet to make itself appear more local than it is. Through its many local virtual and physical affiliates and branches, it is able to localise its global e-commerce model to tie itself more intimately with different cultures and consumption habits across the world.

Third, Dodge stresses that, behind the virtual image of Amazon.com, there is a massive, material, logistical geography. This is distributed in key urban sites across the world and linked together very closely through customised ICT and transport and logistics systems. Behind the familiar 'front end' of the company's web sites, this geography consists of a string of massive warehouses, an extensive array of global logistics and transport networks, a range of data processing centres (including those run by credit card companies), and, at the local, urban scale, a massive system of physical transport flows by couriers to actually deliver the hard manufactured goods that are ordered through its web sites.

Finally, Dodge examines the impacts that the growth of e-commerce are likely to have on traditional retailing spaces in cities. He suggests that prime retailing spaces and malls are likely to survive and prosper by increasing their emphasis on the entertainment and experiential dimensions of their stores and spaces (a trend we examine in Susan Davis, p. 235). The mid-range retailing spaces, which concentrate on standardised and mass market commodities, however, are likely to suffer. Such spaces are being undercut on prices by large, centralised e-commerce services which, freed from the high overheads caused by the need to maintain a range of physical stores, will offer much lower prices.

Amazon.com opened the doors to its online store on 16 July 1995. Today, Amazon.com is the best known e-commerce company in the world, heralded by many business pundits as a paragon of online retailing and a major threat to conventional bricks and mortar store chains. It remains one of few success stories, even though it has only made very modest profits, in the pure-play online retail sector, a sector that is littered with many failures since the collapse of the dot.com boom in 2000 and 2001. In its seven years of existence, Amazon.com has built an enviable reputation as a quality online retailer and an internationally renowned brand. This is all the more remarkable when you consider that Amazon.com lacks the physical presence of stores of the conventional retailers. It has 25 million customer accounts (2001) and rapidly growing sales, reaching $3.12 billion in 2001, up from $1.64 billion in 1999, and is probably the most popular retail-specific web site on the Internet today. The company has also expanded rapidly in terms of the range of goods it sells, moving beyond books to sell music, videos, software, toys, electronics and tools, and thereby making the grand claim of offering consumers the earth's biggest selection.

The geographical dimensions of Amazon.com's online retail business can be considered both in terms of real and virtual infrastructure. Clearly, for any Internet company the principal public face is their web site. Amazon.com's web site is absolutely vital, it is its storefront to the world, accessible to tens of millions of potential customers. Web sites are a form of geography – a micro-geography of the screen. Amazon.com's homepage screen geography is shown in Plate 18. The design is simple and functional, with the overriding aim to make it as usable as possible, and crucially, fast-loading over home modems with only a few small graphics. The site is also backed up with a large amount of computing infrastructure behind the scenes, of course.

The virtual geography of a web site is a world in miniature, a socially constructed world totally under the control of the designers. The design and usability of sites is an area of increasing concern given the growing number of people who are accessing information and services via this medium. This is not necessarily an area of conventional geographical concern being more at the scale of architecture and graphic design. One of the major reasons people shop online is the speed and convenience of the experience; however, too many web site designs seriously fail to meet these requirements. Difficulties with site navigation, confusion over how to order, and most especially, slow-loading pages are some common accessibility problems with online retail. These problems all impact directly on the experience of shopping online. Amazon.com is better than many e-commerce sites and this is reflected in its growing number of customers and high volume of repeat business.

As well as screen space, domain name space is a vital commodity on the Internet. A good domain name is vital to the establishment of an online presence and in many ways can be compared to having the right retail locations in the real world. Domain names are a combination of brands and virtual locations. Their importance is illustrated by the high prices companies are willing to pay to secure the most appropriate and memorable name for their business, particularly at the height of the dot.com boom in the late 1990s. Amazon.com is clearly synonymous with its domain name. But beyond its core web sites (amazon.com, amazon.co.uk, amazon.de, etc.), it has also made a kind of virtual land-grab of its own, registering, by summer 1999, over forty additional names under .com domain alone. Many of these names are clearly related to its core retailing areas of books, videos and music. Other names have been registered for future expansion and also to lock out carpet-baggers.

Another important geographical dimension of Amazon.com is the spatial pattern of growth, expansion and diversification. This has certainly been rapid, as can be seen from the growth in revenues and the growth in new online stores, major investments in other companies and new physical facilities. Figure 6 maps out the extent, and landmarks, of the Amazon.com

Plate 18 Amazon.com's homepage, June 2003.

corporate empire as of January 2000, showing the results of both internal expansion and external diversification through acquisition. At the core of the map is Amazon.com itself, represented by the solid rectangle. In the real world this is spatially represented at the headquarters address at 1516 Second Avenue, Seattle. But it is more meaningfully mapped by its web site location at www.amazon.com. Surrounding Amazon.com are all the associated properties and facilities, both real and virtual.

Obviously, Amazon.com does not have physical, bricks-and-mortar retail stores. Instead it has a growing number of virtual stores on its web site. Amazon.com started with books and moved into music, videos and gifts in 1998, and auctions, toys, software, video games, electronics, tools and zShops in 1999. These virtual stores are mapped as small ellipses physically connected to the Amazon.com core. In some senses these represent natural, evolutionary growth as the main Amazon.com web site becomes a vast virtual department store.

The company has also expanded and diversified, particularly in 1999, through investments and acquisitions of other companies (Figure 6). These additions are shown as small, rounded rectangles ranged along the bottom half of the map. They fall into three broad categories: web retailing properties, strategic technologies and auction related. They represent a sizeable outlay of capital and also highlight the considerable ambition of Bezos, Amazon's CEO, to expand Amazon.com well beyond books. In terms of online retailing, Amazon.com has invested in a range of corporations that operate their own consumer web sites; the most important of these include pharmacy (Drugstore.com), pet products (Pets.com), brand sports goods (Gear.com), luxury gifts (Ashford.com) and grocery shopping (Homegrocer.com). There has also been investment in terms of online auctions with a major alliance with the world's leading auction house, Sotheby's, and the purchase of LiveBid.com. Amazon.com has purchased a number of companies with innovative Internet technologies as well, in order

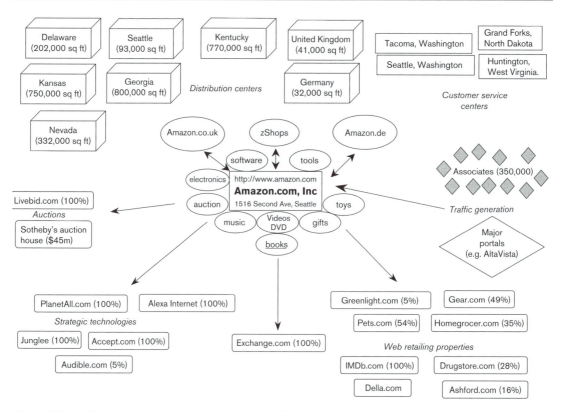

Figure 6 Map of the expanding Amazon.com empire, as of January 2000

Source: Information from company annual reports and press releases

to keep it ahead of the competition. For example, Alexa.com is a leading company in data-mining web traffic to find how people navigate and shop online. Another initiative called zShops provides a sort of virtual flea market, whereby small firms and individual traders can sell to the mass of Amazon.com customers.

Despite the globalisation rhetoric behind much of e-commerce, local identities still matter in much of daily life, including shopping. Even though Amazon.com proclaims the earth's biggest selection, it is interesting that it still felt it necessary to expand overseas by opening subsidiaries for local markets. This was achieved with the national Amazon web sites in the United Kingdom (www.amazon.co.uk) and Germany (www.amazon.de), which both opened in October 1998. (Further national branches have subsequently opened in France, Japan and Canada.) It may appear strange to open local stores. After all, the web suppos-edly transcends terrestrial geography and consumers from around the world, including the UK and Germany, can purchase at the main Amazon.com stores. At issue is the need to build trust with consumers outside the

USA to shop with Amazon.com, when they perceive the store as being American, with prices in US dollars and high overseas shipping charges. The logic here is that customers will be happier buying from a local store, charged in their own currency. Amazon.de is presented in German of course. Clearly, language is a major factor in consumer trust as you want to be able to shop in your own language. Local distinctiveness comes from the domain names, the site design, the language employed and unique editorial content. Another key advantage is faster and cheaper shipping costs for goods that can be sourced and dispatched locally. These national Amazon.com subsidiaries have been successful since their launch and are the leading online bookstores in their respective countries.

Despite the localisation efforts in these national branch stores, the reality is that true power still resides in the Amazon.com head office in Seattle. Like many global operations they are arguably deceiving the public into thinking they are buying locally when in fact the money flows out of the country. In the case of Amazon this can easily be revealed because the web sites of Amazon.co.uk and Amazon.de are actually hosted on powerful servers back in Seattle, despite the appearance of the national Internet domain. Thus the sales, revenues and customer information are automatically transferred back to the United States.

Although Amazon.com is an online company, its principal retail activity is shipping tangible goods to customers, rather than providing purely online information services. To support this activity it needs facilities to store and pack goods. Amazon.com has had to invest significantly in warehousing and the company had (as of 2000) seven major distribution centres strategically placed throughout the United Kingdom to serve different regions (as of 2000). In addition there are two distribution centres in Europe to serve the United Kingdom and German subsidiaries. These are represented by cubes on Figure 6 and are labelled with their location and size. In 1999, Amazon.com also greatly expanded its customer service infrastructure opening three new centres in addition to the original one in Seattle.

CONCLUSIONS

A key question of interest to planners and geographers is what will be the spatial impacts of consumer e-commerce, pioneered by Amazon.com, on retailing in real shops and malls. Will online retailing replace the malls? A corollary of this is the question of whether Amazon.com will still be in the vanguard of change in five or ten years time. Not surprisingly, Amazon's Chief Executive Officer Bezos is very positive about the prospects for his company and also the impacts of online retailing. For example, he is quoted as saying that 'strip malls are history' (Bayers, 1999, 116). However, anecdotal evidence suggests that much of the hype on the impact of online retail from the late 1990s has not come to pass and online retail's impact on the high street has been negligible.

Bezos's argument is that the convenience, wider selection and lower prices offered by online merchants will make the older, general purpose shopping malls uneconomic as increasing market share is taken by cyberspace. However, online shopping is not likely to destroy all retailing in bricks-and-mortar stores, as even the most ardent virtual advocates acknowledge that the web cannot replace the physical and emotional experience of shopping, the pleasure of the hustle-and-bustle of human interaction. For many people shopping is a key leisure activity, where holding the goods in one's hand and the instant gratification of walking out of the store with your purchase are integral to the experience. The point-and-click of web shopping cannot match this experience and often seems unsatisfying.

There are many goods that are not easily amenable to online retailing and there are also practical limits to home delivery of certain goods. The store-based retailers that will continue to flourish are likely to be at two ends of the spectrum – the high-quality 'shoptainment' stores that provide personal service and /or an entertainment experience and the '24/7' convenience store. Retail space in the middle will be increasingly squeezed by the online world, although the degree of impact will vary from product to product. More recently, commentators have begun to argue for the advantages of a combination of web site and stores, in a strategy dubbed clicks-and-bricks. The dominant retailers will be those that can offer an integrated shopping experience by combining the best of online and offline. Indeed Amazon.com is beginning to move in this direction with strategic alliances with bricks and mortar bookstore chains Borders in the United States and Waterstone's in the United Kingdom.

REFERENCE FROM THE READING

Bayers, C. (1999), 'The inner Bezos', *Wired*, March, 115ff.

Editor's suggestions for further reading

Dixon, T. and Marston, A. (2002), 'U.K. retail real estate and the effects of online shopping', *Journal of Urban Technology*, 9(3), 19–47.

Dodge, M. (2001), 'Finding the source of Amazon.Com: Examining the store with the "World's biggest selection"'. In S. Brunn and S. Leinbach (eds), *Worlds of Electronic Commerce*, Chichester: John Wiley and Sons, 167–81.

Leinbach, T. and Brunn, S. (2001), *World of E-Commerce: Economic, Geographical and Social Dimensions*, London: Wiley.

'The Web, the Grocer, and the City'

Andrew Murphy

Editor's Introduction

In the second of our analyses of the urban implications of e-commerce the geographer Andrew Murphy, who works at Birmingham University in the UK, examines another area of spectacular e-commerce growth: online grocery sales. As he argues, in this specially commissioned piece, the mass diffusion of the Internet has allowed a remarkable growth in online food shopping to occur across the world. This has major, but poorly explored, implications for the organisation of food shopping in cities, for urban transport flows, and for the broader structures of urban form.

As Murphy notes, an interesting aspect of the massive growth of food e-commerce has been the emergence of a wide range of business models through which food retailers attempt to organise the delivery of services. Some food retailers (like Tesco in the UK) simply added e-tail deliveries to their existing stores in a relatively low-tech way. Some built massive food delivery warehouses to serve whole urban regions from one site. Still others off-loaded responsibility to an intermediary. The result has been a frenzy of experimentation as food retailers attempt to achieve profits by minimising costs whilst maximising customer satisfaction.

The lessons from this early phase of Internet food shopping demonstrate that achieving success is made difficult by the inherent logistical complexities of food e-tailing. Hugely complex individual food orders, from a vast range of suppliers, must be brought to the consumer, when they are at home (or able to receive it) at speeds that allow the food to remain fresh. This must be done at a low cost. And such deliveries must physically negotiate the extending, and increasingly gridlocked geographies of urban regions.

These complexities point to a process of evolution rather than revolution. They suggest that the future will be about subtle relationships between continuing food stores housed in larger and larger supermarkets complemented by e-tailing operations which link very closely to them. As with all the examples in this book, the reality of cybercity retailing involves complex combinations of face-to-face co-presence, physical movement, and ICT-mediated exchange, all of which work within the complex and extending geographies of city-regions themselves.

A vehicle pulls up outside. Another day, another delivery of food to the home. Another van plying the backstreets, saving customers a trip of their own. This scene fits a number of possible extensions of retail capitalism to the last mile to the customers doorstep: the daily milk round, still available in many streets in Britain, but a dying (if not already well interred) business elsewhere; or a hot pizza delivery. Of interest in this reading is the delivery of food purchased via the internet.

Home delivery of food is nothing new; in fact, many supermarkets offered this service in limited form in their early days, and some still offer home delivery of items purchased by the customer in-store, often to assist the less abled or aged. However, systems predating e-commerce were limited in scale, and were either inherently unresponsive to changing customer needs (by supplying a standing order of staples, for example) or expensive to operate (due to the time required by store staff to take and process orders, largely via telephone). The Internet was therefore heralded in the 1990s as a means to overcoming the problems of order complexity and time costs that beset other home delivery systems. In the same way as Amazon.com promised to be the 'largest bookstore on earth', a virtual grocer could *potentially* offer a wider range of items than even the largest store, at a lower cost than was previously possible, given an order volume sufficient to offer economies of scale (Martin Dodge, p. 221).

E-GROCERY BUSINESS MODELS

A veritable explosion in experimentation occurred in Internet-enabled home delivery of food during the late 1990s. Out of this three distinct types of systems can be discerned, although as always there are numerous variations. The quickest and cheapest to establish is in-store fulfilment, where an existing retailer utilises its established store network (the so-termed 'bricks and clicks' method). The most expensive method is to construct a purpose-built e-commerce fulfilment warehouse (a 'pure-play' if the operator has no 'bricks and mortar' store network). A third option is for (usually small) existing retailers to contract out the web ordering and customer management functions to an intermediary (sometimes termed an 'infomediary') whose core function is to pass stock, order and delivery information between the customer and the retailer.

Numerous 'e-tailers' have tried various forms of these models. Tesco, the largest UK nationwide supermarket chain, began bricks and clicks operations in 1996, and was roundly criticised in the US press for its 'backward' and 'limited' store-based methods (Plate 19):

The Tesco Direct service is so low-tech it's bizarre. Orders are sent to the Tesco store nearest the customer's home, then 'pickers' pull the desired items off the shelves. You read that right — no warehouse . . . Tesco better invest in infrastructure,

Plate 19 An online delivery van from the UK's Tesco supermarket. (Source: FreeFoto.com.)

including 'dedicated picking centres' free of day-to-day shopper traffic . . . [because] the Web is an unforgiving place (*Industry Standard*, February 10, 2000).

Tesco's approach was only half-heartedly defended in the British press. 'There is one big "e-tailer" whose star shines more fiercely than ever', wrote the *Economist* (5 February 2000). 'The surprise is that this is no Silicon Valley digital hotshot, but a dull old British grocer.'

The warehouse model, so lauded by the soon-to-be defunct Internet bell-ringer *Industry Standard*, is best typified by an American example. Webvan's mission was to automate the grocery business by picking from large warehouses in edge-of-town locations. At 300,000 square feet these 'dedicated picking centres' were three times the size of the largest supermarkets, and would hold 'miles' of conveyor belts and computer-controlled rotating racks that would literally bring the goods to the picker. Webvan raised nearly $800 million in private and public capital, and, with the takeover of rival Homegrocer, managed to enter twelve cities around the US, before rapidly mounting losses. Slow growth in customer adoption led to its collapse in July 2001 (Murphy, 2003a).

By contrast, in the 'bricks and clicks' model pickers roam the aisles with computerised order scanners, which direct the picker to the location of the next item for picking, sorted by store layout. Tesco's model is to pick multiple orders at the same time. Pickers are directed by the system to one of the many bins (which could represent several customers) where they place the bought items. Orders are thus divided by the computer system among multiple pickers, and reassembled at the end. While this system is subject to potential pick errors if pickers place items in the wrong box, it is faster than using a single picker to pick the entire order.

Once complete, orders are assembled in the Home Shopping area in the backroom of a regular store, and loaded into custom-built multiple-temperature trucks. Trucks follow a route pre-determined by head office computers, according to a least-cost algorithm. This path is factored into the ordering stage, since customers order goods for particular delivery slots, combinations of space–time which are only available if sufficient other neighbouring slots are also being utilised. Should demand at a particular time lie elsewhere in the delivery area, slots are closed where there is less demand and opened where the demand is greater. Tesco has fulfilment centres in hundreds of its stores across the UK, and boasted of being within economical reach of 90 per cent of the UK population by 1999. It has exported its model to subsidiaries in Ireland and South Korea, and to the United States in a joint venture with Safeway. In the year to June 2002 Tesco.com (the online division) sold nearly £400 million of groceries, appliances, books, DVDs, CDs and clothing. Just before Christmas 2002 it claimed to have delivered to more than one million households around the UK, and to 100,000 homes in one week alone via its fleet of 950 vans. Tesco interpreted this 'as confirmation that Britain has now accepted online supermarket shopping as a normal part of day-to-day life' (Tesco press release, 5 December 2002).

THE E-GROCER AND THE CITY: URBAN IMPLICATIONS

The logistics for groceries is complicated – not as problematic as scheduling aircraft, to be sure, but providing the thousands of grocery lines in-store or online is a lot more difficult than selling books or CDs online (Amazon), or acting as an auctioneer (eBay). Food is sourced on a daily to weekly basis from thousands of manufacturers located all over the world, and distributed on a daily basis to millions of customers through thousands of stores. Moving this convoluted system over to the net was never going to be a straightforward task.

Supermarkets have increasingly paid attention to the logistics of distribution management, realising that this can make or break the supermarkets which survive on razor-thin margins. 'Bricks and mortar' food retailers have developed sophisticated behind-the-scenes grounded infrastructure to act as an efficient intermediary between the many suppliers and many customers. The increasing size of supermarkets during the 1970s required backroom areas to receive, bulk-break and store product inventory prior to shelf restocking. Some of this inventory came directly from manufacturers, but the larger chains began to build centralised distribution centres (DCs) to serve groups of stores, as a means of improving efficiency. The further up-scaling of store sizes during the 1980s and 1990s reduced the backroom space for individual stores, as the DC became more important. It also became larger and more sophisticated, designed to send single truck deliveries to the stores to remove the task of coordinating multifarious manufacturer deliveries to the store. The modern regional distribution centres cater to many stores through automated 'continuous replenishment' systems, and can range up to 700,000 square feet (more than six times the size of the largest hypermarket).

These regional distribution centres are designed for one purpose only: the fastest and most efficient restocking of a chain store. Manufacturers' deliveries are often already labelled with their destination stores and quantities, with the pallet contents bulk-broken and re-routed to the store loading bays without on-site storage, in a process known as 'cross-docking'. Full truck loads are then sent to the store, with goods often already ordered and pre-stacked for easy shelf display in roll-in cages. Goods therefore spend little time in transit or storage, thus quickening the turnover of retail capital and fattening retailer margins. The emphasis of this business model is on scale: large trucks to large stores using as few deliveries as possible at the greatest speed.

Because of this dedicated purpose, the regional distribution centres do not make ideal assembly points for single-customer orders to be delivered to the home. For one thing, a store's product range may come through multiple distribution centres:

> In terms of our logistics model and our supply chain, the only place that all our product range comes together is in a store. In terms of our warehouses that we have at the moment, we've got dry goods warehouses, we've got separate chilled and frozen warehouses, and separate fresh warehouses, and typically they are physically never in the same location. They might be in the same city, but that is about as close as they get (Manager interview, Woolworth's Supermarkets, New Zealand, March 2000; Murphy, 2003a).

Even if ambient, chilled and frozen goods, along with produce, meat, fish and bakery items were all co-located, the breaking of bulk necessary to supply individual customers would require quite different sorting areas, with goods tied up in storage much longer. This would definitely be going against recent retail logistics trends (Murphy, 2003a).

By placing e-tailing fulfilment within stores, existing retailers are able to capitalise on the already grounded infrastructure. Retail grocers operate on very thin margins, and have developed the logistics for the cost-efficient resupply of goods. Bricks and mortar retailers have also off-loaded some of the most expensive parts of the business onto their customers: the picking, packing and delivery. Modern customers do not like to – or perhaps do not know how to – calculate the costs involved in getting to the grocery store, or the opportunity costs of their labour whilst there, which in many places includes bagging the groceries and in some even doing the scanning. As one online grocer put it on their web site:

> In the past grocery stores were built of bricks and mortar anchored to huge parking lots. They had few marketing means to coax you into their stores. The colored insert in our daily paper, the endless jingles and TV ads describing the reasons to visit bigger and bigger food stores. But no matter what, we still had to spend hours each month, waiting in line, while we inched closer and closer to the check out clerk. Once there, scanners have made the clerks' work easier, but we still wait and wait [. . .] now we have to wait until all the products are scanned [. . .] come to think of it, in the last 20 years, not too much has happened to make the grocery shopping experience better for the consumer . . . just better for the store! (www.quick.com, accessed 15 June 2001)

Online grocery retailers aim to change this alleged poor in-store experience, by giving time and energy back to customers in return for regular, larger order sizes and payment for delivery costs. They stand a good chance of revolutionising the home delivery of many products, through diversification in product ranges (as per Tesco) or strategic alliances with other companies. However, as many entrepreneurs have found to their cost, it is essential to avoid over-building for an excessively optimistic rate of customer adoption, and to avoid doggedly following a failing distribution strategy, if online grocery is to avoid the fate of Webvan and the home delivery milkman.

REFERENCES FROM THE READING

Economist (2000), 'Tearaway Tesco', *Economist*, 5 February.

Industry Standard (2000), 'Biggest online grocer not who you think it is', *Industry Standard*, 10 February.

Murphy, A. (2003a), '(Re)solving space and time: fulfil-ment issues in online grocery retailing', *Environment and Planning A*, forthcoming.

Murphy, A. (2003b), 'The Web, the grocer and the city: on the (in)visibility of grounded virtual retail capital'. Working Papers on Services, Society, Space, University of Birmingham School of Geography, Earth and Environmental Sciences WPSSS10.

Editor's suggestion for further reading

Currah, A. (2002), 'Behind the web store: the organisational and spatial development of multi-channel retailing in Toronto', *Environment and Planning A*, 34, 1411–41.

'E-Commerce and Urban Space in Japan: Accessing the Net via Convenience Stores'

Yuko Aoyama

Editor's Introduction

It is increasingly clear that the relationships between retailing, e-commerce and cities vary dramatically across the world. In this specially commissioned piece the geographer Yuko Aoyama, who works at Clark University in the USA, demonstrates this 'path dependency' very powerfully. She examines the extraordinary phenomenon of the growth of e-commerce operated through Japan's massive numbers of neighbourhood convenience stores.

Offering a stark contrast to the home-PC based e-tailing model which tends to prevail in North American and European cities, Japanese e-tailing is dominated by well-established convenience stores working as intermediaries. Such stores link neighbourhood consumers (who have a much lower ownership of PCs, lower PC skill levels, and lower use of credit cards than in the West) with a massive galaxy of online retailers. Through the reorganisation of neighbourhood stores into classic hybrid brick-and-click spaces, Aoyama shows how neighbourhood stores, which have always been exceptionally sophisticated in their use of ICTs in Japan, now dominate e-commerce in that country.

Such stores provide critical social spaces which help many consumers gain knowledge of, and confidence in, online consumption. They add trust and credibility to the consumers' links with online providers. And they help to maintain the economic strength of an incredibly dense and rich network of neighbourhood stores which, in other nations, may actually be threatened, or undermined, by home-based e-tailing.

However, as Aoyama notes, in the absence of cross-national research on the variety of relations between e-commerce and cities, we are all too prone to fall back on the US-dominated stereotypes of purely individualised and home-based e-commerce undermining the sociability, and face-to-face contact that has traditionally occurred in city spaces. Her work is an important counterpoint to that temptation.

In Japan, there are three ways to access e-commerce: via home-PCs, cellular telephones (m-commerce) and terminals located in neighborhood convenience stores. Partly because e-commerce originated in the United States, the home-PC based internet access has assumed the norm as a medium of exchange in its developmental trajectory. The rapid diffusion of home-PC based e-commerce activities requires widespread PC ownership, a level of computer literacy, and a critical mass of consumers who use credit cards actively, but these conditions are often not met by many other societies.

Because online stores are often simply digitized forms of catalog businesses that use internet as a medium of exchange, the US retail sector was able to exploit a long and well-established history of direct marketing, including catalog houses and mail order businesses. In the case of Japan, non-store retailing never amounted to a widespread form of consumption, and both merchants and consumers lacked experience in engaging in non-store retailing. The lack of history and unfamiliarity among consumers contributed to the delay in B-to-C e-commerce development in Japan. Instead, a different model of consumption using the internet is emerging, which uses convenience stores and mobile telephones as access points. In this Reading I describe how neighbourhood convenience stores became players in the fast growing e-commerce business in Japan.

CONVENIENCE STORES AS 'BRICK-AND-CLICK' BUSINESSES

Neighbourhood convenience stores are the pioneers of Japan's 'brick-and-click' businesses (see Plate 20). Convenience stores serve as access points to e-commerce for Japanese consumers by resolving three major society-specific obstacles for e-commerce diffusion: online access, distribution, and secure payments. Because convenience stores are ubiquitous in Japanese urban areas, consumers can access e-commerce without buying a PC, signing up for an internet service provider, or using credit cards.

Japan's convenience stores were particularly innovative on three fronts: in terms of their marketing strategy, their strategy to adopt information technologies, and their locational strategy. In this Reading I focus on the latter two strategies. Japan's convenience stores invest more per unit floor space on information technologies than any other forms of retail outlets. Convenience stores initially invested heavily in information technologies with the sole purpose of improving delivery efficiency. To shorten the time required for restocking, the point-of-sale (POS) system was developed by Seven Eleven Japan in 1982. At the time, information networks were used simply as an instrument for delivery trip rationalization and cost reduction.

Once the information infrastructure was in place, however, convenience stores quickly realized that the same network could be used to gather customer information. The second stage of informatization thus involves the use of the network for market research. Information is used to determine the variety and the amount of products on the shelf for each store. To facilitate market research, special cash registers were developed with buttons that represent gender and age

bracket of customers. Without punching in these buttons, cashiers cannot open to the drawer and provide change or a receipt. The information, including time and item of purchase, as well as the gender/age of the customer, is then transmitted to the headquarters and used for restocking and the tracking of sales by stores. The primary purpose of this system was not simply to identify customer demand, but also to identify products that do not sell. By doing so, stores are able to identify and stock only fast-selling products, increasing space efficiency.

The convenience store chains realized the potential of offering e-commerce through the network only after their information systems were in place. By using their already present Integrated Services Digital Network (ISDN), which allowed data to be sent over normal phone lines quickly, convenience stores began delivery of new online based services. Seven Eleven Japan began experimenting with online utility bill payment systems as early as 1987. Today, it offers online utility bill payment system with bills payable to 126 utility companies including telephone, gas, electricity, and water.

Plate 20 An e-commerce terminal in a Japanese convenience store. (Photograph by Yuko Aoyama.)

In a country where personal checks are non-existent and credit card use is not widespread, utility bill payments were done exclusively through bank transfers until convenience store chains began offering the service. Multiple alliances between convenience stores and e-commerce vendors further increase the product variety. Convenience stores have already begun performing the functions of a travel agency by allowing customers to book a flight, purchase domestic or international package tours, and reserve last minute hotel rooms at a discount rate. Customers can purchase discount airline tickets via the internet, pay cash at a local convenience store, and the flight coupon is delivered to the store for a pickup.

Convenience stores have therefore become an effective intermediary between traditional retailing and e-commerce in a manner that has not been observed in the US or Europe. For Japanese consumers who were not accustomed to the practice of long-distance shopping, convenience stores provide face-to-face interactions at each of their local storefronts, thereby reducing consumer anxiety. This makes the service and product delivery far more user-friendly. With the convenience store functioning as a financial as well as distributional intermediary, it provides easy access and legitimacy to online retailers and bypasses all security concerns that surround online transactions.

The locational strategy adopted by Japan's convenience stores was decidedly urban: the chains initially focused on opening stores with close proximity to each other in the largest of the metropolitan areas. This was done in order to maximize distribution efficiency and minimize delivery routes. Convenience stores' smaller-than-supermarket floor space enabled them to strategically locate stores in high-density urban areas, thus providing advantages over supermarkets in terms of access. Japan's typical urban consumers shop in small volumes and more frequently than American counterparts. Since consumers are likely to walk or bicycle to a nearby grocery store, or drop by a store after work en route from a nearby train or subway station, occupying strategic urban locations is the single most important factor for successful retail performance in Japan. Furthermore, due to limited housing sizes and high energy costs, most households did not own large refrigerators until recently. This necessitated frequent restocking. Lack of storage space provided no incentive for customers to shop at the large-volume, discount stores that are popular in the United States and, increasingly, Europe.

Convenience stores use their ubiquitous and strategic urban locations, which function as access points for e-commerce retailing. In Tokyo's 23 urban wards, for instance, there are 16 convenience stores per square mile. This is compared to 10 stores per square mile in Manhattan, New York, and 1.4 stores in Washington, DC. There is one convenience store for every 2,000 residents in urban Tokyo, while the comparable figures for Manhattan and Washington, DC, are one store over 6,000 residents. This means, for urban Tokyo residents, a trip to the nearest convenience store is often shorter than a trip to the nearest train/subway station. An average Tokyo commuter is estimated to walk by three convenience stores on his/her way to work. Unlike American consumers, who are more willing to travel an extra distance for low prices, Japanese consumers are known to be far more willing to trade after-market service, convenience and proximity for a higher price. Such consumer characteristics, combined with an unfamiliarity with nonstore retailing, provide storefront retailing with an added advantage in densely populated urban areas.

While convenience stores served as pioneers of informatization, this is not without problems. Japanese consumers have not been eager to use e-commerce in general, either at home or at stores. The most frequently used e-commerce kiosk service at convenience stores is the purchase of concerts and other event tickets, followed by in-store payment and delivery arrangement for products purchased online elsewhere (at home using PC-internet access, or mobile telephones). In both cases, however, competition is emerging from mobile telephone-based internet access.

CONCLUSION

The ways in which a society adopts technologies are determined by a complex combination of various factors, involving economic, social, historical, and spatial conditions. The case of convenience stores in Japan suggests that the relationship between online and bricks-and-mortar stores is by no means universal. Nor can such a relationship be technologically determined. The case of mobile telephone use in Japan showed the urban conditions in Japan prompted consumer demand to shift from stationary access to portable access.

More cross-national research is needed to identify how urban residents in different societal contexts respond to various technological solutions. For example, we know little about how wireless and wired communications differ in their impacts on travel patterns in various metropolitan contexts. More specifically, few studies exist of the variations of information technology adoption in non-western societies, including the proliferation of internet cafés in some urban areas, and how they compete against home-based access (Nina Wakeford, p. 263). Understanding the process of technological adaptation would bring further insights into the interactions between consumption, travel patterns, and urban space.

Editor's suggestions for further reading

Aoyama, Y. (2001a), 'The information society, Japanese style: corner stores as hubs for E-Commerce access'. In S. Brunn and S. Leinbach (eds), *Worlds of Electronic Commerce*, Chichester: John Wiley and Sons, 109–28.

Aoyama, Y. (2001b), 'Structural foundations of E-Commerce: a comparative study of the organisation of retail trade in Japan and the United States', *Urban Geography*, 22(2), 130–53.

Hosaya, H. and Schaefer, M. (2001), 'Tokyo metabolism'. In C. Chung, J. Inaba, R. Koolhaas and S. Tsung Leong (eds), *Harvard Design School Guide to Shopping*, Cologne: Taschen, 749–63.

'Space Jam: Media Conglomerates Build the Entertainment City'

from *European Journal of Communication* (1999)

Susan Davis

Editor's Introduction

Martin Dodge (p. 221) predicted that, as e-commerce based retailing grows, premium retailing spaces will be reconfigured to emphasise experience and entertainment over the instrumental transactions of the purchase. This trend has already been widely observed (Hannigan, 1998).

Our final reading on the economic dimensions of cybercities takes up Dodge's point. In it, the US communications scholar Susan Davis examines how the massive multimedia transnationals that increasingly control publishing, film, entertainment, software, music and online content are themselves attempting to reconstruct selected downtown urban retail spaces for entertainment-led retailing. These are attempts to use the power of their brands to develop urban stores which are, in effect, media constructions themselves. Davis highlights how these developments result from a new phase in the ways in which transnational media corporations like Disney, Sony, and Warner see the city.

Moving beyond their obsession with private, home-based, and ICT-mediated entertainment and consumption – with their dominant representations of the public realm of the city as a dangerous place to retreat from – the emphasis of such media corporations now falls on the selective and extremely careful production of what Davis calls private collective spaces in malls and regenerated city cores. As a result, to quote Davis, 'the city (or at least certain districts of the city)', is 'seen as dangerous in its diverse unpredictability, is being made safe imagistically as well as physically.'

By this she means that the transnational media corporations who run the new entertainment-based retail plazas and spaces do everything in their power to maintain the sense of safety that will tempt suburban shoppers into the downtown street or central mall. Thus, in an example that prefigures our discussions in Section VII, spaces are privately policed, regulated by extensive rules, and operate through an intensive screening-out of those deemed to be threatening to the practices, and potential profits, of corporate retailing.

Davis's work highlights that major media corporations, as well as shaping the production landscapes and online exchanges of the cybercity, are increasingly powerful in determining its consumption landscapes as well. Given that tourism, leisure, retailing and urban consumption now underpin large swathes of contemporary city economies, we need to be mindful of this growing influence. For, as the sociologist Mark Gottdeiner (1997) has argued, the physical landscapes of cities are increasingly being shaped directly by large corporations who are keen to model it in their image and tie it seamlessly together with their virtual products, services and brands. As a result, theme parks, themed retailing, malls, resorts and other corporate leisure and consumption spaces need to be seen as hybrid spaces that are being produced together with online and multimedia content.

In the first half of the 20th century, the entertainment conglomerates were central in creating a nearly all-penetrating national and international mass culture, first through film and later through animation, popular music and televised sports. In the second half of the century, they have brought this largely American mass culture thoroughly and extensively into the home, to hundreds of millions of people. At the cusp of the 21st century, they are poised to weave the private realm together with the collective through the creation of dramatic and focused media-filled spaces. In the process, as they further displace smaller businesses and older, heterogeneous uses of the streets, the media conglomerates are changing the relationships between public and private experience. They are in the process of creating public spaces defined by marketing criteria and shaped to the most profitable audiences. These spaces will be devoted to the circulation of well-tested and 'safe' media content and will exclude experimental imagery or oppositional ideas. Privately produced collective spaces based on and filled with familiar mass media content can create a kind of seamless world, one in which the home – currently devoted to extensive consumption of conglomerate culture – is tightly knit to and continuous with the outside. The city (or at least certain districts of the city), seen as dangerous in its diverse unpredictability, is being made safe imagistically as well as physically.

THE SCALE OF THE BOOM

'Entertainment is the hottest topic in real estate circles', as many economic analysts have noted, but the reverse is just as true: real estate is now indispensable to an entertainment company's portfolio, its growth and promotional strategies. While they worry about how the Internet will change their fundamental businesses, the media conglomerates are elevating their corporate vice-presidents for real estate and development. These strategists are less visible publicly than executives in charge of film, television and multimedia, but they are charged with doing more than building offices. They speak of the old-fashioned ideas of place and community as growth opportunities for their companies.

No recent statistics are able to summarize the extent of location-based entertainment projects, but they are ubiquitous. In today's rebuilt city centres, they are important conceptual, architectural and speculative interventions. In Philadelphia, Denver, Baltimore, Atlanta, New Orleans, Cleveland, San Francisco and Washington, DC, these new projects have been born out of the political clout of the developers using state and federal redevelopment funds, the skills of prestigious architects and tourism promotion policies. At their core are buildings that integrate product, most predominantly media product, into space more fully than ever before, using architecture to synthesize marketing goals with the creation of awe and personal identification.

As usual, Manhattan provides striking examples. On the East Side, within the few blocks between East 57th and 55th Streets is a cluster of entertainment retail projects. The Warner Bros. Store features 75,000 square feet of licensed merchandise, 24 video screens, 'a giant zoetrope in a moving picture cafe' (Gragg, 1997, 84) and hands-on interactive animation stations. Nearby, NikeTown features a museum of Olympic medals and trophies; films of US hockey, World Cup soccer, Michigan football; and a three-storey screen that descends from the ceiling every 30 minutes to show an inspirational sports film. NikeTown alone claims 10,000 visitors daily, evenly split between tri-state residents and tourists. A Disney Store is just a few blocks south on 55th Street, again filled with giant video, interactive toys, merchandise and a travel and ticket agency to help customers connect with other Disney products, such as theme parks, cruises and Broadway shows. At Times Square the Disney Company has anchored the area's 'rebirth' as a film, theatre and 'interactive entertainment' district. Forty-Second Street is now a showcase for Disney's endless array of family-friendly products, and the New Amsterdam Theatre is not only the stage for Disney's new live theatrical enterprises, but also a celebration of the company's vast international cultural power. On the site of the New York Coliseum, Time-Warner will build Columbus Center, its new world headquarters. Near Lincoln Center, Sony and the IMAX Corporation have built a large-screen film, cinema and shopping complex. Here, and in San Francisco and Berlin, Sony's entertainment superstores and multistorey movie screens are being used to stoke commercial real estate expansion, bolster property values and create central tourist destinations.

It is not only a question of metropolitan redevelopment. Location-based entertainment is integral to the centrally planned city or the new town built from scratch. These fully designed entertainment and shopping districts, punched up with postmodern pastiche or careful historical reconstructions, aim to attract locals and regional and international tourists to the replication of an older urban world. Their spaces, in the words of architect Jon Jerde, try to 'break down the walls between cultures and between entertainment and shopping, pleasure and profit, the viewers and the viewed' (Iritani, 1996). CityWalk, one of Jerde's expansive projects, is built on an old Universal Studios backlot. It expands the popular studio tour outward, first into a shopping centre and now into a full sized urban-themed destination. CityWalk is designed to communicate architectural chaos and urban unpredictability. Its city/theme park combination has been so successful that Universal (a subsidiary of Seagram) has plans to make it a central tourist destination in Los Angeles by doubling its 5 million square feet of buildings, adding new rides and a convention hotel. As one West Coast observer recently put it, 'Hollywood used to be in the film business', and thus the popularity of Universal Studios tours, where customers can see animators, costumers, actors, directors and editors at work. 'But Hollywood is now in the "place" business, selling [a synopsis of] itself and its labors as an attraction and place sells' (Fulton, 1997, M1–M6).

[. . .]

THE RATIONALE FOR THE ENTERTAINMENT-RETAIL BOOM

There are several powerful and intersecting reasons for the entertainment-retail building boom. Real estate developers are partnering with entertainment corporations in part because retail and office space were overbuilt in the 1980s and remain so today as speculation runs far ahead of demand. Traditional shopping centres and malls, and even Jon Jerde's elaborate pseudo-cities, are facing tremendous competitive pressure as American incomes 'remain flat and retail space per capita increases' (Phillips, 1995). In addition to retail saturation, the development of new shopping media and ways of selling to people in the home via catalogues, cable, infomercials and Internet shopping is also forcing traditional retailers to be ever more inventive to reach the older, affluent though slower consuming segment of the population. At the same time, these consumers are better understood than ever before – they have been subject to more than a generation of traditional and now computerized electronic and video information gathering. Market researchers now speak of knowing most of the consuming population as niches, each with its particular tastes, habits and preferences.

Entertainment retail is a strategy to get people out of the house. From the point of view of landlords, retailers and especially the vast coalitions of institutional real estate speculators called REITs (real estate investment trusts, essentially mutual funds specializing in real estate), the injection of entertainment content into commercial spaces is a coordinated way to differentiate one retail space from another, to bring people out to shop and, in metropolitan locations, to capture the important tourists. In this effort, media conglomerates like Disney, Sony, MCA and Time-Warner – which own the widely familiar film, television and sports imagery – have a long head start. In the risky and overbuilt retail sector, the already tested media content bolsters investor confidence as do Hollywood's corporate deep pockets. Here Pocahontas, Daffy Duck and the Tasmanian Devil provide a kind of insurance that the customers will keep coming. Indeed, they provide sure ways to locate and appeal to important groups within the broad population of customers. [. . .]

Yet, from the media conglomerates' perspective it is not just a question of breaking through the 'cocoon' of home entertainment. Oddly or perhaps not – what is being used to get people out of the house is the material that has been invading it for 40 years. The largest media conglomerates have realized that they need what Disney chief executive Michael Eisner has called an 'inside/outside' strategy, encompassing both the inside (media consumption in private and domestic spaces) *and* the outside (traditional and new forms of media consumption in public). The two spheres must work together and be mutually supportive.

REFERENCES FROM THE READING

Fulton, William (1997), 'From making dreams to concocting reality', *Los Angeles Times*, 22 June, M-1, M-6.

Gragg, Randy (1997), 'Domination by design', *Metropolis*, June, 62–7, 83–5.

Iritani, Evelyn (1996), 'A mall master takes over the world', *Los Angeles Times* 5 July, A-1, A-17.

Phillips, Patrick (1995), 'Merging entertainment and retail', *Economic Development Review*, 13(2), 13–15 (online version, *Abstracts of Business Information*, online pages 1–3).

Editor's references and suggestions for further reading

Chung, C., Inaba, J., Koolhaas, R. and Tsung Leong, S. (eds) (2001), *Harvard Design School Guide to Shopping*, Cologne: Taschen.

Davis, S. (1999), 'Space jam: media conglomerates build the entertainment city', *European Journal of Communication*, 14940, 435–59.

Gottdeiner, M. (1997), *The Theming of America: Dreams, Visions and Commercial Spaces*, Boulder, CO: Westview.

Hannigan, J. (1998), *Fantasy Cities: Pleasure and Profit in the Postmodern Metropolis*, London: Routledge.

Sorkin, M. (ed.) (1992), *Variations on a Theme Park: The New American City and the End of Public Space*, New York: Hill and Wang.

VI SOCIAL AND CULTURAL WORLDS OF CYBERCITIES

Plate 21 Sign to a nearby Internet café, central Amsterdam. (Photograph by Stephen Graham.)

INTRODUCTION

How do people's experiences of their corporeal lives in urban places interact, and combine with, their experiences and use of electronic media and ICTs? How do collective representations of what a contemporary city actually might be relate to portrayals of the city in new media? How can we analyse, to use the words of the British cultural geographer Kevin Robins (1995), the 'state of suspension' between the social and cultural worlds of the city and those supported by ICTs and the new media? Finally, how do the bases of the identities of people and groups – for example, gender, ethnicity, nationality, class, wealth, (dis)ability, religion, profession, and sexuality – influence how they use and construct cities, ICTs, and the interactions between the two? To successfully address these questions research on the social and cultural worlds of cybercities needs to meet three central challenges.

Combining 'proximate distance' and 'distant proximity'

First, integrated analyses of people's experiences of how ICTs and places interact in practice, within their everyday lives, are needed. Only through these can we understand the complex interrelationships between the two (see Wellman and Haythornthwaite, 2002). This is because today's urban social and cultural worlds are about the subtle and ongoing combinations of what Stefen Waltz (2002) calls the 'proximate distance' of face-to-face sociality in places, and the 'distant proximities' of socialities that are mediated by ICTs. The latter occur at a whole variety of telescoping scales, from emails and text messages between adjacent rooms and streets to neighbourhood interactions, city and nation level interactions, and transnational interactions (Smith, 2001). As the leading anthropologist Arjun Appadurai (1996, 3) notes:

> Electronic media mark and reconstitute a much wider field, in which print mediation and other forms of oral, visual, and auditory mediation continue to be important . . . Electronic media (whether associated with the news, politics, family life, or spectacular entertainment) tend to interrogate, subvert, and transform other contextual literacies.

Of course, telemediated exchange and interaction based on corporeal presence in urban places are very different. 'Wandering in a book or computer takes place within more constricted and less sensual parameters' [than walking in a city], writes Rebecca Solnit (2001, 10):

> It's the unpredictable incidents between official events that add up to a life, the incalculable that gives it value. Both rural and urban walking have for two centuries been prime ways of exploring the unpredictable and incalculable, but they are now under assault on many fronts.

It does not follow, however, that face-to-face encounters are always more intense and meaningful that distanciated ones. 'One can be physically close to someone whilst feeling distant,' suggests Waltz. 'And one can be physically distant but feel close' (2002, 6).

We have already explored some of these diverse combinations of connection and disconnection, presence and distanciation, places and mediated environments in cybercities in Part III of this book. It is clear from our discussion there that people and groups now experience, and try to shape, constant interplays between faceless, ICT-based exchange, and face-to-face exchange demanding physical travel and bodily copresence in place. We also saw (Deirdre Boden and Harvey Molotch, p. 101) that the compulsion to be physically proximate, to enjoy the full anthropological richness of human corporeal co-presence, seems likely to continue to fuel urban agglomeration, transport demands and the culture and politics of most cities for a long time to come (even if it may be more restricted and closely surveilled than in the past). In many cases, the very explosion of online, formal information may, indeed, increase the desire to travel to work, to sustain social or family ties, or to travel for pleasure and tourism. These are the staples which sustain many urban economies.

At a finer level of detail, articulations between proximate distance and distant proximity raise a whole host of questions about how the changing social and cultural worlds of cities interact with the use, construction and experience of new media. How, for example, does the mass diffusion of mobile phones and Internet computers change the use and experience of urban places and transport networks? How do people experience the hybrid and recombinant spaces – Internet cafés, cash points, kiosks, and virtual reality installations – that most powerfully exemplify the ways in which ICTs are being co-produced with urban spaces? How do multilevelled constructions of identity surrounding the body (Hillis, 1999), gender (Flanagan and Booth, 2002), age (Holloway and Valentine, 2002), ethnicity and 'race' (Ebo, 1998; Kolko et al., 2000), (dis)ability (Bourk and Worthington, 2001), the home (Morley, 2000), the neighbourhood (Doheny-Farina, 1996), place (Keeble and Loader, 2001), and the city (Boyer, 1996) articulate with online interactions? And how do the variety of human subjects experience the increasingly automated and software-controlled systems that mediate a growing proportion of their relationships to the city (at least in rich, high-tech cities)?

Finally, now that the Internet has been integrated into the daily lives of hundreds of millions of urban dwellers, how do the social networks facilitated by it relate to face-to-face 'communities' and ties at the level of the neighbourhhoood, the city, the urban region, the nation and the transnational? What do these processes imply for our understanding of notions of 'community', 'place' and the social integration (or disintegration) of the world's metropolitan regions (see Calhoun, 1986)? What do Internet-based links imply for the civic and social life of streets and neighbourhoods? For the maintenance of social networks? For the mobilisation of social and political movements and civic activity within and between cities? And for the linkages between the local, urban, national and international scales of social and political agency?

The dangers of universalist thinking

Evidence here suggests a variety of emerging relationships. On the one hand, many social constructions of Internet use sustain individualistic network connections and experiences. These allow people to transcend the barriers of the local and capsularise themselves in the homespace or gated 'community' (see Lieven de Cauter, p. 94). On the other, ICT connections have also been shown, in some cases, to strengthen neighbourhood level interactions (see Wellman and Haythornthwaite, 2002).

A critical challenge here is therefore to maintain an open mind. We must expect diverse relationships and constructions linking ICTs and urban social and cultural worlds both for different people and groups within the same city and between different cities. Given the increasingly multicultural make-up of cities it is particularly important to avoid the temptation to uncritically draw in our analyses and assumptions on western post-Enlightenment traditions of thought. Since the sixteenth century these have stressed a universalising approach to human experience which emphasises mobility and travel as liberating and emancipatory in their own right.

Such western, post-Enlightenment traditions have tended to simply assume a generalised experience of the intersections of mobility, technology and place within a single urban public realm. Most often, this has been done from a white, western, male perspective. Such a tradition is closely associated with the utopian discourses of substitution, power and absolute mobility that were – as we saw in the Introduction – such a dominant part of discussions about cities and ICTs between the 1960s and 1990s.

Universalising notions of the social experience of ICTs and the city have been rightly critcised by a wide range of post-colonialist, feminist and other writers (Sandercock, 1998). Such critiques have successfully revealed the partiality of such approaches, and the roles they have played within the construction of everything from colonial empires and cities, to biased urban planning and media systems (see Kaplan, 2002; Gregory, 1994; Robins, 1999).

Extreme care is needed, therefore, to avoid using terms which imply a universal experience of city–ICT relations. For example, as the anthropologist Carin Kaplan suggests, the very notion of 'virtual travel through cyberspace' that is so common in depictions of the Internet 'proposes a kind of travel that expands on a long history of metaphors of displacement. The emancipation of ideas promised by flows across borders and boundaries is a Western Enlightenment dream – no boundaries for the mind of the subject' (2002, 34).

In response, Kaplan argues that much more grounded understandings are needed. These need to draw from postmodern, post-colonial and anthropological theories which address the diverse ways in which mobilities, displacements and boundary crossings are experienced (see, for example, Bhabha, 1994; Appadurai, 1996; Hannerz, 1996). Whilst focusing on social learning and experimentation, through the use of ICTs in place, such approaches also need to be acutely conscious of the wider contexts of extreme social inequality and the cosmopolitan mixing of vast ranges of people within and between major cities.

Empirical work on the wide variety of ways in which the Internet is now being used in people's everyday lives, in a wide variety of different places, is a real help here (Wellman, and Haythornthwaite, 2002; Jarvis *et al.*, 2002). With a subtle and flexible view of the diverse experiences of the interactions of places and new media, Kaplan (2002) argues that we might begin to understand how the disembodied identities which people and communities construct, through ICTs, at various scales, relate to the changing social and cultural worlds of our planet's increasingly diverse, multicultural, and partitioned cities.

As key sites within the world's media networks, the increasingly cosmopolitan neighbourhoods and spaces of the world's cities are also the drivers of the blossoming range of TV and ICT-mediated 'diasporas'. (This term describes the ethnic or national communities that are now spread across many places in the world – see Hannerz, 1996; Robins, 1999; Staheli *et al.*, 2002). 'Cities', writes the anthropologist Ulf Hannerz, 'are good to think with, as we try to grasp the networks of relationships which organize' the global cultural flows and connections of our planet. 'They are places with especially intricate internal goings-on, and at the same time reach out widely into the world, and toward one another' (1996, 13).

Thus, on- and offline dimensions of individual and collective identity are being made and remade by constructions and mediations which blend urban place, physical travel, and distanciated encounters based on the net, satellite TV, phones, or printed and digital media (see Hannerz, 1996; Eade, 1997; Dürrshmidt, 2000). For example, a big part of the Internet's success has been to tie together transnational diasporas, largely based in the world's main cities, with unprecedented power (Staheli *et al.*, 2002).

Representing cities and cybercities: metaphors and crises

The second challenge is to be conscious of the growing intersections between the ways in which urban life and ICT-mediated life are represented. The idea of the modern city, as we see (Anne Beamish, p.273), has powerfully influenced the development and representation of the Internet – and vice versa (see Boyer, 1996; Vale and Warner, 2001). The Internet is replete with web sites which purport in some way to be 'virtual cities' or 'cyber towns' (see Anne Beamish, p.273; Geert Lovink, p. 371). Electronic analogies of cities have been constructed through the online decisions of thousands of people to 'build' cities in cyberspace. And, as we saw (Thomas J. Campanella, p. 57), a major job of the Internet is to represent specific urban places for the consumption of global audiences through webcameras and travel web sites.

Indeed, It seems paradoxical, given that 'cyberspace' has widely been portrayed as emasculating geography, that the growth of discourses about cities and ICTs – even the very word 'cyberspace' itself

– have actually been dominated by geographical and territorial metaphors since their inception. The British geographer Steve Pile argues, in fact, that the discourse surrounding cyberspace 'is a plurality of clashing, resonating and shocking metaphors' (1994, 1817). The expanding lexicon of the Internet, for example, is not only replete with, but actually *constituted by*, the use of geographical, spatial, territorial and urban metaphors.

Debates about ICTs use spatial metaphors to help visualise what are, effectively, abstract flows of electronic signals, coded as information, representation, and exchange (see Munt, 2001). Thus, an Internet point-of-presence becomes a web *site*. The ultimate convergent, broadband descendant of the Internet is labelled the information super*highway*. A satellite node becomes a tele *port*. A bulletin board system becomes an electronic *neighbourhood*. Web sites run by urban municipalities become virtual *cities*, complete with their '*residents*', '*citizens*', '*districts*', '*squares*', '*ports*', '*cafés*', and even '*mayors*'. The whole society-wide process of technological innovation becomes a Wild-West-like electronic *frontier* awaiting colonisation by heroic frontier dwellers. Those 'exploring' this frontier become web *surfers*, virtual *travellers*, or, to Bill Mitchell (1995, 7), electronic *flâneurs* who 'hang out on the network' (see Featherstone, 1998). The Internet as a whole is variously considered to be an electronic *library*, a medium for electronic *mail*, or a digital market *place* (Stefik, 1996). And, as Robert Luke suggests (p. 249), Microsoft seductively asked in their advertising in the mid-1990s promoting the Internet '*Where* do you want to go today?'

Some argue that the strategy of developing spatial metaphors is 'perhaps the only conceptual tool we have for understanding the development of a new technology' (Sawhney, 1996, 293). Metaphor-making, they suggest, points 'to the process of learning and discovery – to those analogical leaps from the familiar to the unfamiliar which rally the imagination and emotion as well as the intellect' (Buttimer, 1982, 90, quoted in Kirsch, 1995, 543).

And yet, at the same time, the fusing of ICTs into cities provides many challenges to how we imagine the city itself. Drawing on the 'cyborg' theories discussed by Timothy Luke (p. 106), the anthropologist Donna Haraway asserts that the blurring of the body–technology boundary within pervasive cybernetic systems leads to a 'technological polis where machinic-desires drive cybernetic systems by artificial instincts and recursive feedback loops' (Haraway, 1991, 149–81). The cultural critic Frederic Jameson (1984) famously argued that what he called the 'cognitive mapping' of the material geographies of cities is now made difficult because of shifts towards ICT-mediated societies. And the urban theorist Christine Boyer (1996, 133–9) has argued that we now face a 'crisis of representing invisible cities'. This is because a gap exists 'between the city that we can visualise and the invisible city that is constituted in and through its fields of information circulation'. Control, writes Boyer, now 'acts like a sieve (a computer matrix) whose mesh transmutes from point to point, undulating and constantly at work. The code, not the norm, becomes the important device' (Boyer, 1996, 18).

Mike Davis's (1992, 16) work on Los Angeles concretises such arguments. He explicitly argues that the city's highly polarised social order is now clearly founded on extensive webs of high-tech surveillance, linked into both urban simulations in 'cyberspace' and the extending material simulations of theme parks and consumption-driven landscapes (see Fred Dewey, p. 291). Thus, in Los Angeles, Davis – using Jean Baudrillard's idea of a 'simulacrum', a simulation of something without an original – believes that:

> the contemporary city simulates or hallucinates itself in at least two senses. First, in the age of an electronic culture and economy, the city redoubles itself through the complex architecture of its information and media networks . . . Urban cyberspace – as the simulation of the city's information order – will be experienced as even more segregated, and devoid of true public space, than the traditional built city . . . Second, social fantasy is increasingly embodied in simulacral landscapes – theme parks, 'historic districts' and malls – that are partitioned off from the rest of the metropolis.

Between bias and potential in the 'age of access': understanding political and economic power in cybercities

Our third challenge is to understand the interplays in cybercities between the biased technological, political, cultural and economic configurations of ICTs, and their potential roles as spaces of flexibility where social and political action can change and reconstruct city–ICT relations in new ways. Here we pre-empt our discussions in Part VIII.

On the one hand, it is clear that the parallel production of ICT domains and urban places is being strongly shaped by the broader political, economic, and geopolitical shifts which surround internationalising capitalism. It is therefore important to understand that the shaping of both urban places, and the configuration and use of ICT systems, is profoundly biased.

Such biases are becoming ever more important as vast commercial networks deliver wider and wider ranges of tailored services, knowledge and products to users in the form of subscriptions, memberships and leases which are mediated via consumer ICT systems. Jeremy Rifkin (2000) calls this a shift to the 'Age of Access'. For example, millions of people now construct their own cyberspace 'home' portals using Microsoft software. But the idea that they are expressing their personal freedom and individuality in the process is only part of the story. As Robert Luke suggests (p. 249), the software, and the ICT systems used to support this process, are carefully tailored to maximise Microsoft's surveillance and potential profits, whilst exploiting the iconographic symbols of 'home' to allay consumers' fears.

On the other hand, however, we need to be conscious that, within such biases, significant space remains for people and social movements to adapt, and appropriate both, to sustain their own lives, identities and agendas. Whilst biased, ICTs are also 'resources for experiments with self-making in all sorts of societies, for all sorts of persons' (Appadurai, 1996, 3). 'It is patronising', suggests the geographer Andy Merrifield, 'to believe that citizens of cities are easily hoodwinked by a city supposedly comprised of bits and bytes, texts and simulations, semiotics and CCTV systems' (1996, 67).

ICTs are always flexible to some extent – at least within their biased designs and software and access constraints. Users and communities do have real space to socially shape new media applications and technologies in novel and innovative ways – just as they can influence social and cultural change within cities. Ultimately, social movements linking both cities and new media can, unknowingly or deliberately, change things and can thus move the limits and borders set down by wider political, economic and technological change. For example, it is clear that in many cases ICTs may facilitate social and resistance movements that would not otherwise be possible. They can thus help social movements to escape the often dominating shackles of nation states and corporate capital and so allow social action to 'invigorate local tendencies towards more inclusive and democratic societies throughout the world' (Lin, 2002, 389; see Castells, 1999).

Consider some examples. Feminist movements of various sorts have made powerful use of the Internet to politically mobilise at various scales across the world (Harcourt, 1999). The loosely organised anti-globalisation movement also uses the net and mobile phones to parallel the telescoping scales of international capitalism, and so link a myriad of local social movements and campaigns into a powerful international one. This has made it an increasingly powerful political force. At a more micro-scale, people can also significantly affect the shaping, meaning and use of their own technological devices, within the process known as 'technological domestication' which occurs within their homes and everyday lives (see David Morley, p. 252; Silverstone and Hirsch, 1992).

Finally, whilst city and ICT spaces may both be under increasingly powerful gaze of various forms of state and commercial surveillance (see David Lyon, p. 299, and Peter Huber and Mark Mills, p. 418), there is evidence that the people can, in their everyday lives, significantly limit the negative effects of these processes (see Lin, 2002). For example, the US architecture critic John Kaliski suggests that, whilst cityspaces face increasing commercialisation, gating, fortressing and surveillance:

> curiously, many of the social transactions that are shaping the tenor of culture occur in the very places most subject to the scan of globalism. Shopping mall culture, gated enclaves (whether

suburbs or rock houses), omnipresent recording, and surveillance of every aspect of daily life do not seem to limit ever new and evolving cultural expressions and mutations born of unexpected gatherings. The easy reduction of these places to unitary theories or definitions of globalized space overlooks the physical workings of their quotidian elements (1994, 7).

References and suggestions for further reading

Adams, P. (1995), 'A reconsideration of personal boundaries in space–time', *Annals of the Association of American Geographers*, 85(2), 267–85.

Appadurai, A. (1996), *Modernity at Large: Cultural Dimensions of Globalization*, Minneapolis: University of Minnesota Press.

Augé, M. (1995), *Non-Places: Introduction to the Anthropology of Supermodernity*, London: Verso.

Bell, D. and Kennedy, B. (eds) (2000), *The Cybercultures Reader*, London: Routledge.

Bender, G. and Druckrey, T. (1994), *Culture on the Brink: Ideologies of Technology*, Seattle: Bay Press.

Bhabha, H. (1994), *The Location of Culture*, London: Routledge.

Bourk, M. and Worthington, T. (2001), *Universal Service?: Telecommunications Policy and People with Disabilities*, London: Print on Demand Books.

Boyer, C. (1996), *Cybercities: Visual Perception in the Age of Electronic Communication*, New York: Princeton University Press.

Buttimer, A. (1982), 'Musing on helicon: root metaphors and geography', *Geografiska Annaler*, 64B, 89–96.

Calhoun, C. (1986), 'Computer technology, large-scale social integration and the local community', *Urban Affairs Quarterly*, 22(2), 329–49.

Castells, M. (1997), *The Power of Identity*, Oxford: Blackwell.

Castells, M. (1998), *The End of Millennium*, Oxford: Blackwell.

Castells, M. (1999), 'Grassrooting the space of flows', *Urban Geography*, 20(4), 294–302.

Crang, M., Crang, P. and May, J. (eds) (1999), *Virtual Geographies: Bodies, Space and Relations*, London: Routledge.

Crang, M. (2000), 'Public space, urban space and electronic space: would the real city please stand up?' *Urban Studies*, 37(2), 301–17.

Collins, J. (1995), *Architectures of Excess: Cultural Life in the Information Age*, New York: Routledge.

Curry, M. (1998), *Digital Places: Living with Geographic Information Technologies*, London: Routledge.

Davis, M. (1990), *City of Quartz*, New York: Verso.

Davis, M. (1992), 'Beyond Blade Runner: urban control, the ecology of fear', *Open Magazine*, New Jersey: Westfield.

Doheny-Farina, S. (1996), *The Wired Neighborhood*, New Haven: Yale University Press.

Druckrey, T. (ed.) (1994), *Electronic Culture: Technology and Visual Representation*, New York: Aperture.

Dürrshmidt, J. (2000), *Everyday Lives in the Global City: The Delinking of Locale and Milieu*, London: Routledge.

Eade, J. (1997), *Living the Global City: Globalization as Local Process*, London: Routledge.

Ebo, B. (ed.) (1998), *Cyberghetto or Cybertopia?: Race, Class, and Gender on the Internet*, New York: Greenwood Press.

Featherstone, M. (1998), 'The *Flâneur*, the city and virtual public life', *Urban Studies*, 35(5–6), 909–25.

Featherstone, M. and Burrows, R. (1995), *Cyberpunk/Cyberspace/Cyberbodies*, London: Sage.

Flanagan, M. and Booth, A. (eds) (2002), *Reload: Rethinking Women and Cyberculture*, Cambridge, MA: MIT Press.

Gregory, D. (1994), *Geographical Imaginations*, Oxford: Blackwell.

Hannerz, U. (1996), *Transnational Connections: Culture, People, Places*, London: Routledge.

Haraway, D. (1991), 'A manifesto for cyborgs: science, technology, and socialist-feminism in the late twentieth century'. In D. Haraway (ed.), *Simians, Cyborgs and Women: The Reinvention of Nature*, New York: Routledge, 149–81.

Harcourt, W. (1999), *Women @ Internet: Creating New Cultures in Cyberspace*, London: Zed Books.

Hawthorne, S. and Klein, R. (eds) (1999), *Cyberfeminism: Connectivity, Critique and Creativity*, New York: Spinifex Press.

Herman, A. and Swiss, T. (eds) (2000), *The World Wide Web and Contemporary Cultural Theory*, New York: Routledge.

Hillis, K. (1999), *Digital Sensations: Space, Identity and Embodiment in Virtual Reality*, Minneapolis: University of Minnesota Press.

Holloway, S. and Valentine, G. (2002), *Cyberkids: Youth Identities and Communities in an Online World*, London: Routledge.

Holmes, D. (2001), *Virtual Globalisation: Virtual Spaces/Tourist Spaces*, London: Routledge.

Jameson, F. (1984), 'Postmodernism, or the cultural logic of late capitalism', *New Left Review*, 146, 53–92.

Jarvis, H., Pratt, A. and Cheng-Chong, P. (2000), *The Secret Life of Cities: Social Reproduction of Everyday Life*, London: Prentice Hall.

Jones, S. (ed.), *Cybersociety: Computer Mediated Communication and Community*, London: Sage.

Kaliski, J. (1994), 'Liberation and the naming of paranoid space'. Foreword in S. Flusty, *Building Paranoia: The Proliferation of Interdictory Space and the Erosion of Spatial Justice*, Los Angeles: Los Angeles Forum for Architecture and Urban Design.

Kaplan, C. (2002), 'Transporting the subject: technologies of mobility and location in an era of globalization', *PMLA*, 117(1), 32–42.

Keeble, L. and Loader, B. (2001), *Community Informatics: Shaping Computer-Mediated Social Relations*, London: Routledge.

Kirsch, S. (1995), 'The incredible shrinking world? Technology and the production of space', *Society and Space*, 13, 529–55.

Kolko, B., Nakamura, L. and Rodman, G. (eds) (2000), *Race in Cyberspace*, New York: Routledge.

Kopomaa, T. (2000), *The City in Your Pocket: Birth of the Mobile Information Society*, Helsinki: Gaudemus.

Lin, M. (2002), 'Cyber-civic space in Indonesia: from panopticon to pandemonium', *International Development Planning Review*, 24(4), 383–400.

Lootsma, B. (2002), *Body and Globe: Dwelling in an Age of Media and Mobility*, Rotterdam: 010 Publishers.

Mackenzie, A. (2002), *Transductions: Bodies and Machines at Speed*, London: Continuum.

Massey, D. (1993), 'Power-geometry and a progressive sense of place'. In J. Bird, B. Curtis, T. Putnam, G. Robertson and L. Tickner (eds), *Mapping the Futures: Local Cultures, Global Change*, London: Routledge, 59–69.

Merrifield, A. (1996), 'Public space: Integration and exclusion in urban life', *City*, 5–6, 57–72.

Meyrowitz, J. (1985), *No Sense of Place: The Impact of Electronic Media on Social Behavior*, New York: Oxford.

Miles, M., Hall, T. and Borden, I. (eds) (2000), *The City Cultures Reader*, London: Routledge.

Mitchell, W. (1995), *City of Bits: Space, Place and the Infobahn*, Cambridge, MA: MIT Press.

Morley, D. (2000), *Home Territories: Media, Mobility and Identity*, London: Routledge.

Morley, D. and Robins, R. (1995), *Spaces of Identity*, London: Routledge.

Munt, S. (ed.) (2001), *Technospaces: Inside the New Media*, London: Continuum.

Pile, S. (1994), 'Cybergeography: 50 years of *Environment and Planning A*', *Environment and Planning A*, 26, 1815–23.

Rifkin, J. (2000), *The Age of Access: How the Shift From Ownership to Access is Transforming Capitalism*, London: Penguin.

Robins, K. (1995), 'Cyberspace and the world we live in'. In Mike Featherstone and Roger Burrows (eds), *Cyberspace/Cyberbodies/Cyberpunk*, London: Sage, 135–56.

Robins, K. (1996), *Into the Image: Culture and Politics in the Field of Vision*, London: Routledge.

Robins, K. (1999), 'Foreclosing on the city? The bad idea of virtual urbanism'. In J. Downey and J. McGuigan (eds), *Technocities*, London: Sage, 34–59.

Robins, K. and Webster, F. (1999), *Times of the Technoculture: From the Information Society to the Virtual Life*, London: Routledge.

Sandercock, L. (1998), *Towards Cosmopolis*, London: Wiley.

Sawhney, H. (1996), 'Information superhighway: metaphors as midwives', *Media, Culture and Society*, 18, 291–314.

Shiel, M. and Fitzmaurice, T. (eds) (2001), *Cinema and the City: Film and Urban Societies in a Global Context*, Oxford: Blackwell.

Silverstone, R. and Hirsch, E. (1992), *Consuming Technologies: Media and Information in Domestic Spaces*, London: Routledge.

Smith, M.-P. (2001), *Transnational Urbanism: Locating Globalization*, New York: Sage.

Soja, E. (2000), *Postmetropolis: Critical Studies of Cities and Regions*, Oxford: Blackwell.

Solnit, R. (2001), *Wanderlust: A History of Walking*, New York: Viking.

Staheli, S., Ladwith, V., Ormond, M., Reed, K., Sumpter, A. and Trudeau, D. (2002), 'Immigration, the internet and spaces of politics', *Political Geography*, 21, 989–1012.

Stefik, M. (1996), *Internet Dreams: Archetypes, Myths and Metaphors*, Cambridge, MA: MIT Press.

Turkle, S. (1997), *Life on the Screen: Identity in the Age of the Internet*, London: Phoenix.

Vale, L. and Warner, S. (eds) (2001), *Imaging the City*, New Brunswick: Rutgers Center for Urban Policy Research.

Waltz, S. (2002), 'On the move between proximate distance and distant proximity', unpublished thesis, Stuttgart University.

Wellman, B. (ed.) (1999), 'Networks in the global village', Boulder, CO: Westview.

Wellman, B. and Haythornthwaite, C. (eds) (2002), *The Internet in Everyday Life*, Oxford: Blackwell.

Wilson, R. and Disannayake, W. (1998), *Global/Local: Cultural Production and the Transnational Imaginary*, Durham, NC: Duke University Press.

'Habit@Online: Portals as Purchasing Ideology'

Robert Luke

Editor's Introduction

How do the personal home web pages that have proliferated on the Internet since the mid-1990s relate to the territorial spaces of the city? In what ways do the constructions by major software corporations like Microsoft of the personalised, globally accessible, 'home' in cyberspace, relate to the more familiar 'home' in urban space?

In this first, specially commissioned, piece on the social and cultural worlds of cybercities, the Canadian media theorist Robert Luke considers these questions. Luke argues that Microsoft, especially, has powerfully shaped the way in which the web has been 'domesticated'. He analyses the ways in which their appeals to online notions of the 'home' in Canada (as elsewhere) have, in a content of accelerating global flow and real-time accessibility, articulated strongly with the changing nature of cities.

The corporate selling of personal web space now constructs it as a 'home place', rather than as a pure mechanism of mobility. Thus, Luke suggests that the web is being actively 'territorialised' in a way that moves web surfing from the active to the passive. Instead of going 'out' into cyberspace, the implication is that the world now comes to our 'home' cyberspaces (which are overwhelmingly constructed through Microsoft software). The home portal is thus constructed as a commodified space of safe consumption (biased towards Microsoft's affiliate companies). This is done under intensive commercial surveillance.

The 'home' place of the personalised portal is thus a profoundly commercialised and surveilled environment. This resonates with the increasingly commercialised and surveilled urban spaces that it articulates with. As Luke suggests 'the web portal-based territorialization [of the home place] works two ways; Microsoft seeks to bring users under its territorial control, while these users are encouraged to identify msn.ca with their process of online identification'. As users continually move across Microsoft's firewalls to access their home portal, so they continually feed the corporation personal data in exchange for ease of access. Users are continually pushed towards the participating sites offering an array of online commerce opportunities that involve firms with commercial relations with Microsoft. And they are encouraged to customise their home portal so that its space supports their personal identity as a way of encouraging consumer loyalty to the Microsoft brand.

Interestingly, Luke also contrasts the white, pure, advertising style of Microsoft's home software products, which are plastered around the streets of many cities, with the way Microsoft represents the cityscapes that provide the dominant context for the experience of the software. In a clear suggestion of the sanctity of the idea of the 'home' in cybercities, and to the shift of commercial and shopping activity away from bodies and streets, Luke argues that Microsoft selectively appropriates the icons of the city for their commercial purposes. They do this, however, whilst encouraging users to withdraw from the city in their everyday lives.

PORTALS: THE PERSONALIZED PAGE

Within cyberspace, territorialization is never concrete. What dominates the new network paradigm is the positioning of things in process. Online space is defined by movement and the processes of de- and reterritorialization: even home pages are launching points for other things and ideas. In 1996 Microsoft's advertising campaign for the Internet used to use the slogan "Where Do You Want To Go Today?" followed by the line "This way to the Internet." This reflected the premise that we can travel virtually, anywhere, any time. This idea of virtual travel promotes the notion that the online world is itself a portal, an access point for exploring the world from the comfort of our own home, a promise as old as illuminated manuscripts.

Microsoft's advertising campaign for home Internet access has now morphed into "Make It Your Home," a slogan that seeks to construct an identity of place and recognition of Microsoft as *the* place of first access. The change in Microsoft's slogan represents a shift in the conception of web surfing: from active to passive, from going out into the (online) world to having it brought to the home, sold as a place of safe consumption. In Canada, Microsoft wants to make its portalmsn.ca website the place of first access, and by association to make its browser, Internet Explorer, the browser of choice, to create a brand loyalty not just towards its product, but also to the service provided within online commodity flows. Microsoft's intonation to "Make It Your Home" calls us to identify with Microsoft and to use msn.ca as our point of origin: our home within the flows of online capital. This territorialization works in two ways: Microsoft seeks to bring users under its territorial control, while these users are encouraged to identify msn.ca with their process of online identification.

Conditioned to accept the commercial basis for constructing online identity, users enter the habit@online as they would a rigidly controlled border crossing. Microsoft's Passport™ feature, which again involves signing over personal data in exchange for ease of use, borrows the notion of controlled entry with passports as a means to facilitate ease of entry into online spaces. Perhaps more importantly, ease of shopping is facilitated within this online space: You can use the Passport wallet to make fast, secure online purchases without having to type in the same information many times. Entry to this habit@ is strictly controlled and monitored, under the guise of making life simpler. The rhetoric of consumer empowerment, with its sense of belonging and exclusive access to participating sites, authenticates and validates the habit@online identity. The digital citizen carrying a Microsoft Passport™ is (trade)marked as the territory of Microsoft.

Commercial web sites and portals try to hold their users. This is measured by the extent to which those who access a site are retained, preventing them from just clicking through. Instead, users are encouraged to identify with a given portal as their formal gateway to the World Wide Web. The new Microsoft advert – and portals in general – represent the construction of online space as part of identity formation in the new datastructure of postmodernity. Microsoft's advertisement of their current entreaty to "Make It Your Home" refracts representational space:

> space as directly *lived* through its associated images and symbols, and hence the space of 'inhabitants' and 'users' . . . This space is dominated – and hence passively experienced – space which the imagination seeks to change and appropriate. It overlays physical space, making symbolic use of its objects (Lefebvre, 1991, 39).

The abstract whitespace of this advertisement overlays the cityscape and promises a pristine environment that contrasts starkly with the dirty city street and its threat of contact with the "other" (see Plate 22). This datastructural palimpsest has rendered the physical world almost obsolete, appropriating the territory of shopping with the promise of a utopian freedom to purchase from the sanctuary of the home. The shopping space of the city streets of Toronto is deterritorialized and dis-integrated by this advertisement, as the home-as-portal reterritorializes the shopping experience within online space.

Microsoft literally overlays physical space with its appropriation of the urban landscape: the cityscape becomes a backdrop for the datastructure technology and the inherent, mnemonic promise to take you "where you want to go today". Now Microsoft wants to be your home as well, to be associated with the image(s) of home where 'the space of "inhabitants" and "users"' is truly mixed. The inhabitants of the MSN habit@online can passively experience the benefits of the Microsoft utopia. Just as the "Make It Your Home" billboards and posters overlay the physical space of the cityscape, so too does MSN overlay the space of the home with the promise of its ubiquitous computing.

Plate 22 Microsoft Network's advertising on the streets of Toronto, 2001. (Photograph by Robert Luke.)

Now selling itself as "More Useful Everyday" and as "The Everyday Web," Microsoft is insinuating itself within the normalized routines of daily life. Not content to be the signifier of Home, Microsoft now wants utility, to melt into the background. Microsoft's appropriation of home within the online portal is an attempt to project the cultural notion of comfort, to insinuate msn.ca within the process of Internet surfing, and to normalize the presence of Microsoft within everyday life.

REFERENCE FROM THE READING

Lefebvre, H. (1991), *The Production of Space*, Oxford: Blackwell.

Editor's suggestions for further reading

Luke, R. (2002), 'Habit@online: web portals as purchasing ideology', *Topia: A Canadian Journal of Cultural Studies*, 8, 61–89.

Luke, R. (2002) 'AccessAbility: enabling technology for lifelong learning'. In 'Integrating technology into learning and working'. Special Issue of *Educational Technology & Society*, 5 (1), January. http://ifets.ieee.org/periodical/vol_1_2002/v_1_2002.html

'At Home With the Media'

from *Home Territories: Media, Mobility and Identity* (2000)

David Morley

Editor's Introduction

It is important that any understanding of the relationships between ICTs, the home, and urban and suburban social and cultural life, is placed in the context of the experience of how pre-computerised communications technologies were domesticated within the home. Here we must address the cases of telephones, cable TV, video cassette recorders, and, above all, television.

In this second reading on the social and cultural worlds of cybercities, the eminent British media sociologist David Morley reflects, through an analysis of how gender roles shape the use of TV, telephone and computer use in homes, on the relations between media technologies and home life. Starting with a discussion of how the gendered identities within households shape television viewing, Morley stresses the diversity of sites and experiences of TV viewing in most households. TV viewing, and computer use, within the home, he notes, tend to be fragmented in timing, spatial position, and content. This has been encouraged by the growth of personalised technologies.

Morley then analyses some of the tensions which surround the efforts of different members within households and families to use and access computing and TV technologies to support their own uses. He notes that different members within families increasingly identify with different technologies and media altogether. Children and teenagers tend to use the latest technologies to underline their growing independence and to carve out private social and representational spaces that their parents cannot access or understand. Different members of the same family construct and use the same technologies, most notably home computers, in very different ways. As a result, rather than invoking technologies with the ability to have general and universal social 'impacts', it is necessary to understand the plurality of social meanings that technologies are associated with, at different times and positions within households, as they are 'domesticated' in complex ways.

The social worlds of new media are especially influenced by variations in which men and women tend to attempt to shape and experience media and ICT uses within the home. Morley notes that men and women often conflict over their notions of what technologies should be used for, and where they should be placed, in the home. Boys and men tend to try to colonise, and dominate, use of new technologies. Their technology-driven agendas can bring activities into the home that were not previously there, creating gender conflicts. Such conflicts can be exacerbated when new media are used to allow men and women to work at home.

Thus, Morley notes that the different modalities of telephone use in the home – with men tending to emphasise the instrumental uses of the telephone to get things done, and women tending to emphasise the phone's use as an intrinsic means of maintaining social relationships – are now being replayed with conflicts, and contrasts, in the social construction of home computers and ICT use.

The integration of television viewing into the spatial geography of the home constitutes a core part of how household life is organised. Thus when one of Gauntlett and Hill's (1999) respondents reports that, in their household "each family member watches TV in a different place. Mum and Dad in the living room, my sister in the bedroom and I usually watch TV in the kitchen", the authors note that one or another such form of routinised dispersion of viewing is common to 80 per cent of the households in their study (Gauntlett and Hill, 1999, 38). Here we see one aspect of the much-heralded move to the fragmentation of domestic viewing. In line with his earlier argument about the relationship between the development of particular forms of domestic life and broader socio-technological changes, Hirsch observes that in the recent period, in the UK, "new configurations of television (less centred on the broadcast form) in the shape of cable, satellite [and] video . . . and the new technology of personal computers . . . position themselves in a politico-moral environment of 'choice'" which is based on the desires of the private individual, rather than the household as a unit (Hirsch, 1998, 160). Picking up the theme of fragmentation within the home referred to earlier, Silverstone has noted the extent to which the future of the family itself has been widely perceived to be threatened by the arrival of "portable, individual . . . privatising technologies" which function as "isolating and fragmenting machines". In this dystopic vision Silverstone observes, beyond the issue of withdrawal from the public sphere into the home, nowadays, inside the home itself, we face a situation where "parents and children could be seen to occupy separate domestic times and spaces, isolated by personalised stereo systems . . . passing each other like ships in the night in a jamming fog of electronic communication and information overload" (Silverstone, 1994). From this perspective the personalisation of technologies (the Walkman, the Gameboy, Playstation, the multiple television sets in many homes) makes for the fragmentation of the family or household – an effect which as Silverstone notes, "is further enhanced by the 'time-shift' capabilities of both the VCR and the microwave".

NEGOTIATING DIFFERENCE IN THE FAMILY

In many cases the fragmentation of domestic viewing may function to avoid, if not resolve, what would otherwise be conflicts over viewing choices. Certainly conflicts and disagreements about television and other media are often a site of family and household tensions. In some cases it seems that it is hard for people to tolerate the idea of differences of opinion and judgement as between different members of the family. Thus, when one of the respondents in Lelia Green's study of television viewing in the Australian outback states rather defensively, "we are a close family. Generally we all like the same shows", we perhaps glimpse the tip of an iceberg of anxiety about the toleration, organisation and management of difference (and thus potential conflict) within the home (Green, 1998).

These matters are commonly dealt with, or at least demarcated by, systems of differentiation by which household members negotiate the occupation of separate (and often complementary) physical and/or symbolic spaces. Thus, Graham Murdock and his colleagues argue that children in their study used their time on their computer "to win space and privacy within the household and assert their separation and independence from their parents" (Murdock *et al.*, 1995). In a more developed case, an adult young man living at home with his parents is reported by Moores as inhabiting a "masculine world of gadgets, fast cars and sci-fi fantasies" in which he assembles a collection of electrical goods as "signs of a struggle to fashion some limited degree of autonomy in the face of parental authority". For him, as for one of the young men in the study, whose family customarily referred to his room (full of similar gadgets) as his "Womb", this young man's special area of expertise is inscribed in (and demarcates) his particular symbolic and geographical territory within the house. He puts it this way:

> I'm the only one who knows how to use my electrical equipment. Nobody else comes in my room – I think of it as my space . . . Up here I can watch anything I want . . . As soon as I go into my room, it's like I'm on another planet. (Moores, 1996, 36–8)

The son in this family is heavily identified with satellite viewing and dismisses established terrestrial television as traditional, boring and old-fashioned. In a sideways move which produces a differentiated and thus complementary position for himself in the household, in response to the son's increasing interest and involvement in new technologies his father has become increasingly absorbed in restoring antique furniture. Similarly, in another of the families on which Moores reports in the same study, the wife invests her energies and interests in antiques exactly to the (growing) extent to which her husband has become obsessed, as she perceives it, with modern, electronic gadgetry.

Eric Hirsch further reports the tensions in one household arising from the placing of the husband's computer equipment in their lounge, in so far as it impinges on their joint living space. At the point of upgrading the relevant computer equipment, Hirsch reports that each member of the family perceived the new equipment as offering different (and partly contradictory) possibilities. The same piece of technology (a new computer modem) was variously envisaged by the husband of the family as "a tremendous potential"; by his wife as a threat to her use of the telephone line; by the son as an exciting potential for the use of email; and by the daughter as of only "marginal interest in comparison with television". As Hirsch observes, the family faced a very difficult situation, where their variously conflicting interests would need to be reconciled in dealing with the arrival of this new piece of technology and the multiple connections which it potentially offered with the outside world – in a situation where any one of these connections potentially vitiated others. Clearly, the advent of multiple phone lines in many houses, since this time has gone some way to alleviating some of these contradictions, but the essential point remains, so long as the various members of a household still have competing and contradictory priorities for the use of essentially scarce resources such as money, time and space.

[. . .]

THE GENDERED FORMS OF MEDIA CONSUMPTION

In this section I shall turn to some of the questions about the gendering of domestic life and address the question of gendered differences in patterns of media usage and the forms of domestic conflict to which they sometimes give rise. In this context Hirsch reports a struggle between a woman and her husband who kept his portable computer in the living room, so that he could work on it in the evenings. His wife wanted the

computer taken back to his place of work, where in her view it properly belonged, because it took up too much space, claiming that "he has it everywhere where it shouldn't be" [. . .]

In a further striking example of the gendering of (and differentiation of) attitudes to and uses of technology, Moores reports the complementary cases of a woman and a man in his study who both work at home, but take completely opposed attitudes to the difficulties posed for them, as homeworkers, by the telephone. Thus, in one case the wife in the family is so concerned to maintain some boundary round her private space that not only does she use the answering machine to screen calls, but in the evenings and weekends she unplugs both the phone and the answering machine so as to avoid being disturbed. Conversely, the male homeworker has such a "phobia about . . . people not being able to speak to me" that he makes himself available by phone "seven days a week, 24 hours a day", using telecommunications, as Moores observes, "to create a constantly permeable external boundary to the home" as he is so anxious not to lose potential custom.

There is now considerable evidence to support the idea that masculine and feminine modalities of telephone usage exhibit clear differences. Thus in her study of the gendering of telephone usage in Australia, Ann Moyal reports that she discovered "a pervasive, deeply rooted . . . feminine culture of the telephone in which 'kinkeeping', caring, mutual support, friendship . . . and community activity play a central part . . . which . . . contributes substantially to women's sense of autonomy, security, participation and well being" (Moyal, 1995). As Linda Rakow reported from her American study of the same phenomenon, "telephoning is a form of care-giving . . . gendered work . . . that women do to hold together the fabric of the community, build and maintain relationships" (Rakow, 1988). By nature of this telephone involvement, which links together women in different domestic households, a mediated "psychological neighbourhood" is created and sustained. Moyal's study reveals the existence of a "marginalised but vibrant culture of the telephone – a network of callers which constitutes an 'electronic community' of friendships, mutual support and kinkeeping" as the technology "provides access to a 'neighbourhood' structured across space and time". This kind of "kinkeeping" use of the phone is very rarely found among men and then only when men are playing what is, in effect, a feminine role.

REFERENCES FROM THE READING

Gauntlett, D. and Hill, A. (1999), *Living Television*, London: Routledge.

Green, Lelia (1998), 'Communications and the construction of community: consuming the remote commercial television service in Western Australia', unpublished Ph.D. thesis, Murdoch University.

Hirsch, E. (1998), 'New technologies and domestic consumption'. In C. Geraghty and D. Lusted (eds), *The Television Studies Book*, London: Arnold.

Moores, S. (1996), *Satellite Television and Everyday Life*, Luton: University of Luton Press, 37–8.

Moyal, A. (1995), 'The gendered use of the telephone: an Australian case study'. In Jackson and Moores (eds).

Murdock, G., Hartmann, P. and Gray, P. (1995), 'Contextualising home computing'. In S. Jackson and S. Moores (eds), *The Politics of Domestic Consumption*, Hemel Hempstead: Harvester, 255.

Rakow, L. (1998), 'Women and the telephone'. In C. Kramerae (ed.), *Technology and Women's Voices*, London: Routledge.

Silverstone, R. (1994), *Television and Everyday Life*, London: Routledge.

Editor's suggestions for further reading

Morley, D. (2000), *Home Territories: Media, Mobility and Identity*, London: Routledge.

Morley, D. and Robins, R. (1995), *Spaces of Identity*, London: Routledge.

Silverstone, R. and Hirsch, E. (1992), *Consuming Technologies: Media and Information in Domestic Spaces*, London: Routledge.

T
W
O

'Netville: Community On and Offline in a Wired Suburb'

Keith Hampton

Editor's Introduction

Speculations about the ways in which ICTs might affect social networks and the making and meaning of 'community' have concentrated overwhelmingly on the idea of distance transcendence. 'Virtual' communities have been largely portrayed as multi-scaled, and increasingly wide-ranging affairs through which participants engage at transnational scales whilst withdrawing, and 'capsularising', themselves from their local neighbourhoods (see Lieven de Cauter, p. 94). As a result, the possibility of remaking local and neighbourhood-based communities through ICT-based practices has been given relatively little attention.

In this specially commissioned reading the MIT urban researcher Keith Hampton reports on a pioneering empirical research project that he undertook with Barry Wellman. This focused on the ways in which neighbourhood social networks were affected by local broadband connections. Studying an anonymised Toronto suburb equipped new high capacity ICTs (termed 'Netville' for the purposes of the research), their results were surprising. Residents who used the local broadband services actually developed much fuller neighbourhood ties, associations and engagements than those who did not. As Hampton says 'access to the local computer network was a significant predictor of the number of people Netville residents recognized by name, the number of neighbours they talked to on a regular bases, and their total volume of local communication.' In fact, when the services were taken away as the experiment ended, many people experienced a kind of 'grieving' for the lost sociability it sustained.

Hampton and Wellman's work is important for hammering home the need to engage empirically in analysing the complex and subtle remediations of place, neighbourhood, and community that occur when ICTs are introduced to complement other media and forms of interaction and engagement.

'Netville,' a pseudonym adopted to protect the identity and privacy of residents, was one of the first residential communities in the world to be built from the ground up with a broadband telecommunications infrastructure that provided access to the Internet and a series of experimental information and communication technologies (ICTs) (Plates 23 and 24). Netville residents had free access to services that included: high speed Internet (10 mega-bits per second), a videophone, an online jukebox, online health services, local discussion forums, and a series of online entertainment and educational applications. Located in one of Toronto, Canada's outer suburbs, Netville consisted of 109 new, single-family homes. The typical Netville house had three bedrooms and a study and cost about 7 per cent less than the average price for a new home in the same area. Netville was a model for what many envisioned to be the future in residential computer connectivity. A place where work, leisure and social ties could all be maintained from the *smart home*.

New ICTs have made it increasingly possible to socialize, shop, work, learn and participate in leisure activities, all from within the refuge of the private residence. A combination of Internet use and home

Plate 23 Nineteenth-century wagon advertising homes in 'Netville', a Canadian 'smart community' equipped with a broadband ICT system. (Photograph by Keith Hampton.)

computing has increasingly moved activities, once almost exclusively ascribed to the public realm, into the private home. Concerns about the fate of community and social relations as a result of this change have reignited a century old debate into how technological innovation affects community.

Arguments related to the effects of new ICTs on community can generally be divided into two opposing camps. Critics have argued that these technologies contribute to an incomplete lifestyle, which they see as a consequence of turning away from a full range of in-person contacts believed to be a part of our daily lives (Nie and Lutz, 2000). ICTs are said to advance the home as a center for services that encourage a shift toward home-centeredness and privatization, displacing time previously spent on more social activities, and reducing the opportunity, or even the need, for public participation (Kraut *et al.*, 1998). On the other side of the debate, technological utopians have argued that the Internet has created a whole new form of community, 'virtual communities' where relationships extend online and around the world without regard

to gender, race or geography (Rheingold, 1993). What these two perspectives have in common is a focus on the declining importance of local place-based interactions. Neighborhood contacts either are lost through isolation in the home, or are replaced by more distant social ties that are more easily maintained online.

Personal communities extend across multiple social settings and are generally maintained through the use of more than one form of communication. Ignoring these facts has led past studies of Internet use to examine social ties in virtual communities without looking at how relationships extend offline, or examining in-person ties without considering the role of ICTs. It has thus been impossible to determine if the size of social networks, or the frequency of social contact, has decreased as a result of Internet use, or if ICTs have allowed people to shift the maintenance of ties to a new communication medium. The Internet may even have allowed people to reinvest time previously spent on in-person or telephone contact in the maintenance of a greater number of network

Plate 24 Billboard advertisement for 'Netville'. (Photograph by Keith Hampton.)

members online, as was the case with the adoption of the telephone (Fischer, 1992). Only by recognizing that people have ties of various strengths, in multiple foci of activity, and that they rely on multiple methods of communication, can a clear picture be formed of the effects of ICTs on social relations.

The Netville study takes a social network perspective of how new ICTs affect community. This research argues that online social contact encourages additional contact in-person and over the phone. ICTs allow people to bridge temporal and spatial barriers to social contact. In a highly 'wired' residential setting, this has the effect of increasing the amount of contact and support exchanged with existing social ties, and facilitating neighborhood social capital in the form of community involvement and larger, more connected local social networks.

METHODS

Barry Wellman and I began our research on the Netville project in the spring of 1997 and concluded the fieldwork in the summer of 1999. Approximately 60 per cent of Netville homes participated in the technology trial and had access to the network for up to two years. The other 40 per cent of households, for various organizational reasons internal to the telecommunications consortium providing the technology, were never connected to the network despite assurances at the time residents purchased their homes that they would be. While it was unfortunate that not every household in Netville could be connected to the local network, the presence of an internal group of non-wired homes provided a natural comparison group for studying the effects of living in a wired neighborhood.

Shortly after the construction of the first homes I moved to Netville where I conducted ethnography for nearly two years. I worked from home, participated in online activities, attended all possible local meetings (formal and informal), and walked the neighborhood chatting and observing. I made every attempt to share in the life of Netville, making friends and carrying out the daily obligations of life expected of any other resident of the community. The opportunity to live and work amongst Netville residents provided an in-depth understanding of what life was like in a wired neighborhood. The qualitative perspective of the ethnography was reinforced with a cross-sectional survey administered to a sample of Netville residents. Participants were asked to complete surveys about their social networks both within and beyond Netville.

NEIGHBORHOOD TIES

Netville's wired and non-wired residents were very similar in terms of lifestyle, stage in the life cycle, and the length of time they had lived in Netville. What was remarkably different was the structure of residents' local social networks as a result of access to Netville's local computer network. A comparison of wired and non-wired Netville residents is provided in Table 7. Compared to non-wired residents, wired residents recognized three times as many of their neighbors, talked to those neighbors twice as often, visited them 50 per cent more often, made four times as many local phone calls, and further boosted communication with neighbors through the use of email. As the following comments from two Netville residents indicate, wired residents were very aware of the local computer network's impact on neighborhood social capital.

I have walked around the neighborhood a lot lately and I have noticed a few things. I have noticed neighbors talking to each other like they have been friends for a long time. I have noticed a closeness that you don't see in many communities (Netville Resident, Message to neighborhood email list 1998).

I would love to see us have a continuation of the closeness that many of us have with each other,

even on a very superficial level. Do not lose it, we know each other on a first-name basis (Netville Resident, Message to neighborhood email list 1998).

Regression analysis, a statistical way of assessing if variables are likely to be causally related, confirmed that access to the local computer network was a significant predictor of the number of people Netville residents recognized by name, the number of neighbors they talked to on a regular basis, and their total volume of local communication. For the most part North Americans do not have a large number of strong neighborhood ties and there is no indication from this study that ICTs reverse this established trend. However, while people may choose to form their stronger ties with those outside of the neighborhood setting, the evidence from Netville does suggest that ICTs may be particularly useful in encouraging the formation of weak local ties. Weak ties are a form of social capital that is particularly useful for accessing information and resources that are otherwise not available from more densely knit networks of stronger ties. Social capital at the neighborhood level has been shown to increase neighborhood safety and to reduce crime (Sampson et al., 1997). It may also increase housing values and prevent neighborhood decline. Other studies suggest a positive relationship between social capital and health. In general, neighborhoods with high social capital are safer, better informed, higher in social trust and better equipped to deal with local issues.

PUTTING TIES TO WORK IN NETVILLE

Wired Netville residents used their neighborhood networks to organize local events and to mobilize in

Table 7 Neighbourhood networks in Netville[a]

	Wired	Non-wired
Number of residents recognized by name	25.2[.000]	8.4[.000]
Number of residents talked to on a regular basis	6.4[.061]	3.2[.061]
Number of residents visited at home in the past 6-months	4.8[.147]	3.2[.147]
Number of phone calls to other residents in the past month	22.3[.063]	5.6[.063]
Number of private emails to other residents in the past month	4.1[b]	B

Note: Numbers in superscript are p-values (ANOVA). [a] N= 36 Wired, 20 Non-Wired; [b] ANOVA not performed, no variation from zero for non-wired.

dealing with community issues. The local computer network was used by residents as a means to exchange introductions, organize barbecues and parties, search for missing pets, exchange information on local services, share information related to the local town government, and to help children locate potential friends and seek help with their homework. Through online introductions, often consisting of little more than a name, address, and occupation, residents were able to find others at the local level that shared common interests and experiences. In reaction to a perceived threat or problem, or when faced with an emergency, the residents of most communities would need to knock on doors of near strangers to build support for collective action. In a wired neighborhood ICTs could be used to overcome barriers to contact while reducing the costs of mobilization in terms of time and coordination.

The rapid flow of information within Netville helped residents recognize that they shared many common experiences and concerns. Netville residents used their connectivity, on and offline, to act and work collectively on a number of occasions. Two of the most significant examples include their reaction to perceived deficiencies in the construction of their new homes, and protests against the telecommunications consortium providing their technology, when it was decided to end the technology trial, stop providing access to the high-speed local computer network, and to remove the technology from people's homes.

Netville residents used computer-mediated communication to organize offline meetings, and offline meetings to organize online protests. While residents were ultimately unsuccessful in convincing the consortium to continue supplying their free advanced ICTs beyond early 1999, residents were successful in gaining concessions from the local housing developer. The speed with which residents organized was unexpected by the developer, and it pressured him into addressing customer concerns with more resources and with greater speed than he anticipated. Town planning officials were also surprised by the success of wired residents' demands for improved customer service. They noted that the developer had moved a customer service trailer into the neighbourhood; a service that no other development had received in recent memory. Residents also achieved unusual success in preventing the developer from receiving approval from the town to begin work on a second housing development, even though this was a process perceived as

bureaucratic and involving little more than a rubber stamp. Based on his experiences in Netville, the developer acknowledged that he would never build another wired neighborhood.

EXISTING SOCIAL NETWORKS

To build a complete understanding of how the introduction of new ICTs affects social ties, Netville residents were asked a series of questions that compared their relationships with ties before and after their move to Netville. Participants were asked about change in support and contact with network members living at distances of (1) less than 50 km (excluding neighbourhood ties), (2) 50 to 500 km, and (3) greater than 500 km in comparison to one year before their move to Netville.

Moving had a negative impact on the social networks of non-wired Netville residents. As with other studies of suburban movers, moving reduced contact and the support exchanged with social ties. In Netville wired residents fared significantly better than non-wired residents as a result of their move. In contrast to non-wired Netville residents, wired residents experienced a slight increase in contact with members of their social networks who lived more than 50 km away. Although the overall increase in contact was small, there was no indication that the available technology damaged contact with distant ties. This finding contrasts with other empirical studies of the Internet that suggest that contact with existing distant social networks decreases with Internet use (Nie and Lutz, 2000). Contrary to utopian predictions that virtual communities would connect people to ties located great distances away, relative to non-wired residents wired Netville residents experienced the greatest increase in social contact with those who were 50–500 km away. They experienced a smaller increase in contact with ties at distances greater than 500 km and no change in contact with ties within 50 km (but outside the immediate neighbourhood of Netville).

When compared to non-wired residents Netville residents with access to the local computer network reported a slight increase in support exchanged with members of their social networks. Again, the greatest increase in support relative to non-wired residents, although still a small increase, was with those ties within the 50–500 km range. There was no change in support with non-neighbourhood ties less than 50 km away, and only half of the increase in support with ties beyond 500 km was experienced with those 50–500 km away. This finding contrasts with those of Kraut *et al.* (1998), who found a negative, although not statistically significant, relationship between Internet use and social support.

CONCLUSION AND DISCUSSION

In Netville, access to a series of advanced ICTs was associated with increased social capital both within and beyond the neighborhood setting. Contrary to the findings of other studies (e.g., Nie and Lutz, 2000) there was no indication that Internet use inhibited or substituted for other forms of social contact, in-person or over the telephone. In a situation where there was near ubiquitous access to computer-mediated communication, Internet use encouraged visiting, neighbor recognition, collective action, and the maintenance of local social ties. Existing social ties did not suffer as a result of Internet use. But, instead, access to Netville's local computer network helped residents to counter the loss of contact and support usually associated with moving. In Netville contact led to contact. And access to ICTs encouraged additional social contact through multiple means of communication: online, in-person and over the phone.

When social relations are examined in terms of networks and not groups, and when the Internet is not treated as its own unique social system, we find that new ICTs support the growth of social networks, social capital and community well-being. While Netville was certainly a unique situation with its widespread availability of high-speed, always-on Internet access, demographically the residents of Netville are comparable to that proportion of today's population that is most wired. Rather than dramatically altering our way of life, new ICTs may simply help overcome obstacles in existing patterns of relations. People may no longer have to give up ties as a result of geographic mobility, but instead may use the connectivity offered by new ICTs to maintain ties in situations that were previously prohibitive. At the same time, as useful as ICTs are for reaching across great distances, as a communication medium, computer-mediated communication may ulti-

mately be very similar to the telephone. As successful as the telephone has been in facilitating social contact with distant ties, telephone calling is primarily local. Studies have shown that telephones are of most frequent use with those within 8 km of the home (Putnam, 2000, 168). Many North American neighborhoods lack opportunities for social contact and the local institutions that exist to promote local interaction (cafés, bars, community organizations, etc.) are either in decline or are absent from the suburban setting. The introduction of ICTs specifically designed to facilitate communication and information sharing in a residential setting may help overcome existing barriers to social interaction and reverse the trend of neighborhood non-involvement and declining social capital.

REFERENCES FROM THE READING

Fischer, C. (1982), *To Dwell Among Friends*, Berkeley, CA: University of California Press.

Fischer, C. (1992), *America Calling: A Social History of the Telephone to 1940*, Berkeley, CA: University of California Press.

Kraut, R., Lundmark, V., Patterson, M., Kiesler, S., Mukopadhyay, T. and Scherlis, W. (1998), 'Internet paradox: a social technology that reduces social involvement and psychological well-being?', *American Psychologist*, 53(9): 1017–31.

Nie, N., and Lutz, E. (2000), 'Internet and society: a preliminary report', Stanford Institute for the Quantitative Study of Society: Stanford University. Retrieved 24 May, 2001 (http://www.stanford.edu/group/siqss/Press_Release/Preliminary_Report-4-21.pdf)

Putnam, R. (2000), *Bowling Alone*, New York: Simon and Schuster.

Rheingold, H. (1993), *The Virtual Community: Homesteading on the Electronic Frontier*, Reading, MA: Addison-Wesley.

Sampson, R., Raudenbush, S. and Earls, F. (1997), 'Crime: a multilevel study of collective efficacy', *Science* 277: 918–924.

Wellman, B. (1979), 'The community question', *American Journal of Sociology*, 84: 1201–31.

Editor's suggestions for further reading

Doheny-Farina, S. (1996), *The Wired Neighborhood*, New Haven: Yale University Press.

Gurstein, M. (ed.), *Community Informatics: Enabling Community Uses of Information Technology*, Hershey, PA: Idea Group.

Hampton, K. (2001), 'Living the wired life in the wired suburb: Netville, globalization and civil society', Ph.D. dissertation, Department of Sociology, University of Toronto.

Hampton, K. (2002), 'Place-based and IT-mediated community', *Planning Theory and Practice*, 3(2), 228–31.

Hampton, K. and Wellman, B. (2002), 'The not so global village of Netville'. In B. Wellman and C. Haythornthwaite (eds), *The Internet and Everyday Life*, Oxford: Blackwell, 345–71.

Hampton, K. and Wellman, B. (2003), 'Neighboring in Netville: how the Internet supports community, social support and social capital in a wired suburb', *City and Community*. Forthcoming.

Keeble, L. and Loader, B. (2001), *Community Informatics: Shaping Computer-Mediated Social Relations*, London: Routledge.

Rheingold, H. (1994), *The Virtual Community*, London: Secker and Warburg.

Schuler, D. (1996), *New Community Networks: Wired for Change*, New York: Addison Wesley.

Wellman, B. and Haythornthwaite, C. (eds) (2002), *The Internet and Everyday Life*, Oxford: Blackwell.

'Gender and Landscapes of Computing in an Internet Café'

from *Virtual Geographies: Bodies, Spaces and Relations* (1999)

Nina Wakeford

Editor's Introduction

The influence of gendered identities on the shaping and use of ICTs is taken up by our third reading of the social and cultural worlds of cybercities. In this, Nina Wakeford, a sociologist at Surrey University in the UK, develops a sophisticated social analysis of the relationships between gender, ICTs, and the construction of a classic hybrid space in which ICT use blends into an urban place: an Internet café.

Wakeford's starting point is the assertion that studies of Internet cultures need to engage with real urban places rather than merely addressing disembodied and placeless exchanges on the Internet itself. Moving beyond the pervasive fantasies of transcending the body, and engaging with entrenched inequalities in ICT-based social relations, Wakeford asserts that studies of hybrid spaces like Internet cafés are needed. This is because such spaces are what she calls 'translation landscapes' of computing where the Internet is produced and interpreted for people who, in the case of Internet cafés, consume time on the machines and/or food and drink.

Wakeford proceeds to analyse the relations between online and offline interactions within one such anonymised Internet café which she calls 'Net Café'. Spaces such as Internet cafés, she notes, are multidimensional. They are nodes on global ICT-based flows. They are social, public urban places within which the use of machines is structured, produced and contextualised. And they are urban spaces of sociability and the consumption of food and drink. Above all, Wakeford argues that Internet cafés are places of performance where technology is produced, and experienced, in a gendered and corporeal way. Despite efforts to break down gender-stereotypes in Net Café, for example, these were interrupted by old stereotypes.

Wakeford thus demonstrates that hybrid spaces, where distant proximities blend with proximate distance, have complex social and gender ecologies of their own. Above all, she demonstrates that, rather than somehow 'leaving' their bodies behind, hybrid ICT–urban spaces allow web surfers to continuously link between online exchange and the traditional urban biosocial reality of bodies in place. In these, Wakeford suggests that the stifling traditional gender stereotypes of males as technical shapers and fixers of technology, and women as technological supplicants, are being reforged in complex ways. But they are not being thrown away.

RATIONALE FOR STUDYING A 'REAL' PLACE

The early research on gender and Internet cultures was stimulated by the claim that gender and other aspects of social identity might become irrelevant in the new worlds created by information and communication technologies. This belief was built on the premise that computer networks allowed users to be physically invisible to other users. The most transformative visions were offered by futurists and utopian thinkers. Particularly amongst communities of practice inspired by Science Fiction, the deeply entrenched inequalities associated with certain social identities were treated as if they might disappear as the significance of the 'real' body diminished. Amongst virtual world programmers this was frequently translated as a more widespread fantasy of transcending the body. In the new electronic networks which were conceptualised as *spaces* (rather than mere conduits), identities, freed from the restrictions of embodiment, were presented as malleable. Surveying the accounts of cyberspace, Stallabrass concludes:

> The greatest freedom cyberspace promises is that of recasting the self: from static beings, bound by the body and betrayed by appearances, Net surfers may reconstruct themselves in a multiplicity of dazzling roles, changing from moment to moment according to whim (Stallabrass, 1995, 15).

Even in the early literature the ideas of multiple gender roles and 'gender swapping' were used to exemplify this promised escape from body and appearances. Moreover the declining significance of the body was associated with social change beyond electronic environments, a theme of many manifestos of the 'digital age'. Amy Bruckman, a computer scientist and creator of on-line MUD environments reports:

> Gender swapping is one example of how the Internet has the potential to change not just work practice but also culture and values (Bruckman, 1996).

[. . .]

Studying an Internet café builds upon the existing research on gender on-line by exploring how gender operates in a 'real' place where the Internet is both produced and consumed. Observations and interviews during my fieldwork at 'NetCafé' suggested an approach which borrows metaphors of spatiality from cultural geography to explain gender in terms of its production as part of *landscapes of computing*. Landscapes of computing are defined as the overlapping set of material and imaginary geographies which include, but are not restricted to, on-line experiences. The choice of an Internet café was influenced by its role in relation to the Internet and computing more widely. The café is a *translation landscape of computing* where the Internet is produced and interpreted for 'ordinary people' who consume time on the machines, and/or food and drink. My initial question was 'How do the staff achieve this production and interpretation? [. . .]

GENDER AND TRANSLATION LANDSCAPES

The daily activities at NetCafé involve participation in on-line landscapes, but many of the practices do not happen on-line. Unlike many domestic or institutional settings of computers, Internet cafés are locations in which the explicit process of consumption of the machine includes attempts at its contextualisation and interpretation as part of the product which is purchased by a customer. In Internet cafés the product could be conceived as having several interrelated components available for consumption, including:

- the machine as an isolated computer;
- the machine as part of a local network within the café;
- the machine as part of a global network;
- the systems/technical infrastructure (e.g. speed of network connection);
- the staff and their embodied knowledge;
- the café atmosphere/ambience;
- the café decor;
- the café location; and
- the food and drink.

Each of these components is itself complex and may combine incongruous elements. The café decor, for example, indicates the ways in which the café places itself in relation to other local or national businesses and structures of finance (advertising in windows, free postcards and flyers, promotions, etc.) as well as being the partial outcome of staff claiming space to represent their own version of 'the cybervibe'. While not suggesting that every actor in an Internet café will perceive or consume the same product, in general the key feature of these spaces is the combination of a series of familiar experiences (buying coffee, sitting in a café, observing norms of sociability, etc.) with the often more unfamiliar encounters with the computer . . .

REFLECTIONS ON BODIES, GENDER AND THE LANDSCAPES OF COMPUTING

At NetCafé, representations of gender appear to be achieved at least partially through the 'doing' of technology. However technology cannot be equated with the computers alone. Rather, technology is constituted by both discursive practices and alliances of materials and meanings. In NetCafé, the computers function via networks of social relationships which bring together disparate participants from several levels of the building. Put another way, the technology exceeds the boundaries of the machines. It leaks into the 'cybervibe', the interactions between cyberhosts and customers, and even the names given to other products in the café ('cybersalad' for example). One of the achievements of NetCafé as a translation landscape of computing is that it enables participants to scatter discursive representations of the Internet, the 'cyber', and global computer networks upon a range of encounters and artefacts which had not previously been recognised for their alliance with the technological. The processes of enacting the translation landscape of computing are a way of doing the Internet and a way of doing gender. In this section I reconnect the findings of my fieldwork at NetCafé with the previous work on on-line landscapes by returning to the way in which the gendered body is invoked. Framing my thinking is Adam's insistence that the connection of embodiment and technological systems must be taken seriously by feminist theorists and allied critics (Adam 1997). If the question of embodiment for feminist theory rests on the role of the body in producing knowledge, and Adam agrees with many theorists that this is the case, then the task here becomes one of articulating the kinds of bodies which inhabit NetCafé and specifying the knowledges which they produce (or are restrained from producing). The turn to bodies, in fact leads the focus back to the metaphors of spatiality with which I began my discussion of NetCafé . . .

Returning to the existing literature on gender and on-line landscapes, it is clear that bodies also figure in the deliberations of the extent to which physical presence matters in these spaces. However, the kinds of bodies which appear in the discussions of on-line landscapes tend to be restricted to linguistic performance or representations with limited circulation beyond the spaces in which they are created. The knowledges produced by these bodies come into view through a way of 'doing' gender (predominantly via textual input) which cannot be easily equated with that found amongst the bodies in NetCafé. Within NetCafé it is difficult if not impossible to avoid connecting women's bodies to material/biosocial everyday realities of being female and the knowledges produced by experiencing these realities. Whatever version of the café mission or the cybervibe was appropriated for the cyberhosts' own use was also subject to interruption by customers who reinscribed notions of gender and skill on to some bodies but not others in the course of their interactions. Of course, all users of on-line landscapes have material/biosocial realities, unless they are 'bot' (robotic) imitations within the software, and yet in most current accounts their bodies are figured in such a way as to obscure the role of the gendered body in producing knowledges outside that of the on-line landscapes. This is clearly the case in some of the futurist predictions about escape from physical presence, but is also evident in later research which has not questioned the process by which 'virtual' gender becomes 'real' and vice versa.

Another way of explicating this point is to describe an alternative means by which bodies were productive of knowledge at NetCafé. At this site bodies also manufactured knowledge in the course of their *movement through* the physical spaces of the café as well as the on-line landscapes which could be accessed there. Bodies-in-movement produce and incorporate accounts of their journeys as they encounter durable materials and discourses in the landscape of translation. Customers walked around the café floor interacting with both machines and cyberhosts, consuming machines and food, experiencing the decor and music and hearing the history of NetCafé. In this process they generated stories of how to do gender and the Internet through mobile bodies. Cyberhosts and other staff also moved through the building, creating descriptions of the levels of the café operations in terms of gender and technological expertise, including constructing the type of masculinity on the top floor among technical support staff. The female directors' bodily transit between NetCafé and other public arenas was integrated into the meanings of doing technology at the café. When the media attention on NetCafé was at its height, having transportable bodies which represented women and computing was crucial in the way the café was able to become a profitable translation landscape of computing without having a huge advertising budget. Last, the bodies of machines (or their body parts) were carried around the building particularly between the floors on which repairs were executed. As the machines moved around, so gendered meanings were made about who would fix a broken computer and who needed to be flattered or cajoled into doing so. Sustaining an approach of exaggerated gratitude was crucial when the need for those with detailed computer networking knowledge was at its most acute: at exhibitions and off-site trainings. Machines which travelled to participate in such activities were at most risk of suspending the image of NetCafé as a translation landscape of computing by not functioning at all, or, more commonly, by 'almost' working (for example having a very slow Internet connection, or by displaying broken Web links).

In conclusion I would like to suggest that the formulation of bodies as travelling within NetCafé directs us back to the utility of spatial metaphors and to landscapes of computing as ways to focus on specific material and imaginative geographies. For the study of gender in relation to technology it seems particularly apt to follow material/biosocial bodies in order to reach an understanding of how gender might be differentiated from technology in landscapes of computing such as are apparent at NetCafé. It might also be fruitful to follow material/biosocial bodies in

landscapes of computing where it is less obvious to do so, such as on-line landscapes. This approach acknowledges Ormrod's recommendation that the feminist sociology of technology should move away from approaches that isolate technology from patriarchal social relations unless one is involved in the social shaping of the other. In NetCafé, by taking seriously the range of materials and meanings which were being used, a complex interplay of gendered representations and experiences was found which cannot easily be assimilated into the old rubric of technology as inherently masculine. Rather, as NetCafé's daily activities unfolded, the Internet was translated as a place where new alliances for gender were being forged at the same time as these alliances were being interrupted by old stereotypes through which gender and technology are still often understood. These alliances and their interruptions were as dependent on the local cultures of place and space as they were on the landscapes of computing.

REFERENCES FROM THE READING

Adam, A. (1997), 'What should we do with cyberfeminism?' In R. Lander and A. Adam (eds), *Women into Computing: Progress from Where to What?*, Exeter: Intellect.

Bruckman, A. (1996), 'Gender Swapping on the Internet'. In P. Ludlow (ed.), *High Noon on the Electronic Frontier: Conceptual Issues in Cyberspace*, Cambridge, MA: MIT Press.

Stallabrass, J. (1995), 'Empowering technology: the exploration of cyberspace', *New Left Review*, 211, 3–32.

Editor's suggestions for further reading

Cutting Edge: The Women's Research Group (1999), *Desire by Design: Body, Territories and New Technologies*, London: I.B. Tauris.

Harcourt, W. (1999), *Women @ Internet: Creating New Cultures in Cyberspace*, London: Zed Books.

Liff, S. and Steward, F. (2001), 'Community e-gateways: locating networks and learning for social inclusion', *Information, Communications and Society*, 4(3), 317–40.

Wakeford, N. (1999), 'Gender and landscapes of computing in an Internet café'. In M. Crang, P. Crang and J. May (eds), *Virtual Geographies: Bodies, Space and Relations*, London: Routledge, 178–202.

'Speaking Mobile: Intensified Everyday Life, Condensed City'

Timo Kopomaa

Editor's Introduction

Zac Carey (p. 133) analysed some of the implications for the explosive growth of mobile phone use for the streets and spaces of cities. In our fourth Reading on the social and cultural worlds of cybercities the Finnish urban sociologist Timo Kopomaa takes this analysis further. Kopomaa, in a specially commissioned piece, draws on his pioneering empirical analyses of the social uses of mobile phones in that most iconic of mobile phone cities: Helsinki. He is concerned with three questions:

- How does the saturation of the city with mobile phones affect its patterns of socialisation and interaction?
- What does this process mean for notions and experiences of public space?
- How can we conceptualise a city in which urban participants live parallel lives of a 'continuous present' through the real-time interplays of intense physical and ICT-based mobilities?

Kopomaa begins by documenting the extraordinary speed with which mobile phones have become utterly banal, and virtually ubiquitous, features of contemporary urban life in Finland. Even by 1999, he notes, 78 per cent of all Finns had at least one mobile phone. As it saturates the city, the mobile phone is, he argues, simultaneously an object of desire, an instrument which collapses previous forms of public and private life, and a tool of mobile connection and control. Increasingly, as the mobile becomes utterly normalised, it is the spaces and people without mobile phones that become objects of curiosity, not the spaces and people with them. This is reflected in public campaigns across the world to wire up spaces without mobile access, such as urban subway and metro systems (see Plate 15).

Kopomaa's next discussion centres on the ways in which mobile phones are used to reconstruct the time and space coordinates of, in this case, western urban life. In an individualistic, network based, and mobile society, the mobile is a perfect device for shifting from the rhythmic space–time routines of the postwar city (with unpredictable encounters on the street) to a city of flexible, but highly coordinated, encounters. Social, co-present encounters can be organised at any time or place and at extraordinary speed. Notions of a permanent, switched-on accessibility start to transcend the long-standing notions of the regular time–space rhythms of the city. More flexible rhythms of work mean that work becomes spread out spatially and temporally and so work becomes less defined by specific times and places within the day. Finally, the mobile phone space provides, in effect, a 'third place' of encounter and engagement between the traditional separations of (private) homeplace and (public) workplace in the city.

Kopomaa argues that such processes are being pushed further by the deepening functionality of second and third generation mobile phones. The massive explosion in use of SMS (Short Message Service), for example, has led to an extraordinary culture of ongoing real time dialogue which is much less disruptive than full calling. This has deepened the flexible coordination potential within and between cities. It has supported a playful youth culture which has enabled young people to deepen their independence from parents and schools. And it has enabled young people, in particular, to flexibly coordinate the ways in which they use the public streets and squares of the city for social gathering.

Mobile phones, above all, involve the extensions of private worlds, private communications, and private lives into public space. Many social and regulatory conflicts have arisen from this. Transport operators, and organisations operating semi-public spaces like restaurants, have sought to ensure new norms of behavior to reduce the auditory imposition of private worlds into public space. The gaggle of smokers furiously puffing outside the doors of smoke-free buildings is now paralleled by the cluster of mobile phone users outside places where mobile use is prohibited or inaudible. At the same time, Kopomaa argues that mobile phone calls in many case enliven the street life as they bring connections that previously occurred within private buildings or phone boxes into an animated, public space.

As with the web portal analysed (Robert Luke, p. 249), the mobile phone permits users to roam the extending geographies of cities and urban systems at multiple scales whilst always being 'at home'. This is being supported by shifts towards the global standardisation of mobile phone networks. In 2002, for example, Vodafone, the world's largest mobile operator, advertised at airports across the world that it had established a 'right to roam' through 138 countries (at least in the parts of those nations covered by mobile infrastructures).

The mobile phone, yuppie phenomenon of the 1980s, has come to symbolise ubiquitous global communication and new urban culture. As a lifestyle commodity, it is now accessible to most western European people. Finland, and Helsinki in particular, have been international pioneers in the use of mobile phones, but it is also there that the problems engendered by the public use of the mobile phone are felt most acutely. By studying Finnish mobile phone culture (Kopomaa, 2000), we can anticipate wider trends in the practice of this postmodern form of communication. In this Reading, we shall look at the use of mobile phones and its social meaning as part of urban culture and lifestyle through interviews with focus groups. We also present and analyse articles from newspapers, including letters to the editor, assessing the public use of mobile phones. Our aim is to describe the forms of interaction that are engendered when communication activities, which used to be private, are conducted in public space. How do social situations and everyday street practices adapt to the new forms of communication? What kind of problems has the public uses of mobile phones created?

The postmodern information society facilitates living a life both of continuous presence and of complete mobility. The key element in both is the association between communication and information networks. The mobile phone has contributed to the reorganisation of work and leisure. It has created flexible schedules, increased information exchange, with more mobility and increased safety. The mobile phone has become one of the central symbols of the new urban culture and lifestyle. Digital reality offers new islets of communality within the city and between cities.

THE MOBILE PHONE: AN EVERYDAY APPLIANCE

In recent years Finland has been one of the leading countries of mobile telecommunication. It has more mobile phones per capita than most of the countries in the world. The mobile phone has virtually become the symbol and lifeline of Finnish national unity. It is not only the object of individual desire, but also a recognised focus of collective interest and a system of meanings. One of the challenges for mobile phone companies in the late 1980s and early 1990s was the yuppie image of mobile phones. While the manufacturers have worked hard to break down that image, the users have, through their choices, made the mobile phone into a popular tool. Now that the majority of Finns have a mobile phone, they have also made a commitment to it in a way which, at least for the present, is not possible nor perhaps desirable to retract.

According to a consumer survey conducted by Statistics Finland (Nurmela et al., 2000), as many as 78 per cent of 2.35 million Finnish households had at least one mobile phone in autumn 1999. Nearly all Finns aged between 15 and 39 years used a mobile phone, whereas some 40 per cent of the men and 20 per cent of the women aged over 60 had a mobile phone of their own. The question 'Who has a mobile phone?' has been replaced by 'Why doesn't everyone have a mobile phone?'

THE NEW COORDINATES OF EVERYDAY LIFE

The development of mobile phones seems to go hand in hand with the intensive growth and maintenance of mobility. Like the clock, computer and robot, the mobile phone is used to optimise time and performance. It has revolutionised traditional social structures. It is perfectly suited to the ideology of an individualistic networking society. It constitutes an expanding game of social interaction, in which continuous action and unexpectedness are essential elements. The mobile phone challenges its user to engage in real-time participation, which brooks no delays. The mobile phone side-steps anticipatory social arrangements and allows for spontaneous forms of real-time interaction. It offers a tool for a social and practical control of the urban environment.

BREAKING TIME SCHEDULES: RHYTHMIC TIME

The increased speed of connectivity and information transmission, as well as the increased facility of movement, have changed our notions of time, of the nature of the present, and of distance. The present is structured and understood as processes; people are able to react much faster, chains of events have become more compact. Mobile phones change our notion of time as something linear and mechanical, distinct from rhythm of nature and always divisible into smaller parts. When living according to the schedules – or without the schedules – of mobile phones, we can speak about the notion of rhythmic time. It takes into account the cyclical nature of time based on seasons, the times of day, and biology. For people carrying a mobile phone, accessibility begins to take over their waking hours at all times of the day, regardless of their level of activity.

Continuous availability means more flexible working hours; the mobile phone rearranges the division of time into work and leisure, previously dictated by the clock. The reverse side of increased efficiency and flexibility at work is that continuous accessibility may result in the invasion of work into moments that one would like to dedicate to rest and recreation. In some cases, the nature of work itself is such that the mechanical division of time is impossible. In childcare, for instance, the mobile phone offers a tool for organisation, which takes into account the nature of the task at hand.

THE MOBILE PHONE AS A 'THIRD PLACE'

As an instrument for maintaining contact, the mobile phone can be viewed as a 'place' adjacent to, yet outside of home and the workplace, a third place in the definition of Ray Oldenburg (1989). Oldenburg defined his concept in terms of physical spaces, applying it to coffee-houses, shops and other meeting places. In addition, the mobile phone is, in its own way, a meeting place, a popular site for spending time. The reason for effecting a meeting – i.e. calling someone up – is often just the need for contact. There does not have to be any other reason. The urban lifestyle has room for neutral ground, where you can meet a close friend or an acquaintance:

> *Raija, 49*: You can go jogging and talk to those friends you don't necessarily like to talk to at home . . . It's really important, it's one of the most important features of the mobile phone.

The mobile phone offers a 'place' to which you can withdraw when you feel like it. In addition to small-talk and managing everyday chores, the mobile phone also provides an arena for more serious and intimate discussions, which one may not want to do at home in the presence of one's partner, for example. As a third place, the mobile phone tends to be characterised by a certain playfulness in contrast with other spaces, which require more sombre behaviour.

SHORT MESSAGE SERVICE (SMS) MESSAGES: VISIBLE SPEECH

The idea of sharing the rhythm and pace in one's activities is also illustrated by the popularity of SMS messages (Short Message Service). In 2000, around one billion short messages were sent on Finnish mobile networks. These messages have become even more popular among young adults than verbal phone calls.

The SMS messages are a form of processed or visible speech. They are typically read while waiting or travelling, not unlike paperbacks. A message received at work offers a break, a private moment. Young adults sending these messages to each other seem to feel only pleasure at the interruption caused by the arrival of an SMS message. Messages fly back and forth, and the phone gives a beep to signal the arrival of a new message. Of course, messages can also be saved for perusal at a later time, in a more appropriate moment. SMS messages are sent and replied to 'on the fly', since the sending or receiving of a message is less of a disturbance than making a call. In fact, almost any moment can be suitable for sending an SMS message – during the night, at a concert, even at the movies:

> *Minna, 25*: There's actually quite a lot of message traffic. You can send out a message saying that you're changing bars, or going to have a coffee, or something else. It's like unnoticed, continuous use.

An everyday practice, SMS messages can function as invitations, means for keeping contact or transmitting information. When you don't want to have a conversation, the SMS is ideal for conveying a short, concise message; it is an effective way of avoiding the need to talk. The SMS message is a perfect way of arranging a future meeting while avoiding clashes of the coordinates of everyday life. Young people also appreciate the cheap price of the message service, when the alternative, even at its shortest form, is a phone conversation requiring some amount of small-talk. Sometimes, especially when processing a text message takes too much energy, the contrary happens and a phone conversation is the preferred option. An SMS message is often personal or witty. Both characteristics are ways to ensure that the message is noticed. The playfulness of the messages is also a way of reinforcing social ties and to prevent the likelihood of power games.

PUBLIC SPACE IN PRIVATE USE

As nomadic objects, mobile phones are a prime illustration of mobility, which is so characteristic of the postmodern way of life. The world is fitting itself into the mould of a vagabond. On a mobile phone,

answering a call and the preliminaries of conversation call for spatial as well as temporal definition: 'Where are you?', 'Are you in a bad spot?', etc. Mobile phone conversations tend to include a report on the place and the situation where the call is received (Plate 15). The mobiles are used for expressing movement and its direction. A distinctive characteristic of the mobile phone is that during the conversation, two things take place, unlike on an ordinary phone, where the conversation tends to focus, at least in principle, on the subject at hand.

The use of a mobile phone implies a privatisation of public space. It involves a relationship with the milieu, which implies a yearning for privacy and a place for oneself. With the mobile phone, the user separates him/herself from the surrounding space in a 'bubble' that is necessary for the private speech act. A person speaking on a mobile phone withdraws from the social situation or walks about within a circle with a radius of a few metres. The speaker's behaviour is characterised by absence and a certain introversion. Staring into space as well as smiling are both indications of withdrawal into mobile phone sociability.

Who, then, are the people who use mobile phones in public? Our estimates of these agents of the realm of mobile telephony are based largely on photographic material. On the basis of the photos, we can estimate that roughly two-thirds of the users of mobile phones on the street are men. This is not surprising, since proportionately more men than women own a mobile phone (although in the age range 15–29 years girls and young women have used mobile phone and send text messages as much as men). A considerable proportion of the young women would appear to have adopted the text message as their main mode of using a mobile phone. Perhaps men find it also easier to make the public space their own, to privatise it by talking on their phone. Older persons (estimated age over 50) are rare as users of mobile phone in public; this relatively small group was all males (5 per cent, N = 178). The pictures show more young females (< 25 years) than males (62 per cent F; 38 per cent M; n = 47). Girls start hanging out with friends in public earlier than boys, which probably explains their higher visibility as mobile phone users in the streets of Helsinki city centre.

The privatisation of space by mobile phone use is easier in open public places such as pavements than in semi-public places, such as shared premises, the collectivity of which is more tangible owing to the communal use of the space. The significant aspect here is the social situation, the shared interaction, the focused event. Thus, in public spaces and in public social events, such as dining out in a restaurant, mobile phone users tend to go to the door or otherwise withdraw some distance. Calls made from home or the workplace are always made from the home base, but also the user of a mobile phone domesticates the public sphere through the call episode.

THE CONDENSED CITY

The mobile phone offers a new opportunity to stay and be in a public place at the same time. Increased use of the phone and spontaneous phone conversations in public enliven the street scene. Modern forms of public engagement and urban social interaction include mobile togetherness, where people meet each other

Plate 25 The emergence of injunctions to ban mobile phone use in areas where it is deemed offensive: an example from a shop in Amsterdam, 2001. (Photograph by Stephen Graham.)

on the move, and stationary togetherness, where people gather in places, such as waiting halls or public transportation. Such ahistorical, fleeting encounters provide the common stage for mobile phone use as well. The placeless communication has not only engendered fleeting functional condensation, but it has also the decentralisation of mundane interaction. This might especially be the case in the 'non-places' of super-modernity presented by Marc Augé (1995). Non-places – with the ideal type being the traveller's space – are transformed to loci of here-and-now by mobile phone users.

Roughly speaking, mobile phone users can be observed in two types of public spaces. On the one hand, places that attract many people also attract mobile phone users. Such popular loci include pedestrian streets and restaurant terraces. On the other hand, users can often be spotted in the vicinity of places where mobile phones are not allowed, either by specific injunction (Plate 25) or tacit agreement. Mobile phone use is also common in places where people release the need for contact which they have had to repress, such as the arrival lounges at airports.

In relation to calling from phone booths, mobile phone use is mainly substitutive, whereas in relation to private wireline phones, the mobile phone is both substitutive and supplementary. Like phone booths, cars provide a safe bubble of privacy for the mobile phone user. They are popular places for making calls because they provide an opportunity for idleness and the corresponding state of mind. This is also the reason why radios and tape or compact disc players are popular car accessories. Also, the car functions as a place for distance working, enabling one to work while in transit. The possibility of making an emergency call from the mobile phone also increases traffic safety, although the use of mobile phones in traffic has also increased the risk of accidents.

Digital life thrives in public places. These places are attractive because they are stimulating and create pleasurable experiences. People become explorers; to be modern, one must leave home, yet the mobile phone allows one to always be at home as well. The modern subject thrives in the streets, squares, department stores and cafés.

REFERENCES FROM THE READING

Augé, M. (1995), *Non-Places*, London: Verso.

Kopomaa, T. (2000), *The City in Your Pocket: Birth of the Mobile Information Society*, Helsinki: Gaudemus.

Kreitzman, L. (1999), *The 24 Hour Society*, London: Profile Books.

Nurmela, J., Heinonen, R., Ollila, P. and Virtanen, V. (2000), *Mobile Phones and Computer as Parts of Everyday Life in Finland*. Statistics Finland, Reviews 5/2000, Helsinki.

Oldenburg, R. (1989), *The Great Good Place: Cafés, Coffee Shops, Community Centers, Beauty Parlors, General Stores, Bars, Hangouts and How They Get You Through the Day*, New York: Paragon House.

Editor's suggestions for further reading

Brown, B., Green, N. and Harper , R. (eds) (2001), *Wireless World: Social and Interactive Aspects of the Mobile Age*, Cologne: Springer-Verlag.

Fortunati, L. (2002), 'The mobile phone: towards new categories of social relations', *Information, Communication and Society*, 5(4), 513–28.

Katz, J. and Aakhus, M. (eds) (2002), *Perpetual Contact: Mobile Communication, Private Talk, Performance*, Cambridge: Cambridge University Press.

'The City in Cyberspace'

from *Reimaging the City* (2001)

Anne Beamish

Editor's Introduction

How do the 'virtual' worlds and 'virtual cities' that are today so common on the Internet relate to, and influence, the cultural idea of what a 'city' actually might be in the twenty-first century? In our fifth reading on the social and cultural worlds of cybercities, the American urbanist Anne Beamish develops an innovative and ambitious reflection on what makes virtual worlds so appealing to their online participants. She then analyses the ways in which the city is represented, and constructed, in cyberspace. Finally, she looks in the opposite direction to question how 'urban' the virtual worlds of cyberspace are. To do this she uses the traditional criteria used in urbanism to judge whether urban designs and developments support genuinely public, civic interactions and the forging of real 'communities.'

Beamish believes that people participate in constructed online versions of digital worlds, cities and communities because they afford new opportunities for sociability, creativity, and identity play. Howard Rheingold (1994), one of the founders of the virtual and online communities movements, has famously suggested that urban sprawl, motorisation and the decline of informal and public spaces in US cities have directly forced people to search for these opportunities in online worlds. Beamish argues that people are also attracted to online and virtual worlds because they can also actually shape online environments much more directly and clearly than they can usually shape the physical environments of the cities in which they mostly live.

Graphical facsimiles and analogies of cities have had a major influence on the cultural evolution of the Internet. Increasingly, they are also influencing broader cultural notions of the city. The Internet is awash with digital, virtual or simulated 'cities.' There are literal simulations of specific real cities which use digital simulation techniques to create a 2- or 3-D representation of a particular city on the net. Whilst impressive, Beamish notes that these examples tend to be static. Usually, they lack the pulse of human life and human inhabitation.

A second range of virtual and digital cities, not analysed by Beamish, involves the use of a web site by an urban authority or municipality to try to feed back positively on to the economic and social development of a specific city. Reflecting the fact that it may be easier on the net to find a service, a date, or a weather forecast from the other side of the world, these initiatives attempt, in effect, to 'ground' the Internet so that its relations to a particular real urban space become more transparent and less fragmentary. (Geert Lovink (p. 371) analyses an iconic example of such a 'grounded' virtual city: the Amsterdam *Digitale Stadt* (or 'Digital City') (see Aurigi and Graham, 2000, Graham and Aurigi, 1997).)

Beamish is particularly interested in a third type of virtual city: one constructed through mass online participation in a kind of virtual urbanisation through software. Analysing the Alphaworld city, which grew in the late 1990s through the use of Active World software combined with some rudimentary 'building codes', Beamish reflects on the use of explicitly urban simulations and metaphors as frameworks to shape mass participation online fantasies and games. Whist the city metaphor is used to draw participants in from the shapeless infinities of the Internet, Beamish notes that such metaphors rarely develop with any depth. Instead of deeper reflections about the idea of the city in the twenty-first century they offer, rather, a form of online decoration, a familiar 'urban' language that participants can use and relate to which provides frameworks for their participation. The creators of these virtual worlds appear to take the image of the city literally. But they also do so superficially. Their creators and participants generally do not seem to have given much thought to what it is about a 'city' that their visitors would find appealing.

Beamish finishes by analysing virtual cities from the perspective of urbanism. She questions how successful they really are as 'cities', 'public places' or 'communities'. She criticises their pallid, static and anodyne attempts to simulate the vitality and excitement offered by real cities. But she also recognises that sometimes this is a result of the limits imposed by technology, economics or regulation. Beamish acknowledges, however, that many of these virtual cities can actually be classified as new types of civic and public space. Whether they can be defined as that most elusive and contested of urban phenomenon – as 'communities' – she notes, depends, not surprisingly, on what one actually means by that term.

THE APPEAL OF DIGITAL WORLDS

For the uninitiated, it is often difficult to understand the attraction of virtual worlds, because conversation can be superficial, even inane, and the environments can be primitive and crude. In spite of this, many individuals spend countless hours a week building, governing, and participating in their chosen world, and find the experience deeply satisfying and rewarding.

The three main reasons people find these worlds appealing are sociability, creativity, and identity play. Digital worlds can be very sociable places; visitors and inhabitants are more willing to strike up conversations with strangers, since everyone assumes that everyone else is there to socialize. Digital worlds also offer a creative outlet. Users can create their own avatars; many digital worlds allow the building of private rooms, spaces, and objects; others allow users who learn the world's programming language to navigate and express themselves in quite unusual ways. Giving the individual the ability to experiment with identity is an attractive feature of these worlds; participants may alter or hide their gender, personality, or physical looks, a feature that seems to horrify and delight outsiders in equal measure.

There is another important reason these virtual worlds attract people. Participants are not simply consumers of these worlds – they are also their creators, producers, and inhabitants. These digital worlds and communities offer participants the opportunity to create not only public space but *civitas*. And because they are investing their time and energy, users tend to judge their world by its potential and how engaged they are with others, not simply on superficial appearances. Because they are so visual, graphical worlds are the most rewarding ones to explore in terms of how they represent and reflect our images of community, public space, and the city.

Simulated cities such as Planet9, Deuxième Monde, Virtual Whitehall, and Virtual Los Angeles represent the city literally. Unlike other digital worlds, these are not imaginary places – their creators have tried to duplicate the physical world, re-creating some of the larger, more exciting cities such as London, Paris, New York, San Francisco, and Los Angeles. Cities such as Tulsa, Springfield, and Newcastle do not seem to spark the creators' imaginations in quite the same way. They also seem to approve of the city's physical environment, because buildings and streets are painstakingly re-created; but unlike the real thing, they are clean and free of crime. Unfortunately, in their efforts to sanitize and secure the city, the creators also succeed in removing much of its life – often there are no people, no cars, no trees, and the streets are eerily empty.

Plate 26 Active Worlds' Metatropolis. (© Activeworlds .com.)

Virtual worlds have not been successful at creating completely safe and detached realms. In many worlds, deceit and evil deeds are introduced by people and are not necessarily related to the physical environment. As a result, regular participants spend many hours developing social rules for their worlds and defining what is acceptable and unacceptable behavior. AlphaWorld has created a police force to help fight anarchy. [. . .]

Another group of worlds has been created using Active Worlds software, including AlphaWorld, Metatropolis, and Virtual City. Metatropolis is dark and beautiful. Newly arrived visitors see the glittering city in the distance, but as they travel through the world, they realize that they are unable to reach the beckoning lights, which remain just out of grasp on the horizon. In Metatropolis, visitors find themselves stranded in a lonely world of late-night road stops and neon lights (see Plate 26).

[. . .]

AlphaWorld is the largest of the Active Worlds; it grew significantly in the late 1990s (Plates 27 and 28).

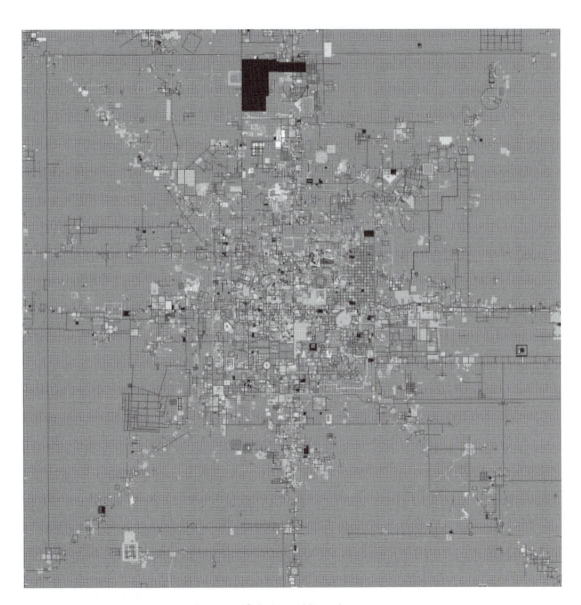

Plate 27 Map of Alphaworld, December 1996. (© Activeworlds.com)

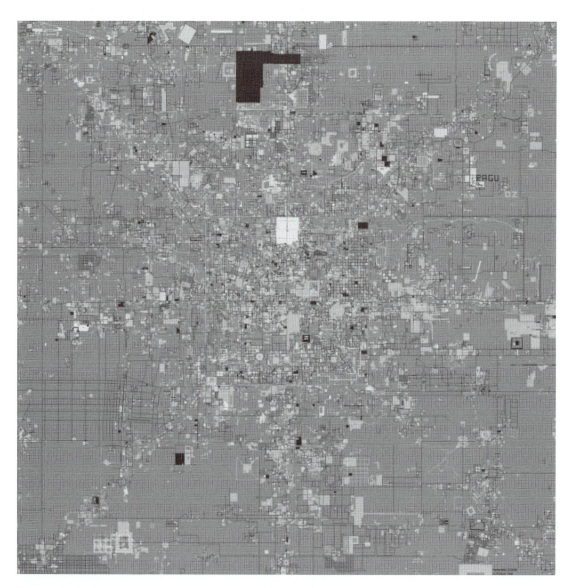

Plate 28   Map of Alphaworld, December 1998. (© Activeworlds.com)

It, like other worlds, has an odd frontier/suburban feel about it, with disconcerting characteristics such as two-lane highways in a world that has no cars and where residents get around on foot or by tele-porting themselves over longer distances. Buildings with sloped roofs and walls are built in a world with no snow, rain, wind, or cold. There are few public buildings and apparently no zoning. There is a building code, however, which can be summed up as follows: Keep it simple, keep it sparse, and keep it spread out.

[. . .]

What do these digital worlds tell us about the creators' image of the city? When digital urban envi-ronments are designed, the downtown is often seen as the Holy Grail – the vivid, exciting, teasing, tantalizing city is held up within sight, but out of reach. The image of the city is used to attract us and to draw us into the world, but it functions mainly as a decoration or marketing technique intended to get the customer in the door. The creators of these virtual worlds appear to take the image of the city literally but superficially, and they generally do not seem to have given much thought to what it is about a city that their visitors would find appealing. They use the image of the city liberally but strip it of meaning.

VIEWS OF DIGITAL CITIES FROM THE PHYSICAL WORLD

If we are to look at the images of physical cities through the lenses of their digital cousins, it is only fair to reverse the view and look at digital cities through the lens of the "real" world. To do so, one must question how successful they are as cities, as public places, and as communities.

Are they able to successfully re-create the urban experience? Too often, rather than mimicking the vitality and excitement of downtown, the digital environment is disconcertingly desolate and empty; the buildings are blandly modern; and it is common to travel around these worlds without meeting another soul.

To be fair, though, the crude and simplistic environment is not always a reflection of the creators' aesthetic taste; it is also a reflection and result of technology, economics, and regulations. The database of Virtual Los Angeles, for example, whose graphics are quite detailed and realistic, is currently hundreds of gigabytes and will soon exceed one terabyte. This much detail requires either massive storage or large amounts of bandwidth to transfer the information. When designers and builders of simulated and imagined cities are forced to choose between speed and detail, the visually interesting details usually lose. Buildings are plain because, as in the real world, complicated, intricate buildings are more expensive and difficult to build. Regulations can also affect the environment. In AlphaWorld, for example, residents are required to pave over land to stake their claim – they may never return to build on it, but it remains out of circulation, resulting in vast paved open areas.

Are these digital worlds public places? Spiro Kostof in *The City Assembled* defines public space as a place that promotes social encounters, ensconces community, and serves the conduct of public affairs. Certainly many other theorists have weighed in with their own definitions: Public space is a stage on which communal life unfolds; a place freely accessible and shaped by its inhabitants; a landscape that reflects us; a world of strangers; a place that hopefully engenders tolerance of diverse interests and behaviors; and a place of social and commercial encounter and exchange. Regardless of which definition is used, many virtual worlds arguably could qualify as public space.

Not only are they public spaces, but many are civic spaces as well, and it is in the realm of promoting *civitas* that these worlds have much to offer. Worlds such as De Digitale Stad (Digital Amsterdam) (p. 371), Municipia, and community networks emphasize civic debate and discussion rather than attempting to duplicate the physical characteristics of a city. With these sites, the boundary between the physical city and the virtual begins to blur, creating a new type of public space.

Are these virtual worlds communities? The verdict is mixed; the answer lies, of course, in how community is defined. Our long history of lamenting the loss of community is probably matched only by our equally long history of not being able to agree on exactly what community is. Definitions of community can include geographic area, common consciousness, shared interests or attitudes, collections of institutions, social interaction, or networks of social relations. We are often unable to separate what we feel it *should* be with what it *is*, we are often unsure whether we're discussing a locality, experience, or relationship; and we aren't always able to distinguish between the familiar community and the nonfamiliar public space.

Editor's suggestions for further reading

Aurigi, A. and Graham, S. (2000), 'Cyberspace and the city: the "virtual city" in Europe'. In G. Bridge and S. Watson (eds), *The Blackwell Companion to Urban Studies*, Oxford: Blackwell, 489–502.

Beamish, A. (2001), 'The city in cyberspace'. In L. Vale and S. Warner, (eds), *Imaging the City*, New Brunswick: Rutgers Center for Urban Policy Research, 283–300.

Bleecker, J. (1994), 'Urban crisis: past, present and virtual', *Socialist Review*, 24, 189–221.

Campanella, T. (2001), 'Anti-urbanist city images and new media culture'. In L. Vale and S. Warner (eds), *Imaging the City*, New Brunswick: Rutgers Center for Urban Policy Research, 237–54.

Doheny-Farina, S. (1996), *The Wired Neighborhood*, New Haven: Yale University Press.

Graham, S. and Aurigi, A. (1997), 'Urbanising cyberspace? The scope and potential of the virtual cities movement', *City*, 7(8), 18–39.

Keeble, L. and Loader, B. (eds) (2001), *Community Informatics: Shaping Computer-Mediated Social Relations*, London: Routledge.

Rheingold H. (1994), *The Virtual Community*, London: Secker & Warburg.

'Identity, Embodiment, and Place: Virtual Reality as Postmodern Technology'

from *Digital Sensations: Space, Identity and Embodiment in Virtual Reality* (1999)

Ken Hillis

Editor's Introduction

Our reflections in this part of the book have, thus far, centred on how identities, bodies and online and urban spaces become remade through the textual, aural and visual technologies of screens and keypads that surround mobile phones, Internet cafés, and PC-based virtual communities and cities. In this final reading on the social and cultural worlds of cybercities we complement these analyses by addressing the links between social constructions of the human body and the growth of completely virtual, immersive realities (known as Virtual Reality or VR) constructed entirely through computer software. In VRs, participants are totally immersed in three-dimensional computerised spaces that link to their senses. VR spaces can thus give the impression of residualising the subject's physical body.

This reading, which draws from the work of the US cultural geographer Ken Hillis, first summarises the utopian dreams of many of the proponents of truly virtual and immersive realities. As cities spread out and become more and more fragmented, Hillis notes that virtual reality technologies have been widely celebrated as a means of achieving a kind of 'virtual reunification': a (largely private and commodified) way of bringing back 'community' in an age of atomisation and fragmentation. But how, Hillis asks, might such a 'digital public sphere' actually work given that the physical separation, and indeed, invisibility, of people's bodies – encased in the capsules of car, home, mall and gated community – continues to prevail, particularly in many US cities?

Hillis is deeply sceptical of such a vision. To him, such a dream, first, ignores the extreme material inequalities of contemporary urban societies. These mean that any truly immersive virtual realities are likely to remain the privilege of those who are already wealthy and mobile. Second, Hillis is troubled by any system in which the agency of human beings is entirely translated into digital symbols that float through ICT systems and become 'reality' in the complete absence of people's real bodies. For him, even when cyberpunk theorists and hackers celebrate the potential of VR technologies for sustaining resistance against deepening corporate control, virtual reality will always be about the denial, and removal, of the embodied physical worlds of people's bodies based in and around places, cities and urban regions.

BODY LANGUAGE

> Travelers on . . . virtual highways . . . have . . . at least one
> body too many – the one now largely sedentary carbon-
> based body at the control console that suffers hunger,
> corpulency, illness, old age, and ultimately death. The
> other body, a silicon-based surrogate jacked into imma-
> terial realms of data, has superpowers, albeit virtually, and
> is immortal – or, rather, the chosen body, an electronic
> avatar "decoupled" from the physical body, is a program
> capable of enduring endless deaths (Morse 1994, 157).

Human bodies form a basis for social relationships.
The poststructural "social body" is determined by
linguistic categories, yet while this social body "may
be named as a theoretical space, it is frequently left
uninvestigated. It is as if the body itself . . . does not
exist" (Shilling 1993, 72). In Virtual Environments (VEs),
however, are users' bodies really dispensed with,
"parked" somehow, or "collapsed"? Or do such metaphors mask a retheorization of the social body by
academics and others more than they uncover any
threat to natural bodies, or how always shifting forms of
power are reconfigured by and in optical technologies?
Although a VE, for example, minimizes ambulatory
experience, users interacting with virtual technology
nonetheless constitute material phenomena engaged
in practices. Users wearing Head Mounted Displays
(HMDs) confirm a sense that technologies such as VR
are able to obtain a grip on human bodies.

We experience place as embodied human beings.
This embodiment is situated somewhere along a continuum alternatively and confusingly conceived to exist
between what are loosely identified as "nature and
culture," or "culture and civilization," or even "nature
and civilization." Embodiment can be considered either
to help link a now-individuated sense of self to a wider
community or, conversely, to contain this self "inside,"
apart from the broader sphere of social relations, which
then comes to be conceived as an organism or entity
somehow apart from individuals. Embodiment is a
leaky concept; it suggests nature, culture, even civilization. "The body and its actions . . . have a richly
ambiguous social meaning. They can be made to
emphasize perceived distinctions between nature
or culture as the need arises, or to reconcile them"
(Marvin 1988, 110).

David Levin asserts that a bodily nature never
encountered except in a historical situation is one
that denies our abilities to resist oppressive uses
of history or even history as epistemology. Bodies resist
history and do so in an inventive manner. For Levin,
save for its extermination, culture and history can do
nothing to the human body. Its physical reality and
form resist the text of history, except for that part
of history that would inscribe itself "biologically" over
nature. This resistance is plausible, in part, because
unless we are to believe that humanly produced
representations somehow might have preceded the
existence of the first human beings, human bodies were
present on the earth before the first story being told
and before its recording in any representational format.
[. . .]

Human bodies, therefore, are an intriguing pivot
for theory, and it is difficult to imagine any geography
that would matter without them. They straddle the
dichotomy erected between nature and culture, their
space both influenced by social relations and influencing what forms these social relations may take. The
degree to which theorists have remained unwilling to
look at our bodies as powerful means of countering the
hegemonizing power released by the nature-culture
dichotomy is perplexing. Incorporating human bodies
into understandings of social relations allows a broader,
more defensible, if continually shifting, material base
from which theory might develop. The concept of *territory*, for example, works well to describe the external
physical reality that results from our crafting of a place.
The agency of both individuals and collectivities is easy
to acknowledge in the act of making a territory, but
the roles of the bodies of a territory's makers remain
implicit and vague. [. . .]

The body I am interested in, however, is not the
obverse of the Cartesian mind or some prepackaged
concept. Rather, bodies are particular *and* plural, have
minds, spirits, and take place in an evolving fashion;
this is a universal that has always been the case and
will continue to remain so. Our bodies are where we
locate individual difference; my body is here, and yours
is there. Our bodies, however, also share operational
similarities, which is why I sympathize with you when
you are ill, as I recognize that my body can operate in
a similar fashion. Nevertheless, it is the establishment
and performance of spatial differences among bodies
that works in unpredictable and relational ways with
other bodies and with places. In material reality,
the subject's embodied agency always locates a place
from which resistance may proceed, from which ideas
presented to her or him may first be tested and evaluated. In a VE, however, this place – the place of the
subject's body – seems to lie physically "behind"

the cyborg agent, who (or which) to date is made to take the ephemeral form of a subjectivity manifested as a cartographized point of view or avatar within invented computer language.

[. . .]

The current popularity of virtual technologies partly reflects the technical elite's hope that these machines might represent, in commodity form, an acceptable commons in which fragmented, "cocooning," but highly individuated modernist subjectivities might achieve virtual reunification with other such selves without having to venture from behind real-life "spatial walls" such as gated communities. In this digital commons, plural identities in flux, feeling overburdened by the weight of their own subjective freedom, might be *pictured* more as one in their search for a transcendent continuity.

[. . .]

A virtual on-line commons, like absolute space, would offer an infinitely extensible grid for potential reunification of fragmented selves, with plenty of room for commerce, too. Such a digital "public sphere" might permit imaginative vaulting of Elias's modern wall. However, the critical spatial separation of users' bodies in "absolute space" would remain unaltered. Each spatially isolated individual would become, as it were, a discrete modern category unto himself or herself. VEs – a vision disguised as space – are the Ideal public sphere for imaginative subjectivities believing "themselves" virtually freed of bodily constraints. These environments are a privileged psychic variation on contemporary physical homelessness. Divorced from the body's constraining intelligence, "the fully extensible self" – assuming sufficient personal wealth – busies itself with the shallow fantasy of building a virtual house in a postsymbolic "environment." At the

same time, this self ecstatically embraces the digital means that serve to extend its own psychic disarticulation, hoping to sharpen its ability to control personal meaning by consuming virtual experience as both leisure and culture. Along the way, it turns its "back" on an intransigent material homelessness embedded in grossly unequal social relations.

[. . .]

When people become symbols, any ethical urgency that they be treated as human beings is reduced. From the fiber-optic privilege of cyberspace, the human body becomes one more thing within a nihilistic perspective of data overload. The temptation to engage with the new virtuality – even, following cyberpunk and hacking, as forms of resistance – must append a caveat that also calls the imagination back to the embodied physical world. An appropriate phrasing for any message recalling the human body in a virtual age might read, "DON'T LEAVE HOME WITHOUT IT."

REFERENCES FROM THE READING

Idhe, Don (1991), *Instrumental Realism: The Interface between Philosophy of Science and Philosophy of Technology*, Bloomington: Indiana University Press.

Marvin, Carolyn (1988), *When Old Technologies Were New*, New York: Oxford University Press.

Morse, M. (1994), 'What do cyborgs eat? Oral logic in an information society'. In Gretchen Bender and Timothy Druckrey (eds), *Culture on the Brink: Ideologies of Technology*, Dia Center for the Arts, no. 9, Seattle: Bay Press.

Shilling, Chris (1993), *The Body and Social Theory*, London: Sage.

Editor's suggestions for further reading

Hillis, K. (1999), *Digital Sensations: Space, Identity and Embodiment in Virtual Reality*, Minneapolis: University of Minnesota Press.

Holmes, D. (2001), *Virtual Globalisation: Virtual Spaces/Tourist Spaces*, London: Routledge.

Robins, K. (1999), 'Foreclosing on the city? The bad idea of virtual urbanism'. In J. Downey and J. McGuigan (eds), *Technocities*, London: Sage, 34–59.

Schroeder, R. (1994), 'Cyberculture, cyborg postmodernism and the sociology of virtual reality technologies', *Futures*, 26(5), 519–28.

VII CYBERCITY PUBLIC DOMAINS AND DIGITAL DIVIDES

Plate 29 Web graffiti on a street in Vienna, Austria, 2002. (Photograph by Stephen Graham.)

INTRODUCTION

The globe shrinks for those who own it; for the displaced or the dispossessed, the migrant or refugee, no distance is more awesome than the few feet across borders or frontiers

(Homi Bhabha, 1992, 88)

In the seventh part of this book we move on from the discussions in the last section to consider some of the ways in which the public cultures and social divisions of cities relate to the 'digital divides' in the shaping, use and experience of ICTs and new media. Understanding the mutual constitution of the public realms and social inequalities of cities and ICTs is the critical, multi-scaled, challenge here. For digital technologies increasingly mediate all social divisions in cities as the ability to access services, exercise rights to political expression, and gain employment and income within cities operate through, and within, ICT and software systems.

Cities, themselves, are now being structured in highly unequal ways through uneven investment patterns by states and firms that are facilitated by intensive ICT-based geographical surveillance and targeting. Infrastructures and public spaces are saturated by electronic means of surveillance which are used by regulators and power holders to try to monitor, and shape, who has access and who does not. And the extremely uneven social architectures of cities are now manifest in, and reinforced by, stark 'digital divides' in the abilities to use, and access, telephones, computers, mobiles, and the Internet.

As we saw in the last section, it is increasingly impossible to separate the public realms of a 'city' from the webs and connections that tie it intimately to the rest of the world. This means that we must consider urban public realms to be both multiple and geographically stretched between many geographic sites (through TV, web, telephone, financial and transport connections). Thus, whilst cities are still 'localities', they are also sources of connection within a myriad of stretched public realms, cultures and social connections – mediated by ICTs and media technologies. These may or may not intersect within the city. Cities are thus not contiguous zones on two-dimensional maps with opaque boundaries. They are not, suggests Doreen Massey, 'areas with boundaries around' (1993, 66). Rather, they must now be viewed as 'articulated moments in networks of social relations and understandings' (*ibid.*). It is how these 'articulated moments' in the diverse circuits and space–times of urban life (some of which are mediated by ICTs) come (and do not come) together within a place that shapes the dynamic nature of that city (Amin and Graham, 1997, 1999).

The imperative now is thus to understand the ways in which this dynamic and multi-scaled 'transnational urbanism' intersects dynamically, through ICTs and other forms of flow, with the sites, places and practices of urban life in particular cities (Smith, 2001). The challenge for urban policy and planning is not to shape some overpowering single public media or urban domain (which will always be biased in practice). Rather, it is to try to use ICTs to help support the emergence of urban places which support a creative, positive and democratic intermingling of public cultures and domains. These must bring together both public spaces and electronically mediated exchanges and articulate with various scales, from the body to the globe. Such a challenge faces extreme difficulties however. This is for five reasons.

First, much ICT use in the city is both invisible and individual. Whilst the mobile phone and internet cabin or café have brought ICT use into the more public parts of cities, in many cases, ICTs have been shaped to sustain an extending urban privatism (with gated communities, ICT-linked automobiles, CCTV-controlled malls, and individualised, firewall-protected, entry points into globally stretched 'cyberspace' domains). In response, many artists and urban activists have toyed with installations which force the private and invisible cultures of 'cyberspace' to be (at least temporarily) public and visible. Rude architecture, for example, placed a screen at a subway station in Berlin to publicly display text messaging from mobile phones (see http://www.rude-architecture.de and Andreas Broeckmann, p. 378). Some agencies and urban social movements have also developed web sites that attempt to mimic in some way the 'public' spaces of the physical city (see Geert Lovink, p. 371).

Second, ICT systems have tended to be appropriated by the most powerful interests in cities to perpetuate their own narrow interests. At least currently, the explosion in the use of ICTs still overwhelmingly represents an extraordinary extension in the social, economic, cultural and geographical powers of those groups and organisations who are best connected, most highly skilled, and most able to organise and configure the shift online to their own advantage. It is those particular groups, organisations and places that are orchestrating the instantaneous and often international mediation of money, work, service distribution, transportation, leisure and media access – as these sectors are reconfigured as flows of electronic signs. So far, they are thus the key beneficiaries of the so-called 'information revolution'. As the geographer Erik Swyngedouw puts it, far from being a universally liberating stampede online, 'the changed mobility, and hence, power patterns' associated with new information technologies 'may negatively affect the control over place of some while extending the control and power of others' (1993).

The third key point that we need to stress here is that there is a close connection between ICTs, global urban inequalities, and the extending power of transnational corporations to shape urban development. Against the rhetoric that cyberspace is a purely virtual and disembodied world, the radical growth of ICTs is closely related to the restructuring of real geographical places at all geographical scales. This is essentially because ICTs offer unparalleled choice and flexibility to mobile firms and socio-economic groups to exploit differences between places and people.

As we saw in Section V, this is being done through the construction of highly elaborate divisions of labour which can then be intimately integrated in real time through ICTs. Currently, international telecoms tariffs are collapsing, capabilities are growing exponentially, and mergers, alliances and acquisitions in telecom and media industries are beginning to offer global one-stop shops for international corporate ICT systems. Such capabilities enhance the ability of corporations to separate out elements within their operations to geographically dispersed cities and settlements, whilst maintaining intimate control and coordination. Thus, transnationals, and their affiliates, can benefit from the seamless and instant integration of plants located in globally spread networks of specialist and very different urban places, with very different labour conditions and costs.

The fourth key point to stress is that ICTs also allow socio-economically affluent urban groups to selectively overcome the barriers of local geography by extending their access to distant places. Increasingly, such groups can use ICTs to bypass the immediate constraints and the real or perceived risks of place in the process (especially those surrounding crime and the fear of crime). Thus, ICTs like the Internet help their users to connect, without risk or fear, to services, social groups and people across local, national or international distances. At the same time, ICTs can be used to support the selective disassociation of affluent groups from exposure to the differences, contrasts, risks and social mixing traditionally associated with cohabitation in urban streets and spaces. Other technologies like the automobile, the privatised mall, the mobile phone, the CCTV system, and the gated condominium complex, can be used to further support this process.

For some residents in all cities, then, instant electronic connection to all geographical scales is thus becoming more and more important. But, from the point of view of human settlements and cities, fracture and social polarisation seems to be the pervasive logic of the intensification and liberalisation of a global capitalist system within which ICTs play such a facilitating role. Intensifying global connections between the valued and powerful parts of cities, and the groups and organisations who control and inhabit them, are tending to be combined with a growing sense of partitioning and disconnection at the local scale within cities.

The uneven growth of ICTs thus become closely embroiled in the restructuring of real urban space. Walls, ramparts, security fences, electric fences, armed guards and defensive urban design are the physical manifestations of this process in both northern and southern cities, as globally connected and ICT-saturated social and economic groups and enclaves strive to insulate themselves from the surrounding landscapes where poverty and exclusion often concentrate. Once again, ICT systems can be configured to help support the power of the well-off to extend their actions in time and space (with 'smart' home technologies, intelligent utility metering, electronic finance and consumption systems) whilst also bolstering the powers they have to maintain safe and secure local spaces (through closed circuit TV-equipped homes, malls and office spaces, electronic alarms, movement and face-recognition sensors, electronic gates and electronically tolled smart highways which filter out the vehicles of the poor).

Our final key point to stress is that dominant applications of ICTs, and the things that flow on them, are currently heavily biased in cultural terms. The consolidation of digital multimedia conglomerates, fuelled by the Internet, the shift towards digital media (CDs, DVDs, digital television, web TV, digital magazines etc.), the volatilities of stock markets, and global liberalisation, are supporting a global trend towards centralisation of electronic power on a smaller number of people, institutions and places. Such organisations increasingly dominate the global flows of technology, capital, infrastructure, and intellectual property rights, transcending the traditional powers of (some) nation states (see Schiller, 1996). Lawrence Lessig (2001) has shown clearly that even the very technical and legal architectures of the Internet – which were originally designed to support decentralised and open communications – are now being reconfigured through growing private control. This is so that the Internet can be turned into a medium which is heavily controlled by groups of large transnational corporations who exploit it for maximising profit and consumer surveillance.

The world is thus facing increasing global concentrations of what we might call electronic economic and cultural power. This process raises many questions about geopolitical relationships, accountability, democracy, global citizenship, the ownership and control of information and the means of cultural expression, and the relationship between the global and local cultures that surround global urbanisation. This is especially so given the extreme asymmetries of north–south relations that surround this process. In many developing nations and cities, TV and Internet media provide an overwhelmingly Anglo-Saxon content as US culture, in particular, is extended, and exported, through the growth of electronic connections. For example, it is far easier for a Russian language speaker with a computer to download the works of Dostoevsky or Chekhov translated into English than it is for them to get the original in their own language.

The capital, technology, skills and finance of the north are taking a dominant role, encouraged by the World Trade Organisation (WTO), in supporting the international liberalisation and consolidation of telecommunications, media and technology. These processes are encouraging the 'opening up' and integration of trade and finance systems. They are leading to the privatisation and internationalisation of scientific research. And they are supporting the imposition of tight systems for allowing corporations to extend their control over intellectual property rights (everything from media content through software, information, knowledge, to human genetic sequences and bioengineered life forms). The industrial strategy of the United States, in particular, as the world's only information superpower, is essentially to extend and intensify its dominance in telecommunications, Internet backbone infrastructure, e-commerce, multimedia, biotechnology, education, research and development and digital content, within widening, liberalising, global markets. With many nations having better and lower cost Internet connections to the US than to adjacent nations and regions, Kenneth Neil (1999, 5) calls the resulting situation 'bandwidth colonialism.'

References and suggestions for further reading

Amin, A. and Graham. S. (1997), 'The ordinary city', *Transactions of the Institute of British Geographers*, 22, 411–29.

Amin, A. and Graham, S. (1999), 'Cities of connection and disconnection'. In J. Allen, D. Massey and M. Pryke (eds), *Unsettling Cities*, London: Routledge, 7–48.

Amin, A. and Thrift, N. (2002), *Cities: Reimagining the Urban*, Cambridge: Polity.

Bhabha, H. (1992), 'Double visions', *Artforum*, January, 82–90.

Calabrese, A. and Burgelman, J.-C. (eds) (1996), *Communication, Citizenship and Social Policy*, New York: Rowman and Littlefield.

Celik, Z. and Ingersoll, R. (eds) (1994), *Streets: Critical Perspectives on Public Space*, Berkeley: University of California Press, 9–22.

Curry, M. (1998), *Digital Places: Living with Geographic Information Technologies*, London: Routledge.

Davis, M. (1990), *City of Quartz*, London: Verso.

Drucker, S. and Guthrie, G. (eds) (1997), *Voices in the Street: Explorations of Gender, Media and Public Space*, Cresskill, NJ: Hampton.

Ellin, N. (ed.) (1997), *Architecture of Fear*, New York: Princeton Architectural Press.

Freeman, C. (2000), *High Tech and High Heels in the Global Economy*, Durham and London: Duke University Press.

Fyfe, N. (ed.) (1998), *Images of the Street: Planning, Identity and Control in Public Space*, London: Routledge.

Goss, J. (1995), '"We know who you are and we know where you live": The instrumental rationality of geodemographic systems', *Economic Geography*, 7(192), 171–98.

Graham, S. (1998), 'Spaces of surveillant-simulation: new technologies, digital representations, and material geographies', *Environment and Planning D: Society and Space*, 16, 483–504.

Graham, S. and Marvin, S. (1996), *Telecommunications and the City: Electronic Spaces, Urban Places*, London: Routledge, Chapter 5.

Graham, S. and Marvin, S. (2001), *Splintering Urbanism: Networked Infrastructure, Technological Mobilities and the Urban Condition*, London: Routledge, Chapter 6.

Hacker, K. and van Dijk, H. (2000), *Digital Democracy*, London: Sage.

Liff, S. and Steward, F. (2001), 'Community e-gateways: locating networks and learning for social inclusion', *Information, Communications and Society*, 4(3), 317–40.

Lessig, L. (1999) *Code – And Other Laws of Cyberspace*, New York: Basic Books.

Lessig, L. (2001), *The Future of Ideas: The Fate of the Commons in a Connected World*, New York: Random House.

Loader, B. (ed.) (1997), *Cyberspace Divides*, London: Routledge.

Lyon, D. (1994), *The Electronic Eye: The Rise Of Surveillance Society*, London: Polity.

Lyon, D. (2001), *Surveillance Society*, Buckingham: Open University Press.

McCahill, M. (2002), *The Surveillance Web: The Rise of Visual Surveillance in an English City*, London: Willan.

Mansell, R. and Silverstone, R. (eds) (1996), *Communication by Design: The Politics of Information and Communications Technologies*, Oxford: Oxford University Press.

Massey, D. (1993), 'Power-geometry and a progressive sense of place'. In J. Bird, B. Curtis, T. Putnam, G. Robertson and L. Tickner (eds), *Mapping The Futures: Local Cultures, Global Change*, London: Routledge, 59–69.

Neil, K. (1999), 'Bandwidth colonialism? The implications of Internet infrastructure on international e-commerce'. Mimeo.

Norris, C. and Armstrong, G. (1999), *The Maximum Surveillance Society: The Rise of CCTV*, Oxford: Berg.

Olalquiaga, C. (1992), *Megalopolis: Contemporary Cultural Sensibilities*, Minneapolis: University of Minnesota Press.

Perelman, M. (1998), *Class Warfare in the Information Age*, New York: St Martin.

Pickles, J. (ed.) (1995), *Ground Truth: The Social Implication of Geographic Information Systems*, New York: Guilford.

Schiller, H. (1996), *Information Inequality: The Deepening Social Crisis in America*, New York: Routledge.

Schön, D., Sanyal, B. and Mitchell, W. (eds) (1999), *High Technology and Low Income Communities*, Cambridge, MA: MIT Press.

Servon, L. (2002), *Bridging the Digital Divide: Technology, Community and Public Policy*, Oxford: Blackwell.

Smith, M.-P. (2001), *Transnational Urbanism: Locating Globalization*, New York: Sage.

Solnit, R. and Schwartenberg, S. (2000), *Hollow City: Gentrification and the Eviction of Urban Culture*, London: Verso.

Sorkin, M. (ed.) (1992), *Variations on a Theme Park: The New American City and the End of Public Space*, New York: Hill and Wang.

Speak, S. and Graham, S. (1999), 'Service not included: marginalised neighbourhoods, private service disinvestment, and compound social exclusion', *Environment and Planning A*, 31, 1985–2001.

Swyngedouw, E. (1993), 'Communication, mobility and the struggle for power over space'. In G. Giannopoulos and A. Gillespie (eds), *Transport and Communications in the New Europe*, London: Belhaven, 305–25.

Winner, L. (ed.) (1992), *Democracy in a Technological Society*, Boston: Kluwer.

Wyatt, S., Henwood, F., Miller, N. and Senker, P. (eds) (2000), *Technology and In/Equality*, London: Routledge.

Zukin, S. (1995), *The Culture of Cities*, Oxford: Blackwell.

'Cyburbanism as a Way of Life'

from *Architecture of Fear* (1997)

Fred Dewey

Editor's Introduction

In our first reading on cybercity digital divides the Los Angeles-based urban critic Fred Dewey analyses what he calls the 'cyburban' condition in his home city. Dewey's critical perspective is based on a critical theoretical framework which suggests a substitutionist relationship between ICTs and embodied, co-present relations in the city (and elsewhere). In many ways his position is similar to that of Paul Virilio (see Paul Virilio, p. 78).

Dewey argues that ICTs are facilitating the restructuring of LA in depressing directions, within the context of deepening corporate control and growing social inequality. Building on the effects of highways and telephones, Dewey suggests that ICTs, and cybernetic techniques, are convincing urbanites to turn inward and away from the exterior urban world. In this infinite regress the physical world becomes ever more problematic and residualised. The control and surveillance capabilities of elites, corporations, and military-commercial complexes are deepened and extended. And the widespread metaphorical representations equating ICT-based spaces with 'highways', 'cities', 'communities' and 'salons' is little less than a giant political erasure which further perpetuates the mass migration inward and away from socially positive public engagements of people and their bodies in cities.

The discourse of the 'information age', to Dewey, thus perpetuates an anti-worldly and anti-democratic agenda which needs to be resisted at all levels. If this near conspiratorial alliance of ICT-celebrating rhetoric, and intensifying corporate and military power is allowed to achieve its ends, argues Dewey, local and personal autonomy will be destroyed in a process of intensifying globalisation.

To Dewey, the prospects of such resistance are not good, however. Already, he notes that the saturation of city life with vast arrays of ICT-mediated exchange threatens to mean that physical gatherings become an anachronism. In LA Dewey believes that any truly public urban realm is already pretty much extinct. He argues that face-to-face and embodied human contact of any description is also endangered. Instead, Dewey suggests, urbanites seek refuge and escape from human co-presence, and all the perceived risk that is now associated with that term. To him, cars, ICT terminals, gated communities, corporately controlled malls, airlines, and so on – the capsularised network-spaces analysed by Lieven de Cauter (p. 94) – are *already* the city – or at least already the *LA version* of the city.

There is certainly much evidence to support claims by Dewey, and others (see Davis, 1990; Flusty, 1997), that transformations to cybercities often involve growing corporate influence, deepening social division, the increasing social control of gatherings in the physical spaces of cities, and the growth of private or semi-private capsularised network spaces (see Garland, 2001). But one problem with Dewey's analysis is its absolute language of substitution and its hints of technological determinism. His analysis fails to address the subtle combinations of identity, place, mobility and power which are always at play in cybercities and which tend to be much more complex than the absolutist, and dystopian, portrayals in his writing. Dewey ignores the ways in which ICTs can support new patterns of embodied sociality, as analysed, for example, by Timo Kopomaa (p. 267) in Finland. He fails to address the continued physical mixing of wide ranges of people which continue in all cities – even in LA. He misses the widening range of hybrid and recombinant spaces, such as those analysed by Nina Wakeford

(p. 263), which transcend the simple duality of public and private, and which deny some absolute logic of privatisation and the inexorable disembodiment of urban life. Finally, Dewey, like so many analyses of Los Angeles urbanists, seems to imply in his work that, to quote Ed Soja (1989), 'It all comes together in Los Angeles.'

In taking the *noir* portrayals of contemporary LA as his sole example Dewey is guilty of what is called synecdoche – that is, taking one small example of the infinite times and spaces of city life and generalising from it (see Amin and Graham, 1997). The reality, however, is that different cities demonstrate a *variety* of relations with ICTs. Clearly – as we shall see in the other readings in this part of the book – there are general structural trends towards privatisation, fragmentation and increasing corporate power. But there are also many examples where ICT use has been shaped to intensify, rather than replace, the interactions of embodied people in real urban spaces.

THE CYBURBAN CONDITION

The new dilemma of our time is hardly merely that of image and information. It is the drama of how these facilitate the metamorphosis of a tapeworm inside the world itself. In Los Angeles, heavily secured and encased environments can be found a block away from intensely vulnerable areas; sophisticated communications technologies rise amid communities without telephones; trains for edge-city "information-age" riders zip by overcrowded, filthy, and virtually unsubsidized buses for the laboring poor. Rising from the ashes of its own mistakes, and most of all propelled by them, Los Angeles has become a laboratory for the new cyburban order.

While the freeway system was kept clear, by Eisenhower, of private co-optation, the massive force it exacted by means of the automobile is still undermining localism and the stability of place. The profound damage of television as a nonpublic mechanism for the creation, transportation, and control of public meaning is lamented, but nothing can be solved by referring to such developments as technological. Indeed, we seem congenitally unable to confront their profoundly political nature. Cybernetics makes this particularly clear. Conceived in government labs and research centers during World War II, the cybernetic signal's central trait, its very core, was exclusion of the world. Things had to be shut out in order to achieve a "steering" of matter and effect. From observing the telegraph, telephone, and radio, early cyberneticists deduced the unsurprising concept that "signals" could regulate and control behavior of both things and people. The scientists, rather than seeing in these prior forms sobering lessons for human decision, action, and assembly, saw that the signal "achieved" aims and goals. Anything that interfered with the signal, that

might block its command, was seen as noise, resistance, and interference. Inside the signal, you enter the realm not only of total fascination but of mobilized immobilization, mind and body separated and reconnected by the coordinating, electrical tug of a mindless, chaotic motion that has no place. With cybernetics, as the physical world is made to disappear, command becomes clearer, more absolute, and less visible. In this way, a cybernetic message becomes the only meaning, disappearing every previous, subsequent, and surrounding possibility, which is to say, context. The result would be a total, unambiguous, immediate "now" of control, corroding and neutralizing the world in perfect, masterly sequence, striving against memory, consciousness, and the meaning the world alone can provide.

The cybernetic signal, like the enclave, theme park, and mall, is a materialization of power pursuing a division between inside and outside, enclosing and removing the world from within. Embodying a politics of invisible buffers and exclusions, this artifice becomes only more insidious with the language used to describe it, as people speak casually of the "space" in cyberspace, the "highway" in "information highway," "on ramps" "salons," "virtual communities," and, most chilling of all, from the founder of Microsoft and software billionaire Bill Gates, "the road ahead." Such metaphors do little more than embed deep-seated confusion, sanctifying a broad illusion of space and possibility.

The notion that electronic signals could constitute space, no less an "urban" or "community" space, consolidates a much larger political erasure. Wonderful interactions may indeed occur on the Net, but that is not the point. The issue is what happens in the world in which we still live. Cyberspace, for all its novelty, represents the suction of the impeding world away into nothingness. By convincing us to turn away from our own world, cyburbia throws us into an infinite regress capable of obliterating not only people but professions, architecture as much as urbanism. Trying to build common space, as well as private space, becomes pointless when people are pulled apart and cannot make it there. Unlike the "steering" we achieve in the material envelope of the car, in the new organizing movement, the world disappears behind us as we move through it. We cannot keep track of the world from whence we came, or indeed, the world to which we are going, which is to say, space and time vanish out from underneath us.

This destruction is far more effective than, but directly linked to, that achieved during the era of nuclear deterrence. One of the first uses of computers was to calculate the complex relationships of explosive charges necessary to create a nuclear chain reaction. Out of this, an extraordinary new power was built through calculation and its electronic extensions, allowing telecommunications to assert regulations and mobilize the ground for it, from multiple sites and variable distances, turning millions of messages into a network of microabsolutisms. We become, in this new world, more and more helpless against powers that operate from further and further away. In a vast system of billions of switches, a single phone can be turned on or off or monitored undetected, while a person's entire history can be tracked, down to even books purchased, through credit hits. Citizen or neighborhood control over such a system becomes unfathomable, and, more importantly, exhausting.

The information age's antiworldly and antidemocratic agenda is reflected in the re-engineering and downsizing of corporations, a savaging Los Angeles is hardly unique in experiencing. As companies turn internally to electronic communication and control over vast distances, the political power of the executive and senior manager grows exponentially while the power of the worker shrinks to zero. "Liberation" through such a space is often celebrated, yet many of the terms used to describe this liberation recall those of the "liberation" charted by Sigmund Freud, Edward Bernays (Freud's son-in-law and a public relations genius), and others via the unconscious. As with the unconscious, the issues of *what* and *whom* are liberated by cyberspace and its zone of free movement remains the central issue.

The consequence of cyberspace may be unfettered movement for some, but for everyone else the outcome is far less certain. The independence of communities and individuals, savaged with the rise of monopolies in the railroad, telegraph, and telephone eras, may slip completely away in the era of computers and telecommunications, as power becomes increasingly the switching of a great calculus, taking from one place and sending to another with ever-greater speed and capacity. When money operates "in" cyberspace, for example, through spreadsheet calculations, exchange rates, and telecommunications management of portfolios, entire communities and whole countries have been sucked out instantly like a horror-movie corpse, then robbed of any consciousness of why or how this happened. For instance at the end of 1994 American government officials and Wall Street executives experimented in currency and securities transfers out of Mexico. They could manipulate from a great distance, unseen, the precise number of millions of people to be plunged into poverty, plague, and death. Michael Camdessus, head of the American-directed IMF, was hardly innocent when he called this "the first crisis of the twenty-first century."

When an architect/architectural critic recently tried to define public space, he noted that "the security of our freedom is dependent on the assembly of bodies." This emaciated notion of the public dovetails well with the prevalent organizing fiction that the world can and must be shut out. Angelenos, typically, can exert unmatched control over how and with whom they interact. Unplanned contact is restricted to the immaterial realm of the phone or, if you are well off, fax and e-mail. This attempt to supersede face-to-face interaction alters even the nature of human friendship. Perhaps someone is not home. You do not just stop by. Perhaps they do not return calls. You do not hear from them for months. Where did they go? No matter. You do not need to see them. They do not even exist.

In cyburbia, conditions of contact that would prevail across continents or thousands of miles become common across thirty miles, ten miles, half a mile, and between rooms in a single home. Life, even in the best of circumstances, becomes the precursor of the computer bulletin board as animated conversation occurs between answering machines and digital signals, forever on the run, from Malibu to North Hollywood, Long Beach to Mount Wilson and beyond. Gatherings become an archaism, allowing us to avert the mess of sweaty, fractious, flawed, and uncomfortably differing humans that constitute the actual common world. In cyberspace, characteristically, we are seemingly unrestrained, able to say things we would never say face to face. It does not matter if the African-American female truck driver you are talking to is neither African-American, a woman, nor a truck driver. You will never see him or her. Does it make a difference? As in totalitarian fictions, the "who" of people ceases to carry any weight.

The glimpse of the void that arrives when fictions do burst, as they must, is terrifying. To think of entering the political realm from such a nonexistent foundation merely reiterates our weaknesses, so we leave everything outside our envelope for others to manage, turning holes of oblivion into zones of helplessness. Such a matrix of envelopes cannot be a city; it ceases even to be a society. People find themselves pulled apart by unseen movers and invisible hands steering from every direction, whether commerce, career, family, self-interest, or the vaguest convenience. "Go on, follow your dream," our confused imaginations urge us. We are left with an endless egotism of relationships without links, privacy without protection, place without neighborhood, identity without otherness, world without worldliness, and homes that make us feel exposed and fundamentally groundless.

THE METAPHYSICS OF EMPIRE AND DISAPPEARANCE

Cyburbanites, even more than suburbanites, are unable to understand or combat arbitrary power; it is their fate to regard everything outside their controllable space – be it their home, office building, mall, car, or identity – through a screen. This screen is part haven and part defensive weapon, creating protection and magical sanctuary even as it gives unseen, effectively occult forces the platforms through which to propel their soft coding. The allure of this is that the few who make it inside can inhabit an environment the majority outside cannot understand, grasp, or challenge. Power becomes encrypted, action indexed by an electronic, blinking hieroglyph. You could trace the origin of this strange and fatal metaphysics to the screen – which before photography and cinema meant to hide and block things – being reconceived as a universal surface on which meaning could appear. Freud spoke early in this century of the "dream screen" and the "mystic writing pad," as if the screen were the site of imagination, apart from others, buried inside the body. While the notion of screening phone calls suggests the

older meaning has not entirely vanished, this cinematic meaning of the screen allows us to interact and appear to imagine while turning away from the world and others. Whereas before we might go into a dark theater or sit with family and watch television, now, in more perfect form, we surround ourselves with the pseudo-meaning of a ravenous cybernetic envelope. We slide eagerly into its permanent, psychedelic refuge, a voyage without end.

This new refuge, unlike the old homogenized refuge of the suburban home, does not even need to materially exist. Inhabitants increasingly traverse a world falling apart yet are just as increasingly oblivious to it. The public nature of the built world is replaced with a great Hindu wheel of suffering. Once at the center, we can flame away in a riotous, soothing Godhead of pure signs; outside, all is war. While cyburbia beckons as the placid, motionless, secure center of this wheel, the new Brahmins cannot afford to arrest its spin. Instead, they turn to the ecstasy of worldlessness, the Rapture and Nirvana of withdrawal; the actual becomes a realm fit for losers, the defective, and expendable, or, as the cybernauts put it, "meat." In this way, the cyburbanite becomes the most devout proponent of simultaneous, total fictionality and utter barbarism.

African-American writer Henry Louis Gates, Jr. described this grim confluence of destruction and ignorance in a recount of Dexter King, the son of civil rights leader Martin Luther King, Jr., speaking of a new memorial to his murdered father. King suggested that interactive, virtual displays "could transport a person, say a kid who didn't know what it was like, to be in a civil rights march." He intoned enthusiastically, "We could actually take you into that experience" (King, cited in Gates, 1995, 6). The manufacture and virtualization of experience pulls people out of the world and politics entirely. It is enough to know "what it is like" – this way, there are no people clubbing us over the head, hauling us off to jail, or murdering us in cold blood. Indentured thinking reemerges through a neo-Hindu fatalism, the new veil, technology. Gates rightly asks, "When King's heirs devote themselves to building a virtual civil rights march, how long before we learn to content ourselves with virtual rights?" (Gates, 1995, 6–7).

The answer is, we already have. Our problem is not merely "silicon snake oil," as Clifford Stoll has aptly called the hype around cyberspace, nor, as Theodore Roszak put it in 1986, the "cult of information." The problem is larger, as the political order, bureaucracies, and cartels have for some time been swarming into the realm of free elections, making the world an increasingly hazardous and despotic realm, replacing the face-to-face with the facade-to-facade. In this new politics of appearance and disappearance, the majority are manacled steerage in ship after ship headed for the new world, their misery exceeded only by that of the rebellious, sick, and silenced who must be tossed overboard to get there.

The clearest indication of the politics of this new cyburban order can be seen in the fact that, since the beginning of the 1980s, public assemblies that do not fit the corporatist vision – those of pro-choice advocates, labor groups, and Gulf War protesters – have been systematically erased. In 1994, a public assembly, quite possibly the largest in Los Angeles's history, against city hall and the anti-immigrant Proposition 187, barely registered. Concern brought people into the streets but could not surmount the webs and waves blocking formation of public space. Framed as "threat," then subject to confusion and dispersal, marchers disappeared in scapegoating, recrimination, doubt, and uncertainty, the event over almost as soon as it had begun. The few demonstrators to actually "appear," to be allowed through the veil, were framed as anti-American, and so repressive legislation was accelerated. Proposition 187 passed "overwhelmingly."

REFERENCES FROM THE READING

Gates, H. L. (1995), 'Heroes, Inc.', *New Yorker*, 16 January, 6–7.

Editor's references and suggestions for further reading

Amin, A. and Graham, S. (1997), 'The ordinary city', *Transactions of the Institute of British Geographers*, NS 22, 411–29.
Davis, M. (1990), *City of Quartz*, London: Verso.
Dewey, F. (1997), 'Cyburbanism as a way of life'. In N. Ellin (ed.), *Architecture of Fear*, New York: Princeton Architectural Press, 260–80.
Ellin, N. (ed.) (1997), *Architecture of Fear*, New York: Princeton Architectural Press.
Flusty, S. (1997), 'Building paranoia'. In N. Ellin (ed.), *Architecture of Fear*, New York: Princeton Architectural Press, 47–60.
Garland, D. (2001), *The Culture of Control*, Oxford: Oxford University Press.
Soja, E. (1989), *Postmodern Geographies*, Verso: London.

'San Francisco: Capital of the Twenty-First Century'

from *Hollow City: Gentrification and the Eviction of Urban Culture* (2000)

Rebecca Solnit and Susan Shwartenberg

Editor's Introduction

Our second reading on cybercity digital divides returns to the theme of Internet and multimedia clustering that we have already encountered in the Readings by Vincent Mosco (p. 199) and Matthew Zook (p. 205). Here, the US writer Rebecca Solnit – writing in the book *Hollow City* that also included stunning photographs by Susan Shwartenberg – discusses the social and cultural implications of the growth of the Internet cluster that mushroomed in downtown San Francisco in the late 1990s.

Solnit argues that the colonisation of certain old-industrial neighbourhoods by dot.com workers, and the upscale services that were attracted by their wealth, had a profound effect on the social ecology, and real estate ecology, of central cities during the 1990s dot.com boom. She demonstrates how the processes of gentrification, exclusion, eviction, and changing land and building use in San Francisco districts like SoMA (South of Market St Area) and the Mission, dramatically transformed these districts. In the process she analyses how the artists, activists and bohemian communities that had long made central San Francisco a mecca for all those resisting the conservatism and commercialism of many other US cities, became increasingly undermined.

The driving force of this transformation was the migration of dot.com billions, and millionaire workers, north from the Silicon Valley technopole, some thirty miles to the south. This occurred as Internet entrepreneurs and start-ups grew tired of the suburban landscape of Silicon Valley and sought to create the urban 'cool' seen to be more relevant to content, rather than hardware or software, production. This migration sparked a brutal, Darwinian, shift as real estate and lease prices rocketed, landlords deliberately evicted low rent tenants to try and cash in, and expensive restaurants, retailers and services sought, in turn, to colonise the newly 'cool' central neighbourhoods.

The result, as in many such gentrifying 'cyber districts', has been an increasingly polarised urban landscape where access to the fruits of the central city is increasingly dependent on the extreme wealth that is only accessible through highly skilled jobs in the new Internet economy. With these jobs being largely white, and the colonised and gentrified neighbourhoods often being largely black or Hispanic, the resulting exclusions and polarisations have had stark racial overtones. Not surprisingly, in a city of healthy levels of social activism, processes of Internet-based gentrification and restructuring in downtown San Francisco have generated massive political and social opposition.

Solnit's work demonstrates that the economic shifts surrounding the growth of Internet industries, far from being some 'virtual' or 'antispatial' phenomenon, involve very profound social, cultural and physical transformations to very real urban places and neighbourhoods. These shifts, although apparently new, force us to add a new high-tech twist to old urban questions. Whose city is it? Can a truly multicultural city be maintained and promoted in a world of intensifying economic neoliberalism and urban gentrification? And can the increasing gaps between rich and poor in cities, fuelled by the gentrification of city cores as some of the key strategic sites within digital capitalism, be managed or even reversed by progressive planning and social policies?

San Francisco has been for most of its 150-year existence both a refuge and an anomaly. Soon it will be neither. Gentrification is transforming the city by driving out the poor and working class, including those who have chosen to give their lives over to unlucrative pursuits such as art, activism, social experimentation, social service. But gentrification is just the fin above water. Below is the rest of the shark: a new American economy in which most of us will be poorer, a few will be far richer, and everything will be faster, more homogenous and more controlled or controllable. The technology boom and the accompanying housing crisis have fast-forwarded San Francisco into the newest version of the American future, a version that also is being realized in Boston, Seattle, and other cities from New York and Atlanta to Denver and Portland.

A decade ago Los Angeles looked like the future – urban decay, open warfare, segregation, despair, injustice and corruption – but the new future looks like San Francisco: a frenzy of financial speculation, covert coercions, overt erasures, a barrage of novelty-item restaurants, websites, technologies and trends, the despair of unemployment replaced by the numbness of incessant work hours and the anxiety of destabilized jobs, homes and neighborhoods. Thirty-five percent of the venture capital in this country is in the Bay Area, along with 30 percent of the multimedia/Internet businesses, and the boom that started in Silicon Valley has produced a ripple effect throughout the region from south of San Jose to Napa and Sonoma in the north. San Francisco has had the most expensive housing of any major American city in the nation for two decades, but in the past few years housing prices – both sales and rents – have been skyrocketing, along with commercial rents. New businesses are coming in at a hectic pace, and they in turn generate new boutiques, restaurants and bars that displace earlier businesses, particularly nonprofits, and the new industry's workers have been outbidding for rentals and buying houses out from under tenants at a breakneck pace. Regionally home sale and rental prices have gone up by 30 percent over the past three years, but the rate of increase is far more dramatic in San Francisco (where rents rose 37 percent from 1996 to 1997, before the boom really hit, and nowadays can go up 20 percent in less than six months in some neighborhoods, vacancy rates are below 1 percent, and houses routinely sell for a hundred thousand dollars over offering price).

Part of the cause is the 70,000 or so jobs created in the Bay Area annually, nearly half a million since 1995.

Evictions have skyrocketed to make way for the new workers and profiteers of the new industries; at last estimate there were seven official evictions a day in San Francisco, and 70 percent of those evicted leave the city. For decades San Francisco has been retooling itself to make tourism its primary industry, but in late 1998 a city survey found nearly as many people were employed in the brand-new Internet/ multimedia industry as in the old hotel industry, 17,600 compared to 19,200, and that doesn't count the huge number of freelancers working in multimedia who bring the numbers to more than 50,000 (in a city whose population is about 800,000). Construction and business services to accommodate this boom have also expanded rapidly, though the construction workers are not building housing they themselves are likely to be able to inhabit. All over the city, buildings are being torn down and replaced with bigger ones, long-vacant lots are being filled in, condos built and sold, old industrial buildings and former nonprofit offices turned into dot-com offices and upscale lofts. As San Francisco's Urban Habitat Program puts it, "The growing gap between low wage and high wage workers and the scarcity of housing, especially affordable housing for low income households, is resulting in the displacement of low income people by middle and high income households in historically urban communities of color." San Francisco and many Silicon Valley cities are exacerbating this housing crisis by encouraging the influx of new enterprises and new jobs without addressing the housing needs such jobs create, thereby ensuring a brutal free-market struggle for places to live and an aggravation of traffic problems that are already among the worst in the nation.

Silicon Valley was the sprawling suburban capital of the first wave of new technology – computers, electronics and software design. In recent years San Francisco has become both a bedroom community for the Valley's highly paid workers and the capital of the next technological wave – the Internet, aka multimedia, with biotechnology about to become a huge presence in Mission Bay. The newness of this new technology is celebrated everywhere, but in some ways it's just continuing by other means an old history in San Francisco: an assault on the poor that began with urban renewal programs in the 1950s and has taken many forms since. And in some ways, the new technology is returning us to an old era, perhaps to the peak years of the Industrial Revolution, with huge gaps between rich and poor, endless work hours and a spartan work ethic, a devout faith in progress and technology. The manic greed at work here also recalls the Gold Rush, another nineteenth-century phenomenon often referenced in the Bay Area; but the differences matter, too. In 1849, California was a remote outpost and prices on everything soared when the world rushed in: laundresses and farmers could charge prices in proportion to the wealth being dug out of the Motherlode and join the boom, a prospect impossible in globalized contemporary California.

The influx of high-tech money is producing a sort of "resort economy" in the Bay Area, with real estate prices so inflated that the people whose work holds the place together can't afford to live in it. In Jackson Hole, Wyoming, the latte-makers and janitors live on the other side of a mountain pass that becomes treacherous in winter; in the Bay Area, the help just faces an increasingly long and hard commute, and air pollution has increased with the sprawl accommodating those who can't live in the most expensive real estate in the country. What Jeff Goodell wrote about the economy of Silicon Valley is coming true here: "The brutality of the Silicon Valley economy is apparent not just to newcomers who arrive here to seek their fortunes

but also to anyone who is so unwise as to choose a field of work for love, not money. Schoolteachers, cops, construction workers, nurses, even doctors and lawyers – as the tide of wealth rises around them, many are finding it harder to stay afloat. Despite the utopian rhetoric of Silicon Valley boosters . . . it's clear that Silicon Valley is developing into a two-tier society: those who have caught the technological wave and those who are being left behind. This is not simply a phenomenon of class or race or age or the distribution of wealth – although those are all important factors. It's really about the Darwinian nature of unfettered capitalism when it's operating at warp speed. And while the divide between the haves and have-nots may be more extreme in Silicon Valley than in other parts of the country right now, that won't last long. 'Silicon Valley-style economies are what we can look forward to everywhere,' says Robert H. Frank, an economist at Cornell University who has long studied the increasing gap between the rich and poor. 'In this new economy, either you have a lottery ticket or you don't. And the people who don't are not happy about it.'"

[. . .]

Editor's suggestions for further reading

Dolgon, C. (1999), 'Soulless cities – Ann Arbor, the cutting edge of discipline: post Fordism, postmodernism and the new bourgeoisie', *Antipode*, 31(3), 276–92.

Graham, S. and Guy, S. (2002), 'Digital space meets urban place: Sociotechnologies of urban restructuring in downtown San Francisco', *City*, 6(3), 369–82.

Pratt, A. (2000), 'New media, the new economy and new spaces', *Geoforum*, 31, 425–36.

Solnit, R. and Shwartenberg, S. (2000), *Hollow City: Gentrification and the Eviction of Urban Culture*, London: Verso.

Zukin, S. (1995), *The Culture of Cities*, Oxford: Blackwell.

'Surveillance in the City'

David Lyon

Editor's Introduction

Social divides and public spaces in cities have always been constructed and regulated through surveillance: the observation of social behaviour at a distance. Indeed, all social relations have always involved an element of surveillance. This has been a constant from the use of the human gaze of those constructing and trying to maintain law and order in medieval cities, through the bureaucracies and welfare regimes of modern national and local states, to the design of the public spaces, and panopticon prisons, analysed by Michel Foucault (and discussed by Gilles Deleuze, p. 73). Analysing the links between surveillance and cities has long provided a fruitful perspective to unearth the linkages between space, technology, power, and vision.

The huge range of ICTs that are saturating contemporary cities are all, essentially, instruments of surveillance. All are about capturing data at a distance to be recorded, processed, and classified using computer hardware, software, and human interpretation. Cybercities are thus intricate and multidimensional sites of comprehensive, and intensifying, digital surveillance. As the digital tracks of people, cars, products, movements, interactions, habits, tastes, wants, desires, risks, and transgressions become coded and captured, it is vital to understand how surveillance shapes the digital divides and public spaces of cybercities.

This third reading on cybercity digital divides, which has been specially commissioned, presents a state-of-the-art review of the inks between surveillance and cybercities from a Canadian sociologist who is probably the world's leading surveillance expert: David Lyon. Lyon's starting point is the palpable sense of intensifying surveillance in western cities since the al-Qaida terrorist attacks in the US on September 11th, 2001. He notes how surveillance practices have been most intensified in the site where, over the past few centuries, anonymity has been most prized – in the city (which was, after all, the target of the terrorists). However, Lyon locates the roots of the intensification of electronic surveillance in the deeper constructions of modern urban life. As modern urban life has grown in scale, complexity and anonymity, so more and more techniques, categories and technologies for surveilling urban citizens, flows and spaces have been invented. These techniques have engaged in the regulation of modern city life to support intrinsically ambivalent notions of public order through social control.

The ambivalence of urban surveillance is heavily stressed in Lyon's work. Rather than simply railing against reductions in privacy or the deepening social controls that emerge through the extension of the surveillance capabilities of ICTs, Lyon urges us to see that surveillance can also have productive power. For example, the human gaze of others can help to ensure that streets feel safe rather than threatening. Remote cameras can save lives through modifying the behaviour of speeding motorists. The logistics systems which support unprecedented levels of personal service for consumers (and which are analysed by Keller Easterling: p. 179), are made possible by massive advances in surveillance (for example bar coding). And new surveillance systems might allow older people to live in their own homes longer because the technology could act as an early warning system if their behaviour or routines showed anything unusual.

Lyon also stresses that surveillance is often about persuasion and seduction as well as coercion. (Recall Robert Luke's analysis (p. 249) of Microsoft's home page software.) Much contemporary urban culture actually celebrates the invasion of previously private lives and spaces by the ever-more revealing gaze of cameras and

ICTs. Think of the Big Brother TV shows. Or the restaurants in London and New York where CCTV cameras link up every table.

Lyon argues that the productive and positive sides of surveillance, however, may arise just for the privileged. Judgements about the right to pass, the right to move, and the right to access everything from Internet sites to urban highways is now being automated by software-driven surveillance systems (see David Holmes, p. 173; Danny Kruger, p. 320; Stephen Graham, p. 324; Lianos and Douglas, 2000). Thus, it is likely that surveillance systems are being used to differentiate people more starkly in cybercities, according enhanced privileges and mobilities to some, whilst reducing the relative or absolute privileges and mobilities of others. And, as we will see (Peter Huber and Mark Mills, p. 418), the most sophisticated and software-driven civil surveillance systems in cities are now being utilised, and diffused, as part of the US-led 'war on terrorism'. This means that essentially military techniques, which use unknowable software to automatically identify and track 'target' populations, vehicles and packages, seem likely to be a feature of urban life.

With such rapid developments the key, to Lyon, is to maintain some degree of transparency in how the judgements of surveillance systems are made and the effects that they have. With surveillance systems so opaque and invisible, and with post-September 11th systems being installed so secretly and quickly, this is, indeed, a difficult political challenge. The instinct to saturate cities and systems of cities with antiterrorist surveillance systems is driving US efforts to establish truly globe-spanning systems which attempt to surveil all electronic interactions across the world. This instinct, to Lyon, results in a depressing, and ultimately futile, spiral of paranoia and combines worryingly with attempts to develop totally secure city spaces (see Peter Huber and Mark Mills, p. 418).

Urban surveillance was not invented on September 11th, 2001, the day New York City received a devastating attack on the World Trade Center twin towers. But that event served to accelerate and widen surveillance processes that were already evident, not only in New York but in major cities throughout the world (see Lyon, 2001). Surveillance is now a commonplace feature of city life, but it is worth considering what is involved in this. Paradoxically, the place often prized for its anonymity in the modern world is precisely where we may expect to be noticed, recorded, identified, checked, categorized, and sorted. Even if, perhaps especially if, we are doing nothing.

Most people today live in cities. Even for those who do not, the urban environment is never far away. Everyday life became more and more urban in modern times and this has strong implications for surveillance. It is in the city that we experience surveillance in ways that are multifaceted, multi-layered, and moment by moment. In the world's affluent societies we take for granted the ubiquitous signs in stores and on streets warning of constant video surveillance. Whether out on the freeway with its camera speed checks or indoors at the computer with its Internet privacy alerts there is no escape from one monitoring device or another.

Urban experience involves the regulation of daily life. Myriad checks are made to ensure that we are in the right place at the right time, travelling at the right speed or carrying the correct items. We are positioned, placed, directed and traced as we travel, buy, study, telephone, find entertainment and work. Sometimes as part of a social category, sometimes as individuals, surveillance sieves our activities in the city. This social sorting process, that depends on surveillance, is based increasingly on attempts to predict and to simulate behaviours.

SOCIAL CONTROL IN THE CITY

Making the city visible so that it could be a place of safety and of public order is nothing new. But in modern times spaces in cities frequently were designed to permit maximum visibility, to discourage deviance and to promote public safety. To see was to ensure social control, to plan for order. By the 1960s, various competing and sometimes complementary models of the city were available, each with its remedy for impersonality or criminality or both. Jane Jacobs (1961) argued classically for cities as places of mutual recognition where neighborliness and communal responsibility would provide whatever eyes on the street were needed. Citizens themselves could provide conditions for crime limitation.

Plate 30 Razor wire and CCTV: surveillance as boundary enforcement in the city. (Photograph by Stephen Graham.)

While this vision does assume a certain cultural homogeneity, it is founded on an important principle of embodied personhood which still is worth seeking in a more electronic era. Other contemporary models, after all, exacerbate rather than ameliorate the segmentation and isolation of some experiences of city life. One of these is the urban fortress, now seen in electronic form in dystopias such as Mike Davis's (1992) *City of Quartz*, which focuses on Los Angeles. This is characterized by physical security based on technical isolation. The other is the idea of defensible space, which sets one area off against another in a struggle for competition for resources and privileges (Plate 30).

It is important to note that surveillance may be seen here as productive power, and not merely in its more paranoid panoptic guise. Safety, security, social order are all seen by most people as positive accomplishments. Who would not wish to walk without fear on a street after dark, to know that terrorism had been kept out, to eliminate recklessly fast driving in urban areas,

or to be alerted if fire broke out or a thief broke in? Many surveillance practices and devices are intended to improve city life in significant respects and are welcomed as such. The question is: what other effects accompany the positive face of surveillance, especially as it is automated and informatized? And, are the effects of surveillance positive for all or just for some?

How is the city made visible, and order constructed, today? Contemporary urban areas are rightly thought of as informational cities (Castells, 1989). The idea of maintaining visibility is present not only in a metaphorical sense of keeping digital tabs on populations but also in a literal sense of their being viewed by video surveillance. Smile! instructs the rear side of all Tasmanian windshield licence stickers, surveillance cameras are everywhere. Construction sites and central streets, shopping malls and department stores are all under constant surveillance scrutiny. At the same time, the information infrastructure allows those visual images to be checked, stored, and compared with other kinds of personal data. Within the city, citizens may expect to be constantly illuminated – made visible – by a multitude of means, not only from dawn until dusk but from dusk until dawn as well.

On a daily basis life in the city spells surveillance in constantly increasing contexts. From the road tolling system to the mobile phone call, the camera in the subway station to the barcoded office door key, the loyalty program in the store to the Internet usage checks at work, surveillance webs are thick in the city. Yet the aim is not necessarily to catch a glimpse of every actual *event* – though that remains an important goal – so much as to anticipate actions, to plan for every *eventuality*. The bustling co-presence of the town, with its square and its marketplace, gives way to both more fleeting relationships and to the absence of persons in their particularity.

URBAN SURVEILLANCE

Why is surveillance important for understanding the city today? Surveillance has to do with focused attention to persons, and in particular the gathering of personal data for specific purposes. The kinds of agencies that collect and process personal data have been doing so for that past two hundred years, to register voters, to record births, marriages and deaths, to enrol subscribers, and to keep track of property transactions. Modernity is characterized among other things by the growth of surveillance. But the means of collecting data altered dramatically with the introduction of the computer, and allowed many others to get in on the act. Today, surveillance has been dispersed, decentred, disorganized, and is a feature of all organization in every city. Different agencies have different purposes in mind, and almost all would recoil in horror if they were associated with Big Brother. But whether their combined efforts add up to something that should still cause apprehension is a question worth exploring.

Note that surveillance is not simply coercive and controlling. It is often a matter of influence, persuasion, seduction. We are all involved in our own surveillance as we leave the tracks and traces that are sensed and surveyed by different surveillance agencies. Our activities may well be orchestrated in a sense, but at the same time our active participation is essential for the music to be produced. We are indeed the bearers of our own surveillance, as the French social theorist Michel Foucault noted, but our activities trigger surveillance mechanisms in different ways, and with varying effects. The power of surveillance is strung out along a broad spectrum from tight, coercive control, through to loose and mild seduction; from obligation to influence. So what I call the social orchestration produced by contemporary surveillance practices is, like music, both soft and hard, both gently wooing and compellingly direct.

Take for example the logic of industrial location. Production may be scattered over a range of different locations but this does not mean the process is unconnected or random. Far from it. Telecommunications links enable constant contact between different aspects of production and also between the highly skilled knowledge-workers in research and development, and the unskilled workers, elsewhere, who are still required for routine assembly and other tasks. The work process is thus tightly integrated and coordinated, using flows of information rather than geographical proximity to hold things together. Similarly, the slowly growing practices of telecommuting, tele-banking, teleshopping (and, for that matter, electronic commerce using the Internet), and distance learning also reduce dependence on old patterns of space and time. Flexibility, mobility, and speed of communication make a huge difference to the way the city is organized.

But these informational cities are also surveillance cities. Whatever the real or imagined benefits of the new flexibility and mobility in what Manuel Castells (1989) calls the spaces of flows, those flows carry data and information about persons and processes so that their activities can be monitored and tracked (Manuel Castells, p. 82). But this is not all. When speed has become so central, not only knowing what is happening in the present, but also anticipating what is about to happen, also becomes crucial. Surveillance overtakes itself, as it were, to produce data on events and processes that have yet to occur in real time. Whereas once surveillance in the city meant the use of street lights and physical architecture to keep watch and to contain deviance, it now means keeping electronic tabs – including camera images – on the population at large as well. It is not that older methods are simply superseded, but that new ones are superimposed. And the newer methods are designed to push surveillance beyond all its old limits.

There are some very mundane aspects to this, but ones that can easily go unnoticed – unlike the cameras that seem to breed on street posts. An example is public utilities, enhanced by connection with the information infrastructure. Residential power lines in some areas are checked by smart meters that permit almost continuous digital sampling. As Rick Crawford (1996) shows, this allows watchers to develop a fine-grained profile of the occupants' electrical appliance usage. While electrical power monitoring has been used in the USA by drug enforcement agencies to root out illicit marijuana gardens that use high-power lights, the more lucrative purposes for smart meters would be market data gathering. These could help share the burden of power at key times, for instance during hot, humid weather, when air-conditioners are running on high. But equally, they alert their users to malfunctioning domestic appliances, providing an opportunity for targeted advertising of new machines.

The space of flows allows the new, information rich, global elites to create their own separate world and even to impose on place their new secluded environments. They are global elites in the sense that they can also create similar symbolic areas throughout the world, in hotels, airports, and transport and communication systems. In their home bases the elites inhabit little fortress zones, the gated communities that further insulate them from other, local, people in the same city. Maximum security is the ever-elusive goal. Surveillance is its necessary partner.

Behind all these moves lies fear. Fear of attack, of intrusion, of violence, prompts efforts to ward off danger, to insulate against risk. Architectural shelters that protect their inhabitants from hazards, are nothing new. But their electronic and video enhancements are, and these are among the latest attempts to cope with intensified fear. According to urbanist Nan Ellin (1997), in the later twentieth century fear generated divisive architectural policies that turned inward and backward rather than facing the actual social challenges of urban life today. Fear may be met with turns to an idealized past, to a fantasy world – such as a themed mall – group cohesion, and privatization, seen especially in gated dwellings. Defensive spaces appear as the way forward, rather than what she seeks: living places.

Surveillance practices are among the flows that enable the increasing division between the global and local in the urban environment. But not only division. The flows simultaneously connect and divide. They hook up the dispersed and the diasporic, uniting diverse groups in terms of shared identities and common interests. They may, for example, provide protection against racist attacks for beleaguered ethnic minorities in the city as well as, on occasion, also strengthening such racism. The London Borough of Newham has installed a Mandrake facial recognition system to locate known racial offenders on the street by comparing a camera image with those on a digitized database. Surveillance systems are a means of maintaining new relations of power and of resistance in the global–local city.

Surveillance flows have a direct effect on place. The new power in the space of flows requires that the city take a certain form. This is very concrete in specific buildings so that the city appears a certain way to outsiders and creates part of what Mike Davis (1998) calls an ecology of fear. Davis sees market-driven urbanization as a transgression of environmental and social common sense. It produces avoidable tragedies from fires and floods to the video-recorded beating of Rodney King. Of course, city planning tries to exclude the very conflicts that it tends to generate. As Sharon Zukin (1991) says, 'the service sector society is a Disneyized dream where conflict [is] designed out and comfort designed in. Surveillance contributes to both features at once.'

Take, for example, Victoria, the capital of British Columbia in Pacific coast Canada. A recent Clean Air by-law banned smoking in all public buildings from bars to homes for the elderly to the provincial parliament. At the same time video cameras were installed in some nightclubs and police renewed their commitment to enforcing a street camping by-law. This means that homeless young people, attracted by its affluence to Victoria, routinely are moved off the sidewalks. The reason is that Victoria is a tourist town with Olde English charm, a Scottish castle and a replica of Anne Hathaway's cottage. The hyperreal city is remade for consumption by tourists. Its poverty can be swept off the streets, its violence contained by cameras and its impure atmosphere cleansed by no smoking signs. Certain symptomatic behavior may even be deterred successfully although its causes remain untreated, unremarked.

Is this an over-polarized view of the city? It is clear that many people in Victoria, and other large cities, support the vision of secure and sanitized public places. It is politically acceptable to do so. Global elites and desperate poverty do coexist in the city, but between them usually lies a large middle class. They may not aspire to the laptop, the satellite phone (or whatever is the current symbol of high-tech status), and VIP lounge culture of the elite, but they certainly wish to avoid the fate of non-consuming – or of choicelessness – that is the lot of the poor. Particularly persons in that fraction of the middle category which David Ley (1996) calls the cultural new class actively imagineer new inner city environments, where they are both producers and consumers.

CONCLUSION

Surveillance is the means whereby knowledge of populations and individuals is obtained in order to minimize risks of many kinds. People provide the data, sometimes knowingly, sometimes unwittingly, as they go about their daily lives in the city. This includes doing nothing: the CCTV cameras will note as a lingerer the person sitting on a sidewalk bench which means that the onus is on ordinary people to prove their innocence. Increasingly, surveillance data are used for simulating, modeling, anticipating situations that have not yet arisen, and decisions are made on the basis of such information. Much planning, policing and marketing are based on such simulations. Moreover, agencies and organizations involved in planning often try to coordinate their activities. In Britain, for instance, Town Centre Management initiatives and Safer City programs bring together those responsible for city centres and urban spaces. Risk management is fed by surveillance.

TWO

Surveillance is generated by the need for tokens of trustworthiness, such as credit cards and driver licenses, in increasingly private and privatized societies. Mobile and flexible lifestyles are individualized, and are ones in which more relationships are fleeting, transitory, inconsequential. While privacy may be seen by some as the need caused by greater surveillance, privacy is actually part of the problem. And the more people are categorized and classified by surveillance systems, the more they are sorted and split up into segments of the population, with whom they have some traits in common. Surveillance often appears to be interested only in those fragmented interests, not in the whole person, let alone the whole community or city. At the same time, the intentions of some surveillance programs and devices are positive. Safety, security, and convenience are sought through surveillance systems, which is why – when we are aware of them – we collude with them so readily.

It comes as no surprise, then, that inequality is basic to such surveillance cities. Some can afford to pay for minimized risks by insuring themselves against them. Others are more likely to be vulnerable to risk just because they do not have the means to pay. The city is divided by simulated surveillance mechanisms into zones, with electronic and armored barriers between them. The Disneyfied city centre shopping malls become no-go areas for those classified as not belonging, by virtue of their credit ratings or their ethnic background.

If surveillance does not actually create inequalities in the city, it certainly tends to reinforce or accentuate them. The city in which mobility is high, where strangers encounter each other, can be a place of stimulation, of excitement, and of the cross-fertilization of customs, styles, ideas, practices. But it is also a place of potential danger, or so it is construed. Meeting others who are not the same as you may be a great source of mutually enriching conversation, but it may be a situation that inspires fear, from the perceived threats of otherness. The capacity to cope with difference is challenged in the city, all the more so in the multicultural metropolitan areas of today's global cities than in those about which the sociologist Georg Simmel (1964) classically wrote in the early twentieth century. The society of strangers, accentuated in the city, was for Simmel a situation in which relationships could both flourish and fragment.

The worsening of social divisions, paradoxically through surveillance for risk management, is one negative outcome of contemporary urban trends. It is one that calls for a renewed sense of what social justice in the city might comprise. But today's surveillance is a mode of social orchestration that operates, not according to some shared standards of morality and of justice, but according to merely utilitarian norms, so it tends to bypass the language of justice. It contributes to the removal of life situations from moral scrutiny. It works by means of sorting mechanisms that are just that, mechanisms. Like bureaucracies, they tend to exclude personal responsibility, feeling, emotion, moral judgement. Yet they work all too effectively, despite their self-referential character. They create complementary computer selves that may not be recognized by, yet profoundly affect the life-chances of, their human namesakes. And as the sorting mechanisms of surveillance continue to be used to manage risks, those human namesakes are more and more at the mercy of a myriad of agencies that, they must fondly assume, have their interests at heart.

It is time to challenge the power of these mechanisms, in the name of a justice that is sensitive to ordinary people living their everyday lives in cities. All the more so since the intensification of urban surveillance consequent on September 11th, 2001. Such a challenge will both highlight the power of surveillance practices and processes to reinforce unequal and unfair social conditions, and at the same time demand proper accountability of those agencies and organizations whose purposes pervade the software running such systems. Unless the ambiguities and subtleties of surveillance in the city are met with analysis and policies to match, the politics of fear and suspicion will continue to outweigh those of care and trust.

REFERENCES

Castells, M. (1989), *The Informational City: Information Technology, Economic Restructuring and the Urban-Regional Process*, Oxford, UK and Cambridge, MA: Blackwell.

Crawford, R. (1996), 'Computer assisted crises'. In G. Gerbner, H. Mowlana and H. Schiller (eds), *Invisible Crises*, Boulder, CO and Oxford: Westview Press, 57.

Davis, M. (1992), *City of Quartz*, New York: Vintage Books.

Davis, M. (1998), *Ecology of Fear*, New York: Metropolitan Books.

Ellin, N. (ed.) (1997), *Architecture of Fear*, New York: Princeton Architectural Press, 44–5.

Jacobs, J. (1961), *The Death and Life of Great American Cities*, New York: Random House.

Ley, D. (1996), *The New Middle Class and the Remaking of the Central City*, Oxford and New York: Oxford University Press, 15

Lyon, L. (2001) 'Surveillance after September 11', *Sociological Research Online*, 6(3), www.socresonline. org.uk 6/3/lyon.html

Simmel, G. (1964), 'The metropolis and mental life'. In K. Wolff (ed.), *The Sociology of Georg Simmel*, New York: Free Press, 409–24.

Zukin, S. (1991), *Landscapes of Power: From Detroit to Disney World*, Berkeley: University of California Press.

Editor's suggestions for further reading

Brin, D. (1999), *The Transparent Society*, New York: Perseus.

Garland, D. (2001), *The Culture of Control*, Oxford: Oxford University Press.

Graham, S. (1998), 'Spaces of surveillant-simulation: new technologies, digital representations, and material geographies', *Environment and Planning D: Society and Space*, 16, 483–504.

Jones, R. (2001), 'Digital rule: punishment, control and technology', *Punishment and Society*, 2(1), 5–22.

Jupp, B. (2001), *Divided by Information?* London: Demos.

Lianos, M. and Douglas, M. (2000), 'Dangerization and the end of deviance: the institutional environment', *British Journal of Criminology*, 40, 264–78.

Lyon, D. (1994), *The Electronic Eye: The Rise of Surveillance Society*, London: Polity.

Lyon, D. (2001), *Surveillance Society*, Buckingham: Open University Press.

Lyon, D. (2002), *Surveillance as Social Sorting: Privacy, Risk and Automated Discrimination*, London: Routledge.

Lyon, D. (2003), *Surveillance After September 11th*, Oxford: Blackwell.

McCahill, M. (2002) *The Surveillance Web*, Cullompton: Willan.

Norris, C. and Armstrong, G. (1999) *The Maximum Surveillance Society*, Oxford: Berg.

Nunn, S. (2002), 'When Superman used X-ray vision, did he have a search warrant? Emerging law enforcement technologies and the transformation of urban space', *Journal of Urban Technology*, 9(3), 69–87.

Thrift, N. and French, S. (2002), 'The automatic production of space', *Transactions of the Institute of British Geographers*, 27(4), 309–35.

Young, J. (1999), *The Exclusive Society*, London: Sage.

TWO

'Defining the Technology Gap'

from *Losing Ground Bit by Bit: Low-Income Communities in the Information Age* (2000)

Benton Foundation

Editor's Introduction

With the city's conduits of opportunity, information and exchange increasingly computerised, exclusion from access to, or ability to use, ICTs is clearly becoming a debilitating condition which is likely to marginalise people, groups and neighbourhoods. Recognition of the growing importance of ICTs to sustain social, economic and cultural participation in cities and societies has fuelled a major wave of research into the nature of the so-called 'digital divide', much of it in the USA.

In this fourth reading on cybercity digital divides the US Benton Foundation – an urban social research think tank and pressure group – attempts to define the nature of the technology gap that structures the digital divide in US cities. Arguing that many low income communities and neighbourhoods in US cities have been largely bypassed by the wealth, opportunities and technologies of the 1990s Internet boom, the report analyses the structural dualisation that helped to produce a profound unevenness in US cities between rich and poor neighbourhoods within very short geographical distances of each other.

The report notes that, with telephone and telecommunications investment increasingly unregulated and privatised, investment is concentrating on 'cherry picking' high-market areas of cities whilst bypassing many poor and African–American neighbourhoods. Household and school access to computers and Internet connections remain low in such poorer districts whilst high income neighbourhoods are becoming saturated with such technologies. And many poorer neighbourhoods remain physically isolated in city cores whilst the main growth areas of the urban economy are in the 'edge cities' of the urban periphery. Often, these can only be accessed by car.

Thus, the urban poor face a multidimensional process of exclusion that has geographical, physical, technological, social, economic and political aspects. Simply wiring up neighbourhoods with Internet-ready PCs is unlikely to make much difference in such a context. The report argues, rather, that comprehensive and multidimensional strategies are required that can address the complex interweaving of the multiple processes that perpetuate exclusion in contemporary cities.

In an October 8, 1996, article describing one of California's technology corridors, the *Wall Street Journal* captured some of the enthusiasm many people feel for the revolution arising from the marriage of computers and communications networks. "Silicon Valley," it said, "is in the midst of an epic boom, opulent even for this glittering edge of America."

But such riches haven't reached many low-income communities – even ones like East Palo Alto, which is right in the middle of Silicon Valley's technological abundance. "Anywhere else in Silicon Valley, your parents use computers, there is a shop down the street to sell you a computer, another to fix your computer another to give you computer classes, (and) there are Kinko's [computer service shops] everywhere," notes Bart Decrem, director of a California youth technology initiative called Plugged In. "In East Palo Alto, there's none of that."

The contrast between affluent and low-income communities may be particularly sharp in places like Silicon Valley, but it exists almost everywhere. The simple fact is that poor communities are entering the Information Age far behind their wealthier neighbors.

"While [middle-class communities] are rapidly approaching the 'next cycle,' the technology of the previous cycle has already bypassed the inner city," says Richard Krieg, executive director of the Institute for Metropolitan Affairs, a public interest organization in Chicago committed to seeking practical answers to problems involving education, health care, and crime. Krieg notes that while families in affluent areas are rapidly acquiring home computers, people in many low-income neighborhoods have little exposure even to earlier generation tools such as laser scanners at supermarkets and bank automatic tellers. "Despite limited empirical study of technology diffusion . . . it is clear that computerization, telecommunications, and mass media applications are dramatically underrepresented in distressed urban areas."

As Krieg suggests, the technology gap is not simply a reflection of the choices made by individual households. The deeper problem is that many poor neighborhoods lack the infrastructure available in affluent areas. Groups such as the United Church of Christ that have studied patterns of telecommunications investment have found that, all too often, telephone and cable companies have moved quickly to wire wealthier suburbs with advanced systems, while poor inner-city neighborhoods aren't upgraded. While public attention is often focused on whether individuals can get a service, the equally important problem is that lack of adequate telecommunications facilities makes an area less attractive for businesses. This can feed a spiral where the lack of investment at the community level leads to fewer economic opportunities for people who live there. As a result, the poverty in the neighborhood makes it a less inviting target for investment, further aggravating the problem.

The same neighborhoods that lack infrastructure are comprised of households that are far less likely to have the tools of the Information Age. In an August 1996 survey of southern Californians, the *Los Angeles Times* found that just 22 percent of households earning less than $25,000 had access to computers, compared to 69 percent of those with incomes over $50,000. "Poor neighborhoods of the region are just totally cut off from the potential benefits of an economy that integrates such vast scientific skill," says Mike Davis, a Los Angeles historian and teacher of urban studies at the Southern California Institute of Architecture.

More recently, according to a Computer Intelligence 1998 Consumer Technology Survey, 80 percent of families making more than $100,000 have computers. By contrast, of those families making less than $30,000 a year, only 25 percent have computers. A 1998 study led by David Birdsell of Baruch College found significant disparities in the area of education: of people with an undergraduate degree or higher, 53 percent use the Web while only 19 percent of people with a high school education or less are Web users.

While demographic trends are changing quickly, there is some evidence that race and income may interact in troubling ways. A 1998 Vanderbilt University study based on Nielsen data from late 1996 and early 1997 indicates that racial inequities in computer ownership and Internet access jump significantly when household incomes drop below $40,000. In such cases, African Americans were less than half as likely as whites to own a home computer and about 60 percent as likely to have Internet access.

Similar trends appear in telephone service, a much older technology that many poor Americans still don't have. While all but 6 percent of U.S. households have telephones, 43.5 percent of families who depend entirely on public assistance and 50 percent of female-headed households living at or below the poverty line lack even this basic technology. And African Americans and Latinos lag about 10 percentage points behind their white counterparts in access to telephones even when income is held constant.

WORRISOME TRENDS

There is no easy way to measure the impact of the current inequitable distribution of information technologies, but it clearly is becoming an increasingly important contributor to inequality in America. The Office of Technology Assessment (OTA) described the effect as "the concentration of poverty and the deconcentration of opportunity."

Email, video conferencing, fax machines, and computer networks are making it easier for jobs to migrate from city centers to suburbs and beyond, the OTA explained in a 1995 report. These technologies are enabling industries that once had to be close to customers and related businesses to operate at greater distances. Similarly, they are allowing distributors and financial institutions like banks and insurance companies to consolidate operations and locate "back room" facilities farther from their customers, eliminating many downtown jobs [see pp. 213, 218].

At the same time, new technologies have led to sweeping changes in manufacturing processes, making old factories in urban centers obsolete. The OTA estimated that the 28 largest counties in the Northeast and Midwest lost one million jobs in the 1980s. The city of Chicago alone has more than 2,000 unused manufacturing sites, according to Krieg.

As employers take advantage of technological advances to relocate to suburbs, the labor market in many cities has become fractured. Many highly skilled managerial and professional jobs remain downtown because they require a great deal of face-to-face contact and networking. But increasingly, the only work for unskilled people consists of low-paying, service sector jobs. Such jobs offer little hope of advancement, and intermediate jobs that would help less skilled workers climb career ladders are hard to find.

"We are witnessing the wholesale disappearance of work accessible to the urban poor," concludes Milton J. Little, Jr., executive vice president and chief operating officer of the National Urban League. His view was confirmed in 1996 by Harvard sociologist William Julius Wilson in *When Work Disappears: The World of the New Urban Poor.*

Editor's suggestions for further reading

Benton Foundation (1998), *Losing Ground Bit By Bit: Low Income Communities in the Information Age*, Washington, DC: Benton Foundation and National Urban League.

Graham, S. (2002), 'Bridging urban digital divides? New technologies and urban polarization', *Urban Studies*, 39(1), 33–56.

Servon, L. (2002), *Bridging the Digital Divide: Technology, Community and Public Policy*, Oxford: Blackwell.

'Bangalore: Internal Disparities of a City Caught in the Information Age'

from *The Information Society* (1998)

Shirin Madon

Editor's Introduction

In our fifth piece on cybercity digital divides the focus shifts from the USA to the high-tech boom town of Bangalore in southern India. A city that is a paradigm case study of the growth of technopoles in the global south, Bangalore is the fastest growing city in India.

In this reading the communications scholar Shirin Madon analyses the ways in which social, geographical and economic divisions in Bangalore have intensified as the city has emerged as a global focus of investment in high-tech manufacturing, research and development, software engineering, and advanced call centres (especially for software support). Madon argues that the growing inequalities that are a key feature of this stage in the city's development have intensified pre-existing tendencies – common in many cities of the developing world – towards extreme polarisation and fragmentation. In contrast to the separation of colonial elites from the colonised masses the new segmentations in Bangalore, as she puts it, are now 'based on participation in information-intensive global economy by a core elite, and non-participation by the masses'.

Madon describes how a whole galaxy of the world's major ICT companies have located major manufacturing, research and development, and software support bases in Bangalore. Massive public investment has been made into increasingly fortified technology and business campuses. These are equipped, at great public expense, with the high quality, and reliable, transport, water, sewerage and telecommunications infrastructure necessary to link them into global production and logistics systems.

At the same time, however, the poor are moving in great numbers to the city's fast expanding informal settlements, to try to benefit from the economic opportunities generated by the growth of the city. Often these people, who are the vast majority of the city's population, are facing worsening conditions. Such areas tend to be starved of public investment. They are often unable to benefit from even rudimentary water, sewerage and ICT services. And, in many cases, they are being forcibly demolished to make way for the technology parks and highways configured for by the technological elite. Thus an all too common form of technological apartheid shapes the internal planning of the city.

In a situation that is repeated across many of the high-tech growth areas of the global south, some non-governmental organisations are struggling to improve the position of the majority of the city's population who live in informal settlements. But they face the overwhelming power of a city council apparently besotted with the clean, modern spaces of global high-tech capitalism on the one hand, and driven by prejudiced notions that residents of informal settlements have no rights whatsoever to the fruits of modern metropolitan life on the other. Bangalore is thus a case study which single-handedly debunks the glib assertions from technological utopianists that ICTs somehow intrinsically make the world a more equal place.

A distinctive feature of most cities in the developing world is the fragmented character of their spatial organization. During the colonial period, this fragmentation was a direct consequence of the need to separate the European population from the indigenous. The few decades that have elapsed since the end of colonialism have not been sufficient to transform the fragmented city into an integrated one. In fact, it appears that the nature of fragmentation has shifted to one based on participation in the information-intensive global economy by a core elite, and nonparticipation by the masses.

Bangalore presents a case study of this type of fragmentation. It is the fastest growing city in India (ranked fifth by size in the country) with a population rising from 1.66 million in 1971 to 2.92 million in 1981 (a 76% growth rate), and a projected population of 3.8 million by 2000. At the turn of the century, Bangalore was a provincial town in Southern Karnataka that had evolved from a medieval temple/fort center and a colonial military encampment. Industrialization until the 1940s centered around textile manufacturing. The city's emergence as a center for information technology stems from decisions in New Delhi shortly after independence to locate strategically sensitive industries well away from borders and coastlands. Bangalore was therefore an obvious place to base the Indian air-force base and other public-sector institutions, which promoted the establishment of a number of universities, institutions, and colleges providing engineering and scientific training.

Between 1951 and 1971, there was a marked correlation between industrial development and accelerating population growth, with figures doubling during the 20-year period. By the 1970s, the occupational profile of the city showed the distinctive patterns of an emerging information society with 10.5% of the population in professional positions, 16.5% in clerical jobs, and 45% in production of which around half were in electrical fields. By the late 1980s, Bangalore included 375 large- and medium-scale industries and had 3000 companies employing 100,000 people in the electronics industry alone. The city contained up to 10,000 small industries and 8 large industrial parks.

Bangalore now plays a prominent role in international electronics, telecommunications, and information technology, contributing almost 40% of India's production in high-technology industrial sectors (hardware, software, telecommunications). The availability of highly skilled technicians, relative political and social stability within the state, the absence of labor conflicts, and an efficient banking network have meant that almost every big player in the information technology scene has its place there. Among indigenous high-technology companies that have established themselves in Bangalore are Wipro Systems (one of the largest Indian hardware and software vendors), Infosys Technologies, and Sonata Software. These, in turn, have been followed by a galaxy of multinational information technology companies attracted by the temperate climate of the city. These companies have adopted varying strategies to gain competitive advantage through offshore software outsourcing using local programmers and satellite links to design and produce customized packages. Some of these transnational companies (TNCs) – Digital Equipment, IBM, and British Aerospace – have formed joint ventures with existing Indian players.

Many of the high-tech companies in Bangalore are growing by more than 50% per year and are employing increasing numbers of graduates. For example, Siemens Communications, set up 18 months ago to build software for the group's digital switching systems, already employs 250 people and plans to grow its workforce of software engineers to 1000 by the end of the decade. The increasing demand for highly skilled "knowledge" workers threatens to outstrip the potential of local institutes in Bangalore to produce such skills, and many of the high-tech companies are having to recruit their workforce from other cities. In recent years, an increasing number of specialized small-scale workshops have been established to supply the high-technology industries. These workshops, most of which are characterized as being in the informal sector, recruit an increasing number of semiskilled migrant workers and help to bridge the gap between the demand and supply of labor.

To capitalize on its emergence as a popular location for software research, Bangalore has established a software technology park at Electronics City, just three-quarters of a mile out of the city. Hundreds of acres of research laboratories are occupied in the park by the likes of IBM, 3M, Motorola, Sanyo, and Texas Instruments. Companies that locate within the park are insulated from the world outside by power generators, by the leasing of special telephone lines, and by an international-style work environment. Bangalore is also in the throes of constructing a new upmarket international information technology park in the suburb of the city. The $250 million project, due to open in

1996, is a joint venture between the government of Karnataka, a Singapore consortium, and the Tata Group. The project aims to integrate advanced work facilities with recreational and residential activities in a single location. The 56-acre complex has already attracted international attention, and thousands of enquiries regarding investment opportunities in the park have flooded in from the United Kingdom, North America, Europe, and Hong Kong. In parallel with the launching of this new park, a new international airport is also being funded by the Indian Tata Group and Singapore International Airlines, which is intended to provide a boost to industrial activity in Bangalore, in particular to the electronics and software industries.

Increasingly, the presence of a sizeable modern industrial sector has brought prosperity to the city and has given its central parts a cosmopolitan outlook. Measured in terms of expensive restaurants and pubs, boutiques, shopping plazas, and other signs of available purchasing power in the context of Western behavior patterns, a middle class is strikingly present in the central parts of the city and in other expensive areas beyond the center. A similar indicator of prosperity is the boom in the construction industry catering for the upper segments of the housing market and for commercial use. Cheap real estate has been one of the reasons why entrepreneurs prefer this city to the choked or very expensive cities like Bombay.

For all the preceding reasons, today Bangalore is perceived internationally as a prosperous and modern Indian city. But this label is misleading. At least three more characteristics have to be added to portray the reality of the situation. First, there exists gross inequality between groups of different socioeconomic status within the city of Bangalore and the region of Karnataka. Second, extreme poverty prevails among many inhabitants in the city. Third, there is an acute problem of civic deficiency both in Bangalore city and in the state of Karnataka, together with poor access to information outside the capital. These three characteristics are described in the remainder of this section.

First, contrary to the international reputation earned by Bangalore, the advent of the information age has yet to make a dent in the overall economic picture of the state, which remains primarily an agricultural state. Out of a total population of almost 50 million people, around 76% live in rural areas, and there is a high incidence of rural poverty in these areas. For example, 95% of the rural poor population in Karnataka have an annual family income of less than $102. A sizeable share of the increase in the population of Bangalore city is related to migration. Many households of predominantly landless agricultural laborers and marginal farmers have been pushed out of their native rural villages by lack of means to survive and have been forced to move to Bangalore to find employment in the informal sectors as unskilled laborers. Tension has escalated within the city of Bangalore with the vast influx of migrant laborers from other Indian states, mainly from neighboring Tamil Nadu. These Tamil migrants seek menial jobs in construction and pose a real threat to local poor inhabitants, as witnessed in the violent anti-Tamil riots of 1991.

Foreign high-technology companies are largely responsible for the rampant wage inflation in Bangalore, particularly among experienced technical staff. For example, these companies typically offer remuneration of about $395 a month, growing to triple the salary over 3–4 years. These discrepancies have led to social tension between the state government in Karnataka, which is committed to supporting the influx of multinational companies into the state, and the lobby of farmers and agricultural workers who are against the establishment of high-technology multinational fast-food chains within the state. The latest of these attacks was recently launched against the KFC (Kentucky Fried Chicken) chain by a lobby of Karnataka farmers.

Second, although poverty could be claimed to be of nationwide concern, in Bangalore the problem is more acute. In terms of absolute numbers, the poor easily predominate over the middle classes and professionals. While the share of urban inhabitants living in huts without access to infrastructure facilities is relatively small in Bangalore (10%) in contrast to other Indian metropoles (25–30%), there has been an exponential growth in this share over the last decade. The condition is continuously deteriorating both in terms of an increase in new slum areas and in terms of an increase in the population density of existing slums. The discrepancy in habitat conditions between the rich and the poor inhabitants is also more extreme and more dramatically visible in Bangalore than in other cities. Due to societal modernization, the state is pressured to cut down trees to accommodate use of central urban land for nonresidential purposes and for expensive housing. Local middle-class residents and the urban poor are driven out of the city because of the rise in price of real estate and rent in the city center. They are therefore forced to squat on the urban fringe and to incur transport costs of commuting to the city each day in search for work.

While it is an open question as to whether the proliferation of shanty towns and environmental degradation is as a result of rural distress and increased population growth, or due to urbanization and globalization, what is certainly noticeable is the increasingly negative attitude of policymakers and planners in Bangalore toward slums. In discussions with the Bangalore state authorities, most policymakers did not officially acknowledge slum dwellers as citizens of the city, even though they clearly constitute the majority of the population. This attitude is reflected in the periodic demotion of any visible manifestations of poverty such as the frequent "clean-up" programs in which squatter settlements are demolished in order to preserve the quality of life of the so-called "modern" sector. Contrary to policy in other Indian cities, an increasing number of slums in Bangalore are located on private land that does not belong to the government and therefore precludes the eventual transfer of ownership to hut dwellers. In many cases, private developers build high-rise blocks so that less space is consumed for housing and the remaining space is allocated to commercial units for which private developers can reap hefty revenues.

A growing number of small, recently established nongovernment organizations such as CIVIC and the Bangalore Poverty Alleviation Programme (BUPP) have taken issue with the claim that slum dwellers have no right to live in the city. According to these groups, the recent internationalization of industrial activity in Bangalore has had a negative impact on the poor since less money is allocated to improving public services and providing urban development programs. These organizations are striving to give poor urbanites some say in the functioning of the city and are currently fighting against the state government's decision to force slum dwellers to share their already scanty land with private entrepreneurs. The BUPP is also developing an information system on slum activities using data compiled by slum dwellers themselves on the status of the land they occupy, the number of people living in each hut, access to amenities, slum dwellers' prioritization of their own problems, and on the nature of information and communication channels between economic agents in the informal sector. The information from the system will be circulated to city authorities for planning purposes and to slum dwellers

themselves to make them more aware of their role in the life of the city.

Third, civic deficiencies caused partially by Bangalore's industrial success need to be faced. Although software companies are not big power users, they need to install voltage regulators, uninterrupted power supplies, and generators to run their computers, because of shortages of power. Similarly, there is a looming water shortage, with municipal pumped water only available 2 days a week and boreholes drying up. The Bangalore Development Authority is trying to encourage more private-sector involvement in the development of infrastructure, especially transport. In terms of industrial development, the city has almost reached saturation point. Today, the state of Karnataka is home to some 114 companies – all of which are located in Bangalore, due to the poor infrastructure facilities outside the city. The development of secondary cities within the state of Karnataka has been envisaged for some time by the Bangalore metropolitan regional development authority, with an ambitious project to develop a mega ring road and a light railway connecting all towns in the state. However, little has been done to date. As a result, thousands of young computer professionals continue to struggle along terrible roads to get to their jobs in software factories.

Information infrastructure in the region remains confined to the business community in the city with poor access to information outside the capital. While the 1965 Public Library Act specified the setting up of 10 district libraries, and 120 subdistrict libraries, by the mid-1980s, resources in the district and subdistrict libraries were suffering from decline in funding resulting in inadequate building maintenance and staffing levels, discontinuation of periodicals, and outdated book collections. In terms of information and communication systems for the public sector and for universal usage, each district in the country has an information center set up by the National Informatics Centre for the purposes of planning and monitoring of development programs and for disseminating information to the public. However, to date, these systems have been largely underutilized, with a few districts performing ad hoc data processing rather than improving productivity of development programs. More recently, the Department of Science and Technology (DST) has launched its Natural Resources Data Management System (NRDMS), with Karnataka volunteering to be the first state in the country to diffuse the system to all district offices by the end of 1996. This system attempts a more sophisticated and integrated areal planning approach through the use of geographic information systems, as opposed to schemewise monitoring of earlier initiatives. It is envisaged that each district in Karnataka will have its own computer system integrating a wide variety of physical and socioeconomic data relevant to the area concerned. However, one of the big problems with the initiative threatens to be sustaining the computerization effort once the pilot phase is over.

To sum up, in recent years Bangalore has grown into a modern, industrial city. Its economy has become more competitive, global, and increasingly dominated by information and communication technology. However, at the same time, the city and region of Karnataka have experienced growing poverty, social inequality, and gross deficiencies in public services.

Editor's suggestions for further reading

Castells, M. and Hall, P. (1994), *Technopoles of the World: The Making of 21st Century Industrial Complexes*, London: Routledge.

Freeman, C. (2000), *High Tech and High Heels in the Global Economy*, Durham and London: Duke University Press.

James, J. (2001), 'Information technology, cumulative causation and patterns of globalisation in the Third World', *Review of International Political Economy*, 8, 147–62.

Madon, S. (1998), 'Information-based global economy and socioeconomic development: the case of Bangalore', *The Information Society*, 13(3), 227–43.

Singhal, A. and Rogers, E. (2001), *India's Communication Revolution: From Bullock Carts to Cyber Marts*, London: Sage.

'Public Internet Cabins and the Digital Divide in Developing World Megacities: A Case Study of Lima'

Ana María Fernández-Maldonado

Editor's Introduction

Our sixth exploration of cybercity digital divides continues the theme of ICT access in the fast growing cities of the developing world. Here Ana María Fernández-Maldonado, a Peruvian urban researcher who works at Delft University in the Netherlands, analyses the complex dynamics of changing ICT access in Lima, the capital primate city of Peru.

In a rare quantitative discussion of digital divides in the megacities of the global south, Fernández-Maldonado concentrates on exploring the ways in which Peruvian telecommunications liberalisation has helped to widen access to phones and the Internet. The growth of mobile phone access has been especially important. She notes, however, that private access to phones, cable TV, mobiles and the Internet is only an option for the city's elite as the very low incomes of the rest of the city's population makes this impossible.

In the light of this, Fernández-Maldonado explores the critical roles played by private and public internet cabins – the *cabinas públicas de Internet*. These are hybrid spaces where Internet access can be rented relatively cheaply by the minute or hour. They provide a crucial link between ICTs and urban space, particularly in Lima's vast range of informal poor and middle-class settlements, or *barriadas*.

Connecting 70 per cent of the city's population, these innovative spaces provide a crucial entry point into the worlds of ICT-based e-commerce, social and community networks, information exchanges, gaming networks, and global diasporas. Without the *cabinas* these would be impossible to access because private telecommunications investors tend to target only the highly paid and formal economic sectors of the city that have done best in the recent period of macroeconomic liberalisation. With over 50 per cent of Lima's population living below the poverty line, home and personal ICT access is possible only for a small minority.

A critical issue, therefore, is the need to support collective Internet access, in public and semi-public *cabinas* spaces, as cheaply as possible. A mass of informal entrepreneurs have met this need with startling entrepreneurial effort. Their *cabinas públicas de Internet* have quickly become crucial hubs of social and economic activity in the *barriadas* and elsewhere. The *cabinas* are especially important as hubs of activity, and sociability, for young people. Fernández-Maldonado urges local municipal and state policy-makers and NGOs to support the growth and extension of the *cabinas* so that they can play a bigger part still in helping to overcome the extremely stark digital divides that characterise Lima's geographical and socio-economic landscapes.

What makes Lima an interesting case in the development of telecommunications is that, despite its huge concentrations of poverty, hundreds of *cabinas públicas de Internet* (public Internet cabins) have emerged in poor and middle-income areas of the city, on a commercial basis and without any support, providing access to ICTs to places where the private sector normally does not invest. They now provide connectivity to more than 70 per cent of the Internet users in Lima (see Plate 31 on p. 316).

Housing eight million inhabitants, Lima concentrates most of Peru's investment, services and economic activity. Since the late 1950s the city transformed itself as a result a strong rural migration flow. Its periphery was gradually occupied by *barriadas*, spontaneous neighborhoods where the newcomers built their homes, which now house 40 per cent of households of Lima. Lima is now famous in the urban literature because of its *barriadas*. It is especially well known because of the level of organization of the residents in building their neighborhoods, and the effort they show to gradually build their own dwellings.

During the 1980s, the emergence of a dense network of 'survival' associations dedicated to food- and health-related issues, and the outstanding growth of informal economic activities, were the outcome of the economic crisis during that period. Meanwhile, the economic, social and political landscape of the city transformed rapidly and new social actors emerged.

Deep economic reforms during the 1990s improved the macroeconomic indicators, and the economic growth was partially restored. But the application of the new economic policies has had tremendous social costs. More than half of Lima's population now lives below the poverty line.

The telecommunications revolution takes place in Lima in this context of socio-economic polarization, increasing 'informalization,' economic scarcity, a pragmatic neoliberal state with low resources, and a dense network of local associations. In Peru there is still no national strategy or policy to guide the transition to the 'information society.' This is linked to the withdrawal of the state from the sector according to the new regulations launched in 1993. These embody the principles of free market and free competition in the telecommunications sector, encouraging international investment. The two public companies were sold to the Spanish telecom company Telefónica in 1994. They agreed to expand the networks with the condition of a period of 'restricted competition.'

The new set of rules have been useful to modernize and expand Lima's telecom networks. But they have not been able to reduce the prices of telephony to levels affordable by most residents of Lima. Telephonic diffusion has greatly improved since privatization but shows no increase since 1998. The use of telecommunications services is highly skewed. There is saturation of the market in the upper niches and low diffusion in

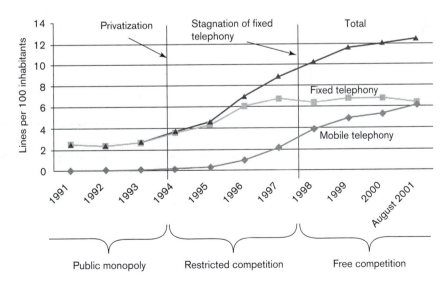

Figure 7 Evolution of telephonic density (lines per 100 inhabitants) in Peru, 1991–2001

Source: El Comercio, 2001b, with data from OSIPTEL

Plates 31a and 31b *Cabinas públicas de Internet* (public Internet cabins) on the streets of an informal district in Lima. (Photographs by Ana María Fernández-Maldonado.)

the lower. The average indicators of telecommunications diffusion are low compared to other countries of Latin America and very low by international standards.

The evolution of the sector can be characterized in three main phases: public monopoly, restricted competition and free competition. They coincide with periods of no growth in telecommunications, the expansion of fixed telephony (from 1994 to 1998) and the expansion of mobile telephony, respectively (see Figure 7 on p. 315).

After the privatization of the public monopoly in 1994, (fixed) telephonic density increased steadily. But since 1998 there has been a stagnation in this segment, linked to the socio-economic circumstances of the country. During the third trimester of 1998, 9 per cent of the total telephone connections were cancelled. This coincided with the beginning of a period of economic recession. The fixed telephone service became too expensive for the average resident. It is the second most expensive in the region after Chile (see Table 8), and more than double the average of the region.

The stagnation of fixed telephony was accompanied by the expansion of mobile telephony. Table 9 shows the clear preferences of the low-income sectors for the pre-paid system, which now provide two-thirds of all mobile phones. During 2001 mobile telephony has become the main way of accessing telephones in Peru, with 1,700 000 users (Roquez, 2001). As in fixed telephony, Telefónica is the dominant operator, with a 65 per cent share of the market (El Comercio, 2001a).

The polarization in telecommunications diffusion in Lima is even more pronounced if we take a look at the figures for cable TV, computer access and Internet connection at home (Table 10). The differences between high and very low strata are remarkable, and the ones in cable TV, PC possession and Internet at home are even more dramatic than in fixed telephony.

The difficult socio-economic situation of the country, added to the high charges for fixed telephony, hinders the diffusion of Internet services in Lima. The average monthly charges for dial-up Internet connection ($40 per month) represents approximately one-third of the

Table 8 Monthly residential subscription charges (in $US) for countries in Latin America in 2000 (adapted from source)

Country	Monthly residential subscription charges (US$)
Chile	16.33
Peru	14.75
Mexico	14.01
Argentina	12.83
Uruguay	8.74
Venezuela	8.2
Brazil	6.7
Costa Rica	4.28
Ecuador	3.67
Colombia	3.29
Average for the region	6.59

Source: ITU, 2000

Table 9 Percentage of mobile phone possession according to socioeconomic sectors in Lima, 2000 (adapted from source)

	Total	A sector	B sector	C sector	D sector
Pre-paid card	16	27	32	17	17
By suscription	8	63	27	6	1
Mobile phone (total)	22	78	53	22	17

Source: Apoyo, 2000

Table 10 Telecommunications diffusion in Lima in 2000, according to socioeconomic stratum (%) (adapted from source)

	Total	High (A)	Middle (B)	Low (C)	Very low (D)
Fixed telephone	46	100	95	62	27
Mobile phone	22	78	53	22	17
Cable TV	23	95	68	23	7
PC	13	75	49	8	1
Internet at home	5	53	21	0	0

Source: Apoyo, 2000

minimal wage in Lima. Despite this huge limitation, Internet diffusion in Lima is considered as notable in the Latin American region, thanks to very widespread public access to Internet in *cabinas*.

THE CABINAS PÚBLICAS DE INTERNET

The *cabinas públicas* are small businesses where computers with Internet connection are rented for a certain period at prices which are between 40 per cent (in middle-income areas) and 70 per cent (in low-income areas) lower than the prices with a dial-up connection. In 2001 the prices fluctuated between $1.20 to $0.40 per hour, according to the location. The low prices have been one of the main factors of success of the model. The *cabinas* began to be visible at the end of 1998 and they are now established in most of the middle and low-income neighborhoods of the city, commercial areas and even in some peripheral areas.

Their size varies from 40 to 3 computers. The large types have many kinds of peripherals and services, while the informal versions located in kiosks inside informal markets can have between 3 to 6 computers and a few peripherals. However, both formal and informal types generally have reasonably good equipment, fast computers and the latest software. Their speed of connection is much better than the speed from a domestic connection.

The users are very young groups, with high proportions of students and persons with higher-than-average education level. Chat rooms, email and instant messaging are the main applications. Surfing the web for information for work, school and academic activities is also popular. Internet telephony is another popular application. Recently, entertainment applications have become enormously popular, among younger users,

and new *cabinas* have been established dedicated only to online games.

The hype of the Internet in Lima is present not only in the media or on the computer screen. It also exists at street level – a simple visit to Lima will give a good impression of how the residents are using and appropriating the new technologies. Growing numbers of *cabinas públicas* are spread all over the city and are full of young people trying to connect with the outside world. Short courses for improving computer skills are advertised everywhere. Street sellers offer manuals for the many computer programs available. And the expanding cluster of informal ICT-related enterprises in Wilson Avenue provides all kind of services, training, selling hardware and software to individuals, firms and institutions.

The introduction of Internet and expansion of telecommunications are having effects on the life of the residents of Lima. Internet and mobile telephony has rapidly changed the means of business communication. Mobile telephony has increased the telephone accessibility of individuals who perform informal services. Computer availability has improved the efficiency of public and private institutions. Although the effects of ICT applications are much more evident in the life of better-off groups, they are also expanding to lower-income groups, especially the more educated and more skilled youth groups. For the youth in Lima the Internet constitutes the easiest way to connect to 'modernity,' to virtually participate in the more advanced parts of the world.

The *cabinas públicas* constitute the most remarkable development in the telecommunications field in Lima. They are playing a significant role in expanding ICT access to middle and low-income groups. More important, at city level, the *cabinas* are providing local facilities: the libraries, recreation facilities, study places, youth centers, etc., that are lacking in the *barriadas* are now present and combined in the multiple services offered in the *cabinas*.

The emergence of *cabinas* in a context characterized by high costs of telecommunications services and scarce economic resources makes this initiative more praiseworthy. These small entrepreneurs have found a way to make profit for their small ICT businesses on a sustainable basis. This confirms that to provide connectivity to people, public access is an alternative that works.

The initiatives presented here constitute an effective answer to the problem of access. But access is not all. ICT activities observed in *cabinas* are basically limited to the social, academic, training and entertainment aspects. There are still no (digital) networks for the local (social) networks: even if access is not an issue, the new technologies are not integrated into the local economy, and are not functional to the objectives of local associations. Cyberspace still needs to be locally appropriated by the thousands of local networks present in Lima. Despite their vitality to claim their rights, they have not reached the level to claim their own portion of cyberspace. This is a step that still has to be taken.

Without a more decisive role of the state to guide the development of ICTs, and a strategy to provide the minimal conditions to promote and improve the existing situation of the *cabinas públicas*, a splendid opportunity for empowering the development of people and communities will be wasted.

REFERENCES FROM THE READING

Apoyo, Opinión y Mercado (2000), *Nivele Socio-económicos de Lima Metropolitana*. Lima.

El Comercio (2001a), *El Momento de la Competencia*, 19 December 2001, Sección Especial, p. 2, Lima.

El Comercio (2001b), *RCP: Internet no Detiene su Crecimiento*, 19 December 2001, Sección Economía, p. B4.

Dupuy, G. (1992), *L'urbanisme des réseaux. Théories et méthodes*, Paris: Armand Colin Editeur.

International Telecommunications Union (ITU), (2000), *America's Telecommunications Indicators*, Geneva: ITU.

Roquez, A. (2001), "El año de la telefonía móvil", *El Comercio*, 19 December, Sección Especial, p. 2.

Editor's suggestions for further reading

Alaedini, P. and Marcotullio, P. (2002), 'Urban implications of information technology and new electronics for developing countries', *Journal of Urban Technology*, 9(3), 89–108.

Fernández-Maldonado, A. (2001), 'The diffusion and use of information and communications technologies in Lima, Peru', *Journal of Urban Technology*, 8(3), 21–43.

Graham, S. (2002), 'Bridging urban digital divides? New technologies and urban polarization', *Urban Studies*, 39(1), 33–56.

Liff, S. and Steward, F. (2001), 'Community e-gateways: locating networks and learning for social inclusion', *Information, Communications and Society*, 4(3), 317–40.

Sanyal, B. (2000), 'From dirt road to information superhighway: advanced information technology (AIT) and the future of the urban poor'. In J. Wheeler, Y. Aoyama and B. Warf (eds), *Cities in the Telecommunications Age: The Fracturing of Geographies*, London and New York: Routledge, 143–60.

'Access Denied'

from *The Wealth and Poverty of Networks* (1997)

Danny Kruger

Editor's Introduction

Our penultimate analysis of cybercity digital divides draws on the work of Danny Kruger, a researcher with Demos, an influential think tank based in London. Kruger synthesises the various dimensions of the digital divide from a UK perspective. Importantly, though, he moves beyond the common preoccupation with access to ICT hardware and software. Instead, he looks at some of the more subtle and invisible ways in which the growing mediation of cities by ICTs may actually damage the life-chances of certain groups, neighbourhoods and individuals.

As well as the obvious costs of non-participation for the poor, old and unskilled groups with lowest access to ICTs, Kruger looks at how the growing use of geodemographic ICT systems by service providers like insurance companies, financial services companies and retailers may, through their categorisation of the neighbourhoods of cities into spaces of differing market potential, have profoundly exclusionary effects. By segmenting poor, old and low skilled people into unattractive categories such systems can further their exclusion from a wide range of crucial private and public services. They may therefore help to perpetuate the very inequalities that are embodied in their crude categorisations through the instrumental rationalities of geodemographic techniques.

Kruger is even more concerned about the probable exclusionary effects of the migration of the whole system of financial transaction within and between cities into electronic systems and spaces. Already, the credit cards that are pretty much essential to participation in e-commerce are the preserve of people who can generate good enough credit ratings. However, Kruger speculates that, as even small transactions are increasingly being made through smart cards, electronic cash systems, and e-purses, so some sections of urban populations may be marginalised in a residual physical cash-only economy. They will thus be ever more distanced from the main economic dynamics of cybercities.

Kruger is concerned that people with physical cash only would be unable to undertake a range of transactions that are critical to daily urban life. They would be charged more for the same service as the infrastructure for processing physical cash is gradually withdrawn. And they would be severely stigmatised as electronic financial transactions become normalised as the way in which urban residents engage with the economy.

It has become one of the orthodoxies of the 1990s that the information and communication technology (ICT) revolution potentially offers one way of tackling almost every social ill. The Information Society Forum's first annual report to the European Commission stated that 'information technology can empower ordinary people and their communities, putting them more in control of their working lives, allowing them a fuller exercise of their rights and an outlet for their creativity'. While the optimism of the rhetoric is often overblown, new technology does indeed hold out great cause for hope in many fields. The opportunities for improving standards in education, for matching people to jobs (or matching employment policy to demographic reality) and for community regeneration in general, are all real. How far can ICT help to combat the problems of social exclusion?

Social exclusion can be defined as exclusion from access to the ladders of social improvement, being cut off from the paths of upward mobility. It involves under-education and unemployment but also encompasses the social pathologies that are generated by such structural forces: disempowerment and a feeling of alienation from wider society. It is linked to the theme of 'network poverty': the state not only of unemploy-ment, but also of lack of access to the informal contacts which provide the most useful paths to decent jobs. The theory holds that it is the 'weak ties' which one makes with acquaintances, friends of friends, and so on that are more useful in connecting us to job oppor-tunities and a sense of civic participation than the 'strong ties' one has with family, close friends and neighbours. It follows, therefore, that the information and communication possibilities of the new technology could offer significant means of escape from the 'ghettos of exclusion' in which so many people live. But, for a number of reasons, things are not working out that way yet. At present, ICT threatens not only to fail in living up to its great potential as an engine of social renewal – it is also threatening to petrify social inequality, to inhibit upward mobility and to consol-idate existing patterns of exclusion.

There are various ways in which the revolution in information and communication technology may be harming the prospects of certain individuals and groups. Most obviously, people's long-term interests are being damaged by their non-participation in an increasingly electronic business culture, whose swift development excludes by virtue of its ever-increasing complexity. Hence, the importance of providing uni-versal access to the key ICT infrastructure, of teaching the skills one needs to use it and of ensuring that the technology can provide information and communica-tion possibilities which are useful to all people, not just professionals and other IT literate groups.

Evidence of ICT use shows fairly predictable patterns of inclusion and exclusion, of 'information haves' and 'information have-nots'. Typical have-nots are the old, the poor, the unskilled. These people have little understanding of the nature, the uses and the benefits of ICT and few opportunities of using it. Furthermore, assuming that they can get access to the hardware and are taught how to use to it, there remains the problem that the content of ICT makes the

electronic age profoundly élitist. It is easier to access the Hong Kong Stock Exchange on the Internet than to get information on local job opportunities in the UK. A system initially designed for and by academics still retains the characteristics of an exclusive ivory tower, despite or because of its appropriation by business. A major shift in the design of ICT applications is necessary if we want to get the British population in general 'on-line' and especially if we seek to include in this enterprise those who suffer exclusion.

Beyond the problems of access, skills and content, there are more insidious ways in which the ICT revolution is undermining the attempts of policy makers to bridge the gap between rich and poor, and confirming people in positions of exclusion. Perhaps most worrying is the development of marketing tech-niques based on geodemographic profiling systems. Geodemographic profiling is the process of finding and identifying 'segments' of the population based on levels of affluence, social group, residential area and previous purchases which enables companies to 'target' indi-viduals with the marketing of products considered appropriate to that 'segment'.

In one form or another this has been common practice for over 100 years in Britain, but there is concern now that the increasing accuracy of the geodemographics industry is threatening to confirm people in the 'segments' in which they have been placed on the basis of their marketing profile. Put simply, it is argued that the wealth of private or com-mercial information which is available to companies enables them directly to concentrate their marketing efforts on those people considered 'good prospects', that is the better-off. Thus some low-income individuals and families are denied access not just to the hardware of ICT but to the benefits that a positive rating in a commercial information system can endow – espe-cially, to attractive and affordable financial services. Thus, there is a fear of what the American sociologist John Goss calls the 'instrumental rationality' of geode-mographics, of a system which 'displays a strategic intent to control social life, and [in which] the ideological conception of identity and social space within the model may become real – in other words, that the assumptions will be validated as the strategies take effect.' The targeting of individuals or families according to the segmentation systems of marketing software could potentially have the effect of *ex post facto* realisation: offered only products – and, more importantly, services – considered applicable to the male, C2, west of Scotland, O-Level only, rented accommodation, thirty-something machine worker 'target group', such an individual is more likely to remain in that group and to focus his consumer power on purchasing products considered appropriate to him by the marketeer. Thus the seller, not the consumer, is setting the agenda, controlling demand as well as supply: geodemographics 'is based upon an instrumental rationality that seeks to bring the processes of consumption further under the control of the regime of production.' Goss' conclusion, in short, is that 'we are no longer confronted with our own will.[1]

The problem of instrumental rationality is related to a phenomenon which has become more marked in recent decades: the polarisation of social space. People of widely varying incomes rarely live in the same areas any more (or, if they do, rarely or never come into contact in civic activities) which further limits the possibility of creating useful networks for the excluded. Not only are young people on council estates deprived of useful contacts, they are deprived of role models, examples of successful people they can look up to. Meanwhile, their affluent counterparts live in suburbs which are equally homogenous in terms of income. The result is less social cohesion and more class rivalry. 'Red-lining', an early form of 'segmentation' practised by insurers in the 1960s and 1970s, played a part in this polarisation, one of the most destructive developments in social organisation in the post-war period; there is a danger that more advanced customer research techniques will compound it.

ICT threatens to consolidate social exclusion in another, equally insidious way. The approach of the cashless society, made possible by the ICT revolution, threatens to petrify economic realities even more. It is becoming vital for an individual to have positive ratings with the companies or banks which oversee one's income and expenditure. It is estimated that around 20 per cent of the British population does not have access to individual financial packages and between 5 and 8 per cent has no financial support at all through the banking system, including current accounts. The reliance on cash, or even on cheques, helps to condemn individuals to life at the bottom end of the spending scale and can help keep them there. Not only

can one not make a large purchase: without access to a complex financial infrastructure one cannot pay for anything by instalments, such as a car or a personal computer, to help one break out of the poverty cycle.

The approach of electronic cash (e-cash) and electronic purses (e-purses) presents a possible scenario in which only those judged to be appropriate by banks, credit companies or the like would be in a position fully to engage in the consumer society. Those who still relied on cash would be – indeed, are already – the victim of a double discrimination. It is already the case in some countries (notably Norway) that organisations refuse to take cash payments or charge for doing so. Electronic transfer is already a common practice in Britain, where at the demand of trade unions all salaries are paid directly into workers' bank accounts and where most bills are paid by direct debit: it might not be long before one is charged for doing otherwise. 'Unbanked' people already pay extortionate fees for cashing cheques. If this were not enough to force people to get a bank account (if they can), there is the possibility that those who persist in using cash will have to pay the costs of doing so, which are presently paid by the collective taxpayer. As argued by Dave Birch, 'the young and techno-hip will eventually tire of subsidising automated teller machines, armoured cars and night safes for their less well-off brethren.' The administrative costs of e-cash will be lower than those of cash (though the Bank of England would lose its seigniorage fee on cash transactions) but it is likely that the better-off will use it at first, because it is they who will be offered the services first, and with most insistence, by the banks which provide them.

Some of these problems, including the basic lack of access and of skills, can and must be tackled head-on with initiatives for providing Internet access for all and to encourage people to take up the opportunities opened up by new technology. Only when people start to use the Internet will it start to become useful to them: only by demanding useful information will useful information be supplied in this unregulated, arch-capitalist sector. The problems of information exclusion are largely to do with the relationship between supply and demand. In all the areas we have looked at – access, skills, content, consumer choice

---

1 Goss J. 1995, 'We know who you are and we know where you live: the instrumental rationality of geodemographic systems' in *Economic Geography*, vol 71, no 2.

and the cashless society – *supply* and not *demand* is setting the agenda. It is necessary to find ways to reinstate demand as the engine of change.

ICT offers great opportunities for 'people power' in trading, education, employment, and community organisation. Most of all, it offers the chance to create synergies between related but separate organisations – most importantly perhaps, between educational and employment institutions – and to invigorate whole communities. But to activate these benefits and thus begin to tackle social exclusion, proactive effort is required, effort which is in keeping with the possibilities of ICT. There are more examples of failed ICT initiatives than of successful ones because people rarely realise what ICT can and cannot do. It can facilitate communication but it cannot create the environment for it out of thin air. It can open doors but it cannot compel people to pass through them who have no reason for doing so. Most of all – in education, in employment policy or in anything at all – it must be treated not as a miraculous elixir, but as a tool in a holistic strategy which involves fresh thinking about demand as well as supply and about access as well as applications of new ICTs.

The benefits of ICT are immense but so are the potential dangers. More hard-headed analysis of what we want ICT to do for us, and how we can implement a coherent and sensible strategy, is required. The unregulated, uninhibited development of ICT and its application by businesses is one of the more disturbing and yet exciting developments of our time. ICT can be harnessed to help overcome problems of social exclusion, bringing the excluded into new networks of opportunity. But it will not do so automatically. Strategies to make ICTs accessible to those currently excluded from the information revolution by low income, poor environments and low expectations can play a major part in the wider campaigns to create the inclusive society.

Editor's suggestions for further reading

Curry, M. (1998), *Digital Places: Living With Geographic Information Technologies*, London: Routledge.

Demos (1997), *The Wealth and Poverty of Networks*, Collection Issue 12, London: Demos.

Goss, J. (1995), 'Marketing the new marketing: the strategic discourse of geodemographic information systems'. In J. Pickles (ed.), *Ground Truth: The Social Implication of Geographic Information Systems*, New York: Guilford, 130–70.

Goss, J. (1995), '"We know who you are and we know where you live": the instrumental rationality of geodemographic systems', *Economic Geography*, 71(2), 171–98.

Jupp, B. (2001), *Divided by Information?*, London: Demos.

Pickles, J. (ed.) (1995), *Ground Truth: The Social Implication of Geographic Information Systems*, New York: Guilford.

'The Software-Sorted City: Rethinking the "Digital Divide"'

Stephen Graham

Editor's Introduction

Our final reading on cybercity digital divides draws on the Editor's research into the automatic sorting of citizens' mobility and access rights within and between cybercities. Stressing that ICTs can be used to inhibit people's freedoms and mobilities as well as to increase them, in this specially written piece Graham analyses a range of examples in which computer software automatically prioritises the access and mobility of certain urban citizens whilst inhibiting that of those deemed by operators or law enforcement agencies to be risky, unprofitable, or undeserving.

Graham reviews a range of cases to help illustrate the spread of such software sorting techniques. He looks at the use of software to socially prioritise traffic on the Internet, in telephone queues at call centres, on electronically tolled highways, and in the mediation of urban social geographies by geodemographic systems. Graham also analyses the biased configurations of ICTs and surveillance systems at international borders. Overall, he stresses that the software sorting of mobilities within and between cybercities is so invisible, so automated, and so normalised, that often neither the resulting losers nor the gainers are actually aware that it is going on. Thus a critical element of the digital divide, that helps to perpetuate intensifying inequalities in contemporary cities, carries on largely unresearched and unnoticed.

Values, opinions and rhetoric are frozen into code
(Bowker and Leigh-Star, 1999, 35)

The modern city exists in a haze of software instructions. Nearly every urban practice is becoming mediated by code
(Ash Amin and Nigel Thrift, 2001, 125)

New Information and Communications Technologies (ICTs) are generally portrayed as means of overcoming the barriers of time and space. The dominant depiction of the so-called 'information society' stresses such technologies as new, friction-free means of connecting people, institutions and spaces which speed up and improve the functionality of all manner of services in the process.

And yet it is increasingly clear that ICTs can create disconnections as well as connections. They can be used to slow down and *add* friction to people's lives, making them logistically more difficult. And they can facilitate the withdrawal of services from people and communities and the worsening of their opportunities (see Lyon, 2002; Graham, 1998). Consequently, urban digital divides are not just about the usual focus of debate – uneven access to the Internet. Perhaps just as important are the powerful and often invisible processes of prioritisation and marginalisation as software and code are used to judge people's worth, eligibility and levels of access to a whole range of essential urban spaces and services.

Users of public and infrastructural services have long been sorted and prioritised by the bureaucracies of providers. But, as such practices are augmented, or replaced, by ICTs, software-based techniques, linked to computer databases, increasingly sort users. These work automatically (i.e. without human discretion), continually (i.e. 24 hours a day), and in real time (i.e. without delay).

New information and communications technologies, and digital surveillance techniques, are now being widely used to subtly differentiate consumers within complex systems of transport, communications or service provision. Here, computer algorithms are being used at the interface of databases and telecommunications networks to allocate different levels of service to different users on an increasingly automated basis. This is being done to overcome problems of congestion, queuing and service quality and to maximise the quality of service, especially for premium, privileged users.

The inherent flexibility of ICT-based sorting can allow enhanced functionality to be offered to those deemed attractive. At the same time, less attractive users and communities, or those deemed to be risky in any way, can be pushed away electronically. Successfully employing software-sorting techniques can clearly advantage service and infrastructure providers as they strive, in the context of widespread privatisation and liberalisation, to maximise profits and returns on investment whilst minimising exposures to risk. They are also fundamentally important in allowing standardised infrastructure services, built up to give the same service to all and to cross-subsidise from profitable to unprofitable spaces and users, to be restructured as markets for mobility and infrastructure services (Graham and Marvin, 2001).

As a result, software-sorting techniques are increasingly being harnessed by managers of public spaces (shopping malls and town centres) and by providers of infrastructure (utilities, transport and telecommunications). Examples of the software sorting of mobility spaces include Internet prioritisation; electronic road pricing; call centre queuing; geodemographic sorting; and the use of biometrics to bypass international passport and immigration controls. It is worth exploring each of these in turn.

INTERNET PRIORITISATION

When the Internet first became a mass medium in the late 1990s it was impossible to give one user a priority

service over another. All packets of data on the Internet were queued when there was congestion in any particular part of the network. Now that the Internet has been commercialised and is dominated by transnational media conglomerates, however, algorithmic surveillance techniques are being embedded into the routers that switch traffic and make the Internet work. New Internet protocols – the software codes that route packets of information around the system – are emerging which actively discriminate between different users' packets, especially in times of congestion. 'Smart' routers can sift priority packets, allowing them passage, whilst automatically blocking those from non-premium users.

Thus, high quality Internet and e-commerce services can now be guaranteed to premium users irrespective of wider conditions whilst non-premium users simultaneously experience 'web site not available' signals. This further supports the 'unbundling' of Internet and e-commerce services, as different qualities can be packaged and sold at different rates to different markets (Graham and Marvin, 2001). As Emily Tseng suggests, 'the ability to discriminate and prioritize data traffic is now being built into the [Internet] system. Therefore economics can shape the way packets flow through the networks and therefore whose content is more important' (2000, 12).

There is therefore no longer a singular 'Internet'. Rather, there are a multitude of 'internets' controlled by the sifting capabilities of software. Elite, transnational groups tend to experience broadband, interactive and relatively empowering models which, to them, remain largely democratic and meritocratic. Relatively unimpeded horizontal networking to peers, both locally and around the world, is the dominant model here. The traffic of such premium users is being prioritised within the technological architecture of the Internet (Calabrese and Borchert, 1996, 250). To Calabrese and Borchert the

> cosmopolitanism of the [transnational elites] will be enhanced by its activity on the superhighway. The high spatial mobility of its members will be mirrored by their high network mobility and activity in the formation and maintenance of political alliances and economic relations on a highly privatised, translocal and increasingly transnational basis (Calabrese and Borchert, 1996, 250).

On the other hand, though, even when marginalised groups and citizens within cities do gain electronic access, telecom and media firms are tending to offer much less capable electronic infrastructures and services than those being 'bundled' into the packaged urban spaces, configured for affluent socio-economic groups. Services for lower income groups are often being configured largely for the passive consumption of corporate entertainment and services. In accessing the Internet through a cable set-top box, for example, opportunities to create multimedia products and connect horizontally to other users are very limited (if not impossible). Bandwidth is overwhelmingly configured 'downstream' so that users can consume corporately produced multimedia products and services. The limited 'upstream' bandwidth from the user to the cable company is configured overwhelmingly to support consumption and 'press now to purchase' buttons. As Calabrese and Borchert argue, from the point of view of the US, the worry is that, as a result, 'wage earners, the precariously employed and the unemployed will interact infrequently on the horizontal dimension, except primarily in commercial modes which are institutionally and hierarchically structured, and controlled for commercial purposes, such as games and shopping' (1996, 253).

In a detailed analysis of how growing corporate control of the Internet is configuring its media spaces, Dwayne Winseck (2003) found that these trends towards the software-sorting of users were highly advanced. In 1999, for example, Cisco, the manufacturer of most of the Internet's routers, advertised to corporate media and Internet firms offering them:

> absolute control, down to the packet, in your hands . . . You can identify each traffic type – Web, email, voice, video . . . [and] isolate . . . the type of application, even down to *specific brands*, by the *interface used*, by the *user type and individual user identification or by the site address* (cited in Winseck, 2003, 183; original emphasis).

As a result, corporate internet firms now routinely prioritise the traffic from the 'premium', selected users that they think will bring the best revenues and exposure to their brands, their services, and their corporate tie-ins. At the same time, they actually downgrade the priority of web and Internet traffic that is deemed to be of marginal profit potential or which will benefit competing firms.

CALL CENTRE QUEUING

The integration of customer data bases within call centres provides another example of the use of digital techniques to discriminate between consumers. Initially, call centres operated through the judgement and discretion of call centre operators. One system installed at South West Water in the UK in the mid-1990s, for example, meant that:

> when a customer rings, just the giving of their name and postcode to the member of staff [a practice often now automated through call-line identification], allows all account details, including records of past telephone calls, billing dates and payments, even scanned images of letters, to be displayed. This amount of information enables staff to deal with different customers in different ways. A customer who repeatedly defaults with payment will be treated completely differently from one who has only defaulted once (*Utility Week*, 1995)

Now that call centres are equipped with Call Line Identification (CLI) – allowing operators to detect the phone numbers of incoming calls – such practices are being automated. Algorithmic surveillance system are emerging which can queue incoming calls differently, depending on inbuilt, algorithmic, judgements of the profits the company makes from them. 'Good' customers can thus be answered quickly whilst 'bad' ones are put on hold. As with Internet prioritisation, neither user is likely to know that such prioritisation and distancing is occurring.

SOFTWARE SORTING IN URBAN ROAD SPACES

In our third example, new algorithmic techniques are being used to reduce road congestion, whilst improving the mobilities of privileged drivers (see David Holmes, p. 173). With road space increasingly congested, electronic road pricing is an ever-more popular political choice. A range of governments have brought in private or public–private regimes to either electronically price entry into existing city centres (for example, Singapore and London), or build new private premium highways that are only accessible to drivers with in-car electronic transponders (including Toronto, LA, San Diego, Melbourne and Manila).

In both cases, access to road space becomes a priced commodity dependent on users having the technology in their cars, and the resources, and often bank accounts, to pay bills. In some cases, systems allow traffic flow to be guaranteed whatever the level of external traffic congestion. On the San Diego I-15 highway, for example, the monitoring of congestion levels on the premium, priced, highway can signal real time price increases when flow starts to decrease because of congestion. Communicated to drivers, this reduces demand and reinstates free-flowing conditions.

Whilst such techniques can be justified through arguments about environmental sustainability, it can also be argued that their implementation is closely related to the changing political economy of cities. This is because, like Internet prioritisation and call centre queuing, highway monitoring and access control systems facilitate the removal of what might be crudely called cash-poor/time-rich users from the congested mobility network. In the process this allows premium network conditions to be experienced by cash-rich/time-poor users. (This, of course, is a gross over-simplification, as we also need to consider the many cash-poor/time-poor people in cities such as single parents or poor families: see Graham and Marvin, 2001).

The government of Hong Kong, for example, recently discussed implementing a city centre road pricing system like that in Singapore. This was not to reduce greenhouse gas emissions, however. Rather, it was a direct response to the lobbying of corporate CEOs who were sick of having to walk the last half mile to meetings in hot, humid conditions because of gridlock. In Singapore, these same executives had grown used to seamless door-to-door service uninhibited by traffic in the priced central area.

THE GEODEMOGRAPHIC SORTING OF CITIES

Our fourth example of software sorting involves the use of computerised Geographical Information Systems (GISs) to sift neighbourhoods and life-chances in cities. 'No longer is the city *visualised* or *composed*', suggests the urbanist Sze Tsung Leong (2001, 766), 'as much as it is empirically *computed*' (original emphases). This has been supported by rapid advances in georeferencing technologies, such as digital cartography, satellite remote sensing, the global web of Global Positioning

Systems (GPS) satellites, and digital telecommunications. Together, these technologies allow the social geographies and patterns of flow within cities to be precisely defined, surveilled, and virtually simulated, against a global geometry of precise, digital, time–space coordinates. 'Currently, data represents the technology by which the city is being reconfigured, regardless of its physical composition,' continues Tsung Leong (2001, 766, 765):

> Information has become the new mapping device that unlocks the city to reveal the inner workings of life, economics, and society in vivid detail. In order to understand, control, and direct market behavior, the proliferation and availability of data on who and where we are has effectively opened up all spaces to statistical and informational analysis, erasing the traditional ways of comprehending space around us.

In locational decision-making and investment targeting, for example, retailers and banks now use GIS and geodemographic targeting systems – which add census and consumer profile data – into store investment and disinvestment decisions (Goss, 1995). A growing industry exists, supporting the sophisticated geodemographic profiling of census tracts and households with the now familiar litany of clusters – 'affluent achievers', 'urban venturers' 'have nots' etc. These are precisely mapped onto urban spaces, to aid locational decision-making, direct marketing and customer targeting. Such consumer profiling information is also valuable in its own right as a commodity to be traded within the burgeoning 'network marketplace'. The largest private 'information bureaux' in the USA, for example – TRW, Equifax and Trans Union – make large profits by maintaining detailed birth, family, address, telephone number, social security and salary history, credit and transactions, mortgage, bankruptcy, tax and legal records of US citizens; in effect, intimate, spatially referenced, software-based sortings of the whole of US society.

Thus the city itself becomes a software-based simulation, a fine-grained dynamic map of consumption and spending potential, as the large geodemographic bureaux now attempt to capture more and more direct consumption information, into GIS-based 'data warehouses', from store credit cards, credit bureaux, direct marketing campaigns, Internet responses and the like. Goss (1995) outlines how such GIS-based systems become transformed from partial representations and simulations of reality, to effectively operate as *reality itself* – the basis for precise locational decision-making and profit-driven targeting within large retailers:

> the GDIS (Geo-Demograpic Information Systems) is literally represented as a construction, a 'built environment' consistent, of course, with the architectonic metaphors so pervasive in the discourse on information technology. This architectonic metaphor effectively gives substance to a language, reifying the binary code that represents information as an alternate world, literally a data 'structure' . . . The abstract data structure is then anchored to a direct representation of reality, which leads to the conceit that the world of the GDIS is itself another reality . . . Here is the perfect edifice for strategy, an ironic doubling of the interiority–exteriority relationship. A representation of the 'exteriority' of the world is interiorized on the computer. The world of the 'other' and its identity have been captured and contained on a spatial grid by the machine technology, where it can be systematically observed and manipulated, by the strategy and power on the other side of the screen (Goss, 1995, 143–4).

The strategies which result from the use of such surveillance-based simulations, more often than not, involve the withdrawal of banks, retail outlets, or financial investment from poorer areas, and the careful 'cherry picking' of the best locations from the socio-spatial matrix of the city (see Speak and Graham, 1999). In the UK, for example, the main retail banks have used GIS techniques to select the least profitable branches as part of their withdrawal of over 50 per cent of bank branches since the 1970s. In the United States, advanced trends towards the 'cherry picking' of lucrative markets and the 'social dumping' of marginal consumers were noted by Susan Christopherson (1992). As US banks have restructured to address profit crises, they have attempted to:

> withdraw or increase revenue from routine transactions-intensive markets and focus on markets with the potential for higher value-added transactions. As banks have established minimum account balances and imposed fees on small accounts, increasing numbers of people, especially in poor neighbourhoods and communities, have had

to forgo basic banking services such as current accounts (Christopherson, 1992, 28).

Similar trends can also be discerned in food shopping (Christopherson, 1992). Wider spatial shifts in shopping towards out-of-town centres reliant on customers having access to cars exacerbate this growing social unevenness in access to basic services. In the USA, careful locational decision-making ensures that new out of town superstores are located as close as possible to the peri-urban zones of maximum market potential. Bloomingdale's and Nordstrom, for example, use GIS techniques to 'look for sites with a 10–15 mile radius that contains at least 10,000 households with an average income of $60,000, or for smaller concentrations of high-income households' (Christopherson, 1992, 282). In the UK, mean while, 10,000 small local shops closed between 1954 and 1994 whilst 750 superstores were opened in carefully selected locations with the highest market potential. Since then these trends have accelerated further.

But GISs also increasingly shape how people consume places and neighbourhoods. In the UK, for example, the web site upmystreet.com displays geographical information, by post code, on social make up, house prices, educational attainment, council services, planning information, and environmental quality. Thus, places become increasingly constructed through consumer decisions which, in turn, are influenced by the web-based surveillance, and sorting, of cities – a process which is heavily influenced by real estate interests. This furthers the advantages of ICT-literate and ICT-connected populations over others. As yet another example of how ICTs are being 'knitted' into places, however, this web site has started to support community emails and bulletin boards for individual post code zones.

BYPASSING BORDERS: BIOMETRIC SURVEILLANCE FOR THE GLOBAL 'KINETIC ELITE'

Finally, algorithmic surveillance techniques now allow highly mobile and affluent business travellers to directly bypass normal arrangements for immigration and ticketing at major international airports. This allows them to seamlessly, and speedily, move through the complex architectural and technological systems designed to rigidly separate air side and ground sides within major international airports (Virilio, 1991, 10).

After a pioneering agreement, for example, hand geometry scans for the most frequent business travellers are now in operation in major airports linking the US, the Netherlands, Canada and Germany and other OECD nations under the Immigration and Naturalization Service Passenger Accelerated Service System (INSPASS). Selected premium travellers are issued with a smart card that records their hand geometry, 'Each time the traveller passes through customs, they present the card and place their hand in a reader that verifies their identity and links into international databases', allowing them instant progress (Banisar, 1999).

By 1999, the scheme had 70,000 participants and the INS were already planning to extend the system globally. Such systems back up the extending infrastructure of highly luxurious airport lounges and facilities only accessible to elite passengers carrying special passes. In the wake of September 11th, 2001 such schemes are proliferating (see David Lyon, p. 299). For example, the Privium system, installed at Amsterdam's Schipol airport in 2001, scans the irises of frequent and business flyers to speed them through immigration controls.

Such configurations of surveillance and ICT systems privilege some users whilst those deemed to warrant less (or no) mobility (especially illegal immigrants and refugees) face ever-increasing militarised control and surveillance efforts to make international boundaries *more* rigid through new border control systems. The world's kinetic elite – to use the term coined by the German philosopher Peter Sloterdijk – thus are able to transgress national borders at the very same time that 'illegal' migrants and refugees find that such borders become more fortress-like because of other configurations of the same ICT and surveillance technologies (see, for example, Verstraete, 2001).

CONCLUSION

Within the context of the privatisation of urban and mobility spaces across the world, ICTs are being socially shaped in cybercities in two very different ways. On the one hand, the surveillance and monitoring capacities of ICTs are being shaped to prioritise and enhance the power and mobilities of privileged human bodies within the many scales of global neoliberal capitalism.

On the other hand, ICTs are being configured to *add* friction, barriers or logistical costs to the mobility and everyday lives of those deemed by dominant states or service providers to be risky, unprofitable, or undeserving of mobility.

Crucially, these techniques of prioritisation and inhibition are often so invisible and automated that neither the losers nor the beneficiaries are even aware that they are in operation within the complex socio-technologies of cybercities (Graham and Wood, 2003). Thus, the broadly standardised infrastructure and mobility services and rights to public space that were a key part of the Keynesian and Fordist welfare state in many nations are often being replaced by consumerist notions of a wide spread of individualised and market-based entitlements to the rights of mobility and of the city. Automated software increasingly polices these judgements as cross-subsidies are abandoned between profitable and unprofitable users and areas in the desire to extract profits only from those who can provide it.

Importantly, the use of software sorting to 'unbundle' public spaces and services is extending rapidly. Public spaces and malls in the UK and USA are increasingly being sorted through face recognition CCTV. Mutual health insurance threatens to be unbundled through individualised assessments of genetic risk based on genomic research. And even retailers have debated the use of software sorting. In 1999 UK supermarkets considered using ICTs to bring in a two-tier pricing policy which would use high prices at peak times to shift the cash-poor/time-rich citizens of cities out of the way of cash-rich/time-poor ones.

As a result of these processes of change, the clear risk is that the previously largely integrated public domains, infrastructures and spaces of cities will become progressively 'splintered' in socio-technical terms (Graham and Marvin, 2001). Rapidly emerging are separated socio-technical realms of premium and marginalised citizens that are structured so that citizens experience totally different life-chances, access rights, and service qualities based on the continuous and automated judgements of vast arrays of (often unknowable) software. Whilst sometimes these splintering domains are clearly visible, more often they are totally invisible. Their agency can even be missed by the people who are sifted by them – either the winners or the losers. Increasingly, then, the politics of the right to the city amount to the hidden politics of code as the agency of software structures urban access and exclusion in subtle but powerful ways (Thrift and French, 2002).

This is reflected in the widening range of local social movements which centre on remaking the local politics of digital information within cities in more egalitarian, democratic and transparent ways. Most notable here is Neighborhood Knowledge Los Angeles (see http://nkla.ucla.edu) which uses progressive GIS-based analyses, accessible over the Internet, to struggle against the ways in which banks and housing suppliers use the very same technologies to systematically disinvest of marginalised neighbourhoods within LA (see Krouk et al., 2000). Here the hope is that urban software applications like NKLA might help to reconnect and unify cities that face splintering via mainstream applications of ICTs. The NKLA project 'aims to demonstrate that technology can help connect and mobilize neighborhoods across the multi-ethnic, automobile dependent municipal region' of LA (see http://nkla.sppsr.ucla.edu/tiiap.htm).

REFERENCES FROM THE READING

Amin, A. and Thrift, N. (2001), *Cities: Reimagining the Urban*, Cambridge: Polity.

Banisar, D. (1999), 'Big brother goes high-tech', *CAQ Magazine* (available at www.worldmedia.com/caq/articles/brother.html).

Bowker, G. and Leigh-Star, S. (1999), *Sorting Things Out*, Cambridge, MA: MIT Press.

Calabrese, A. and Borchert, M. (1996), 'Prospects for electronic democracy in the United States: Rethinking communication and social policy', *Media, Culture and Society*, 18, 249–68.

Christopherson, S. (1992), 'Market rules and territorial outcomes: the case of the United States', *International Journal of Urban and Regional Research*, 17(2), 274–88.

Goss, J. (1995), '"We know who you are and we know where you live": The instrumental rationality of geodemographic systems', *Economic Geography*, 71(2), 171–98.

Graham, S. (1998), 'Spaces of surveillant-simulation: new technologies, digital representations, and material geographies', *Environment and Planning D: Society and Space*, 16, 483–504.

Graham, S. and Marvin, S. (2001), *Splintering Urbanism: Networked Infrastructure, Technological Mobilities and the Urban Condition*, London: Routledge.

Graham, S. and Wood, D. (2003), 'Digitising surveillance: categorisation, space and inequality', *Critical Social Policy*, 23(2), 227–46.

Krouk, D., Pitkin, B. and Richman, N. (2000), 'Internet-based neighborhood information systems: a comparative analysis'. In M. Gurstein (ed.), *Community Informatics: Enabling Community Uses of Information Technology*, Hershey, PA: Idea Group. 275–97.

Lyon, D. (2002), *Surveillance as Social Sorting: Privacy, Risk and Automated Discrimination*, London: Routledge.

Speak, S. and Graham, S. (1999), 'Service not included: Marginalised neighbourhoods, private service disinvestment, and compound social exclusion', *Environment and Planning A*, 31, 1985–2001.

Thrift, N. and S. French (2002), 'The automatic production of space', *Transactions of the Institute of British Geographers*, 27(4), 309–35.

Tseng, E. (2000), 'The geography of cyberspace'. Mimeo.

Tsung Leong, S. (2001), 'Ulterior spaces'. In C. Chung, J. Inaba, R. Koolhaas and S. Tsung Leong (eds), *Harvard Design School Guide to Shopping*, Cologne: Taschen, 765–74.

Utility Week (1995), *Special Issue: IT in Utilities*, 19 November.

Verstraete, G. (2001), 'Technological frontiers and the politics of mobility in the European Union', *New Formations*, 43, 36–43.

Virilio, P. (1991), *The Lost Dimension*, New York: Semiotext(e).

Winseck, D. (2003), 'Netscapes of power: convergence, network design, walled gardens, and other strategies of control in the information age'. In D. Lyon (ed.), *Surveillance as Social Sorting*, London: Routledge, 176–98.

Editor's suggestions for further reading

Curry, M. (1998), *Digital Places: Living With Geographic Information Technologies*, London: Routledge.

Demos (1997) *The Wealth and Poverty of Networks*, Collection Issue 12.

Goss, J. (1995), '"We know who you are and we know where you live": the instrumental rationality of geodemographic systems', *Economic Geography*, 71(2), 171–98.

Goss, J. (1995), 'Marketing the new marketing: the strategic discourse of geodemographic information systems'. In J. Pickles (ed.), *Ground Truth: The Social Implications of Geographic information Systems*, London: Guilford, 130–70.

Graham, S. (1998), 'Spaces of surveillant-simulation: new technologies, digital representations, and material geographies', *Environment and Planning D: Society and Space*, 16, 483–504.

Graham, S. (2002), 'Bridging urban digital divides? New technologies and urban polarization', *Urban Studies*, 39(1), 33–56.

Jupp, B. (2001), *Divided by Information?*, London: Demos.

Lessig, L. (1999), *Code – And Other Laws of Cyberspace*, New York: Basic Books.

Lessig, L. (2001), *The Future of Ideas: The Fate of the Commons in a Connected World*, New York: Random House.

Phillips, D. and Curry, M. (2003), 'Privacy and the phenetic urge: geodemographics and the changing spatiality of local practice'. In D. Lyon (ed.), *Surveillance as Social Sorting*, London: Routledge, 137–52.

Pickles, J. (ed.) (1995), *Ground Truth: The Social Implication of Geographic Information Systems*, New York: Guilford.

TWO

PART THREE

Shaping
Cybercities?

Plate 32 New financial services buildings in the Pudong redevelopment area, Shanghai.
(Photograph by Stephen Graham.)

VIII CYBERCITY STRATEGY AND POLITICS

Plate 33 Kiosk advertising Copenhagen's 'cybercity' strategy, Denmark. (Photograph by Stephen Graham.)

INTRODUCTION

In this final section of the book we explore how the relations between cities and ICTs might be creatively shaped and what futures they might have.

Addressing the issue of cybercity strategies – our concern in this first section of Part 3 – we need to analyse the challenges posed by the need for prescription, policy and intervention at the interface of cities and ICTs (see Manuel Castells, p. 82). A range of key questions emerge here. How can strategies and initiatives best attempt to shape the complex articulations between city spaces and digital spaces? Can they bring in positive, virtuous relationships between these mutually constituting domains, rather than tendencies towards the parallel physical and electronic withdrawal into electronic and physical cocoons and capsular systems? What can local and urban strategies do to harness new media and communications technologies to their policy goals? How can policy approaches be developed which take into account the asymmetric knowledge between urban and ICT industry partners, and which adapt to the particular histories, governance arrangements, and cultures and economies of different cities? Finally, how are urban ICT strategies being shaped by their inevitable embeddedness within broader transformations in the politics, and political economies, of urban governance, planning, social movements and new media?

Difficulties abound in developing crosscutting urban ICT and new media strategies within public policy. As we have seen in the rest of this book, the links between cities and ICTs are slippery and all-encompassing. Everything to do with cities in some way involves information flows, communications, transactions and representations that can be mediated, at least in part, by ICTs. This means that focusing on ICTs as a policy instrument in cities brings with it an extraordinary degree of what sociologists of technology called 'interpretive flexibility' (Bijker *et al.*, 1989). Urban ICT projects can be 'problematised' by technological entrepreneurs to virtually every possible stakeholder within the city because every possible aspect of urban life involves communication, information exchange, transactions, or electronic representations. This means that it is relatively easy to convince a wide range of partners to 'come on board an urban ICT initiative. But the very disparate and wide-ranging coalitions that can easily result can lead to problems of a lack of focus. In addition, the fact that everyone involved thinks the project addresses their needs and priorities, rather than those of other members of the coalition, can create major political conflict.

As a result of these problems, coordinating and developing the sorts of crosscutting perspectives necessary to think about cities and ICTs in an integrated way is extremely difficult. Responsibilities for urban spaces and policies are usually separated institutionally, professionally, and paradigmatically from those which shape and regulate ICTs and new media. In many cases they also tend to operate at different spatial levels of governance. Finding common linguistic, conceptual and policy frameworks to deal with the subtle interplays between urban places and ICTs, in an integrated and imaginative manner, is therefore a massive challenge.

In addition, both cities and ICTs are currently being restructured as key sites within the neoliberalising global economy. The result is that many new media and ICT infrastructures (and urban spaces) in cities are now intensively private and in the hands of ruthlessly secretive transnational companies. Even finding out what a city's ICT infrastructure actually *is* can be impossible for urban public policy-makers as a result. (The Editor knows this from his own policy experience in the UK city of Sheffield between 1989 and 1992.)

The extraordinary rate of change, and the mind-bendingly esoteric language and terminology within the ICT world, makes decisive and effective urban ICT policy-making even more problematic.

Urban policy-makers and planners are usually concerned with the visible and tangible dimensions of cities. But ICTs remain largely invisible and intangible. The explosion of digital exchange particularly challenges traditional urban and planning perspectives oriented towards land use, physical form, urban design, and transportation – ways of thinking about modern, industrial-age cities that have long neglected the importance of electronic communications and technologies in urban life. Furthermore, the prevailing cultures of technological determinism often leave city policy-makers with extremely unrealistic ideas which imply that ICTs autonomously have 'impacts' on the social, economic or cultural development of their cities. This leads to many policy failures. But because these embarrassing developments are rarely evaluated or analysed in detail their lessons are rarely learnt or disseminated. Thus the same traps tend to be fallen into again and again . . .

A final set of problems surround the limits which inevitably exist on the power of even progressive and creative urban policy-makers to alter the deep-seated structural dynamics which are broadly shaping the development of cybercities. The nature, geographies and politics and social relations of cities and urban regions are currently being transformed rapidly under the influences of neoliberal capitalism and the ICT technologies that are intrinsic to it. This limits the 'manoeuvring space' at the local level, to shift to different paths of city–ICT relations.

Broadly speaking, the diffusion of ICTs into urban spaces seem to be involved in trends towards fragmentation, sprawl, suburbanisation, individualisation and a withdrawal of social and political interchange from physical spaces into mediated spaces. However, it is dangerous to generalise because city ICT strategies are shaped by the wider traditions and cultures of state and public policy. These vary greatly between nations and urban areas. As we have seen in this book, in some cities, and some states, we also see trends towards the intensifying use of public space, and a growth of public, social and hybrid spaces where ICTs meet the city. It is therefore crucial to stress that diversity exists in cities' articulations with ICTs. It is also important to realise that the particularities of governance, planning, and urban design, as well as social movements, can significantly shape the detailed ways in which individual cities respond to these deeper structural transformations.

Despite the difficulties involved in developing cybercity strategies, the sheer importance of city–ICT relationships means that there has been a rapid recent proliferation of strategies, policies, initiatives, social movements, analyses and artists' interventions which we could define under the term. Crucially, these attempt to treat the politics, governance, planning or urban design of urban places and digitally mediated spaces of exchange in parallel – often for the first time. As social experiments in technological innovation which are rooted in the particularities of place, these interventions can significantly alter the cultural, social and economic dynamics that bring together cities and ICTs, with potentially positive and negative effects (McGrail, 1999; Wessels, 2000).

In what follows we will encounter a range of such cybercity interventions and strategies. These are drawn from Asia, North America and Europe. The interventions encountered display complex variations. Government-led cybercity strategies, for example, tend to reflect wider variations in the political particularities of nation-states, planning, urban design and media policy in each location. They demonstrate that particular policy imperatives have emerged to shape city–ICT policy discourses in various parts of the world. These tend to emphasise, to take a few examples, integrated market-based city–ICT and transport strategies as means to try and overcome the problems of urban sprawl in North America; public–private urban ICT interventions as means to try and revitalise urban public realms in Europe; and ambitious, developmentalist ICT-based urban development and marketing efforts as means to sustain ambitious 'leap-frogging'-style urban economic development in Asia.

References and suggestions for further reading

Bellamy, C. and Taylor, J. (1998), *Governing in the Information Age*, Buckingham: Open University Press.

Bijker, W., Hughes, T. and Pinch, T. (1989), *The Social Construction of Technological Systems*, Cambridge, MA: MIT Press.

Borja, J. and Castells, M. (1997), *Local and Global: Managing Cities in the Information Age*, London: Earthscan.

Brook, J. and Boal, I. (1995), *Resisting the Virtual Life: The Culture and Politics of Information*, San Francisco: City Lights.

Castells, M. and Hall, P. (1994), *Technopoles of the World: The Making of 21st Century Industrial Complexes*, London: Routledge.

Corey, K. (1987), 'The status of the transactional metropolitan paradigm'. In R. Knight and G. Gappert (eds), *Cities of the 21st Century*, London: Sage.

Curry, D. (1998), *Digital Places: Living With Geographic Information Technologies*, London: Routledge.

Doheny-Farina, S. (1996), *The Wired Neighborhood*, New Haven: Yale University Press.

Dutton, W. (1999), *Society on the Line: Information Politics in the Digital Age*, Oxford: Oxford University Press.

Dutton, W., Blumler, J. and Kraemer, K. (eds) (1987), *Wired Cities: Shaping the Future of Communications*, Washington: Communications Library.

Dutton, W. and Guthrie, K. (1991), 'An ecology of games: the political construction of Santa Monica's Public Electronic Network', *Informatization In The Public Sector*, 1, 279–301.

Edgington, D. W. (1989), 'New strategies for technology development in Japanese cities and regions', *Town Planning Review*, 60(1), 1–27.

Everard, J. (2000), *Virtual States: The Internet and the Boundaries of the Nation State*, London: Routledge.

Gibbs, D. (1993), 'Telematics and urban economic development policies: time for caution?', *Telecommunications Policy*, May/June, 250–6.

Graham, S. (1992), 'Electronic infrastructures and the city: some emerging municipal policy roles in the UK', *Urban Studies*, 29(5), 755–82.

Graham, S. (1994), 'Networking cities: telematics in urban policy: a critical review', *International Journal of Urban and Regional Research*, 18(3), 416–31.

Graham, S. (1995), 'Cities, nations and communications in the global era', *European Planning Studies*, 3(3) 357–80.

Graham, S. and Marvin, S. (1996), *Telecommunications and the City: Electronic Spaces, Urban Places*, London: Routledge, Chapter 7.

Graham, S. and Aurigi, A. (1997), 'Urbanising cyberspace? The scope and potential of the virtual cities movement', *City*, 7/8, 18–39.

Graham, S. and Dominy, G. (1991), 'Planning for the information city: the UK case', *Progress in Planning*, 35(3), 169–248.

Graham, S. and Marvin, S. (1999), 'Planning cyber-cities? Integrating telecommunications into urban planning', *Town Planning Review*, 70(1), 89–114.

Graham, S. and Marvin, S. (2000), *Splintering Urbanism: Networked Infrastructures, Technological Mobilities and the Urban Condition*, London: Routledge, Chapter 8.

Guthrie, K. and Dutton, W. (1992), 'The politics of citizen access technology: the development of public information utilities in four cities', *Policy Studies Journal*, 20(4), 574–97.

Hepworth, M. (1990), 'Planning for the information city: the challenge and response', *Urban Studies*, 27(4), 537–58.

Hepworth, M. (1991), 'Information cities in 1992 Europe', *Telecommunications Policy*, June, 175–81.

Horan, T. (2000), *Digital Places: Building Our City of Bits*, Washington, DC: Urban Land Institute.

Keeble, L. and Loader, B. (eds) (2001), *Community Informatics: Shaping Computer-Mediated Social Relations*, London: Routledge.

Komninos, N. (2002), *Intelligent Cities: Innovation, Knowledge Systems, and Digital Spaces*, London: Spon.

Ishida, T. and Isbister, K. (1998), *Digital Cities: Technologies, Experiences and Future Perspectives*, Heidelberg: Springer.

Laterasse, J. and Pauchard, H. (1995), 'Information systems and territorial administration: a new power struggle'. In P. Healey, S. Cameron, S. Davoudi, S. Graham and A. Madani-Pour (eds), *Managing Cities: The New Urban Context*, London: Wiley, 153–68.

Liff, S. and Steward, F. (2001), 'Community e-gateways: locating networks and learning for social inclusion', *Information, Communications and Society*, 4(3), 317–40.

Loader, B. (ed.) (1997), *The Governance of Cyberspace*, London: Routledge.

Loader, B. (1998), 'Welfare direct: informatics and the emergence of self-service welfare?' In J. Carter (ed.), *Postmodernity and the Fragmentation of Welfare*, London: Routledge, 220–36.

McDowell, S. (2000), 'Globalization, local governance, and the United States' Telecommunications Act of 1996'. In J. Wheeler, Y. Aoyama and B. Warf (eds), *Cities in the Telecommunications Age: The Fracturing of Geographies*, London and New York: Routledge, 112–29.

McGrail, B. (1999), 'Communications technology, local knowledges, and urban networks: the case of economically and socially disadvantaged "peripheral" housing schemes in Edinburgh and Glasgow', *Urban Geography*, 20(4), 303–33.

Malina, A. (2002), 'Community networking and perceptions of civic value', *Communications*, 27, 211–34.

Mandlebaum, S. (1986), 'Cities and communication: the limits of community', *Telecommunications Policy*, June, 132–40.

Obitsu, H. and Nagase, I. (1998), 'Japan's urban environment: the potential of technology in future city concepts'. In G. Gloany, K. Hanaki and O. Koide (eds), *Japanese Urban Environment*, Oxford: Pergamon, 324–36.

Raab, C., Bellamy, C., Taylor, J., Dutton, W. and Peltu, M. (1996), 'The information polity: electronic democracy, privacy, and surveillance'. In W. Dutton (ed.), *Information and Communications Technologies: Visions and Realities*, Oxford: Oxford University Press, 284–98.

Roberts, M., Lloyd-Jones, T., Erickson, B. and Nice, S. (1999), 'Place and space in the networked city: conceptualizing the integrated metropolis', *Journal of Urban Design*, 4(1), 51–67.

Rosenau, J. and Singh, J. (2002), *Information Technologies and Global Politics: The Changing Scope of Power and Governance*, New York: State University of New York Press.

Schmandt , J., Williams, F., Wilson, R. and Strover, S. (eds) (1990), *The New Urban Infrastructure: Cities and Telecommunications*, New York: Praeger.

Schuler, D. (1996), *New Community Networks: Wired for Change*, New York: Addison Wesley.

Southern, A. (2000), 'The political salience of the space of flows: Information and communications technologies and the restructuring city'. In J. Wheeler, Y. Aoyama and B. Warf (eds), *Cities in the Telecommunications Age: The Fracturing of Geographies*, London and New York: Routledge, 249–66.

Strover, S. (1988), 'Urban policy and telecommunications', *Journal of Urban Affairs*, 10(4), 341–56.

Strover, S. (1989), 'Telecommunications and economic development: an incipient rhetoric', *Telecommunications Policy*, September, 194–6.

Tsagarousianou, R., Tambini, D. and Bran, C. (eds) (1998), *Cyberdemocracy: Technology, Cities and Civic Networks*, London: Routledge.

Wessels, B. (2000), 'Telematics in the East End of London: new media as a cultural form', *New Media and Society*, 2(4), 427–44.

Williams, R. (2000), 'Public culture and social learning: the new multimedia technologies in Europe', *Information Society*, 16, 251–62, (introduction to a special issue).

'Planning Cyber-Cities? Integrating Telecommunications into Urban Planning'

Town Planning Review (1999)

Stephen Graham and Simon Marvin

Editor's Introduction

The discipline of modern urban planning developed to concentrate overwhelmingly on physical urban spaces and physical transportation. Until recently it has almost completely neglected the immaterial mobilities and communications flows that are such an intrinsic element of contemporary urban life (Graham and Marvin, 1996, 2001). As a result, the disciplinary foundations of urban planning face a paradigm crisis as its preoccupations with orderly physical spaces and configurations seem less and less in tune with the diverse socio-technical worlds of mobility, flow and multi-scale interconnection found within and between cybercities.

The response of urban planning professions and educators to this crisis has generally been slow. This has not been helped by the largely invisible nature of ICTs and telecommunications infrastructures in cities, or by the lack of statutory policy instruments through which city authorities in many nations can engage in this new field. Nevertheless, a growing range of urban policies and strategies are beginning to emerge within which governance partnerships and urban municipalities are trying to shape the strategic interplay of ICTs and urban spaces.

Our first reading on cybercity politics and strategy draws on one of the results of the long-standing partnership between the Editor and Simon Marvin, an urbanist at Salford University in the UK. This reading attempts to provide an international snapshot of some of the urban ICT strategies that emerged in the late 1990s. The authors explore strategies which treat the interactions of transport and telecommunications in cities; city-level new media and ICT strategies; and strategies designed to boost the development of new or old urban neighbourhoods by developing them as 'Information Districts' and 'Urban Televillages'. In each category Graham and Marvin analyse the ways in which strategies and policies are trying to creatively harness the powers of ICTs to support specific developments within cities or parts of cities.

Graham and Marvin particularly stress the crucial symbolic importance of labelling and branding cities and urban districts with 'cyber', 'silicon', 'intelligent', 'digital', 'mutimedia' or 'information' prefixes (see Jessop and Sum, 2000). For often the symbolic allure of ICTs is as important as real hardware in the development of these initiatives. A classic example of this was the case in the late 1980s when the City of Edinburgh seriously considered building a fake satellite dish on a new business park. After an evaluation they found that, technically, the city's satellite services were already globally 'competitive'. For a while the City Council persisted with the idea of a dish anyway. This was because it was a rare instance of an ICT infrastructure emerging, above ground, to be visible and so demarcate the supposedly high-tech, switched-in modernity of its location. 'The location of a dish on the site', reported a European Commission study in 1991, 'would be symbolic rather than of practical use' (see Graham, 2000).

Whilst broadly welcoming the range of experiments analysed in the Reading, this preoccupation with symbolism, branding, place marketing and the subliminal allure of ICTs also makes Graham and Marvin sceptical. They worry that truly progressive and redistributive urban ICT strategy making may not be possible. Instead, urban ICT strategies may really amount to a range of initiatives which simply add value to already privileged spaces, districts and real estate developments – and so users – within cities. Thus, such initiatives may further marginalise those areas which are already being bypassed and marginalised by contemporary urban trends towards liberalisation, globalisation, and the growth of cybercity economies (see Vincent Mosco, p. 199; Tim Bunnell, p. 348).

Clearly, it is no longer adequate to consider policies for cities and those for telecommunications and new media entirely separately. Only through addressing the complex interactions between cities and telecommunications will the potential of the technology be realised. This realisation is currently leading to a wide range of policy experiments which aim to positively shape how the new media relate to specific cities or parts of cities. Early examples can be drawn from a 'bricolage' of evidence, drawn from cities in the USA, Canada, Malaysia, Europe, the UK and elsewhere around the world. These examples have not yet coalesced around a coherent new paradigm of urban policy. Many can be criticised as technologically determinist, environmentally problematic, or socially exclusionary.

But we would argue that, together, they point to a new style of planning and urban policy. Such urban strategies try to shape face-to-face interactions in place (and the transport flows that sustain these) in parallel with electronically mediated ones across distance. Currently, we can identify three emerging styles of such 'urban telecommunications planning': integrated transport and telecommunications strategies, city-level new media strategies, and so-called 'information districts' and 'urban televillages'.

INTEGRATED TRANSPORT AND TELECOMMUNICATIONS STRATEGIES

The first set of initiatives attempt to shape and manage the relations between physical movement and mobility through the application of new media, combined within particular forms of urban physical development. Each embodies a particular conception of the relations between different forms of communication and their role in the development of the city.

Urban and regional teleworking initiatives

First, there are Urban and Regional Teleworking Initiatives. There are a growing set of initiatives, especially in the USA, that are attempting to grapple with the problems of developing a metropolitan-region approach to teleworking to make a positive contribution to environmental improvement – particularly by reducing vehicle emissions. For example, Telecommunications for Clean Air is a two year programme funded by the Californian South Coast Air Quality Management District to use telecommunications to meet rigorous air quality standards. The main aims of the programme are to identify cost-effective solutions to air quality and congestion, to contribute to economic growth, and to develop a regional approach to problem solving.

The Telework Facilities Exchange is designed to expand telecommuting participation in local government by providing low-cost, flexibly located facilities and marketing these practices to other organisations joining the exchange. A public sector employee would normally commute 35 miles each way to their office. Instead, they commuted to a vacant office a few miles from their home to use a workstation connected to their office. Those workers participating in the programme reduced their normal vehicle miles travelled by 88 per cent; as a result, if 30 per cent of the region's 484,000 local government employees each worked one day a week at the exchange, nearly 500,000,000 vehicle miles travelled could be saved each year.

Communication corridors

Second, there are new forms of communication corridor strategies. These attempt to shape how telecommunications, transport and land use interplay within broader urban commuting corridors. Those based on existing rail/transit networks attempt to manage travel demand both on the road and rail network through the provision of teleworking centres and incentives to travel off-peak. The Metro Net initiative in Los Angeles, for example, involves retrofitting a high capacity fibre optic network alongside the 300-mile regional Metro Rail and Link network. The proposals are designed to achieve three objectives: to generate revenue through leasing capacity, to develop services to enhance ridership of the system, and to improve regional mobility through developing station-based employment and service centres. The mobility strategy has been designed to fit in with the objectives of a series of wider regional communications and land use strategies. More specially, it is hoped that the strategy will contribute towards the implementation of the Air Quality Management Plan, the Development Plan, the Regional Congestion Management Program, and the promotion of regional economic growth. The proposals would develop telecommunications facilities at or near stations for conferences, education, and job-training, to make the Metro system a destination itself.

Road transport informatics

The third type of initiative focuses on the development of Road Transport Informatics (RTI). City-wide initiatives are rapidly emerging here, concerned with the use of RTI systems to more effectively manage transport networks. There are major initiatives in Europe, the USA, and Japan and the UK's National Economic Development Council estimate that the global IT and traffic management market will be worth £29 billion in 2010. However, objectives of RTI are often poorly defined and are not often closely linked with land use and work patterns. These initiatives are more often characterised by a form of 'technical fix', dominated by strong producer-led interests. RTI strategies have assumed importance in the context of EU funding programmes where the technologies are seen as making a major contribution to sustainable development.

There have been a large number of feasibility schemes evaluating the potential of various forms of RTI and electronic tolling in the UK. In February 2003 the world's largest central area ICT-based road pricing scheme was introduced in London. Electronic Road Pricing (ERP) is also being considered more widely in North America. The Highway 407, which is currently being built in Toronto, is billing itself as 'Tolls But No Jams' (*Toronto Star* 29 July 1996). Located in one of the most congested highway corridors in North America, a $1 billion dollar 36 km highway will eventually connect the airport to downtown Toronto. The scheme is being developed by Canadian Highways International Corporation – a private consortium of four companies who are funded by the receipts from electronic tolls. In return for the higher charges, users benefit from higher road speeds than the current limit, and no traffic jams. If demand increases, the highway will be expanded to ten lanes and tolls will be raised to reduce peak travel volumes. The scheme is being marketed at those firms operating just-in-time production methods who require high degrees of certainty in travel times for the movement of goods and services. There is now major development interest in highway land involving commercial, retail, leisure and recreation and housing adjacent to the new road. This initiative is an interesting example of the combined planning of electronic, transport and landuse infrastructure, designed to develop a congestion-free, higher-speed and lower travel-time corridor through the congested region. But this new development trajectory is extremely socially exclusionary. It is very much designed to meet the needs of large international corporations and elite users prepared to pay the premium for increased certainty [see David Holmes, p. 173].

Other RTI initiatives focus on the development of local and regional initiatives in driver information and control systems. These initiatives are based on proposals to carry more traffic by making better use of the existing road network, through pre-trip planning, route guidance, traffic management and control, and network management applications. It is hoped that the provision of information to drivers on road conditions can increase the efficiency of the network to minimise delays, unreliability and environmental damage. For instance, it is estimated that driver information could increase the capacity of the road network by 1 per cent, provide a 10 per cent saving in journey times, and a 6 per cent reduction in mileage. The Scottish consultative document argues that the region is well placed to use these technologies because 80 per cent of the population live within a relatively self-contained belt across the country.

CITY-LEVEL NEW MEDIA AND IT STRATEGIES

The second broad emerging policy area is city-wide new media strategies. IT strategies for community networking, local economic development, and public service delivery have been underway in many UK, European, and American cities for a decade or more. Following American experience, community networks like Free-nets, the Manchester, Kirklees, and Nottingham Host computers, and the Newcastle NewNet system, based on the Internet, have emerged which try to use computer communications to support grassroots, local economic, and voluntary activities. Many local authorities are also experimenting with electronic kiosks and smart-card systems to deliver information on public services, and aim to improve the services themselves.

Local services have developed patchily on the new urban cable networks developing across the UK. Virtually all major UK cities now have a presence on the Internet, where so-called virtual cities range from simple tourist promotion and local databases, to sophisticated spaces which attempt to add coherence to all local activities on the Internet, to widen local access and skills, open up interactive services for local debates, and to develop information and communications services which feed back positively onto the development of the home city. Interestingly, the most innovative virtual cities use the analogies of city spaces, squares and districts, so that the many services they offer relate directly to their counterparts in physical urban space. The most sophisticated of these in the UK currently is Virtual Bristol, supported by a partnership of the City Council, Universities, and Hewlett Packard, and launched in April 1997. Not to be left out, BT is exploring the concept of urban intranets – Internet services that are only accessible to specified local communities.

This disparate range of local new media initiatives have two problems, however. First, they have tended to be fragmented local IT 'islands', largely ignoring each other. And second, they have usually been developed with little or no respect for how they relate to the physical urban realm or to the broader development dynamics and geographies of their subject cities. Thus, the challenge for UK cities is to shape coherent partnership-based strategies aimed at harnessing all types of new media applications – Internet, cable, kiosks, telephone, infrastructure – to

their economic, social and cultural development needs Such issues need to begin with social, geographical and institutional matters and policy needs and move onto how new technologies might meet these needs – rather than the other way round. Institutional solutions need to be found that harness the entrepreneurial energies of the new media industries, and their growing interest in market-based local initiatives (like the booming commercial metropolitan Internet sites in the USA), whilst linking creatively and positively to the fragmented sets of agencies involved, in the broadest sense, in the governance of UK cities (local authorities, development agencies, health, education and information institutions, firms, schools, the community and voluntary sector etc.). Clearly, urban media master plans will be impossible: what is needed are strategic frameworks so that the innumerable local media investments and initiatives emerge to be more than the sum of their parts.

In the UK, the packaging of IT infrastructure with individual land use developments – business parks, telecottages, wired villages etc. – is increasingly common. But efforts are also starting to link broader urban media strategies with urban-wide development strategies. After a period when grant, training and technological support was 'pepper-potted' through the city, Manchester is increasingly gearing its broad telecoms and IT initiatives to specific urban redevelopment and reuse projects, and to strategic discussions about combating social exclusion in the city. A widening range of new physical, IT-oriented projects have emerged, linked into the network services on offer: the electronic village halls (linked to community centres and initiatives through the city), existing managed workspaces in New Mount Street, and proposed ones in the Northern Media Quarter, and a centre for Multimedia development and applications in Hulme.

'INFORMATION DISTRICTS' AND 'URBAN TELEVILLAGES'

The final area where new media policy is becoming directly linked with policies for particular urban spaces is the emergence of information districts and urban televillages. Building on the debates about urban villages in Europe, and the new urbanism movement in the USA, interest is growing rapidly in how media infrastructure and services can be designed and managed, geared to sustaining and feeding back on particular urban districts. In California, the concept of the 'televillage' – an integrated urban place supported by a whole suite of ICT infrastructures and services – is gaining support [Thomas Horan, p. 363; Walter Siembab, p. 366]. The Blue Line TeleVillage, a two square mile area on one of the new public transit corridors in LA, is based on a holistic strategy to manage land use, transport trips and electronic communications so that synergies emerge between the three, creating a liveable community with reduced automobile use, higher community-based activities, and higher urban densities than in the usual LA suburbs.

Physical places for supporting IT training and services – community centres, computer centres, telework centres, IT links in schools, hospitals, transport facilities and libraries, and electronic kiosks in public and semi-public spaces – are integral to the plan which is backed up by a broad, public–private–community partnership, and an extensive array of online public services. In partnership with the public transport operators in the LA region, a new fibre network is being developed to link together whole constellations of televillages across the region. Different packages of IT and telecoms infrastructure and services are being offered for different land uses; distributed organisations are being encouraged; and attempts are being made to include more marginalised social groups. The philosophy is that IT-based retrofitting in existing US urban areas will mean that many urban problems might be addressed with very little new physical construction and no dramatic changes in density (Siembab, 1995).

The other emerging example of combining new media and urban regeneration at district level is the concept of the information district. Here, the emphasis is on creating urban 'milieux' that sustain economic growth in new cultural and symbolic industries, where informal face-to-face contact is essential, whilst also providing high capacity online linkages to the wider world [see Vincent Mosco, p. 199; Matthew Zook, p. 205; Rebecca Solnit and Susan Shwartenberg, p. 296]. Most often, information district strategies emerge organically – as in New York's Silicon Alley, Dublin's Temple Bar, and Manchester's Northern Quarter – where clusters of such industries develop spontaneously in inner urban districts. Then, the challenge is to intervene to further support the growth of small and micro firms in the relevant sectors, whilst also ensuring appropriate property is available and that greater efforts are made to improve the broader urban realm and the contribution of the industries to the economic and social revitalisation of the city as a

whole. Thus, both New York and LA have offered grant schemes and tax exemptions to small and micro firms in the new media sectors. Backed by the powerful New York New Media Association (NYNMA), specialised multimedia centres, offering managed workspaces and high-level telecoms bandwidth, have also started to emerge in Silicon Alley, as have dedicated venture capital funds and orchestrated events and programmes designed to encourage local face-to-face networking. Elsewhere in the USA, the city of Spokane, in Washington State, has wired up much of its downtown to attract multimedia firms.

In Europe, strategies at the neighbourhood and district level have begun to look to coherent interventions in urban design, new media, and, at the institutional level, to try to either sustain, develop or encourage, local clusters of multimedia firms. Manchester has explicitly adopted the Silicon Alley model to support its Northern Media Quarter, on the edge of the city centre. Sheffield's well known Cultural Industries Quarter (CIQ) strategy, aimed at clustering the broadest possible range of media, design, music, film and cultural-related industries firms in one part of the city centre, is now backed up by widening range of online services financed by a public private partnership called NEO. Dublin's Temple Bar district is backing up its physical regeneration efforts, weaving a parallel infrastructure for electronic, multimedia exchange.

In London's Soho media core, meanwhile, a specialised telecommunications network was recently constructed by a consortium of film companies called Sohonet. This system links the tight concentration of film and media headquarters in the district directly to Hollywood film studios, allowing online film transmission and editing over intercontinental scales, via highly capable, digital, broadband connections. The network is seen as a critical boost to the broader global ambitions of the UK film and cultural industries.

CONCLUSIONS

We would argue that the current growth of explicitly urban telecoms strategies and initiatives is broadly to be welcomed. This is for three broad reasons. First, they are acknowledgements that city–telecoms interactions are intrinsically bound up with contemporary metropolitan life. Second, these policies are based on much more sophisticated understandings of the complex and subtle relations between new media and urban life than prevailing 'death of distance' of 'end of cities' rhetoric. And third, these proliferating policies suggest that the articulations between urban spaces and new media technologies are open to innovative, local, and planned interventions which can bring benefits which neither untrammelled market forces nor distant central state hierarchies can deliver.

But what might these nascent policies mean for urban planning and urban development more broadly? Obviously, speculation is difficult in such an embryonic policy arena. In these conclusions we would therefore like only to address two key questions. First, are urban telecommunications initiatives likely to be able to succeed in shaping positive synergies between place-based interactions and development and electronically mediated interactions and development? Or are they merely stylistic and symbolic, aimed at add adding value and high-tech kudos to prestigious real estate developments and urban gentrification processes? Second, what might these initiatives mean for broader notions of the 'city' and for ideas of integrated metropolitan-wide planning? More particularly, are these initiatives likely to reinforce and recreate new forms of socio-economic exclusion and environmental damage, or might they genuinely emerge as useful attempts to develop a more inclusionary and sustainable urban future?

Turning to the first question, we must first sound some notes of caution. For, despite the widening range of initiatives, we remain highly suspicious of their current usefulness, in terms of both their magnitude and direction. In terms of their magnitude, we must raise serious warnings against *overstating* the potential role of telecommunications and information technology in urban strategies. Whilst most approaches to urban strategies are still grappling with new ways of planning for transportation grids and urban places in parallel, it is still the case that electronic interconnections and networks are most often still hidden and taken for granted.

In terms of direction, it is clear that even when land use, transport and telecommunications are considered in parallel, real progress will only come when two further problems are addressed. First, policy-makers will need to fight against prevailing assumptions (which are actually deeply embedded within western culture as a whole) that new technologies can somehow be rolled out as technical 'quick fix' solutions to complex urban problems. In a context where most urban policy-makers and planners lack knowledge and experience of the telecommunications sector, there is the danger

that urban strategies could uncritically embrace the transformational rhetoric that characterises contemporary notions of technology 'impacts' upon the city. New telecommunications initiatives are still often intimately connected with utopian and deterministic ideas of technology's purported beneficial and linear impacts upon the social, environmental and spatial development of cities. Developing more nuanced and sophisticated concepts of the potential roles of telecommunications in urban strategy will require policy-makers to look more critically at the role of technology in contemporary urban strategies.

Achieving this, however, is difficult for another reason. For powerful media and technology firms are exploiting the hyperbolic rhetoric of 'cyberspace', the 'information superhighway' and the 'global information society' to enrol poorly informed urban public policy-makers into making local 'partnerships' to develop new information districts, communications corridors, and 'high-tech' economic development zones of various sorts. Growing interurban competition, and the tendency for urban policy-makers to jump on the latest policy bandwagon, are being exploited by a wide range of consultants and media conglomerates. Such organisations are keen to add value and symbolic kudos to their own efforts to build up demand for new configurations of real estate, ICT infrastructure or developed spaces, and all the associated technological hardware and services embedded in them. Public subsidies, discounted land deals, infrastructural assistance, credibility and the sheer marketing weight of public policy-makers can do much to raise the profile of new, planned 'high-tech' spaces (and, therefore, developer profits).

But the real benefits of such initiatives to localities may be dubious or massively overblown because they remain inappropriate to real local needs. Thus, planners and local policy-makers need to educate themselves as quickly as possible about the burgeoning worlds of new media technologies. They need to be wary of being seduced into expensive partnerships of dubious real local benefit by the lustrous promises of information age hyperbole. It is here that critical local debate about the real communicational needs of urban places, and the policy models that derive from these, is necessary. As Sharon Strover argued way back in 1989 (195):

the ineluctable logic behind the equation of economic development with telecommunications is that if the favourable outcomes of such telecommunications initiatives are so overwhelming that the public at large should be expected to subsidise the costs . . . The seductiveness of the 'becoming modern' economic development and social goods rhetoric is great. And we have heard it all before. The language we might have forgotten was that applied to the cable industry throughout the 1970s and 1980s [see Thomas Streeter, p. 53].

Which brings us on to our second question: what might these initiatives mean for broader notions of the 'city' and for ideas of integrated metropolitan-wide planning? Here, too, there is a need for concern. In the light of the above discussion, we clearly need to unpack the social assumptions and biases built into current urban telecommunications initiatives. We need to ask how urban telecommunications initiatives might link to wider urban debates around social equity, the public realm and culture, economic development, and environmental improvement. With such a strong supply push from powerful media and real estate interests, there are clearly dangers that urban telecommunications strategies are being configured in highly biased ways that might perpetuate and reinforce widespread existing trends towards social and spatial polarisation in urban areas.

The danger is that the foci of initiatives will centre overwhelmingly on configuring new media technologies according to the needs and geographies of affluent, privileged nodes, spaces and corridors in metropolitan regions whilst ignoring and excluding marginalised zones. The former, of course, are already at the vanguard of IT applications and are the 'hot spots' of demand for all forms of telecommunications and media applications and services. In the context of liberalising telecommunications regimes, the risk is that market forces will encourage both corporate and media interests and urban policy-makers to invest their efforts in communications corridors for the highly mobile, information districts for the information elites, and media consumption spaces for affluent professionals with high disposable incomes. In short, urban telecommunications strategies may simply work to extend the existing relational privileges of powerful zones, spaces and interests in the city rather than countering the extremes of uneven development already emerging within ICT-based modes of contemporary urban development.

REFERENCES FROM THE READING

Siembab, W. (1992), *Metro Net, Fibre Optics and Metro Rail: Strategies for Development*, Los Angeles: Siembab Corporation.

Strover, S. (1989), 'Telecommunications and economic development: An incipient rhetoric', *Telecommunications Policy*, September, 194–6.

Editor's suggestions for further reading

Ducatel, K. (1994), 'Transactional telematics and the city', *Local Government Studies*, 20(1), 60–77.

Gdaniec, C. (2000), 'Cultural industries, information technology and the regeneration of post-industrial urban landscapes', *GeoJournal*, 50, 379–87.

Gibbs, D., Tanner, K. and Walker, S. (2000), 'Telematics, geography and economic development: Can local initiatives provide a strategic response?' In M. Wilson and K. Corey (eds), *Information Tectonics*, London: Wiley, 219–36.

Graham, S. (2000), 'Satellite dishes'. In N. Thrift and S. Pile (eds), *City AZ: Urban Fragments*, London: Routledge.

Graham, S. and Marvin, S. (1999), 'Planning cyber-cities? Integrating telecommunications into urban planning', *Town Planning Review*, 70(1), 89–114.

Horan, T. and Jordan, D. (1998), 'Integrated transport and telecommunication planning in Santa Monica', *Journal of Urban Technology*, 5(2), 1–20.

Jessop, B. and Sum, N.-L. (2000), 'An entrepreneurial city in action: Hong Kong's emerging strategies in and for (inter)urban competition', *Urban Studies* (37), 2287–313.

Roberts, M., Lloyd-Jones, T., Erickson, B. and Nice, S. (1999), 'Place and space in the networked city: conceptualizing the integrated metropolis', *Journal of Urban Design*, 4(1), 51–67.

Schuler, R.E. (1992), 'Transportation and telecommunications networks: planning urban infrastructure for the 21st Century', *Urban Studies*, 29(2), 297–309.

'Cyberjaya and Putrajaya: Malaysia's "Intelligent" Cities'

Tim Bunnell

Editor's Introduction

Since the late 1980s Asia has been the global centre for the development of grandiose cybercity strategies. This has been driven by the existence there of extremely ambitious and powerful developmental nation and local states; a widespread eagerness to shift from manufacturing to ICT-based urban service economies; and a deep cultural belief in the transformative powers of ICTs. Combined, these factors have spawned a wide range of massive and expensive ICT-centred urban transformation plans. As a result, some commentators have even talked of the 'Siliconisation' of Asia (Jessop and Sum, 2000).

Our second reading on cybercity politics and strategy is a specially commissioned analysis of the largest of all these strategies: the so-called Multimedia Super Corridor that runs 50 km south from Kuala Lumpur (Malaysia's old capital) (Vincent Mosco, p. 199). In this reading Tim Bunnell, a geographer working at the National University of Singapore, takes a critical look at the ways in which the MSC strategy has been legitimised and constructed. He looks in detail at the strategy's aims and explores its elements (which include the new cities of Putrajaya and Cyberjaya). He maps its gigantic proportions. And he cites the global technopole success stories that the strategy is attempting to imitate.

Bunnell's analysis is especially innovative for the way in which it unearths the discursive and symbolic, as well as the spatial and the material, sides of the construction of the MSC. Bunnell powerfully isolates the ways in which ICT technologies and practices have been invested with particular ideological meanings and symbolism. Bunnell shows how these resonate very powerfully with the aspirations of the ruling Malaysian elite (and especially the long-term Prime Minister, Mahathir). Preoccupied with making the transition from a 'developing' to a 'developed' nation by 2020 Mahathir has, in effect, concretised the national ambition into a vast tract of celebrated new urban landscapes. These have been constructed through a rural and semirural landscape without any thought for the area's residents. The utopian rhetoric that has accompanied the MSC strategy, with ICTs being constructed as transformative motors of change on the way to a universally enjoyed high-tech modernity, has been essential in allowing the strategy to reach implementation.

Thus, the contradictions inherent in the MSC strategy remain largely hidden. How, for example, can ICTs deliver benefits to all Malaysians if ICT-based development is concentrated in a small strip between the new KL airport and the old capital? What about the marginalised indigenous communities and plantation workers that have been forcibly evicted from the MSC site? And what about people who are unable to transform themselves into 'global knowledge workers' and so gain access to the high quality environments within the MSC?

The apparent vogue for 'wired' urban spaces in the Asia–Pacific region in the mid-1990s led two commentators to refer to a 'Siliconisation' of Asia (Jessop and Sum, 2000). Malaysia's contribution to this regional high-tech transformation takes the form of a 50-km corridor extending southwards from the national capital, Kuala Lumpur. At the geographical and imaginative core of the so-called Multimedia Super Corridor (MSC) are two new, so-called 'intelligent' cities, Cyberjaya and Putrajaya. Cyberjaya is promoted as the first major MSC-designated 'cybercity'. Initial development in a 2,800-hectare flagship zone aims to foster an urban environment which is attractive to high-tech investment and conducive for creativity. Putrajaya (Plate 34) has a rather more complex history associated with the relocation of federal government departments outside of Kuala Lumpur. The planning of what is now referred to as Malaysia's electronic federal government administrative centre predates the MSC concept, which was officially announced on 1 August 1996.

Since the early 1990s, the Malaysian federal government has played a key role in a range of large-scale urban development projects, most notably Kuala Lumpur City Centre (KLCC) and the new Kuala Lumpur International Airport (KLIA). The MSC cities thus form part of attempts to augment greater Kuala Lumpur's global visibility, connectivity and competitiveness. The MSC concept, in turn, added an 'information age' dimension to the global repositioning of Kuala Lumpur Metropolitan Area (KLMA). Apart from the launch of Cyberjaya, and as is repeatedly alluded to in MSC promotional material, the entire corridor is served by a 2.5 10-gigabit digital fibre optic and coaxial cable network with direct links to Japan, the US and Europe. The spatial concentration of this material and symbolic infrastructure in and around the country's existing main metropolitan area is, at one level, diagnostic of a broader logic of privileging key urban centres as would-be global nodes.

WHY 'INTELLIGENT' CITIES?

Cyberjaya and Putrajaya cannot be understood as merely part of broader attempts to plug KLMA as Malaysia's national node into transnational urban networks. There are two further ways in which the MSC cities were imagined as appropriate means to national development in Malaysia in the 1990s. The first concerns a broader national information technology (IT) push and attempts to foster technological

Plate 34 New Malaysian government buildings under construction in Putrajaya, 1999. (Photograph by Tim Bunnell.)

innovation compelled, in part, by regional economic competition. Having reached a ceiling position within a labour-intensive phase of economic development, investment in IT was intended to enable Malaysia to move to a higher technological niche from its regional manufacturing rivals. In Malaysia, as elsewhere, the innovation and creativity considered necessary for success in the information economy have been associated with a specifically *urban* complex, the technopole (Castells and Hall, 1994). The stated aim of Cyberjaya was to ferment a culture of innovation in which Malaysians could initially participate and to which they would ultimately contribute. Cyberjaya planners visited and sought to divine the secrets of other innovative urban milieux: Sophia Antipolis in France, Bangalore in India (p. 309), Japan's technopolis programme and, of course, Silicon Valley (p. 296). Cyberjaya, in turn, is intended as a model technopole or cybercity which can be cloned elsewhere in the national territory.

This connects with the second rationale for the MSC cities, concerning their role as testbeds for applications of new technology which, if successful, may be extended nationally. A range of 'Flagship Applications' are cast as experimental ingredients for the realisation of Vision 2020, Prime Minister Mahathir's goal of achieving fully 'developed' status by that year (see Bunnell, 2004). Electronic government trialled in Putrajaya, for example, has been depicted in terms of the irrelevance of physical distance to interaction between state bureaucracy and citizens affected by improved information flows and processes. Similar technological utopian promises were highlighted in relation to health care (through telemedicine) and education (via smart schools). Mahathir has thus depicted MSC as 'a pilot project for harmonising our entire country with the global forces shaping the Information Age' (Mahathir, 1998, 30). The putative potential for technological innovation to integrate the entire national territory was performed in the so-called Teleconferencing Dialogue held in April 1997, in which Mahathir was linked to 13,000 Malaysians in 28 locations across the nation.

These imaginings and representations of developmental areal uniformity did not go uncontested even in the technological and economic optimism of mid-1990s (and pre-economic crisis) Malaysia. Participants in the Teleconferencing Dialogue questioned what, if anything, rural areas had to gain from developments physically sited within high-tech cities in the nation's main metropolitan area. The announcement on the same day that the state of Sarawak in East Malaysia would create its own cyber village also implicitly identified an overdrawn opposition in MSC policy discourse between the informational and the real. On the one hand, universal (national) extension of benefits from applications of information and multimedia technology and teleological imaginings of a multimedia utopia benefiting all Malaysia(ns) is imagined possible in relation to technological utopian notions of liberation from the constraints of space and the frictional effects of distance. On the other hand, electronic technologies are dependent upon complex infrastructure in real places: cyber-development literally takes place. While the place of information economy development is acknowledged by plans to replicate the cybercity model in other parts of the national territory, such new cybercities would clearly be doomed to peripherality (if not failure) in the absence of the state-of-the-art infostructure and infrastructure supporting MSC. It is the relationship between Malaysia's intelligent cities and the national territory to which I now turn.

THE INTELLIGENT CITY – NATIONAL TERRITORY RELATION

The MSC cities' location in KLMA may be said to contradict long-held state regional development goals. Attempts to promote a more balanced regional development have been central to a post-colonial politics of ethnic wealth distribution in Malaysia. Political strategies undertaken by the Malay-dominated state to close the wealth gap with the other (Chinese and Indian) communities especially following the introduction of the New Economic Policy (NEP) in 1970 have been intertwined with efforts to reduce national urban/rural and regional development disparities. Despite these efforts, urban development since 1970 has been skewed towards the so-called western corridor of peninsular Malaysia. Penang has been a major destination for foreign investment in the electronics sector; indeed, prior to the development of MSC, it was Penang that had been dubbed Malaysia's Valley. Since the launch of the MSC, however, national high-tech resources and global high-tech attention have overwhelmingly focused on KLMA.

These urban–regional trends should not be considered symptomatic of a simple erosion of nation-state power in relation to the rise of wired cities as nodes in global networks. Rather than giving way to irresistible external forces and/or a new techno-economic paradigm, the Malaysian federal government has played a leading role in high-tech urban development. While MSC boasts an International Advisory Panel (IAP) comprising some of the most powerful figures in corporate IT, it is the federal government that makes possible liberal financial and labour incentives in the 50 × 15 km corridor. And it was the continued need for national political legitimacy that, in the wake of the economic crisis of 1997, saw Mahathir at once railing against the evils of economic globalisation while pressing ahead with development in the liberal economic MSC testbed. MSC is thus perhaps most accurately understood in terms of the reconfiguration (rather than reduction) of state power and, more specifically, its re-scaling to key metropolitan spaces.

Such re-scaling owes as much to the national politics of development as to supposedly more fundamental new global political economic realities. The location of key projects in and around the federal capital has obvious appeal when compared to more distant and often politically disputatious parts of the national territory. Yet central government has also extended the reach of its direct political authority through its high-tech city investment. On 1 February 2001, control of the 4,581-hectare Putrajaya site was handed over by the Selangor state government to become a new federal territory. In addition, MSC's lead agency, the Multimedia Development Corporation (MDC), is chaired by Prime Minister Mahathir, thus increasing federal government control over other parts of the greater urban region, including Cyberjaya. In the next section, I turn to the implications and effects of the construction of these urban spaces.

INTELLIGENT CITIES AND THEIR OTHERS

No less than other new cities, the construction of Cyberjaya and Putrajaya impacted upon people and places on site. Perhaps also in keeping with urban developments elsewhere, it was already-marginal socio-economic groups that were most adversely affected. These included *Orang Asli* (indigenous) communities in Cyberjaya and plantation estate workers at Putrajaya. Despite the fact that Putrajaya, and MSC more broadly, is officially described as a greenfield project, the Environmental Impact Assessment for the city identified four plantation estates on the site with a total population of around 2,400. After months of wrangling over estate compensation involving plantation companies, the federal government, the Selangor state government (because proposed relocation sites were in that state), the Malaysian Indian Congress (MIC) (because the workers are politically classified as Indians), the National Union of Plantation Workers and a number of NGOs, the four communities eventually moved into low-cost apartment blocks outside the designated Putrajaya area. Certain people and places were cleared to make way for the construction of utopian, urban cyber-futures.

Physical infrastructure and property development at both Cyberjaya and Putrajaya is being carried out by consortia of government-linked private sector companies. Both cities appear to have been exempted from government stipulations that all housing projects include at least 30 per cent low cost (that is not exceeding RM 25,000) units. Yet the cities are also bound up with new socio-spatial dividing practices that extend beyond conventional notions of financial exclusion. Aihwa Ong (1999, 218) identifies the cyber corridor as a privileged zone of governmentality oriented to the nurturing of information age subjectivities. In providing the 'physical and psychological space needed for productive contemplation and creativity' (as one marketing brochure put it), Cyberjaya, for example, is bound up with attempts to 'enable Malay(sian)s to realize themselves as homo Siliconvalleycus'. In addition, the liberalisation of labour market regulations in MSC means that such conditions are applicable to gifted world class citizens irrespective of their nationality. The obverse of this planned massive brain gain, however, is that there is no place for individuals and groups unable or unwilling to realise themselves appropriately for high-tech times (and spaces). 'Non-intelligent' selves are thus marginalised to low-cost blocks outside the MSC urban utopia or as in the case of *Orang Asli* to the aboriginal periphery (Ong, 1999, 218).

However, an oversimplified geography of cybercity inclusion/exclusion here would obscure the working of the information economy and society in Malaysia and elsewhere. Former residents of the Putrajaya plantation estates *have* found employment in Malaysia's intelligent cities, as security guards, gardeners and cleaners (see Plate 35). These low-paid, non-unionised service jobs contrast starkly with the mobile, technologically innovative and fast knowledge workers

Plate 35 Former residents of the Putrajaya plantation estates employed as gardeners and cleaners in the new 'cyber city'. (Photograph by Tim Bunnell.)

imagined in new economy discourse. As Nigel Thrift (2000, 688) has suggested, 'for there to be faster subjects, there have to be slower ones'. It might be less appropriate to speak of the intelligent city and its others than of differential and highly uneven incorporation into a Siliconising economy and society grounded in real, urban places.

CONCLUSION

The Siliconisation of Asia clearly denotes some important new urban formations, expanded virtual possibilities and a reworking of interrelations between urban space and cyberspace. However, I have shown how the construction of intelligent cities in Malaysia can also be an occasion for (re)considering more conventional (and less celebratory) issues of urban–regional and national development. Rather than simply emerging from the urban-centred logics of a supposedly new techno-economic paradigm, Cyberjaya and Putrajaya also form part of evolving national imaginings and prerogatives. While facilitating new translocal and transnational interconnectivity, these MSC cities do not somehow float free from the politics of development at the nation-state scale. And, most importantly, the hypermobility and fast subject-positions associated with 'intelligent' modes of living and working are dependent upon complex assemblages of technological infrastructure in real places. The highly uneven nature of individual and collective incorporation of into urban(e) information society and economy is perhaps the most significant continuity with pre-intelligent urban/technological development.

REFERENCES FROM THE READING

Bunnell, T. G. (2004), *Malaysia, Modernity and the Multimedia Super Corridor*, London: Routledge.

Castells, M. and Hall, P. (1994), *Technopoles of the World: The Making of Twenty-First Century Industrial Complexes*, London: Routledge.

Indergaard, M. (2003), 'The webs they weave: Malaysia's multimedia super-corridor and New York's Silicon Alley', *Urban Studies*, 40(2), 379–401.

Jessop, B. and Sum, N.-L. (2000), 'An entrepreneurial city in action: Hong Kong's emerging strategies in and for (inter)urban competition', *Urban Studies*, 37, 2287–313.

Mahathir, M. (1998), *Mahathir Mohamad on the Multimedia Super Corridor*, Subang Jaya: Pelanduk Publications.

Ong, A. (1999), *Flexible Citizenship: The Cultural Logics of Transnationality*, London: Duke University Press.

Thrift, N. (2000), 'Animal spirits: performing cultures in the new economy', *Annals of the Association of American Geographers*, 90, 674–92.

Editor's suggestions for further reading

Bunnell, T. (2002), 'Multimedia utopia? A geographical critique of I.T. discourse in Malaysia'. *Antipode*, 34(2), 265–95.

Bunnell, Tim (2004), *Malaysia, Modernity and the Multimedia Super Corridor: A Critical Geography of Intelligent Landscapes*, London and New York: RoutledgeCurzon.

Corey, K. (2000), 'Electronic space: creating cyber communities in south east Asia'. In M. Wilson and K. Corey (eds), *Information Tectonics*, London: Wiley, 135–64.

Douglass, M. (2000), 'Mega-urban regions and world city formation: globalisation, the economic crisis and urban policy issues in Pacific Asia', *Urban Studies*, 37(12), 2315–35.

Olds, K. (2001), *Globalization and Urban Change: Capital, Culture, and Pacific Rim Mega-Projects*, Oxford: Oxford University Press.

'Grounding Global Flows: Constructing an E-Commerce Hub in Singapore'

Neil Coe and Henry Wai-chung Yeung

Editor's Introduction

Less than two hundred kilometres down the Malay peninsula from the Multimedia Super Corridor lies another iconic cybercity: the island city-state of Singapore. In no other city in the world has the integration of ICT planning and public policy-making and urban planning and urbanism gone so far.

The Singaporean government, long keen to reinvent the city as an icon of ICT-mediated modernity, has invested massively to orchestrate the island city as a global hub of high value-added ICT-based logistics industries. It has regulated and organised the telecommunications sector to ensure near-ubiquitous social and geographical access to broadband technologies. And it has carefully thought through the strategic spatial planning of the island so that the limitations of geography are to some extent overcome through the externalisation of space-hungry and relatively low value-added activities beyond the national borders, to Malaysia in the north, and Indonesia to the south.

In this third reading on cybercity politics and strategy we address the latest in a long line of sophisticated ICT-based development strategies for Singapore: the plan to ensure that Singapore emerges as a global e-commerce hub. In this reading Neil Coe and Henry Wai-chung Yeung, geographers in Manchester and Singapore National Universities respectively, reflect on how a range of policy instruments are working to try to sustain urban economic centrality in this era of decentred, and geographically stretched, e-commerce activity. Singapore's strategy works against the pervasive rhetoric of the placelessness of the global e-commerce economy. It is based on an explicit recognition that only 'premium' city economies, with the requisite infrastructures, skills, capital, connectivity, economies of scale and scope, innovation potential and image, can emerge as the organisational and logistics centres for e-commerce management, coordination and innovation.

The authors suggest, then, that Singapore's is a strategy to 'ground' the internationally stretched chains of e-commerce into the tiny geographical spaces of the island city-state itself. Recalling the discussions in Section V, the key point here is that strategic centrality amongst the world's privileged cities becomes necessary at the same time as the global, decentralised, spread of stretched e-commerce networks and low value-added functions. The authors stress that pure liberalised markets are unlikely to sustain the achievement of strategic e-commerce centrality amongst competing cities. Instead, coordinating and shaping public policy frameworks will be necessary which combine urban and ICT planning, especially in highly congested cities.

Backed by these ideas, Coe and Wai-chung describe how the Electronic Commerce Masterplan, announced in September 1998, has actually operated in Singapore. The strategy has centred on five interrelated goals: developing an e-commerce infrastructure; ensuring that Singapore emerges as an e-commerce hub; encouraging Singaporean businesses to adopt electronic transaction systems; promoting the acceptability of e-commerce amongst public institutions and citizens; and harmonising the legal and regulatory sides of e-commerce between Singapore and its major trading partners. The authors suggest that very real progress has been made towards reaching these objectives.

In this reading, we use the metaphor of nodes in global flows (Castells, 1996: 82) to explore the potential for particular localities to emerge as hubs within the rapidly developing world of electronic commerce (e-commerce). Arguably, much of the literature on information and communications technologies tends to obfuscate the intricate relations between these technologies, their social users, and material geographies. It is true that advancement in these technologies has greatly facilitated the emergence of e-commerce and the possibility of participation in global flows for individuals and firms all over the world. It is equally true, however, that these virtual flows on a global scale do not preclude the important role of place in organising social life. Here, our analytical task is to show how certain places can enhance their positions as nodes and hubs in the cyberspace of e-commerce through tapping into the social dimensions of global flows, recognising the need for support infrastructure such as logistics and distribution, and developing suitable policy initiatives. We argue it is possible for specific localities to ground global flows in e-commerce by constructing nodes and hubs of a physical, materialised nature.

Singapore provides us with an excellent case study through which to explore these conceptual assertions, due to the wide-ranging efforts of the government to establish the city-state as a premier e-commerce hub in the Asia–Pacific region. Since the early 1980s, Singapore has implemented a series of national IT initiatives and policies, of which the 1998 Electronic Commerce Masterplan is a recent example. The Masterplan takes a broad view of the steps that are necessary to establish Singapore as a premier e-commerce hub, covering issues to do with legal frameworks, the security of online transactions, local and international standards, support and logistics services, e-commerce measurement, and the education of both the private and public sectors, and private individuals. Before we examine the case study of Singapore in more detail, it is useful to establish some conceptual parameters about constructing an e-commerce hub and how it enables a place to become an important node in the global space of flows.

TAPPING INTO GLOBAL FLOWS: CONSTRUCTING HUBS AND NODES IN THE 'VIRTUAL' WORLD OF ELECTRONIC COMMERCE

It is often argued that as information and communications technologies allow virtual access to all forms of information, social and economic actors are becoming increasingly liberated from the constraints of space or the so-called friction of distance. Such technologies are also deemed to be eroding the role of place as the central unit of organising social life. E-commerce is often seen to be placeless in two ways. First, it is argued that, from the perspective of consumers, transactions can be conducted via the Internet from anywhere, and by anyone who has access. Second, online companies are often perceived as embedded in cyberspace rather than in real physical places. We would strongly dispute such views, however. Here, we want to argue that e-commerce is very much grounded in particular places in the global economy, and in particular, in a range of leading or world cities.

How does this grounding of e-commerce take place within specific nodes in the global space of flows? We focus here on three specific dimensions: (1) the social and symbolic dimensions of global flows; (2) the role of logistics and distribution in e-commerce; and (3) the role of state initiatives. First, while it is true that e-commerce is not as restricted by geographic barriers as traditional wholesaling and retailing, the operationalisation of e-commerce requires the mobilisation of significant *social and symbolic capital* in specific localities. This is because the human construction of space and place is seen actually to ground and conceptualise applications and uses of new technologies (Graham, 1998, 172). Nodes in the world of technologically mediated flows exist as places of reflexivity, trust and reciprocity. These nodes have a material dimension because they are embedded in specific physical locations. As such, the social relations of e-commerce cannot exist in cyberspace alone; they need to be grounded in places and localities in which social actors have invested significant meaning through *both* face-to-face interactions *and* material flows. For example, whereas they may be able to provide specialised services for consumers, it is questionable whether e-commerce service providers (e.g. online reservation by airlines) can offer the social atmosphere and support that consumers expect from many traditional channels (e.g. travel agents).

Second, the role of specific localities in the *provision of logistics and distribution* of e-commerce products and services further enhances the importance of place in conducting cyberspace transactions. Every successful e-commerce operation that develops needs the services of global credit companies (e.g. Visa and American Express), global distribution companies (e.g. Federal Express and DHL), global producers (e.g. publishers and entertainment companies), local warehousing and Internet service providers. This, however, does not necessarily imply that every locality can become an important node in global spaces of flows via e-commerce. Two factors explain why certain localities are better equipped than others in engaging e-commerce. On the one hand, pre-existing geographies will shape the e-commerce trajectories of specific localities, as information and communications technologies are not evenly distributed over space, and indeed e-commerce also still requires the physical transportation of goods and services. Thus, to a large extent, the effectiveness of e-commerce as a developmental strategy for specific localities depends on the efficiency of the existing communications and transportation infrastructure of a given locality.

On the other hand, e-commerce allows the decoupling of different functions in a given production chain because electronic markets consist of both physical and virtual channels for marketing and distribution. Sarkar *et al.* (1998) have argued that e-commerce lowers supply chain coordination costs and thus contributes to the separation of physical distribution from other cybermediary functions. In the early stages of the electronic market, many e-commerce companies are unlikely to integrate 'forwards' or 'backwards' along the production chain. Instead, they may make use of services provided by other cybermediaries (e.g. gateways, directories, search services, online malls and electronic publishers) as an efficient mechanism for supporting electronic exchanges and achieving economies of scale and scope. Though their physical distribution channel may be shortened, however, these e-commerce companies are likely to continue to depend on integrated logistics companies (e.g. Federal Express) as their distribution systems. This decoupling process also produces complex and longer networks of information-specific intermediaries (e.g. some firms may locate products,

others provide evaluations of related products, training, and settlement services, and so on). Localities can strive to become an important node if they are able to provide favourable social and institutional contexts for the development of these interfaces of physical and cyber networks bringing together consumers, producers, and their intermediaries.

Third, there is clearly a case for *policy and developmental initiatives* because nodes and hubs in the e-commerce world do not develop randomly. The development of e-commerce nodes and hubs requires sustainable initiatives by various actors from institutions at different spatial scales. These actors may come from business and trade associations, labour unions, civil and voluntary associations, and local, regional and national government agencies which are empowered to foster social and economic development and political stability. How, though, do these institutions explain the rise of certain places as important nodes in the global flows of e-commerce? As e-commerce has the potential to bypass specific localities in its global reach, and an e-commerce hub requires substantial social capital endowments, and physical and logistical infrastructure, holding down global e-commerce flows in specific localities becomes a political initiative that necessitates intense collaboration among public agencies, private business, and social actors. To understand the evolution and success of certain e-commerce nodes, we need to go beyond an analysis of their economic functions and interconnectedness. More importantly, we need to examine the ways in which these economic dimensions are woven together with local politics and institutional relations to form complex power geometries. It is these power geometries that explain the success and failure of specific localities in constructing e-commerce hubs as important nodes in global spaces of information and transactional flows.

POLICY INITIATIVES AND THE RECENT E-COMMERCE DRIVE IN SINGAPORE

The Electronic Commerce Masterplan, announced in September 1998, aimed to develop Singapore into a hub for electronically transacted business through a series of policy interventions, incentive schemes and education initiatives. The initial target was to have S$4 (US$2.4) billion worth of products and services transacted electronically through Singapore, and 50 per cent of businesses to use some form of e-commerce, by the year 2003. The first formal move to prepare for an era of e-commerce was however taken in 1996, when the National Computing Board (NCB) established the Electronic Commerce Hotbed (ECH) Programme, with the aim of jump-starting e-commerce in Singapore. The ECH initiative brought together 38 founding parties, ranging from financial institutions to academic, telecommunications and IT organisations, to pursue two broad objectives; firstly, to make e-commerce widely available, and, secondly, to position Singapore as an e-commerce hub.

In January 1997, the EC Policy Committee was established to further explore these legal and policy issues. In April 1998, the committee completed its review and made its recommendation to the government, listing six guiding principles with regards e-commerce development: (1) the private sector should take the lead; (2) the government should put in place a legal framework to provide certainty and predictability; (3) the government should provide a secure and safe environment for electronic transactions; (4) the government should use joint venture pilots and experiments to expedite e-commerce growth and development; (5) the government should be proactive in pursuing innovative, liberal and transparent policies; and (6) international consistency, cooperation and interoperability should be encouraged (NCB, 1998). After a period of public consultation, these policy recommendations were used to formulate the Electronic Commerce Masterplan of late 1998. The plan had five main thrusts, accompanied by various implementation programmes, which we will now consider in turn:

(1) *To develop an internationally linked e-commerce infrastructure.* This was seen as crucial to strengthening Singapore's position as an e-commerce hub, with the financial and logistics sectors being given a key role. The plan aimed to develop and deploy, in partnership with industry, an efficient settlement system for Internet transactions between businesses, covering international trade payment and multi-currency payment. Certain areas of e-commerce infrastructure development were also singled out for special attention, including trust management, rights management, business to business trading platforms and business to consumer payment systems.

(2) *To jump-start Singapore as an e-commerce hub.* This part of the plan focused on the sectors in which Singapore already had an inherent advantage as a hub, including the financial, transport and logistics, and telecommunications infrastructure providers. Incentive schemes have been implemented to attract international and local companies to base their e-commerce hub activities in Singapore.

(3) *To encourage businesses to use e-commerce strategically.* In this thrust, education and other support programmes were seen as essential to helping businesses exploit e-commerce to enhance their productivity and competitiveness. This has been facilitated in several ways: large-scale training schemes for business people to use simple e-commerce platforms; a usage promotion drive and financial incentives to improve adoption levels among Small and Medium Enterprises (SMEs); manpower development through retraining schemes and new tertiary level courses; and support to help leading local e-commerce providers succeed internationally.

(4) *To promote usage of e-commerce by the public and businesses.* This thrust aimed to extend the benefits of e-commerce to the general public. First, the government itself has to help to promote the use of e-commerce in Singapore through its electronic public services initiatives, with key public services being delivered electronically by the year 2001. Second, mass education schemes provide training for the public. Thirdly, mass media schemes improve the general awareness of e-commerce and its implications.

(5) *To harmonise cross-border EC laws and policies.* This final thrust was designed to develop Singapore as a trusted node for e-commerce. The necessary legislation has been developed, and periodically reviewed, to ensure that Singapore laws are competitive, congruent and internationally consistent. The areas covered include intellectual property rights, data protection, consumer protection and taxation. Bilateral agreements have been sought with key trading partners, and Singapore has looked to participate actively in international fora such as the Association of South East Asian Nations (ASEAN), APEC, UN (United Nations), and the World Trade Organisation (WTO) to promote the harmonisation of e-commerce frameworks.

THE CONSTITUENT ELEMENTS OF SINGAPORE'S MOVE TO BECOME AN E-COMMERCE HUB

On paper at least, there can be no doubt that the Electronic Commerce Masterplan represents a thorough, comprehensive and wide ranging policy intervention in the drive to develop e-commerce in Singapore. We now move on to consider the broad range of elements that constitute Singapore's effort to become an important e-commerce hub.

At the level of the *general e-commerce environment*, Singapore has made significant progress in developing legal frameworks, technical standards and incentive schemes for e-commerce. In terms of legal frameworks, a major step was the enactment of the Electronic Transactions Act, an e-commerce Policy Committee recommendation, which came into force in July 1998. This Act covers a broad range of issues pertaining to e-commerce, including the authentication of the originator of electronic messages, the legal recognition of electronic signatures, the retention of records by electronic means, the formation and validity of electronic contracts, and the legal liability of service providers. With regard to standards, Singapore has established a set of open, industry-led standards in the areas of network protocols, security, email and directories, e-commerce, and information sources and exchange, all designed to facilitate the interconnection and interoperability of businesses via computer networks. Singapore has also instituted a range of incentive schemes to promote e-commerce. For example, the Cluster Development Fund and Innovation Development Scheme can provide funds for qualifying companies to develop their e-commerce projects. In addition to financial incentives, a range of educational programmes has been initiated to raise awareness both among local businesses.

The next domain is that of *infrastructure services*, which is characterised by a high level of collaboration between government and business. Infrastructure services can usefully be subdivided into four components. First, *network services* provide the networks that enable e-commerce. Such services are provided by private sector Internet Service Providers, of which there are three in Singapore, and broadband network access providers. Second, there is also a wide variety of *directory services* on offer in Singapore, allowing customers to search for information and web sites. The third infrastructure services component is that of

security services. For example, a company, Netrust, was established in July 1997 as a joint venture between the NCB and the Network for Electronic Transfers (Singapore) (NETS) to issue and manage digital keys and certificates. *Payments services* constitute the fourth and final element of infrastructure services. Several secure online payment systems are already in place in Singapore. An Electronic Commerce Hotbed initiative started in 1996 has made significant progress towards developing a Secure Electronic Transactions (SET) protocol to support secure credit card payments.

The third level of the e-commerce infrastructure is that of *commerce solution providers* (CSPs), of which there is a wide variety in Singapore. Such CSPs offer complete e-commerce solutions to businesses that do not have the capability, or choose not, to implement their own e-commerce services. Overall, there already appears to be a well-developed and sophisticated infrastructure for e-commerce in Singapore, supporting a wide range of specific e-commerce applications, ranging from the government sector (e.g. electronic filing of tax returns), through retailing and leisure (e.g. online shopping with cold storage supermarkets) to the financial sector (e.g. Internet banking and share trading).

CONCLUSION

We propose that the emergent global trend toward electronic commerce will not have the spatially homogenising effects that some technology commentators predict. Instead, as with previous techno-economic systems, the world of electronic commerce will be characterised by distinctive and constantly evolving patterns of uneven development. Place, in the abstract sense, and cities, in their material form, will remain of the utmost importance, as there will continue to be a need for centres of coordination and control, or hubs and nodes, in the intensifying world of electronic flows. At a conceptual level, we have proposed three arguments why this is so. First, there are socio-cultural aspects to e-commerce that are embedded in particular places and contexts. Second, the effective operationalisation of e-commerce does not simply require an electronic link between two parties, but in many cases also relies upon a whole range of e-commerce service providers and cybermediaries. Third, the development of e-commerce also requires a particular policy context that ensures stability, reliability, security and transparency.

Our analysis of Singapore's policy drive to become an e-commerce hub has shown several aspects of policy formulation and implementation that may be suggested as being indicative of best practice. We make four main points in this regard. First, and most simply, we argue that there is a powerful argument for coordinated national policy intervention to stimulate and facilitate e-commerce development, and that increasingly such policies will need to be oriented towards international harmonisation and standardisation.

Second, the Singaporean case illustrates that a broad-based and integrated set of e-commerce policies is desirable. This goes way beyond providing a legal and regulatory framework for electronic transactions, and includes areas such as developing secure payment systems, coordinating standards, nurturing IT skills, building business and public awareness, attracting both foreign investors and local uptake through incentive schemes, and fostering support services.

Third, we suggest that Singapore illustrates the kind of progressive institutional support that is necessary to mobilise such initiatives. The success of e-commerce policy formation in Singapore arguably rests on the large number of parties that have been consulted, covering a broad range of ministries, statutory boards, foreign multinationals, local businesses and educational establishments.

Finally, we suggest that the Singapore case indicates that e-commerce hub development should not purely be left to either state institutions or the private sector, but ultimately relies on sustained and effective collaboration between the two domains.

REFERENCES FROM THE READING

Castells, M. (1996), *The Rise of the Network Society*, Oxford: Blackwell.

Graham, S. (1998), 'The end of geography or the explosion of place? Conceptualizing space, place and information technology', *Progress in Human Geography*, 22(2), 165–85.

Knox, P. L. and Taylor, P. J. (eds) (1995), *World Cities in a World System*, Cambridge: Cambridge University Press.

NCB (1998), Press release, 13 April, available from http://www.ncb.gov.sg/

Sarkar, M., Butler, B. and Steinfield, C. (1998), 'Cybermediaries in electronic marketspace: toward theory building', *Journal of Business Research*, 41(3), 215–21.

Editor's suggestions for further reading

Corey, K. (1991), 'The role of information technology in the planning and development of Singapore'. In S. Brunn and T. Leinbach (eds), *Collapsing Space and Time: Geographic Aspects of Communications and information*, London: Harper, 217–31.

Corey, K. (1993), 'Using telecommunications and information technology in planning an information-age city: Singapore'. In H. Bakis, R. Abler and E. Roche (eds), *Corporate Networks, International Telecommunications and Interdependence*, London: Belhaven, 49–76.

Corey, K. (2000), 'Electronic space: creating cyber communities in south east Asia'. In M. Wilson and K. Corey (eds), *Information Tectonics*, London: Wiley, 135–64.

Grundy-Warr, C., Peachey, K. and Perry, M. (1999), 'Fragmented integration in the Singapore– Indonesian border zone: southeast Asia's "growth triangle" against the global economy', *International Journal of Urban and Regional Research*, 23(2), 304–28.

Ho, K. C. (2003), 'Attracting and retaining investments in uncertain times: Singapore in South-East Asia', *Urban Studies*, 40(2), 421–38.

Van Grunsven, L. and Van Egeraat, C. (1999), 'Achievements of the industrial "high road" and clustering strategies in Singapore and their relevance to European peripheral economies', *European Planning Studies*, 7(2), 145–73.

Vinh, T. (2001), 'Cooperation'. In C. Chung, J. Inaba, R. Koolhaas and S. Tsung Leong (eds), *Harvard Design School Guide to Shopping*, Cologne: Taschen, 204–24.

Yeung, H. (1999), 'Regulating investment abroad: the political economy of the regionalization of Singaporean firms', *Antipode*, 31(3), 245–73.

'Cybernetic Wal-Mart: Will Internet Tax Breaks Kill Main Street USA?'

Adbusters (2002)

Richard Sclove

Editor's Introduction

Our fourth reading on cybercity politics and strategy returns to the implications of e-commerce for local economies that were analysed earlier (Martin Dodge, p. 221; Andrew Murphy, p. 226; Yuko Aoyama, p. 231). Rather than being concerned only with the economic dynamics of e-commerce, however, this reading explores the political and policy options that arise at the interface between the regulation of online and offline retailing.

Here Richard Sclove, founder of the Loka Institute in the USA (http://www.loka.org) and author of *Democracy and Technology* (Guilford Press, 1995), analyses the political biases that are being afforded by governments to e-commerce retailers. As e-commerce transactions grow in scale, Sclove stresses the ways in which the transnational corporations that dominate them are managing to avoid the sales or value-added taxes that their offline competitors in 'brick' rather than 'click' retailing face.

Sclove argues that this biased treatment of e-commerce suppliers by governments is likely to further devastate locally owned and diverse retailing centres in downtowns and high streets across the USA. To Sclove, this effect will amount to an intensification of the damage already done by massive 'big box' retailing transnationals like WalMart, who have been shown to deliberately destroy local retailers when they first colonise a city's geographical market.

This 'cybernetic Wal-Mart effect' is, to Sclove, a reflection of the widespread assumption that policy towards e-commerce supply tends only to account for the supposed wishes of consumers for maximum value and minimum price. This is a result of the denial that e-commerce has any real implications for the complex economic ecologies of urban places. And it is a reflection of the intensification of transnational control over local retailing economies.

Finally, Sclove argues that real political options exist to treat e-commerce differently. 'There is a simple way to maintain a healthy balance between e-commerce and local business', he writes. A simple tax-based cross-subsidy, for example, would allow municipal finances to benefit from e-commerce transactions in the same way as they do from local face-to-face ones. Without using such creative policy solutions to address the damaging effects of e-commerce on local economies, local democracy, and local quality of life, Sclove suggests that e-commerce may become little more than a steamroller driving over local places with extremely damaging, and often ignored, social and economic effects.

The case that Sclove analyses is actually only one of the ways in which large, multisite and transnational firms are using ICTs and e-commerce to bypass geographically based regulation and restrictions as they operate around the world. In the age of the Internet, especially, the extreme difficulties of achieving real corporate regulation and transparency at the transnational scale are likely to make tax-avoidance an ever-more central element within corporate operations. For, increasingly, transnational corporations:

will install their web servers where taxes are lowest, disguise their trade in goods as a trade in services, and even launch their own virtual currencies. The tax burden, in other words, is shifting to those who are unable to move their assets offshore or out of the old economy into cyberspace. With little else to offer, poor countries [and spaces] end up giving everything away in a desperate attempt to attract 'investment'. If taxation is not to become wholly regressive, we will have to revolutionise the means to which the rich are charged (Monbiot, 2000, 13).

By 2003, annual online sales in the United States are projected to soar above $1.4 trillion, and a politically loaded question is now unavoidable: do we tax online shopping? Already, the US Congress has resolved – 423 to 1 – that there should be a worldwide ban against levying special or discriminatory taxes on electronic commerce. The next step, according to such powerful figures as Senator John McCain and house majority leader Dick Armey, is to permanently exempt e-commerce even from *existing* sales taxes.

Thus far, the debate encompasses the perspectives of public servants, businesses, and consumers. But how about that of citizens? What would tax-free e-commerce mean for democracy and civic life? Very possibly it could mean the same thing the proliferation of Wal-Marts and megamalls has meant for Main Streets: a demise that no one intended.

Suppose a Wal-Mart store locates on the outskirts of a town, and half the residents start to do one-third of their shopping there. That means they do two-thirds of their shopping downtown, while the other half of the population continues to do all its shopping downtown. Although all the residents still patronize Main Street, downtown retail revenue drops about 16.7 percent – enough to start killing off shops.

It's a perverse market dynamic: a loss to the entire community that not a single person wanted. It is also self-reinforcing; once the downtown starts to shut down, people who preferred to shop there have no choice but to switch to Wal-Mart.

Systems theorists explain this kind of unwelcome, coercive and extreme outcome as the result of a "positive feedback loop." That is, the output of a process (some residents opting to shop Wal-Mart on occasion) feeds back into the original process as input (a smaller, less diversified local economy), generating more output (more people compelled to patronize Wal-Mart). In other words, a little generates more, more generates a *lot* more. Systems with positive feedback loops can easily burst limits.

To social scientists this is a "collective action problem:" an example of reasonable individual actions that together add up to a socially irrational outcome. As more commerce goes online, an emerging Cybernetic Wal-Mart Effect threatens to aggravate this dynamic. It works just like the regular Wal-Mart Effect, except more powerfully and pervasively.

Brick-and-mortar Wal-Marts mainly threaten mom-and-pop retail shops. But online commerce pits local businesses not just against a mall on the outskirts of town, but against the entire global marketplace. The Internet is spreading into every sector of the economy, from local manufacturers and suppliers to service providers such as travel agents, lawyers, and stock-brokers. A few of them may thrive by going online themselves, but they are the exceptions. In general, the economies of scale involved in enticing a viable customer base to a website will overwhelmingly favor a few deep-pocketed, very un-local enterprises.

If we think of ourselves solely as consumers, this isn't necessarily a problem. While local economies wither, the Net should enable consumers to enjoy access to a wider range of goods and services, in some cases at lower cost. The catch is that we're not simply consumers. We're also family members, friends, local community members, and workers. From the standpoint of democratic society, above all we are citizens.

As consumers, we always ask, "Is this the best deal for me?" But as citizens we must ask, "Does a Cybernetic Wal-Mart Effect serve the common good? Does it further our fundamental interest in preserving and improving the character of our democracy?"

From a democratic citizen's perspective, e-commerce with its coercive Cybernetic Wal-Mart Effect is problematic. My online shopping contributes to shrinking the local economy. Eviscerating a local economy weakens local cultural and community vibrancy. That's bad in its own right, but worse for democracy. As social bonds weaken, people relinquish mutual understanding and the capacity for collective action. Those are essential conditions for a workable democracy.

At the same time, undercutting local economies increases local dependence on national and global market forces and on decisions made in faraway corporate headquarters – powers over which communities have little or no control. As the locus of political intervention shifts to distant centers, the influence of everyday citizens declines.

A refusal to tax e-commerce amounts to a public sanction of this anti-democratic shift. But there's a simple way to maintain a healthy balance between e-commerce and local business, between sometimes perverse market forces and the social good. First, tax online and mail-order catalog sales. Second, grant some of the revenue back to municipalities to invest in their local economies and community life (e.g., sidewalk benches, parks, playgrounds, public toilets, public music and theater, local meeting halls, and soon). If necessary, another portion of the revenue could be rebated to low-income citizens.

Our judgments as citizens need to consider but also transcend our narrower interests as consumers. When it comes to public policy and the common good, our citizen-selves ought to be sovereign over our consumer-selves. And so the unavoidable question changes: If our consumer-selves say "yes" to sheltering e-commerce from taxes and shrug at the Cybernetic Wal-Mart Effect, are our citizen-selves prepared to live with the consequences?

Editor's references and suggestions for further reading

Guy, C. M. (1996) 'Corporate strategies in food retailing and their local impacts'. *Environment and Planning A*, 28, 1575–1602.

Klein, N. (1999), *NoLogo*, New York: Picador.

Monbiot, G. (2000), 'Devil take the hindmost', *Guardian Weekly*, 30 March–5 April, 13.

Sclove, R. (1995), *Democracy and Technology*, New York: Guilford Press.

Wrigley, N. and Lowe, M. (eds) (1996), *Retailing, Consumption and Capital*, London: Longman.

'Recombinations for Community Meaning'

from *Digital Places: Building Our City of Bits* (2000)

Thomas Horan

Editor's Introduction

How can the design of urban places address the pervasive sense that cybercities are being fragmented into a series of capsular physical and technical network spaces which undermine truly public mixing and often lead to social alienation? Can ICTs be part of the solution to these problems as well as part of their cause?

In this fifth reading on cybercity politics and strategy the US urbanist Thomas Horan concludes the discussions in his book *Digital Places: Building Our City of Bits*. He offers some principles which, he argues, can be used to harness the capabilities of ICTs and new media to develop truly recombinant urban spaces in which vibrant place-based and ICT mediated relations work positively together, in a virtuous circle, rather then negatively in a vicious one. Railing against the suburban sprawl, automobilism, social isolation and fortification that characterises so much of suburban USA, Horan suggests that strong public policy and planning frameworks, based on principles of recombinant planning, might help to reverse current trends and so bring about a return to accessible, civically engaged communities.

To Horan, the key is to rethink the role and design of the key public institutions in US cities which anchor the urban public realm: schools, libraries, museums, and community centres. With good funding and innovative ideas such places, he argues, can be turned into the recombinant hubs of the cybercity. They can thus combine strong, accessible and well-designed urban spaces which fuse seamlessly into a relevant range of high quality ICT services, networks and spaces. As with homes and workplaces in cybercities, then, the creative urban challenge is to fuse the design of urban spaces with the design of cyberspaces.

If such recombinant urban strategies are developed with the financial and creative resources necessary, Horan argues that the economic, cultural, social and environmental qualities of US suburbs might be dramatically improved. Single-use public spaces, employed for small parts of the day, might be restructured as multi-use spaces animated throughout the day. Fortified and defensive neighbourhoods with dead centres and a total reliance on car traffic might, combined with the broader principles of new urbanism, emerge as more 'liveable' communities where a face-to-face street life remerges. And the culture of fear which drives people into their capsular lives, based on interlinked complexes of cars, malls, homes, personal computers, mobile phones, and gated communities may, at least to some extent, be undermined.

RECOMBINATIONS FOR COMMUNITY MEANING

During the closing decades of the twentieth century there was an erosion of traditional public space due to a number of technological, sociological, and economic trends. Malls replaced town squares; gated communities replaced traditional neighborhoods; parking lots replaced open spaces. Writing about these trends in *Suburban Nation*, architect and new urbanism advocate Andres Duany and his coauthors summarize the situation we now face: "The choice is ours: either a society of homogenous pieces, isolated from one another in fortified enclaves or a society of diverse and memorable neighborhoods, organized into mutually supportive towns, cities, and regions" (Duany *et al.* 2000). Contrary to the notion that technology can only foster isolationism, innovative electronic networks can encourage the development and growth of accessible, civically engaged communities. They can add a new dimension to public space, one that interacts with and supports physical space.

While much has been made of recent commercial interests in digital communities (see John Hagel and Arthur Armstrong's *Net Gain* for creating loyal online customers), the creation of vibrant place-based communities should not be left strictly to the market forces of the private sector (Hagel and Armstrong, 1997). Rather, local communities can play a crucial role in defining the nature and types of electronic-community services available to their citizens. The major thrust of this book has been to demonstrate how local public institutions such as schools, libraries, and community centers can play a key role in creating digital places at the community level. While communities of interest will continue to thrive on the Internet, this will not erase the need to carve out free digital space to facilitate the interactions of local communities.

While the exact contours of specific new digital communities depend on local circumstance, the strands of recombinant design suggest several important features: they can represent innovative recombinations of uses (*fluid locations* for learning, culture, and health care); they can retain the traditional role of civic institutions (*meaningful places* such as libraries and schools); they can have synergistic virtual and physical dimensions (*threshold connections* to allow access from anywhere); and they should be crafted in a manner that facilitates local and user access and participation (*democratic designs* of community access).

RECOMBINANT DIGITAL COMMUNITIES

Like homes, workplaces, and commerce, community functions are recombining. In designing and redesigning digital communities, some items to consider include:

- *Schools.* School officials can explore the community connections that can be enhanced by combining school network development and lifelong learning to support educational and training needs throughout the community. Teachers can exploit new bandwidth connections in the home to encourage enhanced connections with children and parents in the community. Higher education (both community colleges and universities) can investigate new technology-infused partnerships with businesses and students.
- *Libraries.* Local libraries can consider how new and innovative designs can help them reassert their spatial and electronic presence in the community, including universal access to all community members. New partnerships can be explored with other community resources – museums, schools – to devise innovative electronic and physical "third-places."
- *Museums and Cultural Centers.* Community and cultural groups can develop new interactive dimensions to their exhibitions and programs to enhance local electronic cultural presence and community. These can include taking an active role in getting local artists represented as well as connecting citizens and visitors to local cultural museums and organizations.
- *Community Institutions.* Nonprofit community groups can work with local schools, libraries, city officials, and the private sector to evaluate access needs and devise training and learning programs. Local governments can consider those government services that can be provided electronically and play a role in ensuring access to residents. Local officials can also consider mechanisms for using the electronic medium to enhance face-to-face participation in local debates and decisions.

Recent experience suggests that communities will benefit by leveraging the activities of traditional civic enterprises, such as libraries, schools, and museums, through electronic networks. These institutions provide points of entry and access. The school is no longer just a school but a learning center with strong electronic connections to the home. The museum is no longer a self-contained institution, but can provide a network for artists throughout the community. The library is not just a repository, but a dynamic "third place" for communicating and learning. Taken together, new connections can create a web of activities that, in sum, create community.

REFERENCES FROM THE READING

Duany, A., Plater-Zybek, E. and Speck, J. (2000), *Suburban Nation: The Rise of Sprawl and the Decline of the American Dream*, New York: North Point Press.
Hagel, J. and Armstrong, A. G. (1997), *Net Gain: Expanding Markets through Virtual Communities*, Cambridge, MA: Harvard Business School Press.

Editor's suggestions for further reading

Horan, T. (2000), *Digital Places: Building Our City of Bits*, Washington, DC: Urban Land Institute.
Horan, T. and Jordan, D. (1998), 'Integrated transport and telecommunication planning in Santa Monica', *Journal of Urban Technology*, 5(2), 1–20.
Mitchell, W. (1995), *City of Bits: Space, Place and the Infobahn*, Cambridge, MA: MIT Press.
Robbins, E. (1998), 'The new urbanism and the fallacy of singularity', *Urban Design International*, 3(1), 33–42.
Roberts, M., Lloyd-Jones, T., Erickson, B. and Nice, S. (1999), 'Place and space in the networked city: conceptualizing the integrated metropolis', *Journal of Urban Design*, 4(1), 51–67.
Smith, N. (1999), 'Which new urbanism? The revanchist '90s', *Perspecta*, 30, 98–105.

THREE

'Retrofitting Sprawl: A Cyber Strategy for Livable Communities'

Walter Siembab

Editor's Introduction

Thomas Horan's discussion (p. 363) of the challenges and opportunities of recombinant urban design in the USA merges neatly into this fifth reading on cybercity politics and strategy. Here, the US planning consultant Walter Siembab goes into more detail about how the planned installation of a range of high-quality ICT infrastructures and services might, in combination with new transport strategies, and denser standards of urban design, lead to a 'retrofitting' of sprawling US suburbs.

Like Horan, Siembab hopes that broad-scale recombinant planning may help to significantly restructure the physical, social and technological configurations of US suburbia. To accommodate population growth, and to address crises of environmental sustainability and petroleum supply, Siembab argues that the challenge of such retrofitting must be recognised as a central imperative in contemporary US urban planning.

Siembab's innovative contribution is to analyse how a combined strategy of retrofitting may work to incorporate public transport, ICTs and urban design so that these operate together to create more than the sum of the parts. He envisages key nodes in retrofitted suburbs as combining transit stations, district service cores, and 'Network Stations' – publicly accessible ICT hubs based on broadband networks. Such an approach may, Siembab argues, foster a more distributed urban economy which would also help to improve the resilience of cities to shocks and attacks – such as those on September 11th, 2001.

Siembab believes, however, that to achieve such an approach, a major shift in the conceptual and political paradigms and orientations of US cities will be required. 'Just as the livable communities strategy is a political option to sprawl', he writes, 'so the cyber strategy for livable communities is a political alternative to a pure bricks and mortar approach.'

The chance of such a major paradigm shift happening, however, remains small – at least in the short term. The problem is that US urban planning practice is very localised and dominated by coalitions of real estate companies and boosterist politicians and planners. Combined with the failure by planning educators and practitioners to really engage with a recombinant view of urban planning and design, or with the potential of ICTs as policy tools, this situation seems unlikely to lead to the sorts of innovations urged by Siembab. At the very least, successfully retrofitting suburbs is going to take a very long time, given the obduracy of urban form, the costs of transit investments, and the broader culture of capsularisation within US suburbia which combine to configure defensive households, neighbourhoods, automobiles, and ICT networks

As we have rediscovered in our own day, there is an inescapable lag between the new urban concept and its actualization. Buildings are far more durable than paintings, sculpture, fashions of dress or thought.

(Crouch, Garr, and Mundigo, 1982, p. xv)

Retrofit. To modify equipment that is already in service using parts developed or made available after the time of original manufacture.

(*Webster's English Dictionary*)

There is a growing awareness among urban theorists, practitioners, and politicians that the sprawling metropolis built throughout America and many parts of Europe over the past fifty years needs to be spatially restructured. Large scale tracts of single function developments (residential, retail, office, industrial) have produced a number of unintended consequences.

The automobile has been the key technology enabling the scale of specialized developments. As expressed by architect Moshe Safdie, 'as cars shaped the city, so the city itself is now shaped to require cars' (1997, 127). As a result of the sprawl, urban form and its auto dependence, society absorbs poor air quality and high costs of highway infrastructure construction and maintenance. The underclass is spatially isolated, unable to easily access jobs and services. Consumers pay the myriad costs associated with private transportation including time wasted in congestion. Civilization faces global warming, fossil fuel depletion and threats of petroleum wars.

The most troubling consequence may be that the viability of modern metropolitan regions depends on an abundant supply of ubiquitously available, affordable petroleum. Not only are those conditions likely to become impossible to guarantee in the future, but the act of pursuing them may contribute to the difficulty of achieving them.

URBAN SPACE NEEDS TO BE RESTRUCTURED: THE QUESTION IS HOW?

A candidate to correct sprawl is a reform movement that has been gaining support among metropolitan policy-makers and practitioners throughout the United States. It is often referred to as the movement to support Transit-Oriented Development (TOD), smart growth, traditional neighborhood development, or the livable communities strategy.

The livable communities strategy can be summarized as building compact, mixed-use town centers that feature walking distances to public transit which link residents to job opportunities and social services in order to reduce dependence on the automobile – all under the guidance of a community planning process. (See Characteristics of Livable Communities at http://www.fta.dot.gov.)

The strategy is being driven by population growth. For example, by the year 2020, California's population of 33 million is projected to reach 45.3 million – an increase of 37 per cent. At the current rate, the state is adding nearly 4 million people, or the equivalent of the population of Los Angeles, every seven years. The livable communities strategy provides a political alternative to more suburban sprawl.

There is no doubt that the livable communities strategy is a prudent approach to new development. Its strength is that it can mitigate some of the impacts of metropolitan population growth. However, its most significant shortcoming is that it cannot be implemented at the scale needed in the time available before the costs of the current urban form threaten regional stability and sustainability. In other words, metropolitan regions lack the money and the time for overcoming the legacy of the past by relying on bricks and mortar alone.

Brookings Institution Senior Fellow Anthony Downs provides the numbers. The livable communities strategy is partly confounded by the fact that 85 per cent of the developed portions of the nation that will exist in 2020 were already in place as of 2000. Even if radical changes in the form of the to-be-added 15 per cent could be achieved – which is not likely – that would not substantially change the patterns already in place today, which will necessarily dominate the overall picture in 2020 (Downs, 2001). Basically, the best that the livable communities strategy can accomplish is to slow the pace at which conditions worsen.

Actually *improving current conditions* will require a retrofit of the 85 per cent of the future that is here now. The retrofit must make metropolitan regions capable of quickly adapting their spatial structure by changing the *software* of behavior rather than the *hardware* of buildings. A *cyber strategy* can provide an alternative path to a livable community for all places, not just those expecting development.

CYBER STRATEGY FOR LIVABLE COMMUNITIES

A cyber strategy uses digital network policies and initiatives to change the location of a wide range of urban functions, thereby enabling new transportation options and introducing new economic opportunities. It can also change the social context of technology access, use and control, thereby enabling a strength of place and introducing new civic opportunities.

Cyber strategy is a relatively new option for transportation authorities, local governments and regional planning agencies. It is a phenomenon of telecommunications market liberalization and the resulting emergence of globally integrated, broadband digital networks.

Before the AT&T divestiture in 1984, the regulated public telephone utility provided relatively homogenous service to every region. The subsequent competitive market resulted in regionally uneven levels of investment, services, prices, access, and utilization. At the same time, the flood of innovation spawned by competition provides the opportunity to apply entirely new tools to previously intractable problems like congestion, spatial isolation and urban decline. To realize those opportunities requires three closely linked initiatives moving in parallel:

- Retrofit buildings and existing centers so that their functions are easily re-programmable,
- Modernize the business practices of public and private enterprise so that their functions are easily transportable, and
- Re-invent mobility around the one-mile trip to and between multi-functional centers.

RETROFITTING BUILDINGS AND EXISTING CENTERS

One structural problem with the sprawl model is the large single-function districts that form a complex checkerboard of specializations. The livable communities strategy addresses that problem by mixing land uses, often mixing functions within a single building (e.g., housing over ground floor retail). The idea is to build offices, storefronts and housing close together, even in the same building if possible, in order to fashion pedestrian or short transit access from home to jobs and some services.

Cyber strategy can contribute to that target proximity by providing virtual access to some jobs and services for all residential clusters, regardless of their density. There are two ways to do this. Adding functions virtually to previously single function buildings is one approach. For example, common technologies such as kiosks, the Internet, audio conferencing and interactive video conferencing over Integrated Services Digital Network (a souped-up basic phone line) can help city halls, county buildings, state facilities and federal buildings become multi-government employment and service centers. Public schools can become multi-grade community resource centers. Mini-malls can include a business conferencing center and distance education classroom next to the laundromat and convenience store.

The second approach is based on introducing a *Network Station* to an existing center. A Network Station is a public, shared-use, mixed-use telematics facility that can add functions to a typical village center, retail mall, or office park. A Network Station can be equipped, furnished, and staffed to function as a private office, meeting space, training center, public service counter, classroom, medical examination room, retail shop, etc. for a robust array of public, non-profit and private organizations with their nearest bricks and mortar presence miles away.

In addition to a wide array of functions programmed to satisfy the particular destination-needs of the proximate community, each Network Station will offer public access to a robust suite of technologies, and serve as a portal to e-commerce.

Network Stations can be developed at scales that include the neighborhood, village/town, city, and region. The total number and location of each should be determined by the economic geography and transportation infrastructure of the region. The facility should be established in existing vacant space, allowing it to be deployed in a relatively short period of time.

FOSTERING DISTRIBUTED ORGANIZATIONS

In order for the retrofit of existing centers to function successfully, a second initiative for modernizing the business practices of a critical mass of enterprises (from micro business to large public and private corporations) must be accomplished. For example, colleges must begin to offer distance education classes or there won't be content for distance education classrooms. Demand for shared work stations depends in part on the transition of the regions, public and private enterprises from centralized to distributed structures.

Gartner, Inc., published a *Strategy and Tactics/ Trends and Direction Memo* on January 3, 2002 that addressed many of the issues involved in the transition to what it refers to as the e-workplace. The memo was entitled "Creating Resiliency" with the e-workplace utilizing the theme that collaboration applications and knowledge management systems provide the technology basis for resiliency in the distributed organisation. The memo begins with the following:

> In the aftermath of the attacks on the World Trade Center and the Pentagon, the benefits of distributing operations were readily apparent. The trend toward greater distribution of operations has been evident for almost two centuries and has been accelerated by almost every new emerging information technology since the telegraph. However, until recently, the technology was used to communicate between distributed facilities or to collect and centrally analyze data. It was not until the emergence of PC technology that the physical disaggregation of the workplace itself became possible, first through the exchange of data disks, then through Local Area networks (LANs) and Wide Area Networks (WANs), and finally through IP virtual private network (VPN) technology. Now, new smart, wireless personal devices are extending the disaggregation of the workplace to anywhere that a knowledge worker may be.

In a distributed enterprise, the network is used to coordinate complex problem solving in distributed work teams, and to serve as the main customer service and service delivery channel. Cyber strategy provides a plan for guiding the transition from organizational hierarchies to *network* enterprise structure throughout the region.

REINVENTING MOBILITY

As trip lengths become shorter, markets for new private, short range, low speed, lower cost, *low impact* transportation options should grow. These options include walking, cycles, scooters, motor scooters, golf carts, station cars, and neighborhood vehicles.

Neighborhood trams and smart shuttles are among the public possibilities.

In order to encourage the use of low impact vehicles, Neighborhood Transportation Zones (NTZ) should be established with a one-mile radius around the Network Stations. This will promote mixed-vehicle streets analogous to mixed-use buildings and projects. The innovative public transit options should be initiated within these zones. Rapid bus service routes can connect the largest or most significant Network Stations.

As a greater portion of household trips occurs within the NTZ, a greater portion of trips will be taken via a low performance vehicle or by innovative public transit service. This system of mobility could compare favorably with today's high performance automobility which has the capacity to offer a no-wait, no-transfer, door-to-door service.

As the vehicles get smaller and go slower, consistent with trip limitations within NTZs, less land will be needed for the high performance automobile. The capacity of existing parking lots will appear to increase without any investment as the vehicles parked there become smaller. A low impact vehicle takes-up approximately 20 per cent of the volume of a contemporary 4x4 sports utility vehicle (SUV).

Eventually, some of the land devoted to the high performance automobile can be reclaimed and used for infill development. These infill projects could be designed to provide in bricks and mortar the functionality needed to complete the match between jobs, housing, and services within a one-mile radius. Incentives can be offered for developers to build housing on surface parking lots in retail centers and in office parks. Landscaped walkways, community gardens and even housing projects may be possible in the middle of 128 feet or even 113 feet wide streets found throughout communities built on the sprawl model.

THE LIVABLE COMMUNITIES STRATEGY REVISITED

The relatively low cost and short time to deploy a cyber strategy means that every region, growing or declining, high or low density, with or without rail transit can immediately begin to develop a livable communities plan. A system of Network Stations, mixed-use buildings, and mixed-use streets can be deployed to retrofit a county or region so that a target of satisfying say 75 per cent of today's trip purposes within one or two miles from home can be approached. Under those cyber assumptions, the growth increment could then be more surgically used as a way to add the bricks and mortar facilities that are needed to achieve functional integration at the neighborhood or village scale.

The combination of both bricks and mortar and *bricks and bits* (Siembab and O'Brien, 1999) can create the *match* (not just a balance) between residents and their particular work opportunities and service needs This same combination can also match resources with need in support of urban revitalization and economic development policies. Universal propinquity and economic equity may be achievable. Based on the experience with a cyber strategy for livable communities (Siembab, 1994, 1997), it should be possible to substantially implement the initiatives in at least some regions within five years.

However, to do so requires that transportation agencies, rail transit authorities, local land use planners, air quality regulators, private developers, and especially livable community advocates be capable of engaging in what Graham has called cross-sectoral thinking (Siembab, Graham and Roldan, 2001). To cite just one example of a missed opportunity at cross-sectoral thinking: A survey of rail transit authorities in the US and Europe to determine the strategic uses of fiber networks found virtually no best practices. Rail transit authorities on both continents persistently treated fiber networks developed in the authority's rights of way as a revenue generating asset for the authority rather than as a fundamental element of a cyber strategy for the region.

Just as the livable communities strategy is a political alternative to sprawl, so cyber strategy for livable communities is a political alternative to a pure bricks and mortar approach. Will practitioners and politicians master the required level of cross-sectoral thinking in time to avoid instability and make metropolitan regions more livable in the short run and sustainable in the long run? The race is on!

REFERENCES FROM THE READING

Crouch, D, Garr, D. and Mundigo, A. (1982), *Spanish City Planning in North America*, Cambridge, MA: MIT Press.

Downs, A. (2001), 'Testimony to the house subcommittee on highways and transit: the future of US ground transportation from 2000 to 2020'. Published in *Metro Investment Report*, Abel Publishing, Los Angeles, April.

Safdie, M. (1997), *The City After the Automobile*, New York: Basic Books.

Siembab, W. (1994), *Telework Facilities Exchange: Final Report*, Sacramento, CA: Institute for Local Self-Government.

Siembab, W. (1997), *Blue Line TeleVillage: Final Report*, Sacramento, CA: Institute for Local Self-Government.

Siembab, W. and O'Brien, T. (1999), 'Digital broadband networks for economic development and mobility: a bricks and bits strategy for retrofitting cities', *Journal of Municipal Telecommunications*, 1(1).

Siembab, W., Graham, S. and Roldan, M. (2001), *Using Fiber Networks to Stimulate Transit-Oriented Development: Prospects, Barriers and Best Practices*, Norman Y. Mineta International Institute for Surface Transportation Policy Studies, Report 01–16, October.

'The Rise and Fall of the Digital City Metaphor and Community in 1990s Amsterdam'

Geert Lovink

Editor's Introduction

Can the Internet ever genuinely reflect the complex public cultures of an individual city in a transparent and democratic way? Can the electronic domains of new media relate positively to the public domains of a city? How can public policy-makers, social movements, and activists within a city work to ensure that the potential of the Internet for relating creatively with a city's unique cultural politics is actively explored in ways that challenge the increasingly extreme commercialism of the Internet itself?

In our penultimate reading on cybercity politics and strategy our focus moves from Asia and the USA to Europe. In it the Dutch media theorist and Internet critic Geert Lovink considers the story of Amsterdam *De Digital Stadt* (DDS or The Digital City). This initiative represents an iconic example of the promises and pitfalls that emerge when so-called 'digital' or 'virtual' cities are constructed. The story of DDS is an especially useful way of understanding the difficulties faced when attempting to relate public web services in some explicit or metaphorical way with the social, cultural or political worlds of a real city within the context of increasingly diffused, and commercialised, Internet access.

Analysing the complex history of DDS, Lovink argues that it relates strongly to the broader urban social and political context in 1990s Holland. Founded as a community internet provider in 1993, DDS grew spectacularly and explicitly adopted the interface metaphor of a city to structure its innovative range of free services. As Lovink suggests, the fascination of early Internet users was all about the spatialisation of cyberspace around familiar motifs and metaphors.

DDS took the metaphor of the city further than most. It was constructed complete with a 'town map' (web interface), 'town squares' (thematic zones), 'home districts' (spaces for personal web pages) and a 'mayor' (manager). As a space to support, sustain and reflect the broad spread of social activism, both within and beyond Amsterdam itself, the key aspect of the urban metaphor however, was one of political space. As Lovink suggests, momentum for DDS came from the idea that both the city and cyberspace needed strong electronic public realms to foster democratic discourses in the face of a creeping urban privatism and the increasing corporate takeover of the Internet. The lack of censorship of web content on DDS proved a major attraction to the site as were the operator's financial independence and democratic principles.

The analysis of the decline of DDS, and its restructuring as a normal Internet Service Provider (ISP), is an especially interesting feature of Lovink's analysis. He shows how the success of the DDS city interface, in stimulating, and organising, a wide range of 'virtual' communities, was difficult to sustain financially as the Internet became a mainstream medium. Gradually, the truly local content of DDS – that which originated from, and related to, Amsterdam itself – became swamped by the growth of the more usual, placeless and deterritorialised English language content from around the world. Thus DDS became less 'grounded' in Amsterdam itself (Aurigi and Graham, 2000). In addition, many of the net activists who had fuelled the growth of DDS became successful web entrepreneurs in their own right as Amsterdam took off as a major new media capital, and its urban and media spaces succumbed to gentrification (see Rebecca Solnit and Susan Shwartenberg, p. 296).

As Lovink concludes, the story of DDS brilliantly demonstrates the tensions between local, place-based new media activism and the delocalising logic of the Internet. It hammers home the difficulties of starting and sustaining genuinely independent and non-commercialised spaces for democratic exchange on the Internet in times of intense commercialisation. And it underlines the ambivalence of deliberately using the familiar icons of modern urbanism as metaphorical instruments to organise, and visualise, the chaotic and distributed worlds of electronic domains.

Information wants to be space.

(Erik Davis)

The Amsterdam Digital City, founded in 1993, was one of Europe's largest and best-known community Internet providers. It was a 'freenet', offering free dial-up access, free email and webspace, within which many online communities formed. Over a period of eight years the Digital City (www.dds.nl) went through phases of spectacular growth and change, antici-pating and responding to Internet developments at large. Reflecting its actual and symbolic significance, research about DDS communities and the history of DDS also expanded. The privatisation of its online community services in late 2000 sparked a fierce debate amongst users. Attempts were made to keep the public domain community parts of DDS out of the hands of commercial interests. By mid-2001 the turbulent history came to an end with the closure of the free access services. As of 1 August 2001 DDS trans-formed into an ordinary ISP, reselling broadband DSL services to a largely reduced customer base. Even though the company retained its name, no traces were left of the once mighty interfaces and web pages of tens of thousands of users.

Considerations presented here are to be understood within the specific Dutch context of the 1990s, a period of fierce neo-liberalism in a country once known for its opulent welfare state. Dutch independent Internet culture, driven by a demand for public media access, grew up in the economically fragile post-recession years of the 1990s in a climate of permanent budget cuts in the state funded cultural sector. Non-profit Internet initiatives therefore had to find new ways to operate in between the state and the market. The Digital City story tells of the difficulties in building up a broad and diverse Internet culture within a Zeitgeist of the 'absent state' and the triumph of market liberalism.

Early Internet users were fascinated by the spatial-isation of cyberspace. In this mythological-speculative stage of the Internet a name like Digital City appealed to the imagination of many. Although maps were made to help navigate the vast system, the city metaphor was used in a restrained way. The spatialisation (see Plate 36) was neither a representation of a computer network nor a simulation of an actual city. 'City' in this context was used as the house of political culture. This was by definition both local and global. The two were not seen in an antagonistic relationship. The same could be said of the conceptual pair 'real' and 'virtual'. During the roaring nineties the one was fuelling the other. The facilitation and mapping of a rich collection of virtual communities was exciting enough. It was all about fuelling the imagination and opening up new spaces, not in terms of colonisation; more along the lines of Italo Calvino's invisible cities. The economic factor only came in later, once the caravan was well under way. For a brief moment in time there was the collective illusion that the Internet was a temporary autonomous zone (a term coined by Hakim Bey), not yet occupied by the powers that be.

By the early 1990s the (in)famous Amsterdam squatters' movement, which had dominated the social and cultural (and law-and-order) agenda of the pre-vious decade, had petered out in the city's streets. But its autonomous yet pragmatic mode of operation had infiltrated the workings of the more progressive cultural institutions. The autonomous movement of the 1980s had successfully occupied both urban spaces and the electronic spectrum (free radio and even a brief chapter of pirate television). The movement had built a sustainable alternative infrastructure beyond street riots and political conflicts. It was the time that the cultural centers Paradiso and De Balie, which were both at the vanguard of local cultural politics, embraced the 'technological culture' theme in their programming. In the beginning, this took the shape of

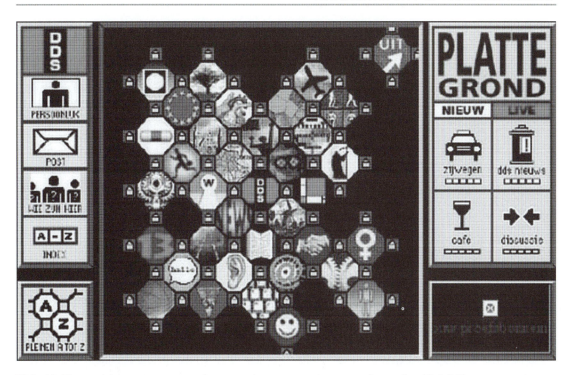

Plate 36   The spatial town square metaphor as used to structure content on Amsterdam Digital City.
(Source: Geert Lovink.)

critical, if somewhat passive, academic observations of the technologies surrounding us and of their risks, but it quickly evolved into a do-it-yourself approach. Technology was no longer seen as the exclusive preserve of science, big business, or the government. The Amsterdam agenda was one of activism, not research. Why wait? It was time to shape the information society. Technology could become the handiwork of average groups or individuals. Mass availability of electronic hardware and software had created a broad user-base for 'low-tech' applications, something that in its turn spawned feasts of video art, robotics and other forms of 'industrial culture', free radio and public access television and well attended cultural events where technology was rearranged and playfully dealt with. The 'home neighbourhoods' within the DDS were a manifestation of this culture (Plate 37).

Elected politicians meanwhile were struggling with another 'situational' problem: that of their very own position amidst fast dwindling public support and sagging credibility. This was – not surprisingly – blamed on a 'communication deficit' for which a substantial application of 'new media' suddenly appeared to be an instant antidote. The clue was not lost on De Balie cultural center which approached City Hall with a freenet-based proposal to link up the town's inhabitants through the Internet so that they could 'engage in dialogue' with their representatives and with the policy-makers. The system itself was to be installed by the people at HackTic Network, the only group of techies at that time that was readily available – or affordable. The Digital City was launched in January 1994 as a ten-week experiment in electronic democracy. The response from the public was overwhelming. And, in no time, 'everybody' was communicating with everybody else. With one exception though: the local politicians never made it to the new medium.

Amsterdam Digital City was an initiative of Marleen Stikker, the later director of the Society for Old and New Media (www.waag.org), then a staff member of De Balie. Before 1993 Marleen had organized projects on the crossroads between theater, new media arts and public debate in which technology always played a key role. In the festivals she got artists to work with interactive television, voicemail games, live radio and video-conferencing systems. The freenet model was imported from the USA where early citizens' networks, such as the one in Cleveland, were already operational. The independent or tactical media element of freenets, run as non-profit initiatives, was combined with another

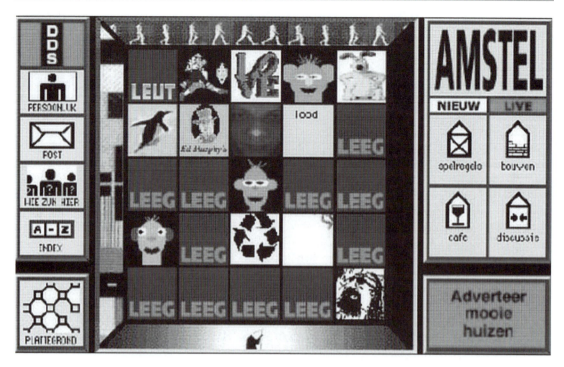

Plate 37 A 'home neighbourhood' within Amsterdam Digital City. (Source: Geert Lovink.)

rumour that had blown over the Atlantic, the 'electronic town hall'.

The idea in Amsterdam was that only an independent public domain could guarantee 'electronic democracy' (comparable to the role of the print media). It was not up to the state or local governments alone to decide how the political decision-making mechanism was going to be transformed in the future network society. In order to get there the citizens themselves had to be empowered to use technology in their own, often weird and seemingly irrelevant ways. What had to be prevented, in the eyes of the Digital City founders was a 1:1 copy–paste from the 'old' days of mass democracy – with its political parties, television and the power of media moguls – into the new electronic era. In order to prevent this from happening the Amsterdam group decided not to write manifestos or reports with recommendations but to take the avant-garde stand and move into the terrain as soon as possible: establish a beach head, land as many troops as possible and occupy the entire territory.

Around mid-1994 a legal structure was formed: a non-profit foundation with a board consisting of experienced administrators, all of them neutral outsiders. The foundation had no legal ties to the users and the employees were not represented either. The Digital

City freenet was founded as a cultural organisation, not as a business. In 1994 the dot.com years were still a way off, despite *Wired* magazine giving a glimpse of what was about to happen. The Digital City had other ambitions – political ones. It was important to get normal citizens involved in shaping the medium which, until then, had only been used by academics and hackers. The architecture of the network society was at stake but who was going to design it? The commercial tidal wave was about to happen, that much was clear. But would commerce really empower average users and guarantee the cyber rights US-American libertarian gurus so heavily promoted? With the history of radio and television in their minds, the fight over a public domain within cyberspace couldn't start early enough.

The prime cause of the Digital City's success was the freedom it granted to its users from the very beginning. This may sound trivial, but it is not, especially if you take the increasing control over net-use in universities and corporations into account (especially outside the Netherlands). Awareness of privacy issues, corporate media control and censorship was high and the necessity to use cryptography was felt early, as was the right to anonymity while communicating via the Internet. The Digital City did not turn into a propaganda mouthpiece for the City Hall, under the guise of 'bringing politics closer to the common people thanks to information technology'. The DDS-system was not the property of the municipal corporation, even though many people assume this to be the case. In fact, DDS never received substantial subsidy from the municipality (the city council was one its biggest customers, though). In the end the 'netizens' were far more interested in dialoguing among themselves than engaging in arcane discussions with closed-minded politicians.

DDS looked for a balance whereby virtual cultures grew optimally without politics being discarded altogether. A precondition for this was the community system's independence. But that was costing money, and quite a lot to boot. By 1998 DDS had increasingly grown into a business with 25 employees and 70,000 regular users while wishing to retain its not-for-profit character at the same time. The management under Joost Flint was pursuing a policy of courting a handful of major customers who brought some serious money in. It was all about attracting projects which would fit into the DDS set-up, but that wasn't a totally frictionless process. DDS was divided into three components: a commercial department that hunted for the hard cash, an innovation wing that developed applications for corporate customers, and the community aspect.

The 'virtual community' image was never really appropriate in this case. After a few years of hypergrowth DDS had turned into a multifaceted amalgam of small communities who shared the intention of perpetrating the DDS system as an 'open city'. If anything DDS was a facilitator for communities, not a community itself. It is there that the central interface of the DDS played a key role. The graphic user interface (designed mainly by Marjolein Ruyg – Plate 36) was made in such a way as to provide an overview of the mass of information on offer. In keeping with the name of the system, the DDS web interface was built around the notions of 'squares', 'buildings/homes', and '(side-) streets', but it did not show pictures or simulations of the actual (Amsterdam) city-scape, as many people expected. There were, for instance, 'squares' devoted to environmental issues, sports, books, tourism, European affairs, women, gay and lesbian issues, information on drug use, social activism and both local and national government. In between the squares there were tiny house icons pointing at the thousands of homepages (Plate 37). DDS also had its own cemetery, a web memorial for those who had passed away. Unlike Yahoo-type web directories,

the interface was not pretending to give a full representation of the underlying activities. The central interface worked more like a guidance to give the vast project a look and identity without presenting itself as a portal.

Another question pertained to the much-vaunted urban metaphor of the Digital City. What about its strictly local role – would that dwindle into insignificance? As a free community service provider DDS faced the paradox that the local significance and the global 'non-located' online components were both growing exponentially. A few years after its launch no more than a quarter of the 'inhabitants' actually lived in Amsterdam, yet DDS remained a Dutch-language site. The management for a long time maintained that upholding the Dutch language was a legitimate aim. For many users it was difficult to express themselves in English. The Internet was increasingly used in a very local or regional context, for example one could go online to check out the program of the nightclub next door, or when the movies would start, etc. At the same time DDS never tried to impose its own (local) metaphor onto users.

By the late 1990s, Amsterdam, long known for its large and diverse alternative social movements, faced some major shifts in its cultural landscape. Dot.com baby suits took over the public image of the gentrified town. The once solidly unconventional activists had in large numbers relocated themselves as creators and managers in the so-called new media culture, which was largely (though not exclusively) ICT-driven. For quite a time after it started to come into its own, this new cultural landscape had remained remarkably free of influence from mainstream or commercial interests. With squatters outlawed, politics declined. The new media scene morphed into something very different from what the Amsterdam model of public digital culture with Digital City as one amongst many projects had become famous for.

In itself the notion of a public sphere within the media has already been solidly entrenched, thanks to the policy of the municipality to cable nearly every household by the early 1980s, and to manage the system as a public utility like the water or electricity supply. So this approach was expanded into the realm of Internet access provision and associated new media facilities without much difficulty. However, the ongoing onslaught of 'the market', and of its attendant ideology of commercialism and privatisation proved increasingly difficult to resist. As in many other global cities Amsterdam in the late 1990s got into the firm grip of 'dot.com mania'. With hindsight, what was actually amazing was how long the new media culture had remained nearly immune to the dictates of the corporate sector. Partially, this had been due to the fact that the traditional elite took a fairly lenient and sometimes even supportive view of this state of affairs. But at the same time they kept resolutely clear of any involvement in it, this according to the hallowed Dutch 'polder model', which established a delicate consensus between state, business and trade unions on the basis of non-regulation.

Five years after its founding, at the height of dot.com-mania, the Digital City had evolved from an amateur, low-tech, non-budget grassroots initiative into a fully professionalised, technology and business driven organisation. And this culminated recently in its transformation from a non-profit foundation into a private sector ICT venture. Come December 1999, the astonished 'inhabitants' learned that the directorate of the DDS had opted for a corporate framework, and that community building and support were no longer a paramount objective. By 1998–9 the free DDS facilities were available everywhere. Scores of new commercial providers and services had popped up all over the place (such as Hotmail, Geocities and even free dial-up providers), offering the same services (often more extensive, better ones) than the DDS was able to provide. The free Internet services advertised massively and attracted a customers pool far removed from the idealistic concerns that used to inform the original Digital City. This resulted in a substantial quantitative, but more importantly, qualitative erosion of the DDS user base. Even if the absolute number of accounts had risen to reach an all time high mark of 160,000 in early 2000, an analysis of the use patterns showed that these could no longer be considered conducive to community building or even to socio-politically relevant information exchange – homepage-building and upkeep for instance, no longer attracted much interest. The once so valuable web site had turned into empty lots. Despite an overall growth of Internet use the Digital City had lost its attractiveness for users.

The strategic issue raised here relates to the problem of local and global. Net activists and artists are confronted with the dilemma between the presuming friction-free machinic globality and the experience that social networks, in order to be successful, need to be rooted in local structures. Internet culture pops up in places where crystals of (media) freedom have been found before. At the same time the net is constantly subverting the very same local ties it grows out of while creating new forms of 'glocality'. The choice global or local is a false one. Even though urban and spatial metaphors in general may have exhausted themselves there is little to be found in the mathematical emptiness of 'pure' disembodied virtuality. Discontent within the Digital City project in the spatial metaphor existed right from the start. Due to Dutch pragmatism no 'metaphor police' was established to look into identity, language and nationality. In that sense DDS was, more than anything, a rich social experiment in Internet freedom with only a few hints of what political liberty in the technological future could look like. The lesson of the Amsterdam Digital City is, if anything, an economic and legal one, and deals with internal democracy: the high art of staying independent in an increasingly commercial environment, no longer being able to rely on government support in matters of public interest.

Editor's references and suggestions for further reading

Aurigi, A. and Graham, S. (2000), 'Cyberspace and the city: the "virtual city" in Europe'. In G. Bridge and S. Watson (eds), *The Blackwell Companion to Urban Studies*, Oxford: Blackwell, 489–502.

Beamish, A. (1999), 'Approaches to community computing: bringing technology to low income groups'. In D. Schön, B. Sanyal and W. Mitchell (eds), *High Technology and Low Income Communities*, Cambridge, MA: MIT Press, 349–70.

Beamish, A. (2001), 'The city in cyberspace'. In L. Vale and S. Warner (eds), *Imaging the City*, New Brunswick: Rutgers Center for Urban Policy Research, 283–300.

Brants K., Huizenga M. and van Meerten, R. (1996), 'The new canals of Amsterdam: an exercise in local electronic democracy', *Media, Culture & Society*, 18, 233–47.

Graham, S. and Aurigi, A. (1997), 'Urbanising cyberspace? The scope and potential of the virtual cities movement', *City*, 7/8, 18–39.

Ishida, T. and Isbister, K. (1998), *Digital Cities: Technologies, Experiences and Future Perspectives*, Heidelberg: Springer.

Lovink, G. and Riemens, P. (1998), 'The monkey's tail: the Amsterdam digital city three and a half years later'. In INURA (International Network for Urban Research and Action) (eds), *Possible Urban Worlds: Urban Strategies at the End of the 20th Century*, Berlin: Birkhauser-Verlag, 180–5.

Lovink, G. (2002), *Dark Fiber: Tracking Critical Internet Culture*, Cambridge, MA: MIT Press.

Tsagarousianou, R., Tambini, D. and Bran, C. (eds), *Cyberdemocracy: Technology, Cities and Civic Networks*, London: Routledge.

Van Lieshout, M. (2001), 'Configuring the digital city of Amsterdam: social learning in experimentation', *New Media and Society*, 3(2), 131–56.

'Public Spheres and Network Interfaces'

Andreas Broeckmann

Editor's Introduction

Our exploration of cybercity politics and strategy closes with an examination of how artists across the world are experimenting with ICTs as means to try and revitalise, or at least augment, urban public domains. This reading, by Andreas Broeckmann, the Director of the Transmedia ars festival in Berlin, links neatly with Geert Lovink's analysis of Amsterdam Digital City (p. 371).

Like Lovink, Broeckmann points out some of the problems involved in using ICTs to revitalise, or merely identify, the multifaceted public domains of contemporary cities. He begins by analysing the deep transformations which are currently underway in urban public domains, as classical notions of public space-based interactions give way to plural, hybrid and technologised ideas of what urban public realms actually are (see Section VII). The key here is that the city, as a site where streets, spaces and face-to-face interactions blend with ICT-mediated ones, can be seen as both an interface to, and the generator of, a wide range of different publics at a range of scales. As economic, social and cultural interchanges become closed-off into ICT systems, however, so the risk of privatisation and commercialisation threatens to undermine the public roles of urban physical spaces.

Broeckmann notes, in this context, that a wide variety of artistic interventions are now emerging at the interfaces between urban public spaces and ICTs. These are attempts to work against the dominant logics of fragmentation, surveillance, and privatisation. They are assertions of the uncontrollability, and democratic potential, of hybrids which blend ICTs and city public spaces. And they subvert, and usurp, technologies of surveillance and capsularisation for other uses.

These interventions and installations thus celebrate the tactics and technologies of conflict and participation. In the process they seek to reclaim public domains with, and for, multiple heterogeneous groups in cities. Crucially, they do this by forcing the often invisible and ICT-mediated interactions in the public domains of cybercities out on to the political visibility of city streets.

Broeckmann finishes by reflecting on some of the most influential examples of ICT-based artists' interventions in urban public realms, imagining these projects as new types of urban interface, or border zones, between embodied presence and mediated exchange. Rafael Lozano-Hemmer's installation in the main Zócalo square of Mexico City for the City's millennial celebration, for example, allowed people from all over the world to programme a series of huge searchlights over the net to light up the city's sky in the pattern of their choice. Other interventions project the texts of ICT and mobile phone-based exchanges on to city streets, walls or subway stations (see, for example, http://www.rude-architecture.de).

The key in such interventions is to connect many layers, many scales, and many publics into a visual hybrid of interface, public space, and urban design. The challenge, Broeckmann notes, is to do this in a way that celebrates the unpredictability and the essential uncontrollability of the social and cultural worlds of cybercities. Such interventions by artists are important in helping us to imagine public domains for cybercities which amount to more than a retreat from the spontaneous interactions of embodied subjects in physical space (other than for surveilled shopping in themed consumption zones).

It is imperative, however, that these temporary experiments emerge not just as one-off demonstrations by artists. The real challenge now is to develop ambitious and imaginative urban policies which will ensure that hybrid installations of ICTs and public spaces become normal, taken for granted, elements within the everyday landscapes of cities.

THE PUBLIC DOMAIN

The public domain is conventionally viewed as an assemblage of urban social spaces – the market place, theatres, libraries, cafés, etc. – where ideas and practical concerns of society can be voiced and discussed in an open dialogue (Habermas, 1989). Even here, the public sphere that is constituted at these sites is not of itself neutral and democratic, but invested with political and economic power. Think, for instance, of the exclusive English clubs and coffee houses of the eighteenth century where a new class of industrialists and traders negotiated their political influence, or the large city squares where the architectural and ritual presentation of power determines the horizontal expanse of the public site.

The notion of what is 'public' is currently undergoing a deep transformation which is brought about by a variety of geographical, economic, geopolitical, technological and discursive drivers of social change. The public sphere, and the way in which individuals and groups participate in it, is taking on a whole different set of meanings, compared with the classical forms of the civic public that emerged in the eighteenth and nineteenth centuries. Most importantly, it seems to be impossible today to speak of only one public domain, but we have to think of the public domain as a heterogeneous, at times hybrid, at times fragmented dispersion of physical and virtual spaces populated by different types of 'publics'.

The public domain is a composite of many stages and layers, open and closed physical spaces, media channels, forms of communication and cultural expression. The city is still a prime site of experimenting with the new public domain; it is a zone of tension in which social conflicts and instabilities are made productive. The city can be seen as both an interface to, and the generator of, new interfaces to the different publics. At the same time, the changing media sphere (telecommunication, broadcasting, WWW, etc.) is creating a translocal topology of the public domain which is both embedded in, and which reaches beyond, the urban territory (Section VII).

Media technologies play a crucial role in the current curbing of the urban public domain. Developments such as the transfer of economic activities into closed electronic networks, the privatisation of public areas, the homogenisation and the separation of functions by allocating specific zones to shopping, entertainment, housing, traffic, etc. – seek to make public space more secure and more efficient. But they undermine the essence of the public domain as a zone of uncontrollability. The instability of the public domain is a condition of its democratic potential. Artistic urban interventions strive to counteract the safely surveilled and appeased urban terrain of the transparent city with its technologies of security and privatisation – zoning, surveillance cameras, biometric systems, etc. – by means of tactics and technologies of conflict and participation, reclaiming the public domain with and for multiple heterogeneous groups (Marchart, 1999; Deutsch, 1996).

The challenge in the creative use of media technologies is to foster the diversity of public actors and terrains and to develop strategies of articulating the new public domains that connect physical urban spaces and the potential public sphere of the electronic networks. This public sphere will only come into being if there are complex forms of interaction, of participation and learning, that use the technical possibilities of the networks and that allow for new and creative forms of becoming visible, becoming present, becoming active, in short, of becoming public.

VISIBILITY, PRESENCE AND ACTION

Visibility, presence and action define three levels of being in public. They are also the main fault lines of public conflict. Invisibility is one of the stigmata of modern society. Whether in the case of minorities, social movements or special interest groups, achieving visibility, the visible factuality of a particular social group, can already be an important achievement towards the goal of political effectiveness. However, visibility is always already connected to the possibility of control. Like other forms of public policing, the increasing use of surveillance cameras in public and private spaces indicates that visibility is a matter of concern for the public order. Whatever is publicly visible always already bears the potential of being illegal, indecent or otherwise unacceptable. Becoming visible in an urban environment is therefore often conducted as a clandestine activity, as in the illegal pasting of posters, the placing of graffiti and tags, pirate radio broadcasts, taking on an artificial identity on the Internet, or the elusive appearance in larger masses of people. In these medial strategies, the regimes of visibility, anonymity and identity are locked in a problematic imbalance (Turkle, 1997).

The notion of visibility is connected to a fleeting, impermanent perception in time and is therefore, at best, a tactical element of being in public. In contrast, the notion of presence – not necessarily coupled to a visibility – makes a claim to 'being here, now' and is an affirmation of a public status. Presence is immediate, which means that it cannot be realised through medial representation. At the same time, presence is affirmed in the form of a mediality, it always manifests itself in a medium: graffiti have presence as signs and images in the urban space, radio programmes manifest a presence on the air waves, squatting is the physical occupation of a space declared empty and unused. Presence is the affirmation of an identity and of a Now in a specific medial environment.

Neither visibility nor presence can in themselves provide the potential for becoming publicly active. This notion of public action relates to a form of political agency which aims to effectively transform a material and symbolic situation by means of argument and through more or less immediate performative acts. Possibilities for such actions are on the decrease where symbolical representation, mediated participation and the equation of consumption and democracy, have replaced a sense of active presence and involvement in public matters. Being in public is identified with potential illegality on the one hand, and with the danger of personal harm on the other. Reclaiming public domains as sites of constructive conflict, and developing democratic forms of agency for the new intersections of virtual and physical public environments, are therefore imperatives of the current situation.

THE INTERFACE

The interface is the connecting zone of two separate elements in a networked system. Most commonly, the interface is understood as the tool through which the communication between the human user and a computer is facilitated, or the connecting plugs or software tools that allow for different technical apparatuses to communicate with each other. More generally, the interface is the border zone, the in-between shared by different systems where the exchange of signs and data translates virtualities into potential effectiveness.

The interface is the site of potential agency, and it has therefore been of central concern in some of the artistic experiments that have sought new articulations of media technologies and human actions in urban and translocal environments. Think of the inner-city projects by the German artist and architect Christian Möller (http://www.arc.de/cm) in which the combination of chance or uncontrolled natural effects with the concrete yet uncoordinated actions of multiple users, creates surprising aesthetic results that oscillate between ambient noise and sublime expressiveness. The interfaces that connect the different levels of agency and make them visible and audible, facilitate an experience in the participants of a creative collectivity in an orchestrated, yet open and unpredictable located process.

The projects of Mexican-Canadian artist Rafael Lozano-Hemmer, in contrast, pay great attention to the action of the individual, both in relation to the technological system and in relation to other participants. The installation *Alzado Vectoriel* (1999–2000) created a setting in which the choices of individual Internet users were immediately translated into a widely visible configuration of powerful spotlights over the central square in Mexico City (see Plate 38). The local experience of the urban space was co-determined by the tele-absent net users who, in turn, received a stream of images back as an indication of the feedback on-site.

A very different structure of mutual influence and interlocked communication was developed by Lozano-Hemmer in the project *Relational Architecture #3 – Re:Positioning Fear* (1997). Here, an online discussion about different aspects of fear was used as the – still legible – visual material for a large-scale projection onto the historical building of a military arsenal in Graz (Austria). The projection of the unfolding discussion text was placed in the shadows cast by

Plate 38 'Vectorial Elevation, Relational Architecture 4', 1999–2002. Large-scale interactive intervention with eighteen robotic searchlights controlled by the public over the Internet at http://www. alzado.net. First installed at the Zócalo Square in Mexico City. (Photograph by Martin Vargas.)

visitors of the square in front of the arsenal. This effected a tight visual and metaphorical relationship between local and online participants. The presence and the movement of the on-site visitors became both the condition of the presence and legibility of the online discourse, and its projection screen, turning the on-site participants into the interfaces, the virtual objects and subjects within this translocal situation (see http://xarch.tu-graz.ac.at/home/rafael/fear).

Successful artistic interfaces like these connect different publics and urban layers, articulating physical spaces with networked communication spaces and facilitating what the Amsterdam-based group De Geuzen call 'social interfacing', i.e. a fragmented and heterogeneous system of engaging different publics in a variety of specific ways. Such interfaces are themselves dynamic and have the potential for being transformed and re-invented in relation to the inherent changeability of the social and cultural environment in which they operate. A prime example for this is

the artistically inclined anti-corporate campaign of the US-based online group RTMARK whose web site is a highly dynamic creative and interactive tool through which they intervene into specific, socially and politically agitated situations (see http://www.rtmark.com).

The artistic research into interfaces of translocal urban interventions, then, is also an investigation into the actors and the agents of the new public domain that emerges in the overlapping zone between physical spaces and electronic networks, their medial tools and the terrains where their agency can unfold. A project like Marko Peljhan's Makrolab, a platform for research into global communications systems, marks the fact that, very often, such artistic projects no longer represent and no longer allow for passive consumption by an audience (see http://makrolab.ljudmila.org). Active participation is not an option, but a condition for the aesthetic experience, an aesthetic experience which is not to be found in an objectifiable processuality, but in the ongoing, transitory process of communication and exchange itself.

THE SUBJECTIVATION OF GROUPS

The subjectivation of groups is one of the prerequisites for an active appropriation of the new translocal public domain that connects distributed physical and virtual spaces. Artistic projects like those of Denis Beaubois, who ironically plays with and reverses the subjectifying gaze of public surveillance camera systems, deal with the question of potentials for agency on an individualistic level (see Beaubois, 2001). They follow the Situationist tradition of urban intervention with its key strategies of estrangement, drifting and camouflage. However, artistic interventions which seek to deal with the translocal urban environment on a more political level require the involvement of larger groups of people and their engagement in communicative and cooperative processes.

Closely associated with the current internal crisis of the western democratic systems and the social effects of globalisation is the demise of collectivities which represented the hallmark of social renewal of modernity. From the public demonstrations of the bourgeois revolutions of the eighteenth and nineteenth century, through the formation of the workers' and trade union movement, to the development of the new social movements in the 1960s and 1970s, the more or less organised collectivity of independent and ideologically united groups constituted a crucial political factor in the public domain of western societies. In the face of a globalised economy and the concomitant ideological changes, this modern collectivism seems to have come to an end, having been replaced by ideologies of culture, identity, life-style and consumption.

In the year 2000, the media channels and interfaces available for bringing forth new forms of communication and cooperation still hardly exist. Multiple experiments for 'group-ware', online communication environments, 3-D multimedia spaces in virtual reality, etc., are underway, but neither the tools nor the necessary changes of mentality in the users seem to be in place for effectively using these. Electronic mailing lists and chat communities remain the most vibrant, distinctively low-tech cooperative environments, and a web-based collective memory project like akaKURDISTAN, initiated by Susan Meiselas and

Picture Projects in 1997, deploys very simple upload and commentary tools for fostering a joint historical identity of the worldwide Kurdish diaspora (http://www.akakurdistan.com).

For the evaluation of the political effectiveness of networked media tools it is important to realise that they rely on connectivity in the sense of the quality of an individual's access to the networks, a factor that defines the degrees of freedom to act in the translocal public domain. The forms of agency that emerge in such networked environments are neither individualistic nor collective, but rather connective. Whereas the collective is ideally determined by an intentional and empathetic relation between different actors, the connective rests on any kind of machinic relation and is therefore more versatile, more open, and based on the heterogeneity of its components or members. The politics of the connective is, as yet, not understood, though experiences like those of the online developer communities of UNIX, Linux and other operating systems, or the global protests in such recent cases as the World Economic Forum (WEF) and World Trade Organisation (WTO) meetings, the legal attacks on E-toy, or the discussions around the ICANN organisation, form useful material for initiating such research (see the Net time mailing list at http://www.nettime.org).

An experimental set-up for the investigation of such new forms of connective agency and their technical and social conditions, was created by the German–Austrian artists group Knowbotic Research in the IO_Dencies project series (1997–9) with groups of people in Tokyo, São Paulo, Venice and the German Ruhr Area, and other nonlocated Internet users (Knowbotic Research, 1997, 1998; Lazzarato et al., 1999). Collaborative online tools were specially designed to help stimulate and articulate current discursive interests of the participating individuals, mostly related to questions of urban development and the possibilities of public action. The online tools combined the possibilities of direct intervention into a medial environment through texts and images, with the manipulation of material provided by other participants, and the machinic reconfiguration of the intentional designs on the basis of multiple, tendential parameters of usage and the intensity of interests for certain materials or topics. In these projects, the connective relation between the participants emerged as an effect of the comparability and vicinity of individual strategies and interests, rather than on the basis of an a priori ideological agreement. What can develop from such assemblages of interests are social formations which are formally 'machinic', yet which have the potential for becoming active in a medial and urban environment that prioritises tendential over intentional forces.

THE EFFECTIVENESS OF CONNECTIVE FORMS OF PUBLIC AGENCY

It is one of the presuppositions of Knowbotic Research and others investigating this terrain, that digital interfaces should neither be seen as tools for mere representation, nor that the action in the present which they facilitate can follow the binary logic of the 1/0, on/off. The interfaces challenging these regimes of signification are intermediate zones, open force fields, in which potentials for action emerge as machinic, i.e. technical and user-dependent tendential shifts, intensifications, breaks, etc. Strategies of effectiveness will emerge, as François Jullien maintains, not on the basis of subjective intentions, but by paying attention to the circumstances which make a certain effect possible: 'The potential depends on the circumstantial conditions . . . therefore it is paramount to use the potentiality of the circumstances' (cited in Julien, 1996). The strategic exploitation of such potentials of agency will be a decisive factor in the struggle for defending the latitude of medial and public spaces.

REFERENCES FROM THE READING

Beaubois, D. (2001), 'In the event of amnesia, the city will recall.' Accessed at: http://www.dirtymouse. net/emg/amnesia.htm

Deutsch, R. (1996), *Evictions: Art and Spatial Politics*, Cambridge, MA: MIT Press.

Habermas, J. (1989), *The Structural Transformation of the Public Sphere*, Cambridge: Polity Press.

Jullien, F. (1996), *Traité de l'efficacité*, Paris: L'Éditions.

Knowbotic Research (1997), 'The urban as field of action'. In V2_Organisation (ed.), *Technomorphica*, Rotterdam, S.59–75.

Knowbotic Research (1998), 'IO_dencies: Questioning Urbanity'. In V2_Organisation (ed.), *The Art of the Accident*, Rotterdam, S.186–206.

Lazzarato, M., Reck, H.-U. and Hardt, M. (1999), 'IO_lavoro immateriale'. In P. Weibel (ed.), *Offene Handlungsfelder – Open Practices*, Köln, S.167–213.

Marchart, O. (1999), 'Art, space and the public sphere(s)'. In A. Lechner and P. Maier (eds), *Stadt Motiv*, Vienna: Silene.

Turkle, S. (1997), *Life on the Screen: Identity in the Age of the Internet*, London: Phoenix.

Editor's suggestions for further reading

Lozano-Hemmer, R. (2000), *Vectorial Elevation*, Mexico City: Conaculta.

Weibel, P. and Druckrey, T. (2000), *Net_Condition: Art and Global Media*, Cambridge, MA: MIT Press.

IX CYBERCITY FUTURES

Plate 39 Internet café in central Amsterdam, 2001. (Photograph by Stephen Graham.)

INTRODUCTION

> Cities are embodiments of desire, they are complex intersections of hopes and fears. Multi-layered and multifac-
> eted, they direct our attention to future and past. They are expressions of the geography of imagination; they are
> represented in imaginary geographies. They are tapestry-like fragments of past, present and future. It is here – in
> cities, in daily metropolitan life – that the unknown desires of anonymous strangers mingle.
>
> (Lipman and Harris, 1999, 729–30)

How can one engage in a sophisticated way with the future nature of city–ICT relationships when the utopian
and technologically determinist dreams of transcendence that we encountered in this book's introduction
underline the very severe risks of such engagement? How can a collective imagination of a desired and
normative future for city–ICT relations be constructed whilst avoiding the dangers of falling into the traps of
rationalism, a depoliticised technological determinism, or assumptions of a universal human subject?

The futures industry

Predicting the future of society, technology, and cities is a major industry sustaining a vast output of films,
books, magazines, journals, advertisements, video games and new media. The Internet is saturated with
futuristic speculations about the ways in which biometrics, surveillance, personal computing and ICT-linked
cars, software, social practices and urban spaces will develop in the future within an increasingly
internationalised and urban capitalist society. Indeed, as we saw back in the introduction, the painting of
vivid scenarios about the futures of cybercities is actually a key element in the construction and constitution
of the hydra-headed monster of 'information age' rhetoric itself. Such rhetoric draws inputs from architecture,
urbanism, science fiction, futures studies, geography, urban planning, engineering, art, business studies,
economics, anthropology, sociology, cultural studies and journalism.

Given the near impossibility of actually predicting future directions of social and technological change,
futuristic speculation about technology and cities tend to say as much about our own times as it does about
the nature of future worlds. Future cybercity scenarios and portrayals tap into, and feed, the Zeitgeist of urban,
technological, and media culture within contemporary societies. They resonate closely within a long line of
now classic utopian and anti-utopian predictions and scenarios about the future of cities, politics, culture,
industrialism, surveillance, the state and technology such as the classic novels *The Shape of Things to
Come* by H. G. Wells (1933), *Nineteen Eighty-Four* by George Orwell (1949) and *Brave New World* by
Aldous Huxley (1932) , and the film *Metropolis* by Fritz Lang (1926) (see Kumar, 1987; Schaer *et al.*, 2000).

As with these classics, contemporary cybercity futures give space to the dreams, fantasies, and, very often,
nightmares that stalk and shape both collective memories of the urban and technological past, and collective
imaginations of contemporary and future urban and technological worlds. They tend to project perceived
current trends onto future scenarios, often in an extreme form. The glamorous, futuristic technological visions,
or dark, dystopian portraits that have tended to dominate scenarios of cybercity futures therefore tend to
reflect, as Joseph Corn puts it, 'the experience of the moment as well as memories of the past. They [are]

imaginative constructs that [have] more to say about the times in which they were made than about the real future' (1986, 219).

Not surprisingly, future cybercity scenarios also tend to draw directly on the pre-existing ideological and political leanings of their creators. Drawing on Enlightenment notions of progress, mobility and the universal subject, for example, technology producers and their representatives have worked very hard to portray ICTs as largely benign, positive and egalitarian forces bringing liberation and emancipation for all in their wake. Here, as with Bill Gates's manifesto of a 'friction-free capitalism' (1995), cyberlibertarianism has tended to blend seamlessly with neoliberal discourses of macroeconomic management, 'globalisation', and restructuring which have dominated the world since the 1980s.

At the same time, however, producer-led discourses of the futures of cybercities have also tended to deny the social, economic and environmental problems and crises that ICTs are associated with. Problems and crises of social polarisation, economic collapse, social isolation, information overload, net addiction, intrusive surveillance, intensifying corporate power, job shedding, vulnerabilities to failure, increased transportation, poor labour conditions, and the environmental crises caused by dumped, obsolescent technologies in the global south have been conspicuously absent from such scenarios.

Finally, as we saw (Tim Bunnell, p. 348) in Malaysia, national and local state policy-makers are also major players in constructing positive and messianic discourses about cybercity futures. Many also put massive physical, financial and discursive resources into projecting harmonious and utopian visions of their future 'information age', 'intelligent' or 'siliconised' city. This is being done as a way of associating their regimes with the glamour and lustrous modernity of ICT technologies, whilst helping to ensure public support for vast and often costly public ICT policies.

Back to square one? The persistence of end of city scenarios

We saw in Chapter 1 that, until very recently, dominant discourses about cities and ICTs have stressed the 'impacts' of new media technologies as a motive force to effectively 'unglue' the city. Whilst these are on the wane, they are far from gone. In fact, 'what is surprising is that we keep repeating the argument' about whether ICTs will undermine cities in and of themselves (Wigley, 2002, 107). Some futuristic imaginings about ICT-based societies – represented here by Martin Pawley, page 402 – remain preoccupied by the idea that new media technologies are a forceful and inevitable centrifuge for the city's contents – a means to restore a purported pastoral, decentralised or smaller scale society.

The resilience of such ideas should not be surprising. Anti-urban ideas run right through the roots of modern and utopian urban planning, particularly in the Anglo-Saxon world. As Leo Marx argued in his book *The Machine in the Garden* (1964), the representation of new technology, especially in North America, has long emphasised the need to support a pastoral ideal by emptying-out the dangerous, destructive and 'unnatural' central city. Such ideas strongly influenced urban planning visionaries in the late nineteenth and twentieth centuries such as Ebenezer Howard and Frank Lloyd Wright (see Fishman, 1982). All such visionaries were fascinated by the potential of then new network technologies like radio, the telephone and the automobile to sustain radical shifts in the social order. These became embodied in the transformed physical landscapes of their preferred, decentralised urban utopias. 'Modern technology', they believed, 'had outstripped the antiquated social order, and the result was chaos and strife. In their ideal cities, however, technology would fulfil its proper role' – namely, integrating the planned metropolis into a functioning, efficient, harmonious, and modern city designed according to their personal blueprints (Fishman, 1982, 13).

Anti-utopian ripostes: cyberpunk science fiction and critical urban social science

For every celebratory and messianic future, however, there has been a dark and anti-utopian parallel. The antecedents here, amongst many others, are the powerful tales of centralised state power based on absolute surveillance, or bioengineering and consciousness-altering drugs, found in Orwell's *Nineteen Eighty-Four* and Huxley's *Brave New World* respectively.

The most important anti-utopian genre in shaping imaginations of both future and contemporary cities has been cyberpunk science fiction. Centred around the writings of authors like William Gibson, Neil Stephenson and Bruce Sterling, cyberpunk has had a massive impact on the cultural imagination of both new media and the meaning of the urban. It has also been a central genre in the debates surrounding wider cultural and technological shifts towards postmodernity.

Cyberpunk novels offer bleak portrayals of ultra-technologised urban life in a near-future world where globalisation, commodification and corporate control have accelerated immeasurably. They are replete with cyborg 'posthumans', absolute surveillance, invasive bio and molecular engineering, and immersive virtual realities which blend seamlessly into 'real' urban landscapes. Gibson's (1984) *Neuromancer*, Neil Stephenson's (1993) *Snow Crash*, and Bruce Sterling's (1988) *Islands in the Net* have been particularly influential (see Cadigan, 2002; Rob Warren *et al.*, p. 395). When translated into films – as in movies like *Blade Runner* or *The Matrix* – cyberpunk sci-fi has produced perhaps the most powerful and influential scenarios of cybercity futures currently circulating. Paula Geyh points out, however, that cyberpunk visions of future cities, far from being radically new, are remarkably similar to

> the great modernist 'Cities of Tomorrow', 'Futuramas', and the 'radiant City' dreams of Le Corbusier and Oscar Niemeyer and their many imitators. It is as if the architects of cyberspace were driven by the same utopian imagination, which, having failed in its real world incarnations, now sought 'realization' in the ethereal realm of cyberspace (2001, 103).

Cyberpunk hammers home both the astonishing rate, and the deeply politicised nature, of technological change in computing, the new media, and biotechnology. It also underlines the way these processes of change come together in reshaping the spatialities and politics of cities and urban life. In achieving this, cyberpunk writers have played a major role in forcing researchers, activists, social movements to take their scenarios seriously (Burrows, 1997).

A particular emphasis in the genre has been on the importance of subversion, activism and direct action. Network-based mobilisation and computer hacking are widely portrayed as means to assert individual and collective liberties and identities within the wider dystopian and bleak contexts portrayed. Such portrayals resonate strongly with the tactics now employed by the anti-globalisation movement. As cyberpunk ideas about the future of cities and technology have taken hold, so 'it has been re-appropriated by the mainstream to refer to a diverse set of cultural forms based around futuristic ideas in computing and communication' (Dodge and Kitchen, 2001, 187).

Major crossovers have also occurred between cyberpunk and critical urban social science (Timothy Luke, p. 106). Writers like Mike Davis (1990, 1992, 1998, 2002) – an extremely influential analyst of the social and environmental crises afflicting US urbanism – have expressed a major debt to cyberpunk imaginings of near-future urban conditions. The British sociologist Roger Burrows has even argued that:

> The themes and processes which a symptomatic reading of cyberpunk reveal are a good deal more insightful than those offered by what now passes for the theoretical and empirical mainstream [in urban studies] . . . I think that one gets a clearer analytical understanding of contemporary urban processes from a reading of Gibson or Stephenson than one does from a reading of Sassen or Castells (1997, 38–45, cited in Dodge and Kitchen, 2001, 186).

This interchange has gone both ways. For example, William Gibson has written about the nature of contemporary cities, focusing especially on fast-emerging paradigmatic cybercities such as Singapore (Gibson, 1994). This fertile crossover has shown considerable promise in helping to develop creative imaginings for the future condition of our urbanising planet (see Kitchen and Kneale, 2002).

However, as a genre cyberpunk does have major key weaknesses. Three, in particular, can be identified. First, cyberpunk writings tend to recreate patriarchical notions of gender relations. Second, they are often guilty, in their obsession with technology, of suggesting a simple technological determinism, in which the entire global society responds in a simple and linear way to new technical innovations.

Third, as Phil Agre argues (p. 415) much of cyberpunk science fiction can be criticised – along with most depictions of city–ICT relations – for its obsession with portraying the 'virtual' world as a domain totally separated from the corporeal and physical domain (Dodge and Kitchin, 2001, 188). As we have seen in this book, such a perspective is unhelpful. It does not allow us to grapple with the hybrid, fused, recombinant and remediated realities of cybercities in the context when ICTs are increasingly becoming embedded, taken for granted, and fused invisibly into the physical and corporeal worlds of the city. Rather than a space apart, ICTs, and the electronic realms they mediate, now need to be viewed as part of the very fabric of everyday urban life itself. As we have seen in this book, they are a critical urban political terrain which is centrally important as local, national and transnational connections fuse together in the remaking of urban places in a wide variety of ways around the world.

Cybercity futures: the normative challenge

> Recently there has been much said about the digital city, the information city, the TeleCity, the cyborg city, the figured city, the cybercity, the virtual city, the virtual urban area, the virtual metropolis, the transphysical city, the city of bits, the imploding city, the overdimensionalized city, the ageographical city, the SimCity, the real-time city, and even the end of the city (although not in Pol Pot's sense!). However, as appealing as it may be to imagine living 'without cities', or in cities transformed by information technologies, the danger is that such ideas tend to marginalize the 'urban question' from political discourse (Smith, 2003, 40).

The overwhelming challenge, in discussing and debating the futures of city–ICT relations, is to debunk the powerful myth that the issues they raise are somehow technocratic, value-free, non-political, or only amenable to urban and 'technological experts' (see Vincent Mosco, p. 199). This is especially important given wider transformations in the politics of cities and urban planning. For the ideologies supporting modern urban planning practices across the world are currently in crisis. They have been severely undermined by the increasing illegitimacy of the notion that the technological configuration of cities and urban life is somehow a 'technical' question that can be 'rationally' shaped by experts for some putative 'public good'. This crisis of technical rationality has been paralleled by shifts away from mechanistic thinking and great embarrassment over past efforts by planners to predict, and creatively influence, city futures.

Tim Marshall (1997) points out that British transport planners, for example, have virtually given up forecasting the future. This is because past experiences were so embarrassingly inaccurate. But it is also because of transport 'planners' sense of being adrift in a confusing and uncontrollable flux' (31). Marshall cites recent attempts in the UK at 'foresighting' (rather than forecasting) technological shifts in transport which confronted the essential problem that 'Technology Foresight is something of a contradiction in terms, because the future development of the transport sector is determined as much by economic, social, political and environmental factors as it is by the availability of technology' (Technology Foresight Panel Transport 1995; 59, quoted in Marshall).

Such crises in the technical legitimacy of urban forecasting raise the need for social movements, and critiques, to assert the politicised and biased natures of contemporary, and future, cybercity configurations. There are many challenges here. First, the complex ways in which ICT systems maintain, support and configure social power relations within and between cities at various scales need to be exposed and analysed. They are, as Richard Smith (2003) hints above, very much part of the inherently political 'urban question' (see Amin and Thrift, 2002).

Second, the dynamics of change in the political economies of contemporary cities need to be revealed in ways that enable the multifaceted roles of ICTs to be understood in a sophisticated and non-deterministic way.

Finally, the practice of developing normative and even utopian thinking about cities and ICTs needs to be further developed. This must be based on the idea that the links between cities, everyday urban life, digital technologies, and the softwares that make them work, are some of the key political nexus points that will powerfully influence the nature of future societies, cultures, and natures.

This book has forcefully demonstrated that it is not sufficient to ignore reconfigurations in either the technological bases of urban society, or the urban bases of our technologised societies. At the same time, futures and strategy for cybercities must not be preoccupied in a fanatical way with technological practices. We must resist the 'monocausal fetishization of the electronic' that remains so common in popular portrayals on new technology (Appadurai, 1996, 3). At the same time, political engagements with the future of cybercities must also not be afraid of being normative – despite the many disasters of prescriptive urban and technological planning, forecasting and futures analyses in the past five decades (see Pitkin, 2001).

Above all, ways of imagining, and shaping, city–ICT relations need to be devised, and creatively implemented. Such strategies must be seen as processes of social learning and experimentation which attempt to meld the inherent flexibilities of ICT–city relations in reflective, purposeful and diverse ways. Rather than static and technocratic they need to be dynamic, performative, and concerned with experimental shifts in the urban as a *process*. As Andreas Broeckmann discusses (p. 378), such normative and experimental interventions must attempt to exploit the largely untapped potential of addressing the fluid and telescoping scales of interaction that link the urban and the telemediated in an integrated way. Rather than being narrowly besotted with technology *per se*, however, such experimentation needs to be fuelled by critical appreciations of the challenges of our urbanising planet. The considerations here are pressing: neoliberal economic restructuring, migration and multiculturalism, sprawl and environmental crises, the privatisation of public space, endemic inequality, social fragmentation and capsularisation, and the widespread disillusionment with mainstream democratic politics.

This, above all, is a practical political challenge for multi-scaled social movements. It is unlikely to be met through even the most innovative mainstream urban and municipal institutions on their own. 'Movement toward a normative vision of the city requires the development of counterinstitutions capable of reframing issues in broad terms and of mobilizing organizational and financial resources to fight for their aims' (Fainstein, 1999, 268).

Finally, it is important to stress that the imagination of cybercity utopias can – under certain conditions – be a useful part of this process. This is provided that the imagined utopias are not realms of fantasy but that they begin by addressing desired futures the role of ICTs within the fine grain of everyday urban life. Utopian thinking thus needs to be reconnected with the process of social change. 'The solution', suggests the British sociologist Ruth Levitas, 'is not to call for more and better utopias, more and better images and maps of possible futures. These will follow if we have better analyses of the present which identify possible points of intervention, paths and agents of change' (1993, 265).

References and suggestions for further reading

Amin, A. and Thrift, N. (2002), *Cities: Reimagining the Urban*, Cambridge: Polity.

Appadurai, A. (1996), *Modernity at Large: Cultural Dimensions of Globalization*, Minneapolis: University of Minnesota Press.

Ascher, F. (1995), *Métapolis ou l'avenir des villes*, Paris: O. Jacob.

Bird, J. , Curtis, B., Putnam, T., Robertson, G. and Tickner, L. (eds) (1993), *Mapping the Futures: Local Cultures, Global Change*, London: Routledge.

Burrows, R. (1997), 'Virtual culture, urban social polarization, and social science fiction'. In B. Loader (ed.), *The Governance of Cyberspace*, London: Routledge, 31–42.

Cadigan, P. (ed.) (2002), *The Ultimate Cyberpunk*, New York: iBooks.

Campanella, T. (2001), 'Anti-urbanist city images and new media culture'. In L. Vale and S. Warner (eds), *Imaging the City*, New Brunswick: Rutgers Center for Urban Policy Research, 237–54.

Corn, J. (ed.) (1986), *Imagining Tomorrow: History, Technology and the American Future*, Cambridge, MA: MIT Press, 219–29.

Corn, J. and Horrigan, B. (1984), *Yesterday's Tomorrows: Past Visions of the American Future*, Baltimore: Johns Hopkins University Press.

Davis, M. (1990), *City of Quartz: Excavating the Future in Los Angeles*, London: Verso.

Davis, M. (1992), 'Beyond Blade Runner: Urban control, the ecology of fear', *Open Magazine*, New Jersey: Westfield.

Davis, M. (1998), *The Ecology of Fear: Los Angeles and the Imagination of Disaster*, New York: Metropolitan Books.

Davis, M. (2002), *Dead Cities*, New York: New Press.

Dodge, M. and Kitchin, R. (2001), *Mapping Cyberspace*, London: Routledge.

Eaton, R. (2002), *Ideal Cities: Utopianism and the (Un)Built Environment*, London: Thames and Hudson.

Fainstein, S. (1999), 'Can we make the cities we want?'. In R. Beauregard and S. Body-Gendrot (eds), *The Urban Moment: Cosmopolitan Essays on the Late 20th Century City*, Thousand Oaks, CA: Sage, 209–41.

Featherstone, M. and Burrows, R. (1995), *Cyberpunk/Cyberspace/Cyberbodies*, London: Sage.

Fishman, R. (1982), *Urban Utopias in the Twentieth Century: Ebenezer Howard, Frank Lloyd Wright and Le Corbusier*, Cambridge, MA: MIT Press.

Gates, W. (1995), *The Road Ahead*, London: Hodder and Stoughton.

Geyh, P. (2001), 'The fortress and the polis: from the postmodern city to cyberspace and back'. In S. Munt (ed.), *Technospaces: Inside the New Media*, London: Continuum, 99–112.

Gibson, W. (1986), *Neuromancer*, London: Harper Collins.

Gibson, W. (1993), *Virtual Light*, New York: Bantam.

Gibson, W. (1994), 'The trouble with squeaky clean', *HQ*, Jan./Feb., 116–22.

Graham S. and Marvin S. (1996), *Telecommunications and the City: Electronic Spaces, Urban Places*, London: Routledge, Chapter 10.

Graham, S. and Marvin, S. (2000), *Splintering Urbanism: Networked Infrastructures, Technological Mobilities and the Urban Condition*, London: Routledge, Chapter 8.

Hayles, N. K. (1999), *How We Became Posthuman: Virtual Bodies in Cybernetics, Literature, and Informatics*, Chicago: University of Chicago Press.

Huxley, A. (1932), *Brave New World*, London: Chatto and Windus.

INURA (International Network for Urban Research and Action) (eds), *Possible Urban Worlds: Urban Strategies at the End of the 20th Century*, Berlin: Birkhauser-Verlag.

Kitchen, R. and Kneale, J. (2002), *Lost in Space: Geographies of Science Fiction*, London: Continuum.

Kumar, K. (1987), *Utopia and Anti-Utopia in Modern Times*, Oxford: Blackwell.

Lessig, L. (2001), *The Future of Ideas: The Fate of the Commons in a Connected World*, New York: Random House.

Levitas, R. (1993), 'The future of thinking about the future'. In J. Bird, B. Curtis, T. Putnam, G. Robertson and L. Tickner (eds), *Mapping The Futures: Local Cultures, Global Change*, London: Routledge, 257–66.

Lipman, A. and Harris, A. (1999), 'Fortress Johannesburg', *Environment and Planning B: Planning and Design*, 26, 727–40.

Marshall, T. (1997), 'Futures, foresight and forward looks', *Town Planning Review*, 68(1), 31–50.

Marx, L. (1964), *The Machine in the Garden: Technology and the Rise of the Pastoral Ideal in America*, London and New York: Oxford University Press.

Miles, M. and Hall, T. (eds) (2002), *Urban Futures*, London: Routledge.

Moylan, T. (2001), *Scraps of the Untainted Sky: Science Fiction, Utopia, Dystopia*, New York: Westview.

Orwell, G. (1949), *Nineteen Eighty-Four*, London: Martin Secker and Warburg.

Pawley, M. (1997), *Terminal Architecture*, London: Reaktion Books.

Pinney, C. (1992), 'Future travel: anthropology and cultural distance in an age of virtual reality; or, a past seen from a possible future', *Visual Anthropology Review*, 8(1), 38–55.

Pitkin, B. (2001), 'A historical perspective on technology and planning', *Berkeley Planning Journal*, 15, 32–55.

Robins, K. and Hepworth, M. (1988), 'Electronic spaces: new technologies and the future of cities', *Futures*, April, 155–76.

Rutsky, R. (1999), *High Techné: Art and Technology From the Machine Aesthetic to the Posthuman*, Minneapolis: University of Minnesota.

Schaer, R., Claeys, G. and Tower Sargent, L. (eds) (2000), *Utopia: The Search for the Ideal Society in the Western World*, New York: Oxford University Press.

Shiel, M. and Fitzmaurice, T. (eds) (2001), *Cinema and the City: Film and Urban Societies in a Global Context*, Oxford: Blackwell.

Smith, R. G. (2003), 'World city-actor networks', *Progress in Human Geography*, 27(1), 25–44.

Stephenson, N. (1992), *Snow Crash*, New York: Bantam.

Sterling, B. (1988), *Islands in the Net*, New York: Arbor House.

Wells, H. G. (1988), *The Shape of Things to Come*, London: Phoenix.

Wigley, M. (2002), 'Resisting the city'. In A. Mulder (ed.), *Transurbanism*, Rotterdam: NAi Publishers, 103–22.

'The Future of the Future in Planning Theory: Appropriating Cyberpunk Visions of the City'

Journal of Planning Education and Research (1998)

Rob Warren, Stacy Warren, Samuel Nunn and Colin Warren

Editor's Introduction

Our first reading on cybercity futures assesses the usefulness of cyberpunk science fiction for social scientific analyses of contemporary and future cities. In it, the US urbanists and planning researchers Rob Warren, Stacy Warren, Samuel Nunn and Colin Warren argue that cyberpunk novels provide a useful stimulus to urban researchers attempting to understand and explore the options for theory and practice in cybercities.

To the authors the genre of cyperpunk portrays the possible forms of the hypermodern and hyper-urban societies of the near future. It creatively imagines how contemporary cities and societies might be reconstructed through yet unheard of (but fast advancing) mutations in ICTs, nanotechnologies, and surveillance. Politically, the authors argue, cyberpunk science fiction portrays extreme versions of today's headlong movement towards the privatisation of public functions and spaces and what they (controversially) believe to be an effective disintegration of national states under the pressures of neoliberal capitalism.

But it is in its provision of imaginative fuel for the understanding of the geographies, socio-technologies, economies and politics of future cities that Warren and colleagues find most value in cyberpunk. In the Reading they review some of these speculations and ideas. They concentrate, in particular, on reviewing cyberpunk speculations about the centralisation of corporate power; the intensifying privatisation of urban public space; the growth of organised crime networks; the sprawl and increasing gigantism of cities; and the extreme volatilities of the digitally mediated global economy and financial system. Recurring themes which also run through cyberpunk renditions of the urban future, they suggest, are the intensifying commodification of everyday life; the spiralling inequities between cheek-by-jowel urban neighbourhoods; and, most tellingly, the shift of politics and power into both the coded spaces of software and the molecular constructions of nanotechnology.

The reading by Warren *et al.* is innovative because it manages to bridge science fiction depictions, based on exaggerating and projecting perceived current trajectories of urban technological change, with debates about how today's cities should actually be managed and planned. This is a crucial exercise which requires much more work. For – as the authors suggest – our imaginative constructions of future cities invariably help to shape the discourses and practices which actually influence the ways in which cities change.

THE WORLD OF CYBERPUNK

Cyberpunk appeared in the mid-1980s. William Gibson is most commonly identified as defining the genre with *Neuromancer* (1984) and the rest of his "Sprawl" trilogy, *Count Zero* (1986a) and *Mona Lisa Overdrive* (1988). It is distinguished from the main body of science fiction by a stylistic shift in the way that it moves traditional settings and themes forward in time as "metaphorical evocations of life in the present" (Fitting, 1991, 299). *Cyber* broadly relates to cybernetics – the science of communications, control, and feedback and to the merger of humans and machine components creating cyborgs. *Punk* situates the voice as coming from a rough, alternative edge, a "hip self-marginalization in opposition to dominant life-styles" (Fitting, 1991, 296). The scientist and explorer heroes of traditional science fiction are replaced by anti-heroes: out-of-work pizza deliverers, machine-augmented adventurers, bicycle couriers, and data hustlers who are swept into seemingly unfathomable intrigues in a *noir* world where the boundaries between real and virtual experience and human and artificial intelligence dissolve.

As McCaffrey (1991) notes, cyberpunk writers were "the first generation of artists for whom the technologies of satellite dishes, video and audio players and recorders, computers and video games, digital watches and MTV were . . . part of a daily 'reality matrix'" (12). In coming together, these authors presented themselves as "techno-urban-guerillas" with the manifesto that:

> both the technological dreams and nightmares envisioned by previous generations of SF artists were already in place, and that writers as well as the general public needed to create ways of using this technology for their own purposes before we all became mere software, easily deletable from the hard drives of multinationalism's vast main-frame (1991, 12).

Cyberpunk's social, political, and economic structures are an extreme version of today's headlong movement to the privatization of public functions and evidence that the national state can disintegrate. An increasingly small and powerful set of multinational corporations and crime cartels become the controlling institutions. Class-based representations of the industrial age are masked or replaced by fragmented cultural and ethnic formations at the local and regional scale that have tribal characteristics. Some urban landscapes are

recognizable and viable; others are mutations; and there is a capability of constructing whole cities overnight with new technology. An attendant array of advanced computer and communications technologies evolves into cyberspace through the merger of the Internet, virtual reality, and a three-dimensional database "containing all a culture's deposited wealth, where every document is available, every recording playable, and every picture viewable" (Featherstone and Burrows, 1995, 7).

[. . .]

SITUATING THE TERRESTRIAL WORLD

Terrestrial geography in cyberpunk, although detailed and complex in description, is inexorably entwined with cyberspace. Urban landscape and the city are extreme but recognizable projections of current trends of privatization and fracturing of nation-states. Bruce Sterling, in *Heavy Weather* (1994, 185), envisions the economic and governmental structure becoming a "raging non-linear anarchy." Once the economy went digital, "the very nature of money mutated beyond repair" (186) and private corporations and organized crime stepped in to electronically mint their own private currency. A "global ocean of black money" emerged that caused stock markets to crash and national currencies to collapse.

These money problems are symptomatic of broader restructuring of the relative roles played by government and private interests. The most extreme cases are countries in the periphery seeking to stay plugged in with obsolete silicon: "Nations so benighted that the concept of nation was still taken seriously" (Gibson, 1986a, 121). In this world, for example, one cyber-cowboy punched through several "African backwaters" like "a shark cruising a swimming pool thick with caviar." In a week's work, he was able to sack several million "laughably tiny bank accounts," which resulted in "the collapse of at least three governments and . . . untold human suffering" (1986a, 121).

In advanced industrial nations, with economic control functions undermined, public authority eroded, and the continuing privatization of public services, there is little to nothing left that governments can do. The operational headquarters of the U.S. government, in *Snow Crash*, has been downsized to "Fedland," a small complex of buildings adjacent to an exit off a Los Angeles freeway. The monuments and museums in Washington, D.C., are now all in a privately managed tourist park that generates a tenth of the government's revenue (Stephenson, 1992, 164). Northern and southern California are formally divided (Gibson 1993). Multinational corporations grow to unprecedented power and supplant the nation-state in controlling the use of force on a global scale (Sterling, 1988, 179).

Crime cartels rather than governments are the only viable rival to corporate power. Their roles in industrial espionage, facilitating the defection of top scientists from one corporation to another and holding stolen data for ransom (Gibson, 1986b, 17), are recurring themes in Gibson's depictions of "the secret skirmishes of the zaibatsus [and] the multinational corporations that control entire economies" (103). More whimsical but no less possible, Stephenson (1992) has converted the Mafia's main business into a chain of pizza outlets with competition conspicuously absent. The only reason its methods of operating are legal now is that past laws that would make it a criminal organization no longer exist (234).

Macro and micro geography

Against this backdrop a wide range of urban environments can be found. Gibson's most straightline vision is at the macro scale. The boundaries of BAMA (Boston-Atlanta Metropolitan Axis) or the *Sprawl* are defined in terms of data exchange rather than population, government, or traditional economic factors. On a computerized map programmed to display frequency of data exchange with each pixel representing a thousand megabytes, "Manhattan and Atlanta burn solid white . . . At a hundred million megabytes per second, you begin to make out certain blocks in midtown Manhattan, outlines of hundred-year-old industrial parks ringing the old core of Atlanta . . ." (Gibson, 1984, 43). However, BAMA receives far less attention and is less instructive than Gibson's descriptions of the urban future at more micro scales.

Some world cities, such as London and Paris, have charm and vary little from their present structure. In the New York portion of the Sprawl, San Francisco, and Tokyo, the marginalizing effects of multiple economies are more evident. New York's skyline is dominated by "unfinished Fuller domes" that produce unplanned microclimates. There are places no bigger than a few blocks where "a fine drizzle of condensation" falls continuously from the geodesics and other "sections of high dome are famous for displays of . . . a particularly urban variety of lightning" (1986a, 114).

Vast horizontal ghettoes exist with poverty, gangs, and illegal toxic waste dumps scavenged by children, along with vertical arcologies, former urban malls, that have become village-like enclaves for marginalized groups.

One of Gibson's most striking images of the urban future is the "Bridge" in 21st century San Francisco described in *Virtual Light* (1993). The city's function was to serve the wealthy in the midst of a growing number of people attempting to survive on the streets. One day thousands of the homeless converged spontaneously on the Oakland Bay Bridge, which had been left for years unrepaired and unused after serious earthquake damage. They appropriate the Bridge and without any set plan use "every imaginable technique and material" to slowly turn it into a "startlingly organic," almost psychedelic commune, largely outside the reach of the law, with residences, shops, arcades, herbalists, tattoo parlors, and bars (70).

Gibson's version of an enterprise zone in Japan is equally interesting in what it presages. Ninsei, a shadowland of black medicine, is a zone in the heart of Night City, "a deranged experiment in social Darwinism, designed by a bored researcher who kept one thumb permanently on the fast-forward button" (1984, 7). In explaining why Chiba City, the area's formal unit of government, tolerated this enclave, one of Gibson's characters concludes that there is "a certain sense in the notion that burgeoning technologies require outlaw zones . . . Night City wasn't there for its inhabitants, but as a deliberately unsupervised playground for technology itself" (1984, 11).

Stephenson plays out the Los Angeles psyche, so demonized in planning, in his characterization of that region. The central portions of LA have been abandoned by anyone with "normal" demographics:

> The only ones left in the city are street people, feeding off debris; immigrants, thrown out like shrapnel from the destruction of the Asian powers; young bohos; and . . . young smart people . . . who take the risk of living in the city because they like stimulation and they know they can handle it (1992, 179).

The suburbs and beyond are dotted with burbclaves, the cyberpunk evolution of the gated community – "a city-state with its own constitution, a border, laws, cops, everything" (1992, 6). A wide variety of lifestyles are accommodated – including White Columns, the classic middle-class enclave carrying on the tradition of Apartheid, and Nova Sicilia and Narcolumbia, respectively developed by the Mafia and Columbian drug money (both very safe places).

[. . .]

Opportunities for entrepreneurship and surveillance

Privatization is carried to its logical conclusion in *Snow Crash* (1992). The CIA and the Library of Congress have merged as one seamless private corporation; UCLA is owned by the Japanese, a private suburb, and a few big American corporations; pay-per-use toilets are the only facility available in some neighborhoods; two private firms control the highways in California and engage in armed conflict where their roads intersect; police services come in no-frills to deluxe packages from different firms and anyone can "Dial 1-800-The Cops – all major credit cards"; jails and the court system are privatized as well. Gibson also has private-sector "armed response" companies, using state-of-the-art military technology, evolving as a dominant provider of security services. However, the traditional concept of policing tends to lose all meaning with the pervasiveness of surveillance systems in this digital world.

Snow Crash offers a variety of anticipations about who will have access to information about whom in the future. The protagonist in the book is, among other things, a part-time stringer for the merged CIA and Library of Congress (CIC). He, along with millions of other stringers, gets information, "gossip, videotape, audiotape, a fragment of a computer disk, a xerox of a document" and directly uploads it to the CIC database (1992, 20). The stringers get paid if one of the CIC's many private clients buys any of their information. Stringers are laggards compared to "gargoyles," who have "their computers on their bodies . . . recording everything that happens around them." Staggering amounts of useless data are dumped on the CIC. "It's like writing down the license number of every car you see . . . just in case one of them will be involved in a hit-and-run accident" (115).

The use of surveillance by others is far less random. All interactions between the Mafia's pizza deliverers and customers are video-taped and piped in real time to CosaNostra Pizza University's management science laboratory for analysis. A real screw-up gets the person fired or worse, and provides a video casebook example for class use (1992, 10). Y.T., a "kourier" who delivers documents in LA, needs a chest full of ID bar codes

to get into business complexes private highways, and burbclaves. But providing access to this data may not be optional. It took a hovering Mafia helicopter's crew half a second, after enveloping her in a laser beam, to know "everything about Y.T. – where she lives, what she does, her eye color, credit record, ancestry, and blood type" (1992, 32).

Individuals are coded into data systems so many times and in so many ways that the possibilities of surveillance and control by others are almost limitless. In a Gibson (1986b) short story, Johnny Mnemonic speculates that a Japanese criminal cartel he is hiding from, the Yakuza, is "settling its ghostly bulk over the city's data banks, probing for faint images of me reflected in numbered accounts, securities transactions, bills for utilities." He goes on:

> We're an information economy. They teach you that in school. What they don't tell you is that it's impossible to move, to live, to operate at any level without leaving traces, Bits, seemingly meaningless fragments of personal information. Fragments that can be retrieved, amplified . . . (16–17).

Citizens, cyborgs, and beyond

Citizens, in the sense of politically competent and active persons, are in short supply in cyberpunk urban life. As noted, large numbers in LA, San Francisco, and the Sprawl are marginalized and alienated from the information society. Many others have become addicted to television or immersed in virtual-reality soap operas. The poverty-ridden mother of a main character in *Count Zero* has brain surgery to implant a new cranial socket. It is designed to reduce static and sensory bleedover when she plugs into her soaps on virtual-reality TV for hours at a time every day. The "medical" work is done at the kind of place "where you don't even get an appointment for the operation. Walk in and they just slap it in your head" (1986a, 33).

Vast numbers of people across class lines use *simstim*, a technology that allows the user to literally crawl inside someone else's body – see what they see, hear what they hear, feel what they feel – as multiple sensory input channels transmit a cascade of data back to the "visitor." These commodified experiences are a dominant product of the media industry. Crafted as simulacra for originals that never existed, millions of simstims are purchased by eager fans who can slip into the identify of their favorite star, who is surgically and chemically augmented to perfection. The viewer becomes *the* star in soap-opera style love stories in exotic settings – with all headaches, bad moods, and other aberrations edited out.

In a sense, people using simstim have become "recreational cyborgs." An important part of their lives is experienced through the flesh-machine linkage required for virtual reality. However, cyberpunk also provides a window on far more systemic possibilities of the blurring of humans, machines, and bioprocesses. Direct input and output through brain sockets are a common feature. If you have the money, knowledge of such things as operating complex airplanes and language fluency are no further away than the right cable jack. Gibson's character, Molly Millions, has been to Ninsei for body extensions to enhance her career as an adventurer/bodyguard. She acquired optical implants with mirrored silver surfaces the size of glasses for lenses and retractable blades in place of fingernails, "each one a narrow, double-edged scalpel in pale blue steel" (1986b, 8).

Gibson's most extreme examples of boundary blurring are found in the cases of Virek and McCoy Pauley (known as Dixie Flatline). Virek is reputedly the wealthiest individual in the world, ruling over a vast corporate structure. He also has a body whose cells "have opted for the quixotic pursuit of individual careers." This requires Virek to spend a tenth of his income annually to maintain his life in a vat in a Stockholm industrial suburb (1986a, 13, 16). However, technology allows him to exercise total control over his corporate and financial operations and interact directly with anyone anywhere in the world in holographic and other forms. Virek is clearly something other than simply a human being. Brande (1996, 91) observes that Virek's successful acquisition of a new super-high-capacity medium for information storage and processing, which he intends to use to escape from the vat and be free "to inhabit any number of real bodies" (Gibson 1986a, 219), makes classifying his form of consciousness even more problematic.

[. . .]

Nanotechnology

Cyberpunk's themes in relation to telecommunications and information technology are hyper-exaggerated visions of the familiar. The incorporation and naturalization of nanotechnology into the socioeconomic structure of the city moves further out from the

recognizable. The proposed technology is based on theorization that has advocates but is far from accepted in mainstream science. But if it does emerge in applied forms, the potential consequences could be extraordinary for urban form and processes.

[. . .]

Cyberpunk explores possible effects of molecular manufacturing on urban economic and social organization, control of ecological processes, and transformations of the built environment.

In one version, nanotechnology makes it possible to supply the population of the planet with food, clothing, transportation, shelter, and almost any other consumer good on demand at minimal cost. Stephenson's *Diamond Age* (1995) presents a detailed picture of a society and its cities in which nanotechnology has been fully articulated. Every home has a "matter compiler" (38). Low-cost raw materials are delivered to a home's compiler through a utility-like feed line to be converted by it into whatever is needed – a bed, a meal, medicine.

In another application of Stephenson's nanotechnology, cities and even customized topography are manufacturable, literally overnight. Imperial Tectonics Limited's (ITL) geotects (programmers) physically *make* real estate with a very high profit yield (Stephenson, 1995, 15). In a crowded world, a few strategically sited islands that had been created by ITL with "smart" coral are "the most expensive real estate in the world outside of a few blessed places like Tokyo, San Francisco, and Manhattan" and all that was needed to produce them "was a hot young geotect, a smart matter compiler, and a jumbo Source" (15). Given the "sheer clunkiness of old cities," Stephenson's protagonist could conclude that, with nanotechnology available, existing "cities of the world were doomed, except possibly as theme parks, and that the future was in new cities, build from the bedrock up one atom at a time" (62).

The plot of *Virtual Light* centers on efforts to recover the stolen specifications of a covert project to use nanotechnology for the reconstruction of San Francisco. This "Sunflower" plan would cultivate a new city layer by layer over the remains of the old beginning with a grid of 78-story office-residential, retail-residential complexes. The structures would be completely self-sufficient and be *grown*, in the same way buildings were being created in Tokyo, by feeding on their own waste (250, 251). The project would make a fortune for corporate interests who know the plan's "footprint" while displacing thousands in a reprise of 20th-century urban renewal.

As Gibson's San Francisco example indicates, the distribution of the benefits of nanotechnology is problematic. Yet, at the same time, wide use of the technology would have far-reaching consequences for the existing socio-economic order. In urban form, nano-communities could be marketed that change design and even location on demand to suit the shifting preferences of their residents. The capacity of matter compilers to meet material needs without depleting natural resources or causing pollution would radically change the economic infrastructure. Retail stores, distribution networks, wholesalers, shippers, and importers, for example, would only be used for limited or specialized purposes, if at all. Even more basic, the relation of work to consumption would require sweeping redefinition.

REFERENCES FROM THE READING

Featherstone, M. and Burrows, R. (1995) 'Cultures of technological embodiment: an introduction'. In M. Featherstone and R. Burrows (eds), *Cyberspace/Cyberbodies/Cyberpunk*, London: Sage, 1–19.

Fitting, P. (1991), 'The lessons of cyberpunk'. In C. Penley and A. Ross (eds), *Technoculture*, Minneapolis: University of Minnesota Press.

Gibson, W. (1984), *Neuromancer*, New York: Ace Books.

Gibson, W. (1986a), *Count Zero*, New York: Arbor House.,

Gibson, W. (1986b), *Burning Chrome*, New York: Arbor House.

Gibson, W. (1988), *Mona Lisa Overdrive*, New York: Bantam.

Gibson, W. (1993), *Virtual Light*, New York: Bantam.

McCaffrey, L. (1991), 'Introduction: the desert of the real'. In L. McCaffrey (ed.), *Storming the Reality Studio*, Durham, NC: Duke University Press, 1–16.

Stephenson, N. (1992), *Snow Crash*, New York: Bantam.

Stephenson, N. (1995), *Diamond Age*, New York: Bantam.

Sterling, B. (1988), *Islands in the Net*, New York: Arbor House.

Sterling, B. (1994), *Heavy Weather*, New York: Bantam.

Editor's suggestions for further reading

Bukatman, S. (1993), *Terminal Identity: The Virtual Subject in Postmodern Science Fiction*, London: Duke University Press.

Burrows, R. (1997), 'Virtual culture, urban social polarization, and social science fiction'. In B. Loader (ed.), *The Governance of Cyberspace*, London: Routledge, 31–42.

Cadigan, P. (ed.) (2002), *The Ultimate Cyberpunk*, New York: iBooks.

Davis, M. (1990), *City of Quartz: Excavating the Future in Los Angeles*, London: Verso.

Davis, M. (1992), 'Beyond Blade Runner: urban control, the ecology of fear', *Open Magazine*, New Jersey: Westfield.

Featherstone, M. and Burrows, R. (1995), *Cyberpunk/Cyberspace/Cyberbodies*, London: Sage.

Fritz, S. (2002), *Understanding Nanotechnology*, New York: Brown.

Gibson, W. (1986), *Neuromancer*, London: Harper Collins.

Gibson, W. (1993), *Virtual Light*, New York: Bantam.

Gibson, W. (1994), 'The trouble with squeaky clean', *HQ*, Jan./Feb., 116–22.

Hayles, N. K. (1999), *How We Became Posthuman: Virtual Bodies in Cybernetics, Literature, and Informatics*, Chicago: University of Chicago Press.

Kitchen, R. and Kneale, J. (2002), *Lost in Space: Geographies of Science Fiction*, London: Continuum.

Kumar, K. (1987), *Utopia and Anti-Utopia in Modern Times*, Oxford: Blackwell.

Moylan, T. (2001), *Scraps of the Untainted Sky: Science Fiction, Utopia, Dystopia*, New York: Westview.

Robins, K. and Hepworth, M. (1988), 'Electronic spaces: new technologies and the future of cities', *Futures*, April, 155–76.

Stephenson, N. (1992), *Snow Crash*, New York: Bantam.

Sterling, B. (1988), *Islands in the Net*, New York: Arbor House.

'Terminal 2098'

from *Terminal Architecture* (1998)

Martin Pawley

Editor's Introduction

Our second Reading on the futures of cybercities draws from the first chapter of the book *Terminal Architecture* (1998). This is a speculation on the ways in which architecture is changing in response to ICTs by the influential British architectural critic Martin Pawley. A devoted modernist, Pawley has had a long career writing in the British press on the negative aspects of continued agglomeration in the cores of old cities. As a believer in the need to usher in an age of total sprawl, based on an honest and functional utilisation of the powers of ultra-high capacity ICTs, and other logistics and transport systems, Pawley here offers us a scenario of urban life in the year 2098.

A clear Virilio-esque logic of simple substitution pervades Pawley's scenario (see Paul Virilio, p. 78). The writer assumes that the raison d'être of economic and social agglomeration in cities will completely collapse over the next century. This will happen, he implies, because the bandwidth of ICTs will start to support mediated interchanges which begin to equal the power, quality and anthropological meaning of face-to-face interaction. To Pawley, the compulsion of proximity (see Deirdre Boden and Harvey Molotch, p. 101) is about to evaporate completely over the next hundred years.

The resulting landscape in 2098 reveals the author's love of the 'big sheds' and 'big box' architectures of contemporary logistical spaces. This is little more than an extended field of capsular spaces hooked into ultra-high capacity ICTs. These terminals, pods and sheds, ranged across the semiurban plain of what once was rural Dorset, presents a chaotic, nonlinear and illegible landscape to people trying to negotiate them in the way that cities are walked through today. However, this is not a problem as, clearly, the dominant way in which their inhabitants engage with each other, and the world, is through the infinitely extendible and malleable virtualised and holographic interchanges available through the terminals' ICTs. 'We don't say houses any more, we say terminals', argues one of the story's characters. 'Houses is a shit word these days. It means something bad.'

As a result of such extreme 'terminal urbanism' all concerns for the immediate physical environment have evidently collapsed. The world physically outside the capsule or the shed is relationally more distant − both imaginatively and socio-technically − than are other terminal-citizens on the other side of the planet. Similarly, any notion that one actually even *has* a geographic location has also evaporated. Pawley implies that, in allowing physical transportation and congested, central, urbanisation to wither and die, the terminal-based societies of 2098 have also overcome some of the main financial and ecological crises of the early twenty-first century.

In a sense, Pawley's scenario of terminal urbanism brings us back to the discussions in Chapter 1 about the preoccupation in discourses about cities and ICTs with ideas of the substitution of cities and transport flows. Whilst an engaging scenario, his speculations can be criticised for their extreme technological determinism. In particular, Pawley is guilty of simplistically resorting to the now extremely fanciful equation that more bandwidth will necessarily mean the collapse of both physical movement and concentration in urban centres. In doing so he denies the very real social, economic, physical, political and anthropological forces that continue to drive urban agglomeration in many cities across the world. The population of Pawley's home city, London, for example, is now growing at a rate faster than at any time since the 1950s. Ironically, this is partly because it has global pulling power in new media and ICT industries that drive the compulsion of proximity that Pawley so simplistically denies.

If you want a guided tour of the house of the future you have to book early, said the message on the monitor. The photographer and I looked at one another. There was an error of 100 years in the programme. Laughing, I punched the keys and asked for a booking. The monitor didn't even blink. It just asked matter-offactly for our ID numbers. We punched those in too and waited again. After a moment the fax started groaning.

'It took the booking!' we exclaimed, half amazed, half amused. It was like keyboarding your way into the computer system of your own bank by accident, and then wondering you really had done it. But there were the press passes coming off the fax. 'Good for twelve hours only,' said the overprinting. 'Please enter and leave by the car park entrance at 836559. Have a nice day.'

836559 was a map reference in Dorset. It took a day or two to set up our trip, and an overnight drive to get there. By the time we arrived we were already deeply sceptical. But there it was, an empty carpark with an entrance and the ticket booth illuminated by our headlights. It must have been about 5 a.m. and still dark when we fed our passes into the machine. We half expected them to be rejected, but no. The turnstile hummed and we pushed our way through. As far as we could see, we were the only visitors. Nobody was on duty at the ticket booth. There was only a big illuminated signboard saying 'To the terminals'. We stood there, one foot in each century so to speak, not knowing where to start. In the end we decided to wait for it to get light to see whether we were being taken for fools.

'Great shapes,' said the photographer, finally. He had been peering through the gloom using his 600mm lens like a telescope. I swung my binoculars in the same direction. The sun was just rising. We watched as it slowly illuminated a cluster of huge, low buildings. As soon as it was light enough we began to make our way on foot towards them. The ground was hard and dusty, sculpted into inexplicable ruts and ridges that ran like railway lines in all directions. Here and there bundles of cables snaked across our track, half buried, half exposed. We must have stepped over a thousand cables that morning, all just lying on the ground.

There were no trees between us and the settlement. Where we were, on the flat Dorset plain, there was just greyish vegetation in sprawling mounds that we instinctively avoided even though it meant detouring from our path. Once we stopped and listened carefully near one of these belts of weeds. It seemed to us we could hear a deep humming and wailing coming from it like a kind of music. There was even some suggestion of movement underneath. The photographer made as if to step into the biomass and suddenly a dog barked. Then others joined in. All their barking was identical and there seemed to be dogs all around us. Synchronized dog-barking. A strange and eerie sound. We backed away from the weed belt and it stopped; we advanced again and it started. The photographer noticed something like a lawn sprinkler sticking up from the edge of the weeds. One by one the dogs stopped. The lawn sprinkler retracted. Then the photographer reached down towards it and it shot out again and started to bark. We looked at one another and smiled nervously. A security system. After that we gave the weed belts a wide berth.

Twenty more minutes of tediously circuitous walking brought us to another obstacle, a series of low mounds of discarded containers and packaging arranged by the wind into fantastic sand dunes of waste flagged by torn and fluttering plastic sheets. Once again it was possible to move through, rather than over, these man-made barkhans by snaking along the ruts and ridges scored into the ground. Treading with special care over the uneven surface, we picked our way among the mounds like holidaymakers avoiding rocky outcrops on a beach. As we got closer to the big sheds we had seen from afar, the mounds of waste seemed to get bigger. Bleached and desiccated by the sun, we could see that they actually reached right up the shed walls and in some cases swept over the shallow slopes of their roofs.

Another five minutes found us walking over rutted ground between the sheds themselves; great monsters with sides 200 metres long, all arranged in a cluster around an access route marked out by an even greater concentration of ruts and ridges than we had seen before. It would be hard to exaggerate how difficult it was to walk in that territory. In the distance the access route led out over an elliptical concrete bridge under which lay a collection of what looked like shipping containers surrounded by more heaps of empty beverage cans, bottles and packaging, and partially hidden from view. Nearer to us, on the exposed side of the nearest shed, some huge lettering could still be seen. 'Z..US..' could still be made out in flaking black patches worn grey by the wind.

'Hi there,' said a voice suddenly, 'are you the long-distance bookings?' We looked around but could see nobody. The photographer shouted out that yes we were. At this a figure disengaged itself from the shadows that still lingered around the corners of the big sheds.

'My name's Carlos,' he said. 'I'm your guide to Ideal Home 2098. Any questions you have, just ask me. It's my job to make sure you see everything you want to see and still have time to have some fun.' Although he must have been 100 metres away Carlos's voice sounded almost intimately close. We later found out it was electronically projected. He advanced towards us with an eager stride. When he arrived I was the one who asked him the obvious question.

'Where are the houses of the future?'

'Houses? Houses is a bad word,' he frowned. 'We don't say house any more, we say terminals. Ter-min-als. Houses is a shit word these days. It means something bad. What you called debit in your day?'

'Debit? You mean debt?'

'Yeah, debt, debit. You know, when they take it all away like in Charles III's time? We call that the Age of Boxes. All live in boxes. Now we have terminals. Terminals are much better than houses. Nobody can take your terminal away. You take your terminal away yourself.'

While he spoke we studied Carlos. He looked normal enough but he wore strange clothes. A one-piece garment that looked like an Airforce mechanic's overalls in shiny grey. The photographer took a picture of him. He laughed.

'No big deal,' he said.

'What do you call those clothes?'

'Terminal,' he replied, smiling. And while we watched he inflated his suit like an airbag and rolled over onto his back on the ground. The suit completely enclosed him but we could still hear his voice as clear as ever: it never changed, far or near, outdoors or in. It always had that boom of presence, as though the loudness button was always on. It always made him sound impressive. Carlos deflated his terminal airbag to its original size and he stood before us again.

'This is the smallest terminal we have. You can live in here for a day or two, maybe a week. A hundred channels HDTV, 3,000 hours minimax video. 300 Gigabytes of ROM, 150 RAM. 25 kilowatts of heat if you need it. Low radiation fog gun. Fully recyclable. Bet you never saw anything like it.'

We did not speak. 'Any questions, you ask me,' Carlos added.

The photographer said: 'That thing with the suit. Will you do that again?'

When Carlos had obliged I had a question too.

'Does everybody live in . . . suits like this?'

'No way. Everybody lives in a terminal, that's for sure, but you can set up a terminal anywhere and have it look like anything. The only requirement is that it should fit into the pattern of the image cascade during daylight hours.'

'What's the image cascade?'

'Well, you know, currently it's Heritage holograms, Listed style, per cent for art, Hammer films, anything in the Bible's OK.'

'You mean it has to conform to some sort of regulation appearance?'

'Yes, but that's no problem. It includes most things.'

'But this stuff round here doesn't look like anything. It's just heaps of rubbish and sheds.'

'It's still pretty early, right? You wait a bit, then you'll see. It won't look like this for long. At least not to us.'

'Can you show us a bigger, er, more typical terminal than your suit?'

'Well, like I said, we're still early, but I guess some of the demonstrators will be around.' He motioned us to follow him and led the way towards what looked like a bizarre cluster of giant wasps' nests clinging to the wall of one of the sheds. Bizarre, because when you got closer you could see they were a collection of miniature historic buildings welded to a collection of cars. Carlos stopped just beneath a 1:72 scale model of the Temple of Abu Simbel that looked as though it was indissolubly joined to a type of car we did not recognize.

'Jim lives here, he's pretty active.'

As we were later to learn, Jim's terminal was indeed bigger than most, but even so it was smaller than his car. It was stacked up off the ground on columns of old CD players glued on top of one another. When Jim's terminal was connected to his car like this it looked as though he might be able to stretch out full length in it, just.

Despite being active Jim seemed uncommunicative at first. He let us look into his dwelling, which resembled the cockpit of a fighter aircraft with a water bed in it. 'How can you live in here?' I asked. Jim looked mystified. 'Excuse me?'

'How can you live in here, there's no space.' Jim picked up an umbilical keypad and punched a button. Suddenly the vast Dolby hush of a great cathedral descended over the interior of his tiny capsule. The terminal's envelope had become a screen and a panorama of open desert stretched away in all directions. It was astounding. It might have been a hundred miles from wall to wall. If we had shouted, we would have heard echoes. 'Plenty of space,' said Jim phlegmatically, and eclipsed the image with his finger.

'Can you do that again?' asked the photographer.

'Let me get this straight,' I said to Carlos. 'Does everybody live alone like Jim, in these little trick boxes?'

'More or less. You have to remember everybody really wants to live alone, they always have. Only traditional houses made it impossible. Big houses, all heavy eh? Cost a fortune, can't get rid of it, can't sell it, can't even throw it away! I know the history! Everybody had a bad time. People without houses were the poor ones in those days. No homes. They had to live in boxes. But if you look back with hindsight, those boxes were the first terminals. All they did was take all the technology you had in your houses or flats a hundred years ago and make a house out of them. Not put the technology in a house, but make a house out of the technology. It's obvious really. That's how you get from a heavy house to a light terminal.'

'Are they called terminals because they are really just like big satellite TV sets, connected like computers to energy, information and . . .'

'Well, really the people are the terminals. But OK, yes, the terminals get all that and nutrients and fluids. Brain packs, virtual worlds, endless interactive movies, cable living, on the wire, that's what we've got.

'Jim here lives in this Egyptian terminal and his car and it's like a vast cinema when he wants. He can drink in the Jumbotron in full sensurround.

'Sometimes he spends days on end in here popping pills, watching and listening. He can paint out any outside noise with white sound, graze in the refrigerator, keep the picture steady, bring it all back, images of everything he ever did – or wanted to do. Better than sharing and arguing, killing and crowding. Think about it. What a rush! To have filmed your whole life. To remember and relive all the things you have ever done. To visualize and display all the things you have ever thought of. It beats cut-throat competition or sharing with other people, I tell you. Nearly nobody lives with anybody any more. You should know that. It started in your time, before your time, more and more people living alone, now everybody does.'

It turned out that Jim's terminal was nearly eleven years old. He had helped the video rental company sling it at night, fixed with tension cables into a tiny gap between two big storage sheds at Exit 16, and propped up with a load of obsolete electronic gear. He had been one of the first to set up his own terminal in the science park. Some planners had come once in an armoured car to look for infringements, but by then there were already 30 or 40 terminals plugged in there, feeding off the electricity the distribution centres generated to keep their warehouse-loads of food and perishable goods from rotting. The terminals did not interfere with the coming and going of the freezer trucks, they just clung like parasites to the big structures that defined the landscape, making use of the air rights, as they used to say. You could tell the planners weren't exactly happy about Exit 16 but there wasn't a lot they could do, not without the army. It was happening everywhere. They say the video companies financed a lot of it. In the end the planners gave up and told the terminal dwellers they were going to set up a light village there to hide the place so that the terminals could stay. Now it was a quiet place – anyway it was at weekends. In the week you wouldn't really be able to see it.

'What's it like living in Dorset, Jim?' I asked. Once again he looked dazed. Carlos intervened.

'Terminal people don't think like that,' he said, giving Jim a wink. 'They've left all that stuff behind them. Time is a vast continuum for them. Just like space. Jim doesn't think about living in Dorset, he lives everywhere. Everywhere he can get sounds, images and temperatures of on his screen. It's what you used to call the top shelf.'

'How does Jim vote?'

Neither Jim nor Carlos answered. Then suddenly Jim remembered. 'The 24-hour consumption environment party,' he said proudly.

'There was one round here once. You just push a button.'

Then he began to try to tell us about one of the last documentaries he had ever seen, squeezed between the commercials at four o'clock in the morning, that had explained everything. The way he remembered it now, the whole continuum of life, space and terminals must have started in Charles III's time when they closed the motorways and evacuated the cities. But he was not used to talking much and he could not manage any more.

A lot of our time in 2098 had already been spent in and around Jim's capsule. The photographer and I were not eager to waste time. Standing outside I looked at the odd arrangement of Jim's porticoed tent, its exterior looking exactly like stone except that it flapped in the breeze where it was velcroed onto his car.

'How do you get your car out?' I asked finally. But Jim had had enough. Carlos explained that in 2098 cars and houses were part of the same mechanism. When you went out in your car, part of your house went too. 'Mobile satellite of the home,' I wrote in my notebook.

'Time to come with me,' said Carlos suddenly. 'They are all leaving now anyway.' We followed him to a black-windowed van with six big wheels to handle the rough terrain.

'Carlos, why did they close the motorways?' I asked.

'They were a mess,' said Carlos. 'Like houses, too big, too expensive, too old. It all got out of control. They just closed them down in 2025 and everybody drove directly to where they wanted to go.'

'What? Across country?'

'Of course, we all do that now. No houses. No roads. Terminals. Four wheel drives. All terrain vehicles. You know it makes sense.'

There was a long silence as we watched what was happening at Exit 16 that morning. Elsewhere in the science park other residents had started to stir. It was like a Western movie when the Indians suddenly rise up from the ground where they have hidden to ambush the cavalry. Now we could see what they were: all around the big sheds were strung bizarre collections of dwellings, part trucks or cars, part tents, part houses. As the sun strengthened they began to bulge and tear; the vehicle parts left the house parts and the house parts left the ground parts. Some whole houses lumbered across the rutted tracks leading to the bridge, their windscreen washers sweeping away the night's accumulation of grey dust. Other vehicles seemed no more than skeletons, dune buggies with refrigerators fixed to them and monster trucks with enormous tyres, waving satellite antennae. Thin monopostos (single-seat cars) with their drivers laid almost flat.

Then as the sun rose in the sky and the early morning departures petered out, with the click and rattle of a suddenly energized electric circuit the walls of the sheds began to change. Images danced and focused on their corrugated surfaces, working the waste dunes and the weed belt into a complex of high-resolution holograms that represented something we were utterly unprepared for: a scene of rural life out of Thomas Hardy. There were hayricks, fields of corn, grazing cows, and old cottages and barns. The desolate scene we had witnessed at dawn was now invisible. We could not see it. Nobody could see it. So it wasn't there. There was a light village there instead.

After the exodus, Carlos turned his 6x6 van out of the complex and began to head west, straight across countryside that had become a vast prairie. Occasionally we saw other vehicles travelling in other directions, but not often. The bits of road and motorway we crossed had small settlements of terminals on them. There was no through traffic any more. Yet the terminals all looked in the main like cars. In fact I noticed early on that Carlos always referred to the van as a terminal, like his suit, like Jim's capsule, like Jim. There were terminals everywhere, like digital clocks back in 1998. After a time our questions petered out. Our conversation was slowly dying of a surfeit of amazement.

As we rumbled along over wasted country from one nest of terminals to the next, we came to see that a great entropy had taken place in the hundred years since our own time. Many great problems that taxed our leaders in 1998 had just disappeared. The housing problem was no more because there were no houses; the population problem seemed to be unimportant because there were no marriages any more, no cohabitation and no children. The sex war was over because there were no competing genders, only individuals bound up in the study of their own person-alities. As far as I could tell, you could not distinguish between women and men, they all looked like terminals. There was no society. No landscape to speak of. The way Carlos talked, it seemed as if machines produced everything all the time. Nobody bought anything, they just took it and then dumped it and took something else. Despite the ozone shields here and there nobody appeared to care much about pollution. There was no unemployment problem because nobody worked. When I asked about food, Carlos thought for a bit and then said that there were special farms in Ireland that fed the whole of Europe. As I stared out of the tinted windows I realized that our progress over this bland but uneven landscape was a paradigm for the future, as purposeful yet as mean-ingless as the noise of a tank, car or plane in a video game. Everything that had happened in a hundred years had already begun happening in our own time. Nothing was new. Everything was different.

In the whole of our journey we only saw the edge of one small town and that was Dorchester. Carlos did not think it was interesting.

'No one has driven in the towns for fifty years,' he scoffed, and that was that. All we could see from a hill to the west of the town were rows of houses that might have been 200 years old apart from the black glass in their windows and armoured front doors like submarine hatches. Among the houses an ancient tower with four pinnacles supported a giant dish antenna. Around and between the old houses crawled brightly coloured concertina tubes and roof-mounted pods and aerials. Cross streets were blocked off with more black glass and corrugated plastic, shipping containers were dumped here and there. Everywhere those symbiotic car/terminals seemed to have plugged themselves onto the sides of houses like twentieth-century sculptures. Carlos said that all the old towns and cities were like this: decayed, ruined, their public buildings, office blocks, palaces and museums either abandoned or used as natural outcrops or caves to hang the ubiquitous terminals on. Because of high-definition interactive TV everybody could see everywhere and be everywhere. So everybody lived everywhere, or thought they did. Apart from Carlos at the very beginning of our visit, we never saw a single person on foot outdoors. But then, what we saw was not necessarily what was there, as the other 'light villages', with their spurious identities that we occasionally glimpsed – and which Carlos insisted were much more interesting – reminded us.

After what must have been half an hour of travel, Carlos stopped his van by a stainless-steel sign that we would not have noticed ourselves. It was decorated with a rose symbol that reminded us of our own time. The incised lettering beneath it read 'Grade II* Heritage attraction ahead. Digital security system operating.'

'This is a big house,' said Carlos triumphantly. 'This is what I brought you to see. This one is rich. He is Nick the driver. He lives in the house of MGBs [a type of British sports car].'

This implausible assertion proved to be true for, when we headed in the direction indicated by the sign, we soon saw a great mound of metal ahead of us. To us it did not look like a house at all, more like a car breakers' yard from our own time, but Carlos insisted that this one really was a house, not a terminal, one of the very few houses left.

When we got out next to the mound we met its owner, Nick. Carlos treated Nick with great respect because, he said, Nick held the title of works driver. Works driver of Dorset. His grandfather had driven on the motorways before they had been closed down, and his great-grandfather had actually driven cars in the old cities. It was his grandfather who had begun the great collection of MGBs that Nick's father had allowed to decay. Some said that there had been more than 300 of them early in the century, but by the time Nick had the MGBs hauled into a large circle and piled up, a lot of them had rotted away. The ones that were left Nick had stacked up three storeys high, corbelling inwards like an Indian hogan. He had made windows out of collections of windshields glued together and had waterproofed the lot with transparent polymers sprayed from a mobile crane. Inside, Nick had built a kind of terminal stage-set of a home that was one of the largest in the West of England. He stood proudly inside his heap of MGBs, winding the window in the door of one slowly up and down while the photographer shot off a roll of film.

'This', said Carlos confidentially, 'is really historic, eh? This is the Ideal Home of 2098.' And he laughed as though he had rewarded us with the surprise that he knew we had been waiting for.

Editor's suggestions for further reading

Pawley, M. (1997), *Terminal Architecture*, London: Reaktion Books.
Robins, K. and Hepworth, M. (1988), 'Electronic spaces: new technologies and the future of cities', *Futures*, April, 155–76.

'Sustainable Tourist Space: From Reality to Virtual Reality?'

Tourism Geographies (1999)

Jean Michel Dewailly

Editor's Introduction

In our third reading on cybercity futures Jean Michel Dewailly, a French geographer of tourism at the Université Lumière, Lyon II, France, offers us another interesting speculation on how future ICTs might complement place-based experiences. Dewailly attempts to identify the future implications of immersive, three-dimensional virtual reality (VR) technologies for the development of tourist spaces and cities (see Ken Hillis, p. 279). He sees such technologies as a positive policy tool in the struggle to deal with the catastrophic environmental impacts of the explosive growth of global tourism. Dewailly argues that VR merely amounts to the latest in a long line of innovations which progressively reduce the 'authenticity' of tourist experiences (although see Thrift 1997; Nigel Thrift, p. 98, for a critique of the notion that cities are becoming less 'authentic'). Dewailly believes that tourist VR, which would allow people to 'go on holidays' and 'vacations' without physically moving, is the logical next step from the theming and packaging of simulated tourist experiences so common in the architecture of consumption spaces today (see Susan Davis, p. 235).

Dewailly speculates on how tourist VR spaces might relate to constructed, physical ones. Whilst stressing the substitutionist potential of tourist VR, Dewailly does not believe that this will lead to the collapse of tourist cities or the cessation of all tourist-related travelling. Rather, he speculates that tourist VR spaces will complement physical tourist spaces, just as IMAX theatres do today. He also admits that there is a risk that tourist VR will merely induce more trips to go and see the 'real' place. Perhaps, Dewailly argues – in a twist which recalls Boden and Molotch's discussion of the uneven social access to 'ordinary talk' in cities (p. 101) – we may see a stark social and wealth divide in the experience of 'real' and 'virtual' tourist spaces. For example, the wealthy may be allowed to experience actual travel and real visits whilst lower, and even middle-income groups are forced to make do with the VR simulation consumed at a distance.

Virtual reality is increasingly allowing interactive and three-dimensional experiences. Large theme parks, like Disneyland and Futuroscope (France), offer powerful experiences transmitted by giant screen or, better still, by head-mounted display units. These include interplanetary voyages or trips to fantasy worlds, as well as sporting exploits that are inconceivable in reality for most individual spectators. Visitors experience sensations 'as if' they were at the heart of the action. The emotional reaction to these virtual experiences is physically very 'real' (expressed by shouts and instinctive body movements), which is why young children and nervous people are advised against participating in them.

Certainly, we could debate the quality of various touristic experiences for the individual. Tropical paradises and artificial ski runs, for example, satisfy some but not others. Why does virtual reality not satisfy the aspirations of certain potential tourists? When virtual reality allows individuals to 'practice safe travel . . . and safe sex', Inayatullah (1995, 413) suggests, it 'will have finally captured nature – possibly making it redundant. Why travel, when reality and imagination are blurred anyway?'. According to psychologists, transitioning to actual behaviour kills fantasy. But would an individual believe that he or she has seen enough of a tourist site by 'visiting' it virtually, or does a virtual visit provide further incentive to go to the site in question so as to enjoy fully its richness?

In other words, is the effect of virtual reality persuasive or dissuasive? Undoubtedly, it can be both, depending on the places and the circumstances, but also on the temperament, age, health, tastes and financial means of individuals. In any case, the connection between the visitor and the area is noticeably changed: the virtual is no longer just the vehicle for tourism, but also becomes the object of tourism. Is it, then, a question of a simple continuity in the construction and practice of tourist space, or an abrupt break induced by the radical modification of access to reality facilitated by technological progress?

It is reasonable to suppose that the creation of virtual worlds will cause tourist flows to increase. We do know that the perception conveyed in the media of a place can be a decisive element in its visitation rates, as is shown by the attraction of sites where motion pictures or television series are filmed, even if they are mythical. Now, virtual reality permits any place to be shown in its best light and thereby to generate maximum interest in it. Moreover, it opens vast new fields of application, which have hitherto undoubtedly been perceived with a degree of uncertainty: 'the prospects and areas of virtual reality applications are unlimited' (Cheong 1995, 416). Virtual reality technology is constantly perfecting the sensory interaction that it brings into play and we can anticipate considerably more progress, notably in the wake of military-oriented research.

Under such circumstances, taking virtual vacations might be plausible, thereby avoiding any loss of time, long queues, traffic jams, risks of accidents and sickness, bureaucratic and health formalities, language problems, climatic hazards, environmental degradation, public access limitations (particularly for the handicapped and disadvantaged) and various cumulative expenses. For example, Williams and Hobson

(1995: 425) note that it will soon 'be possible for a British VR [virtual reality] participant to take a holiday in Spain, even though they have never left the UK'. Virtual theme parks, on a small scale and indoors, have already appeared in the United States, Japan, Australia and France, allowing several tens of people to immerse themselves in a virtual world at one time.

However, it appears that virtual reality will never fully replace 'real' tourism to the point of becoming a 'substitute' for it. For example, immersion into heavy doses of the virtual may be harmful to the individual's physical and mental health. Furthermore, it does not replace direct social and cultural experience, nor unpredictable and memorable interactions with the environment (such as an unexpected conversation). The ability to reproduce virtual worlds at will clearly gives them an artificial character that undoubtedly would not suit everyone. It will probably be considered more as an additional invitation to a journey than as a journey of its own, but it could also appear as a complement to that journey, of which it could constitute one element. An example of the latter is the multi-storeyed IMAX movie theatre located outside the entrance to the Grand Canyon National Park, where visitors tend to spend more time experiencing hang-gliding and rafting in the canyon than they do viewing the Grand Canyon itself.

A MORE COMPLEX TOURIST SPACE

We could, therefore, try to introduce this new dimension into the modelling of tourist space, beyond that which some authors have proposed regarding the development of classical tourist areas. The model in Figure 8 synthesizes the major phases in the development of tourist areas, wherein the arrival of virtual reality constitutes perhaps a new, evolutionary step. The evolution of tourist areas in this model is placed in the context of their degree of authenticity and the pressure of tourism on them. Globally, these two elements evolve in inverse relation to each other: the more advanced the evolution in time, the more the pressure grows and the more authenticity diminishes. The model allows us to distinguish several major phases, corresponding to different types of destination areas, but keeping in mind that, depending on the scale of analysis, not all the constituent elements of an area necessarily belong to the same phase. These types

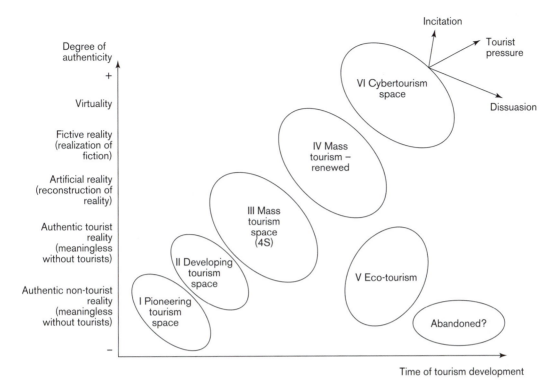

Figure 8 Tourism space: authenticity, reality, virtuality and tourist pressure
Source: Dewailly, 2000, 50

of areas are separated here more for convenience and to indicate more clearly the characteristic stages of development. In reality, there is a more or less rapid spatio-temporal transition from one stage to the next, without any break in continuity as a new phase absorbs various elements of the ones before it (see Figure 8).

The first stage of this development (I) is a pioneer type, marking the beginnings of tourism. It has a very high degree of authenticity, which, however, diminishes sharply to the next stage (II), where tourism acquires a more important structuring role in a destination area. That leads to a stage of mass tourism (III), which could be schematically designated the '4 Ss' (sun, sea, sand and sex). Forced to renew itself in response to a growing and strongly competitive demand, this type of tourism modernizes itself by incorporating even more artificial elements (theme parks, tropical paradises, artificial ski runs, etc.) (IV), while at the same time the environmental excesses and resulting awareness that they create lead to the emergence of an ecotourism concern for authenticity and resource conservation (V). But this ecotourism in turn becomes more massive and

tends to become only a niche sub-product, somewhat denatured, of traditional tourism.

At the same time, in the present state of evolution, cybertourism appears (VI). But this form of tourism is of a different nature because, for the first time since the beginnings of tourism, it does not depend upon real, direct contact by the tourist with the external world, but only upon the perception of the latter through synthetic images permitted by technological advances. If these images induce the individual to become an 'active tourist' only within the virtual realm, we are witnessing a paradox in that the implementation of virtual technology is proven to be very satisfactory from the viewpoint of sustainable tourism. However, it can also produce an opposite effect by serving as an inducement to real tourism. The real and the virtual together can create a growing diversity and complexity, which adds to a destination's richness in terms of its tourist product and the range of choices for its clientele. Can we conclude that, in the end, virtual reality is merely adding a new layer of experience to tourist areas?

REFERENCES FROM THE READING

Cheong, R. (1995), 'The virtual threat to travel and tourism', *Tourism Management*, 16, 417–22.

Inayatullah, S. (1995), 'Rethinking tourism: unfamiliar histories and alternative futures', *Tourism Management*, 16(6), 411–5.

Williams, P. and Hobson, P. (1995), 'Virtual reality and tourism: fact or fantasy?', *Tourism Management*, 16(6), 423–7.

Editor's references and suggestions for further reading

Couclelis, H. (2000), 'From sustainable transport to sustainable accessibility: can we avoid a new *Tragedy of the Commons*?'. In D. Janelle and D. Hodge (eds), *Information, Place and Cyberspace*, Berlin: Springer, 341–57.

Dewailly, J.-M. (1999), 'Sustainable tourist space: from reality to virtual reality?', *Tourism Geographies*, 1(1), 41–55.

Gottdeiner, M. (1997), *The Theming of America: Dreams, Visions and Commercial Spaces*, Boulder, CO: Westview.

Hannigan, J. (1998), *Fantasy Cities: Pleasure and Profit in the Postmodern Metropolis*, London: Routledge.

Hillis, K. (1999), *Digital Sensations: Space, Identity and Embodiment in Virtual Reality*, Minneapolis: University of Minnesota Press.

Pinney, C. (1992), 'Future travel: anthropology and cultural distance in an age of virtual reality; or, a past seen from a possible future', *Visual Anthropology Review*, 8(1), 38–55.

Thrift, N. (1997), 'Cities without modernity, cites with magic', *Scottish Geographical Magazine*, 113(3), 138–49.

'A Letter from the Future'

from *OpenDemocracy.Net* (2002)

John Adams

Editor's Introduction

In our fourth speculation on cybercity futures the British geographer John Adams writes a letter from the future. Adams pretends to write from the year 2021 to the famous British transport and environmental campaigner Mayer Hillman, on the occasion of Hillman's ninetieth birthday. In so doing he retrospectively challenges the assumptions about technology, mobility, cities and society that, he argues, prevailed in the late twentieth and early twenty-first centuries.

Adams is particularly scathing about the dominant narratives of technological 'progress' that existed in this period. Mixed with endlessly repeated utopian mantras that ICTs would help make cities more sustainable through reducing physical movement, Adams reveals how such belief systems merely presided over a catastrophic shift towards societies of 'hypermobility'. Arguing directly against speculations such as those of Martin Pawley (see p. 401), Adams suggests that all current indicators in the early twenty-first century reveal an exponential growth in all scales of physical transport flows *as well as* a growing use of ICTs. Projecting these trends to 2021, he sees a world of hypermobility running out of control.

Adams suggests that, as technology is used to speed up flows and intensify demands on workers and services, so people's lives accelerate and their appetites for transglobal virtual and physical movement grow ever upwards. The world of saturated movement and flow which results leads to a pervasive, ever-present rush hour. This supports the intensifying shift outward from central cities, as people struggle to live in places where auto-travel is still physically possible. It enables a paradoxical shift to a world where people engage with superficial information from all over the world whilst uprooting themselves from real emotional engagement with local places. And it drives globalised societies of intense alienation and anonymity, as people capsularise themselves through technology and engage in social and political contact only when it is instrumental for them to do so personally, as individuals (what Haythornthwaite and Wellman (2002) call 'networked individualism').

Adams's vision of the future – based on the projection of current trends as he sees them – thus stresses the paradoxical nature of mobility, freedom, openness and closure, in emerging cybercities. People may be able to flit around the world's virtual spaces at the click of a mouse. But, increasingly, they struggle to physically move a few blocks within many cities. Developed economies may, according to conventional indicators, have unprecedented material prosperity. But, behind this material standard of living lurks a collapse of the quality of many living environments and social and ecological support systems. Such collapses are especially dramatic in the global south, as the costs of conventional growth and mobility (pollution, poverty, resource extraction, habitat destruction) are externalised at a global scale, from the north to the south. Finally, children may make friends in web chatrooms on the other side of the planet. But the street is now often inaccessible to them because of traffic and the paranoia that comes from living in a capsular, and automobilised, civilisation (see Lieven de Cauter, p. 94).

In Adams's vision, opening up to the growing universe of telemediated experiences and flow is all too often associated with a closing off of technological, social and physical space from any notion of unpredictability or serendipitous contact. He diagnoses an explosion of the quantity of contact but suggests that the quality of that contact is often diminishing.

**Put in time capsule on 30 October 2001.
To be opened 30 October 2021**

Dear Mayer,

Tell me it isn't so! First, congratulations on your venerability. I credit all the cycling you used to do before you were driven off the roads by the fearsome traffic and took to your rowing machine.

Secondly, I must express my regret that your ninetieth birthday party will not be an occasion where one can press the flesh. Raising a glass to your 3-D laser hologram will just not be the same. But with all your now-ancient friends dispersed to the four corners, it is certainly better than nothing.

Thirdly, at the time of writing, only qualified optimists such as me, would have believed that so many of your friends would still be alive to attend your virtual birthday party let alone in robust good health, owing to spectacular advances in the biosciences.

Fourthly, and now we get to the main point of this letter, I describe myself as a qualified optimist because the progress in science and technology that I can imagine at the time of writing will, by the time you read this, have created a world that I won't much like. I promise to be delighted if my pessimism proves ill-founded. But my method of forecasting is the common one; you pick trends that are running strong and project them into the future, unless and until you see something that might impede them.

So much was promised. Telephones, fax machines, videoconferencing, email and the Internet were to be our salvation. They would de-materialise the global economy. They would transform it into an information economy that would tread more lightly on the earth. Electronic mobility would substitute for physical mobility. Red-eye flights to business meetings would be replaced by a saunter into the virtual meeting room. Most of us would compute to work rather than commute to work. Urban rush-hour traffic jams would melt away. We would be better informed, and democratically empowered by the Internet's ability to register our views by means of online voting. We would be liberated to spend more quality time with our families and neighbours. Year by year we would become richer, more convivial and kinder to the planet. Moore's Law stipulated that these trends would be exponential;

the power of computers, the key drivers of all this progress, would continue to double every eighteen months for as far as the eye could see.

CAVERNS OF LONELINESS

This is the vision that was promoted by government and industry with generous quantities of both money and exhortation. Even at the time it was obvious to some (including you and I) that the exhorters were either deluded or (more often) cynically deluding.

The historic evidence was clear. While electronic mobility could on occasion substitute for physical mobility it was overwhelmingly a stimulus to it. Just as with the telegraph and the telephone before them, email and the Internet provided people who were physically remote from each other with new reasons to meet, and facilitated their doing so. The eyes of the increasing numbers of business travellers will have become even redder as they are expected to travel with their computers downloading messages and updating their spreadsheets as they go and attending endless time-zone-spanning virtual meetings, all computers now coming with tiny built-in cameras as standard.

The urban rush-hour traffic jams will indeed have disappeared, to be replaced by pervasive all-day congestion. With the exception of a few small islands of urban café society focused on Starbucks, the search for somewhere to park, and liberation from the morning commute to work, will have encouraged the dispersal of further millions to exurbia, and their settlement at densities unserviceable by public transport let alone feet or bicycles.

Better informed our world will not be. As we roam wider, we spread ourselves thinner. The greater the geographical extent of our known world the lower the level of resolution at which we can know it. As the volume of information overwhelms the human capacity to make sense of it, people are compelled to adopt ever cruder filters. On contentious issues genetic engineering and global warming are current examples. Information overload combines with crude but highly selective filters to polarise global public opinion; people abstract from the Internet only what they want to know.

And convivial it will not be. Hypermobile societies are anonymous societies, and anonymity breeds crime, fear and paranoia. The biggest losers will be children. You will recall our study of children's independence over thirty years ago, in which we discovered that hardly any children were being allowed out of the house anymore unless chaperoned by an adult; and the answers we got when we asked parents why they were denying their children the freedoms that they (the parents) had enjoyed as children: fear of traffic and fear of strangers.

By 2021 there will be many more strangers. The 'Stranger Danger' campaigns run in schools were a late twentieth century symptom of the social stress caused by hypermobility. Neighbourhood Watch was another; neighbours who did not know each other put stickers in their windows and pretended to look out for each other. By the time you read this such schemes will have been abandoned as the ineffectual shams that they were, and replaced by CCTV cameras capable of recognising faces, backed up by data bases storing personal information about every thing from DNA to body odour and, for those who can afford them, private security services and gated communities. The ultimate gated community at the time of writing offers a foretaste: it an ocean-going cruise liner, under construction in Norway, with one hundred and ten residential apartments (at $2 million to $5 million each), combining high security, the ultimate in video conferencing and satellite communications, and the financial advantages of living offshore.

THE DEATH OF DISTANCE

The British Government's Crime Prevention Panel acknowledges the existence of trends that are fostering crime: easier access, reduced geographical barriers, increased size of the rewards and increased anonymity. It characterises the crime problem as a technological arms race and proposes fighting high-tech with higher-tech. And in the threat they spot an opportunity: the potential for the UK to excel in world-leading technology and solutions. Annex 1 lists some of the exciting possibilities. They include: genetics, biometics, biomimetics, sensors technology, magnetic materials, smart materials, encryption, propinquity tagging, nanotechnology, wearable technologies, intelligent alarms.

The report does acknowledge, in passing, the existence of the problem of social exclusion, those outside the gates of the gated community, and proffers a technological solution: new technology can reduce crime and fear of crime by being universally available across society, strengthening communities and reducing social exclusion. But this solution is oblivious to the nature of arms races; universal accessibility to the technological weapons about which the panel enthuses would make them available to the enemy as well.

Their method for anticipating the future, like mine, projects ongoing trends indefinitely in the absence of perceptible impediments. I would be intrigued to know how many of the Crime Prevention Panel have read George Orwell's *1984*, or Aldous Huxley's *Brave New World*. Their future is Orwellian with a brave new high-tech gloss. Can I see any impediments to its realisation in a hypermobile world? None!

What drives the forecasters' models of increasing mobility? Economic growth. Are there any governments anywhere resisting economic growth? No. Is their Orwellian future democratically resistible? No. Participation rates in democracy are decreasing. Increasing mobility increases the scale of problems that need governing, moving the locus of political power from Town Hall, to Whitehall, to Brussels and, ultimately, to completely unaccountable organisations like the World Bank and the International Monetary Fund. As effective accountability decreases, so does the incentive to vote.

The electronic referenda proffered by the proponents of cyber-democracy will routinely leave people frustrated by the simplistic propositions on which they are invited to vote, and disenfranchise large resentful minorities. Nowhere in the genre of science fiction in which distance has been conquered by science and technology can one find a plausible example of effective democracy. The scale of the problems that need governing precludes it.

While writing this letter I took a break to listen to a BBC programme on virtual tourism: without touching fragile environments it will simulate not only the view, but also the noises, smells and even the weather of remote parts of the world which will be spared an invasion by real tourists. I confess, Mayer, that like your virtual birthday party, I find the idea less than completely satisfying (see Jean Michel Dewailly, p. 407).

Perhaps I will log on to your birthday party twenty years from now to discover that your arguments have prevailed that everywhere governments are esteeming the local above the national and global; that walking and cycling have been granted priority over every other form of transport. But, as I write, the tide appears to be running strongly in the opposite direction. I hope you will be able to convince me it isn't so.

I hope to come and see you soon in the flesh. Be prepared. I will need a drink.

Very best wishes,

Your young friend, still imbued with your irrational belief in the efficacy of rational argument.

John Adams

Editor's suggestions for further reading

Couclelis, H. (2000), 'From sustainable transport to sustainable accessibility: can we avoid a new *Tragedy of the Commons*?'. In D. Janelle and D. Hodge (eds), *Information, Place and Cyberspace*, Berlin: Springer, 341–57.

Haythornthwaite, C. and Wellman, B. (2002), 'Moving the Internet out of cyberspace'. In B. Wellman and C. Haythornthwaite, (eds), *The Internet and Everyday Life*, Oxford: Blackwell, 3–44.

'Life After Cyberspace'

Philip Agre

Editor's Introduction

What happens now that the Internet is becoming a taken-for-granted feature of everyday life, at least in the more affluent parts of the world's more affluent cities? With Internet access increasingly ignored and assumed in such places, ICT technologies seem to be in the process of becoming as banal and ubiquitous as electricity supplies amongst many communities in advanced societies (although, we must remember, of course, that even in wealthy cities the Internet is not as ubiquitous as is often suggested).

This is an important stage of development because technologies often have their biggest effects on society when they become, in a sense, invisible because they are taken for granted and assumed. In many cases this shift to invisibility is both a metaphorical and a physical one. People tend to fail to notice technological artefacts and connections because they cease to be novel or exciting. But those technological artefacts and connections also tend to become more hidden, more miniaturised, and more embedded into the everyday environment of homes, workplaces, transport systems, artefacts and cities.

Both such trends can be observed with ICTs in contemporary cybercities. On the one hand, ICTs and Internet connections are becoming so normalised amongst many communities, user groups and spaces, especially in the rich, high-tech cities of the global north, that attention is shifting to those users and spaces that do *not* have ICT access. (Thus, news reports talk about underground trains finally being hooked up for mobile phones, or of cities or neighbourhoods that do not yet have broadband, wireless area networks, or CCTV 'security'.)

On the other hand, computing is becoming increasingly ubiquitous. Instead of obvious, large ICT equipment standing on desks, high-capability ICT systems are being miniaturised and invisibly embedded into homes, cars, buildings, infrastructures, consumer equipment and even clothes and toys (Thrift and French, 2002).

In this fifth reading on cybercity futures the US communications scholar Phil Agre seeks to move away from broad brush, dramatic speculation to focus on the implications of 'life after cyberspace' (see also Wellman and Haythornthwaite, 2002). Agre argues that most futuristic speculations about the implications of cyberspaces and ICTs for the future of cities and societies over the past three decades or so have been well wide of the mark. He argues that cyberpunk science fiction, for example (see Warren *et al.*, p. 395) has been obsessed by the fantasy of plugging ICTs directly into the brain. Agre argues, however, that instead, ICTs are being embedded, invisibly, into the banal artefacts of everyday urban life: cars, clothes, toys, buildings and public spaces. To Agre, the challenge – which echoes the motivation of this book – is not to study 'cyberspace' as some disconnected, disembodied, virtual world. Rather, we must understand how the mediation of social relationships, economic systems, physical movements, corporeal lives, cultural meanings, and human relationships with places, are influenced by ICTs in practice in real lives and in real cities.

This recombinant and hybrid world of cybercities, argues Agre, is also developing directly counter to another long-standing assumption in cyberspace debates: decentralisation. Whilst the architecture of the Internet may be decentralised, Agre points out that the economics and politics of current processes of global economic change favour the centralisation and the consolidation of economic, technological, and political power.

As the Internet matures and becomes integrated within the institutional world around it, it is becoming increasingly clear that science fiction has dis-served us. Although networked computing had already been familiar to many academic and military people for several years, it was taken up by popular culture in the context of the virtual reality craze whose canonical text was Gibson's [science fiction novel] *Neuromancer* (1984). Compelling though *Neuromancer* was as myth, as a forecast it was quite backward. Gibson famously defined cyberspace as a space apart from the corporeal world – a hallucination. But the Internet is not growing apart from the world, but to the contrary is increasingly embedded in it.

Forecasting is a hazardous occupation, and among its many hazards is the mistake of over-generalizing from transient aspects of one's current-day reality. One prominent reality of computer use in the 1980s was the cumbersome nature of interfaces. The paradigm of computer use then, as for most people still, was the box: the desktop terminal, attached by wires to a processor, with a display screen and keyboard that were useless unless the user's body was immobilized in a narrow range of postures. The desire to cast off these chains is widespread, and Gibson spoke for many in imagining that the constraints of the box could be cast off by plugging the darn thing directly into one's brain.

Computer science, though, is headed in an entirely different direction. The great fashion in user interface research is to get out of the box, as they say, and to embed computers in the physical environment: in clothing, architecture, automobiles, and public places, letting the devices talk to one another wirelessly. Computing is to become ubiquitous and invisible, industrial design is to merge with system design, and indeed the very concept of computing is to give way to concepts such as writing reports, driving to work, and keeping in touch with one's family. Computing, in short, is increasingly about the activities and relationships of real life, and the boundary between the real world and the world of computer-mediated services is steadily blurring away.

The early visions of cyberspace have dis-served us in other ways. Certain aspects of Internet architecture and administration are decentralized, and this led to hopes that everything else would become decentralized as well. Information connoted freedom, and networks connoted Adam Smith's market of artisans. Economics, however, has taught us that each of these

associations is misleading. Information as an industrial input and output exhibits vast economies of scale, and economies of scale frequently lead to industry concentration – witness the vast wave of merger activity currently going on in many industries and especially those related to information. Economics also teaches of network effects, which can persuade the whole world to adopt an open protocol like TCP/IP [the software protocol that makes the Internet work] but which also create natural monopolies for any private good whose value to customers lies mainly in the number of other people who have it.

In the case of the Internet we can see this embedding in many ways. I will describe three of them here.

First, we can look at some of the institutional preconditions that almost miraculously made the Internet possible. Contending theories of social order emphasize centralization and decentralization, but in fact the Internet grew out of a special combination of the two. The Advanced Research Project Agency (ARPA), the US State organization which developed the ancestor of the Internet in 1968, was located at both the center and the periphery, in different ways, of a powerful institutional order. Its centralized support enabled a loosely organized process of producing public goods, namely technical standards.

A second aspect of the Internet's embedding in the social world pertains to its user communities. So long as we persist in opposing so-called virtual communities to the face-to-face communities of the mythical opposite extreme, we miss the ways in which real communities of practice employ a whole ecology of media as they think together about the matters that concern them. And so long as we focus on the limited areas of the Internet where people engage in fantasy play that is intentionally disconnected from their real-world identities, we miss how social and professional identities are continuous across several media, and how people use those several media to develop their identities in ways that carry over to other settings. Just as most people don't define their activities in terms of computers, most people using Internet services are mainly concerned with the real-world matters to which their discussions and activities in the use of those services pertain.

A third and final aspect of the Internet's embedding in the real world is the process of social shaping through which the Internet's architecture on various layers evolves. This is a large and complex story, but I want to draw particular attention to the ideas about people's lives that are inscribed in the code. For example, the Internet was originally designed for the scientific community, and its architecture reflected a whole set of background assumptions about that community, for example its high capacity for self-regulation, its openness, and its relative lack of concern with exchanging money. As new communities began to appropriate the architecture for their own purposes, all of those background assumptions came to the surface in the form of security holes and other problems.

All of this is most unfortunate in a way. If we could escape into a parallel world of cyberspace then we could ignore the emerging sprawl of nontransparent and undemocratic institutions of global governance that increasingly order our electronic and non-electronic lives. But that's how it is, and we need to deal with it by recommitting ourselves to the values of democracy here in the real world.

Editor's references and suggestions for further reading

Armitage, J. and Roberts, J. (eds) (2002), *Living with Cyberspace: Technology and Society in the 21st Century*, London: Continuum.

Hunter, S. (2002), *World Without Secrets: Business, Crime and Privacy in an Age of Ubiquitous Computing*, London: Wiley.

Lessig, L. (2001), *The Future of Ideas: The Fate of the Commons in a Connected World*, New York: Random House.

Saracco, R., Weihmayer, R. and Harrow, J. (eds) (1999), *The Disappearance of Telecommunications*, London: John Wiley & Sons.

Thrift, N. and French, S. (2002), 'The automatic production of space', *Transactions of the Institute of British Geographers*, 27(4): 309–35.

Wellman, B. and Haythornthwaite, C. (eds) (2002), *The Internet and Everyday Life*, Oxford: Blackwell.

Woolgar, S. (ed.) (2002), *Virtual Society? Get Real!*, Oxford: Oxford University Press.

'How Technology Will Defeat Terrorism'

City Journal (2002)

Peter Huber and Mark Mills

Editor's Introduction

Our final reading on cybercity futures presents the reflections of Peter Huber and Mark Mills, who both work for the consultants Digital Power Capital, on how intensifying surveillance, and high-tech practices of 'network-centric' warfare, will allow the United States to overcome perceived risks of mass terrorist outrages like the September 11th, 2001 attacks (although we should remember that, cataclysmic as they were, these events involved only 1 per cent of global deaths through political violence annually on our planet).

Use and control of information and ICTs is a key strategic site within the globe-spanning struggles that have occurred since September 11th, 2001. Indeed, ever since the first Gulf war in 1991, the use of ICTs on the battlefield has blended more and more with the use of ICTs for the surveillance of urban civil societies and the 'information war' of propaganda and media output (on the net, on TV, and in newspapers). In that war, for the first time, 'front and home lines were interconnected in a near-seamless regime of information control' (Dyer-Witheford, 1999, 142).

Huber and Mills argue that this seamless linkage needs to be deepened further. To counter the risks of globe-spanning terrorist networks attempting to destroy urban and symbolic targets, the authors suggest that the complex and immense flows and mobilities that characterise cybercities need to be saturated with the very highest possible levels of automated, 24-hour, electronic surveillance (see David Lyon, p. 299). Their analysis underlines the fact that, 'amidst privatization and deregulation, one of the few aspects of the capitalist state generally reinforced is the security apparatus' (Dyer-Witheford, 1999, 141).

Huber and Mills believe that many of the emerging high-tech systems which now sustain mobility and consumption within and between cities – such as ICT-based highways, biometric scanning in airports, and bar-coded logistics systems – provide early models that must now be 'rolled out' and standardised across all walks of urban, social, and technological life. These civil technologies need to become a central part of the ongoing 'war on terrorism'. Huber and Mills have faith that such a technology-based re-ordering of urban society will reduce the vulnerabilities of the vast urban regions of the United States to further terrorist attacks – whether from explosives or biological, chemical or nuclear 'weapons of mass destruction' (WMD).

Huber and Mills's notion of 'threat recognition' implies a belief that urban civil societies need to be reconstituted in an essentially militarised way. This must be based, they suggest, on the experiences and capabilities of the US military in high-tech surveillance, scanning, targeting, tracking, and weapons systems. They suggest that this blending of the civil and the military, the domestic and the geopolitical, will allow 'non-cooperative targets' to be easily separated from 'cooperative' ones. Such a challenge is very problematic within the massive and intrinsically open flows and complexities of cities and city systems.

The key, for Huber and Mills, is thus to saturate cities – and the mobility channels that link them together – with a wide range of anticipatory, software-based surveillance systems. These, they argue, will warn law enforcement, 'homeland security' and military agencies when events, behaviours, people and profiles do not 'fit' the predicted signatures and patterns offered by 'cooperative targets'.

In the face of the risks of urban mass terrorism Huber and Mills readily admit that they 'really *do* want an Orwellian future' (original emphasis). The authors are sceptical about arguments that these shifts will erode civil liberties. They suggest that many of the systems they envisage will rely on the voluntary 'opting in' of participants.

However, many of the examples Huber and Mills describe suggest that scanning will be invisible and ubiquitous. To be effective, such systems will obviously have to be all-inclusive. In addition, it seems very likely that, to be at all effective, such anticipatory scanning systems would have to rely on crude racial and ethnic profiling. The danger here is that those racial groups deemed by the Bush regime's 'Homeland Security' department to be most likely to harbour 'uncooperative targets' will be the focus of much more detailed and invasive scrutiny than those deemed to be 'friendly' (see Stephen Graham, p. 324).

Such an approach is extremely problematic in dealing with the policing and social order issues generated by the complex and cosmopolitan social worlds of US cities. In fact, since September 11th, 2001 there have already been massive interrogations, arrests, indefinite detentions without trial, and intrusive extensions of surveillance. US Muslims and Arabs, in particular, have faced the full wrath of these measures. So much so that many lawyers involved in the civil rights and civil liberty movements now allege that the US is, technically, already a 'police state' (Ahmed, 2002, 266).

Huber and Mills's reading shows how the thinking of US military and 'homeland security' advisors is becoming completely integrated within the neo-conservative 'war on terror'. It demonstrates the ways in which a focus on ICTs integrates the efforts of such planners both to deal with the civil societies of US cities, and to mobilise unprecedented 'network-centric' military power around the world. As is common in the history of US approaches to both civil and military conflict, Huber and Mills are, in effect, hoping for some technological 'silver bullet' or 'superweapon' that will simply translate into US informational, geopolitical, economic and military supremacy (see Marx, 1995; Franklin, 1988).

At the same time Huber and Mills strive to use US supremacy in ICTs to minimise US casualties and exposure to risk and challenge. Their vision of a global 'panopticon' thus blends seamlessly in to the 'homeland panopticon' of surveillance in US regions, cities and city systems. 'Smart' weapons merge with 'smart' sensors which track the world's flows. A crude technological and economic triumphalism – which simplistically separates the world into 'good' guys and 'bad' guys, rich and poor – drives their perspective. The authors display a utopianism and technological determinism which ignores the history of many failed attempts to technologically 'solve' political and military conflicts. More worryingly still, Huber and Mills hint at the deep connections between US supremacy in ICTs, their globe-spanning military power, and the geopolitics of their national economic aspirations (which include the need to control Middle East oil reserves because of increasing home oil consumption).

Huber and Mills's vision of the future is a bleak one. Sadly, at the time of writing (June 2003) – with Iraq in chaos after the second Gulf war – it seems all too likely to emerge. Certainly, there seem few limits to the extension of George Bush's ongoing global 'war on terror' or to the intensification of homeland surveillance. As Robert Warren writes, 'rather than a cause, the "war on terrorism" has served as a prism being used to conflate and further legitimize dynamics that already were militarizing urban space' (2002, 614).

The Editor would argue that the continued US-led bombing of civilian areas in the Muslim world – Afghanistan, Iraq, Syria? – is actually occurring to support the United States' wider, and predefined, geopolitical objectives, and the security of their oil reserves. Within two months of its invasion of Iraq in 2003 the US announced that it would maintain four major bases in the country. Plans for an oil pipeline from Iraq to Israel were released. Syria – another country within George Bush's 'axis of evil' – was coming under the gaze of US military planners now that they had secured a central position within the geopolitical structure of the Middle East.

Such actions seem very likely to generate a *growth* in terrorist attacks against the US and its western allies, rather than the reverse. Without political initiatives to undermine the root causes that generate fanatical terrorism, the Editor would argue that we are unlikely to see it stop in the near future – no matter what high-tech military and surveillance gadgetry is thrown at the problem. In fundamentally open, mobile and interconnected urban societies, dreams of high-tech 'safety' are merely fantasies. Indeed, terrorism often works, as on September 11th, by simply turning the massive technological complexes that sustain everyday life in cybercities into killing machines.

On the knee-jerk resort to mass bombing as a response to terrorist actions, the mass killing of civilians as 'collateral damage' since September 11th, 2001, in Afghanistan and Iraq, reveals the lie of so-called 'surgical strikes'. The mass bombing of Afghanistan in late 2001, for example, directly and indirectly killed around ten times the numbers of innocent civilians who died as a result of the September 11th attacks (5,000 directly killed and 20,000 indirectly through displacement and famine) (Steele, 2003). As a result, many commentators have defined such attacks as little more than state-instigated terrorism. With the help of defence spending which almost amounts to the rest of the world's combined, the US strategy is clearly aimed at 'demonstrating to the rest of the world that America's military superiority is now unprecedented and unduplicable' (Davis, 2003, 3; see Herold, 2002). The world faces a truly globe-spanning 'hyperpower' which has few geopolitical checks and balances constraining its actions.

The bombing of civilians by US and other western air forces as part of the 'war on terror' is based on a double-standard which many allege to be essentially racist. The home, western, and usually white body of the pilot is priceless, inviolable, subject to massive media coverage and must be withdrawn from risk wherever possible. The non-western, usually non-white body, who dies under an air strike as 'collateral damage' is anonymous. It is worthless, unheard, unworthy and uncounted – a non-detected presence within the pixellated and distanciated landscapes of war. To the Editor the stark asymmetries perpetuated by the continued resort to the bombing of civilians to minimise western casualties, will simply destabilise the world further, stoking the resentments and hatreds that fuel terrorism in the first place.

Huber and Mills would disagree with such a position, I am sure. In their piece they readily admit that their suggested programme of a global technologised and militarised struggle against terrorism is 'a horrible vision'. They say that it gives them 'no joy to articulate it'. But, to them, the overriding imperative is to mobilise the United States' unprecedented strategic domination of ICTs to the combined imperatives of homeland security, global military offence, and international economic mastery. To them, this will inevitably mean that 'at home and abroad, it will end up as their sons against our silicon', where the 'our' refers to the US and the 'them' to 'terrorist' forces and those who support them. The result, to them, is not in doubt: 'Our silicon will win.'

◼ ◼ ◼ ◼ ◼ ◼

In transit across Manhattan on any given day are some 4 million letters, 3 million people, half a million motor vehicles and their contents, and half a million parcels – any of which may be carrying something lethal. Step by step, cities like New York must now learn to watch and track everything that moves. Airport screening is coming to much of the rest of civilian life; but it will have to be much smoother, faster, more accurate screening than airports have today, or life will just grind to a halt.

However much local governments may yearn for a one-stop federal solution to their new security problems, much of the initiative and innovation will almost certainly be left to them. And for obvious reasons, the challenge will be especially great in places like Manhattan, where the largest number of potential targets attracts the largest number of cars, packages and people – every one of which could be a bearer of destruction.

In the post–September 11th world, we know we have to see the plastic explosives in the truck before they detonate, the anthrax before it's dispersed, the sarin nerve gas before it gets into the air-conditioning duct – and not just see it but recognize it. Our imaging systems will have to distinguish between the scaffolding on a Wonderbra and the wiring on a bomb. We have, of course, slow, conventional ways of doing all this in a forensic laboratory. But the challenge now is to do it in bulk and in real time. How do you even begin to do this?

Threat recognition is, fundamentally, an exercise in the analysis and transmission of information. Imaging systems and transponders will generate huge amounts of digital data; intelligent networks, pattern-recognition software, and remote databases will make sense of the data as it comes in, chiefly by comparing the image at hand against large databases of stored images. We can't begin to screen the flow of people and goods properly without a massive deployment of digital technology to perform these functions for us.

The first step is to divide the civilian world into two, separating the trustworthy cooperators from the noncooperators, so we don't have to search *every* car, package, and pocket. The airlines already make this separation. They already have detailed profiles of most travelers readily at hand – from credit cards, travel agents, and travel histories – and they effectively enlist us as deputy security agents when they ask us about the packing and custody of our bags after establishing our identity. The Federal Aviation Authority uses such data, in a limited fashion, with the Computer Assisted Passenger Prescreening System that helps target passengers whose checked baggage should get a thorough examination. The same principles can work to screen vehicles, packages, letters, and most everything else that moves.

To understand how, consider the way quite similar tools are currently solving the problem of highway congestion. The tollbooth transponders now spreading rapidly across the United States – New York's E-Z Pass system is an example – track cooperative targets very effectively and save willing cooperators a lot of time, compared with the toll collectors or coin baskets of old. The relatively primitive friend-recognition network to which E-Z Pass belongs now extends across all of New York and from New England down to the mid-Atlantic states. Other networks are more sophisticated, like the new beltway around Toronto, a toll road without tollbooths: transponders, cameras, and license-plate numbers provide the billing information (David Holmes, p. 173). Three years ago, Singapore implemented a much more comprehensive version of the same thing – a card-based toll system designed to curb the massive traffic jams that choke that tiny island-state. Electronic devices mounted on gantries above the road deduct the appropriate charge as the cars pass under them. Even systems as simple as these can be linked up to security networks, too, and can do much to enhance safety, because so much of security comes down to establishing identity and tracking patterns of conduct – just the sorts of things that the automatic toll collectors already do.

Quite similar technologies that address the congestion of people – and in the much more demanding context of maintaining security – are now up and running. The new iris-scanning system at Schiphol Airport in Amsterdam is a dramatic example: you register, and then stare briefly into a lens instead of presenting your passport. The iris is the most personally distinct feature of the human body, even more distinct than a fingerprint. And yes, the toaster-size camera can see through your glasses and contact lenses. As with E-Z Pass, nobody *has* to participate; you're free to wait in line instead. An affront to civil liberties? Hardly. You have to volunteer – and then pay about $100 – for the privilege of getting your eyeball

into their database. The plan is to let you shop the airport stores that way, too, rather than having to swipe your credit card.

Similar technologies can separate innocuous packages from the suspect ones that need closer inspection. Most everything that's shipped by UPS, FedEx, and even the Postal Service (if it's dispatched through a postal meter rather than with ordinary stamps) is already bar-coded and repeatedly scanned while in transit. This is how the package companies maintain web sites that can tell you exactly where your inbound package is as it moves through their system. But technology is now set to move far beyond the printed code and the optical scanner. Motorola's BiStatix radio-frequency identification technology, for example, combines silicon with printed ink to embed smart electronic tags ubiquitously in packaging materials, so that packages automatically announce their whereabouts, without any human having to scan them. There are profound economic advantages in being able to 'talk to' your packages the entire length of the supply chain – this is the heart of shrinking your warehouses and your inventory, and building a business around just-in-time deliveries.

The civil defense advantages are equally apparent: a package that begins in the hands of a trustworthy shipper like Amazon and is tracked carefully while in transit from Newcastle, Delaware, to New York City can be admitted without more ado into a downtown office building and opened with confidence. The by-mail terrorist invariably prefers stamps, a handwritten address, and a late-night drop into an anonymous mailbox.

All of these tracking and screening systems that identify cooperating and friendly targets are prey to sabotage, of course. Smart cards can be stolen. With all these systems, you have to have intelligent networks and databases behind the wall to keep watch for wolves clothed as sheep. The machines have to look for patterns of activity, for daily habits – and for breaks in them – just as Citibank already does when it spots what looks like an unusual spending pattern on your credit card. Try to buy four cheap color TVs in a store in a high-crime neighborhood you never visit, and a computer somewhere deep back in the network is almost certain to initiate an immediate inquiry from your bank.

Today we use these systems, which are ubiquitous and affordable, chiefly for mundane economic screening, but they inevitably will be extended to track security as well. In response to the anthrax scare, public health officials are already taking steps to tap into the networks and databases that pharmacies use to track drug purchases and restock their shelves. When unusual numbers of people in a neighborhood begin feeling lousy, they'll begin buying all sorts of conventional medicines; those purchases may signal a biological attack well before hospital labs can spot the pattern.

Thus, while talk of national ID cards inches forward in the federal government's marble halls, identification systems are already evolving organically, and very much more rapidly, throughout the marketplace. Local governments should help set standards and facilitate collaboration with the private sector, so that building owners, garages, toll collectors, banks, package-delivery companies, and many others already in the track-and-screen business can effectively coordinate and integrate the identification technologies that they're deploying anyway. Yes, to be sure, the American Civil Liberties Union (ACLU) will protest; but like it or not, this is the way things should go and inevitably will go, too.

Once you can identify the vast majority of people, cars, and packages that are innocuous and can be waved quickly through the gate, you can inspect more closely the few that may not be, and if necessary interdict them.

Dealing with the noncooperative targets is a five-step process. First, you project electromagnetic waves across a wide range of frequencies, from radar, millimeter-waves, and infrared heat through visible light, and X-rays. There are other alternatives, too – magnetic pulses and acoustic waves, for example. Second, you carefully look for what gets through or bounces back. Third, you intensely analyze the same, crunching the numbers to turn the massive stream of return data into a coherent image. Fourth, you make sense of it, generally by another massive round of number crunching for pattern recognition, comparing the image at hand with huge databases of images stored previously. Fifth, if you don't like what you see, then you kill it, disable it, or at the very least shunt it aside for closer and more leisurely scrutiny. If you see anthrax, say, kill it with a burst of gamma rays or an electromagnetic pulse intense enough to shatter DNA.

In its most familiar form, 'sight' begins with sunlight and ends with eyeballs. But how do you 'see' the molecules in plastic explosives, in nerve gas, in cocaine, or in anthrax DNA? A dog's nose is pretty good at this

form of 'vision,' which is why places like airports are filled with canines these days. Chemical laboratories can work out the molecular composition of anything, given a big enough sample and enough time to run it through a lot of bulky hardware. But for mass screening, in real time, you project photon power – electromagnetic power – across a wide range of different frequencies, and if you do that cleverly enough, you can see most anything.

Plain old light will take you a long way in distinguishing shiny aluminum from dull wood. But if you push the frequency up quite a lot further, through visible light and on up into X-rays, you can see broken bones; and if, in addition, you also analyze what bounces back, you can get quite a good profile of molecular composition, too. 'Light' in millimeter wavelengths penetrates all sorts of clutter – like foliage, clothes, boxes, and Sheetrock – yet bounces off things we do want to see, like metals, plastic, and flesh, much like X-rays, but at much lower energy levels that pose no threat at all to human health. (A company called Millivision has even developed a millimeter-wave 'flashlight' that detects respiration at a distance – a concept originally pursued to locate wounded soldiers but equally useful in searching collapsed buildings.)

Use still more of the electromagnetic rainbow – infrared bands, ultraviolet, and so forth – and you see still more. Aim the 'light' from multiple angles – as a medical CAT scan does, for example – and you can build complete three-dimensional images. From airport scanners to on-board radar for cars, these things aren't matters of theory or laboratory speculation any more. A whole constellation of imaging products is now maturing into commercial viability. All it takes is enough different kinds of 'light,' and electronic eyeballs to match – and powerful number-crunching microprocessors to make sense of the huge amounts of data generated from these multi-spectral, multi-dimensional scanning systems – and you can see through just about any truck, car, pocket, suitcase, or package you want, looking for just about anything you choose.

If the seeing and recognizing of threats isn't highly automated, the prospect of gridlock will prevent screening from happening at all. So real-time networking will be crucial, as in the federal program that provides law-enforcement officers with fast, nationwide access to fingerprint databases, together with pattern-recognition software that fully automates the process of finding matches. Right now, real-time fingerprint recognition is moving out into the field as a wireless application, so that an officer who makes a stop can check prints in three minutes. All wireless, all fully automated. It's headed for three seconds, and the faster we get there the better. Now just do the same for irises. And the backscatter X-ray profile generated by semtex. Or sarin gas. And dozens of other threat signatures.

The data volumes are staggering. The automated luggage scanner in the catacombs of the airport today, as it spins a highly advanced CAT scanner around your toothpaste and dirty linen, generates so much data that it would take about 4,000 dial-up modems to push it down regular phone lines. The task is so challenging that, so far, only a handful of US airports do it.

The computing requirements are staggering, too. While it takes a fair amount of a highly paid physician's time to interpret a hospital CAT scan, the Federal Aviation Authority insists on a suitcase automatically scanned and analyzed by computer every six seconds. It will be the same for buildings, stadiums, bridges, and tunnels.

A decade ago, none of this would have been economically feasible. It is today. It will entail a lot of new investment; but the technology is there, or very close to there – real, commercial, functional. It's going to get deployed, not only at airports but even more widely on private premises and, later, for municipal uses.

So what happens when you see and recognize something you don't like? You kill it. Many new options are materializing here, too, more than a few of them employing technologies very similar to the sensor technologies.

Here's just one example. Pump up millimeter-wave power high enough – at present, this takes something more like a television tube than a semiconductor chip, but arrays of chips will get there soon enough – pump up the power, and you can cook things, or people, or hostile microorganisms, at quite a distance. If you choose, you can make anyone within several hundred meters feel like his whole body is touching a very hot light bulb, encouraging him to run away, fast. The Pentagon calls this 'active denial technology.' If you shift the frequency a bit and go for a longer pulse, you begin doing more lasting damage: 'terminal denial,' the Pentagon might call it.

Such high-tech power-projecting gizmos may seem to have little relevance for maintaining security in the heart of New York – until you begin worrying about

anthrax, say. Killing biologics is already a familiar objective in the biopharmaceutical industry. Blood extracts, for example, provide proteins that are widely used for health care (as in vaccines) – but possible viral contamination is a major concern. Technology companies have found ways to tune pulses of power to destroy any DNA that may find its way into such products, without destroying the proteins. Similarly, the food handling and processing industry projects carefully controlled pulses of power to kill biological causes of spoilage without damaging the rest of the food too much.

But if we deploy all this screening and tracking technology right here at home, the ACLU will doubtless argue, aren't we stepping into a horribly Orwellian future, a future utterly devoid of any privacy? In fact, we are not, and here's why.

Most of the screening of the future will be entirely by machine, and the machines can be set up to respect a whole lot of privacy. As our pattern recognition gets better and better, more and more can be waved through the checkpoints without any human involvement at all. Most information need never be pulled out of the digital loop for human scrutiny. When you get good enough at detecting threats, then you invade privacy only when you should, and at no other time.

And even then, the surrender of privacy can remain, by and large, an opt-in choice. Nobody is forced to step through an airport metal detector, or send luggage through the CAT scanner in the basement. It's a choice that goes with stepping into an aluminum tube called an aircraft – and the same choice will probably have to come with stepping into a cement tube under the Hudson River or onto a steel-and-cement edifice that spans San Francisco Bay. Civil libertarians won't like the argument at all, but bridge and tunnel tolls already have the effect of excluding those who can't or won't pay, and government buildings already make screening a condition of entry, as do a growing number of public schools and museums. The wave-through screening systems, intended to speed passage through an otherwise resistant – properly resistant, lawfully resistant – gateway will survive the inevitable constitutional challenges.

The equally inevitable health and safety objections won't stop the new screening technologies, either. Some are indeed dangerous – airport luggage screeners have to be carefully shielded, because X-rays are up in the ionizing frequencies that can tear electrons off atoms and rip molecules apart – most notably our own DNA. But despite frequent theorizing to the contrary, the lower-frequency beams don't ionize; they pack just enough punch to penetrate and reflect but not to separate electrons from atoms entirely. And happily, the new focus on altogether real threats appears to have made the public a good bit more sceptical of junk-science health scares.

Anyway, we really *do* want an Orwellian future – not in Manhattan, but in Kabul. We all look for a quick end to al-Qaida, and we hope anthrax is over. And we certainly hope that a sufficiently quick, unambiguous, and violent victory in Afghanistan will impel other governments that might wish us ill to rein in their homegrown terrorists themselves and so avoid the fate the Taliban met in backing al-Qaida. Perhaps some of this will happen, but ultimately wars like this one can't really be won, in the conventional sense of the word. Terrorist 'wars' will continue, in one form or another, for as long as we live.

Many pundits insist that technology won't win such wars; only ground troops will. They have it exactly backward. It is *only* our technology that will let us survive the enduring war against terrorism.

We are destined to fight a never-ending succession of micro-scale battles, which will require us to spread military resources across vast expanses of empty land and penetrate deep into the shadows of lives lived at the margins of human existence. *Their* conscripts dwell in those expanses and shadows. Our soldiers don't, and can't for any extended period of time. What we have instead is micro-scale technology that is both smarter and more expendable than their fanatics, that is more easily concealed and more mobile, that requires no food and sleep, and that can endure even harsher conditions.

Technology in hand today fundamentally changes the calculus of war, especially wars that in any other era would have become wars of human attrition, wars that favor the side most willing to dispatch young men into the line of fire.

The next-generation sensing technologies are far smaller, cheaper, and more powerful than today's spy satellites or SR-71 Blackbird spy plane. The old ones were fine for spying on the military-industrial complexes of nation-states. But now we have sensing technologies that bring to the battlefield abroad, and to the vast arena of civilian defense here at home, the same wizardry that transformed the mainframe computer into the Palm Pilot, the television tower into the cell phone.

Equipped with such sensors, the Predator 'remote-piloted vehicle' (RPV), about the size of a pterodactyl, can see through fog, foliage, and snow, can see objects buried underground, can see and track bullets back to their source, and can shoot Hellfire missiles. The Prowler, roughly half the size, and the high-altitude (65,000 ft) Altus boast similar capabilities. Military RPV development programs are now focused on fully functional bat-size and even butterfly-size RPVs, which have already been built. AeroVironment's electric-powered Black Widow typifies a new family of tiny fliers, with two-mile range and live color video downlink. The company is now developing a wing-flapping, dragonfly-like Microbat that weighs half an ounce, including camera and telecom downlink.

These devices and many others like them are already well past lab-bench theory; they are close to the point where they can be churned out at low cost and in large quantities, like artillery shells. Video cameras are down to lipstick-size; million-pixel deerfly-size cameras are under development. Until quite recently, the best infrared detectors were the size of a coffee can, and they strained to detect an engine's hot exhaust, but today's thimble-size units can now detect the microscopic heat difference between a red and a white stripe on an American flag from four miles out. Sensors to pick up light or X-rays or millimeter waves – the same sensors that can 'see' threats in airports or at the gateways to bridges and tunnels – are now built from the same materials as microprocessors, in the same factories, using the same core tools. Advanced radar will soon be far cheaper and more ubiquitous than binoculars and telescopes once were. It will be built from arrays of single-chip integrated transmitter-receivers, which defense contractors can incorporate by the hundreds or thousands into different platforms, weapons systems, and munitions.

Manufacturers are now etching sensors alongside microprocessor, memory, and transmitter on a single semiconductor chip, and before long they'll be able to build by the bucketload complete sensor modules – with built-in laser, memory, and Central Processing Unit (CPU) – that are no larger than a grain of sand. Dispersed along roadsides, hills, and trails, they will report just about anything that may interest us – the passage of vehicles, the odor of explosives, the conversations of pedestrians, the look, sound, weight, temperature, even the smell, of almost anything.

The cost of such chips is plummeting, as semiconductor costs generally do – down tenfold in the last decade, with another threefold drop projected in the next few years. Whereas yesterday's military technologies always grew more expensive, today's get progressively cheaper, even as their performance doubles and redoubles every few years.

A decade ago, most of this would still have sounded far-fetched. But today we have in place a trillion-dollar infrastructure of semiconductor and software industries, with deep roots as defense contractors. Two years ago, the Department of Defense established the Sensor Information Technology Program to develop the software that will manage distributed networks of communicating micro-sensors. Does anyone now doubt American code writers' ability to write such software? Or their motivation to do so?

Can the other side turn these same technologies against us? No chance. Building and using them requires a digital infrastructure and a digital mindset. The very notion of someone waging a 'digital *jihad*' is oxymoronic. Even if our enemies steal these devices, they can't get them running, or keep them running, without developing a tech-savvy population. And in any event, it's a pretty straightforward matter to lock up digital technologies from the inside. You just build in a secure, Y2K-like bug, with whatever terminal date you like, and right on schedule the weapons will turn into plowshares, or sand itself.

However advanced our power to destroy, we obviously can't bring down the other side's skyscrapers – they don't have any. This kind of enemy never will. So our longer-term objective must be to infiltrate their homelands electronically, to the point where we can listen to and track anything that moves. We can then project destructive power precisely, judiciously, and from a safe distance – week after week, year after year, for as long as may be necessary.

Properly deployed at home, as they can be, these technologies of freedom will guarantee the physical security on which all our civil liberties ultimately depend. Properly deployed abroad, they will destroy privacy everywhere we need to destroy it.

It may seem anomalous to point to micro-scale technology as the answer to terrorists who brought down New York's tallest skyscrapers. But this is the technology that perfectly matches the enemy's character and strength. It can be replicated at very little cost; it is cheap and expendable. Small and highly mobile, it can be scattered far and wide – across Manhattan, the richest place on earth, and also across the Hindu Kush, the poorest. It can infiltrate, image, track, and ultimately destroy at the peasant-soldier's scale of things. It can win a war of attrition.

It is a horrible vision. It gives us no joy to articulate it. But at home and abroad, it will end up as their sons against our silicon. Our silicon will win.

Editor's suggestions for further reading

Ahmed, N. (2002), *The War on Freedom*, Joshua Tree, CA: Tree of Life Publications.

Arquila, J. and Ronfeldt, D. (2001), *Networks and Netwars: The Future of Terror, Crime and Miltancy*, Santa Monica: RAND.

Booth, K. and Dunne, T. (eds) (2002), *Worlds in Collision: Terror and the Future of Global Order*, Basingstoke: Palgrave.

Davis, M. (2003), 'Slouching toward Baghdad'. Mimeo.

De Landa, M. (1991), *War in the Age of Intelligent Machines*, New York: Zone.

Der Derian, J. (1990), 'The (s)pace of international relations: simulation, surveillance, and speed', *International Relations Quarterly*, 34, 295–310.

Der Derian, J. (2001), *Virtuous War: Mapping the Military-Industrial-Media-Entertainment Network*, Boulder, CO: Westview.

Dillon, M. (2002), 'Network society, network-centric warfare and the state of emergency', *Theory, Culture and Society*, 19(4), 71–9.

Dyer-Witheford, N. (1999), *Cyber-Marx: Cycles and Circuits of Struggle in High-Technology Capitalism*, Chicago: University of Illinois Press.

Franklin, H. (1988), *War Stars: The Superweapon and the American Imagination*, Oxford: Oxford University Press.

Graham, S. (2001), 'In a moment: On glocal mobilities and the terrorised city', *City* 5(3), 411–15.

Graham, S. (2002a), 'Urbanising war / militarising cities: the city as strategic site', *Archis*, Part 3, 25–35.

Graham, S. (2002b), 'Special collection: reflections on cities, September 11th and the "war on terrorism" – one year on', *International Journal of Urban and Regional Research*, 26(3), 589–90.

Graham, S. (2003), 'Lessons in urbicide', *New Left Review*, 19, Jan./Feb., 63–78.

Graham, M. and Marvin, S. (2004), *Cities, War and Terrorism*, Oxford: Blackwell.

Graham, S. and Wood, D. (2003), 'Digitising surveillance: categorisation, space and inequality', *Critical Social Policy*, 23(2), 227–48.

Herold, M. (2002), 'US bombing and Afghan civilian deaths: the official neglect of "unworthy" bodies', *International Journal of Urban and Regional Research*, 26(3), 626–34.

Hookway, B. (1999), *Pandemonium: The Rise of Predatory Locales in the Postwar World*, New York: Princeton Architectural Press.

Kraska, P. (ed.) (2001), *Militarizing the American Criminal Justice System: The Changing Roles of the Armed Forces and the Police*, Boston: Northeastern.

Lianos, M. and Douglas, M. (2000), 'Dangerization and the end of deviance', *British Journal of Criminology*, 40, 261–78.

Lyon, D. (2003), *Surveillance After September 11th*, Oxford: Blackwell.

Mahajan, R. (2002), *The New Crusade: America's War on Terrorism*, New York: Monthly Review Press.

Marx, G. (1995), 'The engineering of social control: the search for silver bullets'. In J. Hagan and R. Peterson (eds), *Crime and Inequality*, Stanford: Stanford University Press, 225–46.

Nunn, S. (2002), 'When Superman used X-ray vision, did he have a search warrant? Emerging law enforcement technologies and the transformation of urban space', *Journal of Urban Technology*, 9(93), 69–87.

Rattrey, G. (2001), *Strategic Warfare in Cyberspace*, Cambridge, MA: MIT Press.

Rose, J. (2003), 'We are all afraid, but of what, exactly?', *Guardian*, 20 February, 14.

Steele, J. (2003), 'Counting the dead', *Guardian*, 29 January, 23.

Warren, R. (2002), 'Situating the city and September 11th: military urban doctrine, "pop up" armies and spatial chess', *International Journal of Urban and Regional Research*, 26(3), 614–19.

ILLUSTRATION CREDITS

COVER PHOTO

Times Square, Manhattan, New York. © Andrew Mould. Reproduced with permission.

PLATES

1 Mobile citizenship: commuters text as they walk in Tokyo's Shinjuku district, June 2003. © Stephen Graham. Reproduced with permission.

2 New media complex in the new Sony Plaza at Potsdamer Platz, Berlin, 2002. © Stephen Graham. Reproduced with permission.

3 Domestic satellite TV installations in Dubai, the United Arab Emirates. © Stephen Graham. Reproduced with permission.

4 Blizzard of 1888, New Street, New York. Used with permission of the Museum of the City of New York.

5 The Florence Duomo Cupola webcam. Public domain.

6 Webcam from the Place de La Bastille, Parispourvous.com. Public domain.

7 The Upper West Side webcam in New York City. Public domain.

8 Web graffiti in Berlin, 2001. © Stephen Graham. Reproduced with permission.

9 Travellers using an Internet and information kiosk in London's Heathrow Airport, 2002. © Stephen Graham. Reproduced with permission.

10 Transarchitectures: Marcus Novak's outline of complex spaces. © Marcus Novak. Reproduced with permission.

11 Mobile phone users on a central London street. © Zac Carey. Reproduced with permission.

12 The reinvention of street stalls as communication stalls in central London. © Zac Carey. Reproduced with permission.

13 Telecommunications workers install a new fibre grid in downtown Manhattan, 1999. © Stephen Graham. Reproduced with permission.

14 Physical and virtual: global port and airport infrastructures in Tokyo Bay, June 2003. © Stephen Graham. Reproduced with permission.

15 Reaction of the London *Evening Standard* cartoonist Patrick Blower to the news that the London Underground was being wired to allow travellers to use mobile phones (24 November 1999). Used with permission of Patrick Blower.

16 The Eurotéléport Roubaix. © Stephen Graham. Reproduced with permission.

17 A sign announcing the northern entrance to Silicon Alley in Manhattan's Flatiron district, New York. © Stephen Graham. Reproduced with permission.

18 Amazon.com's homepage, June 2003. Public domain.

19 An online delivery van from the UK's Tesco supermarket. Photograph: used with permission from FreeFoto.com.

20 An e-commerce terminal in a Japanese convenience store. © Yuko Aoyama. Reproduced with permission.

21 Sign to a nearby Internet café, central Amsterdam. © Stephen Graham. Reproduced with permission.

22 Microsoft Network's advertising on the streets of Toronto, 2001. © Robert Luke. Reproduced with permission.

23 Nineteenth-century wagon advertising homes in 'Netville'. © Keith Hampton. Reproduced with permission.

24 Billboard advertisement for 'Netville'. © Keith Hampton. Reproduced with permission.

25 The emergence of injunctions to ban mobile phone use in areas where it is deemed offensive. © Stephen Graham. Reproduced with permission.

26 Active Worlds' Metatropolis. © Activeworlds.com. Reproduced with permission.

27 Map of Alphaworld, December 1996. © Activeworlds.com. Reproduced with permission.

28 Map of Alphaworld December 1998. © Activeworlds.com. Reproduced with permission.

29 Web graffiti on a street in Vienna, Austria, 2002. © Stephen Graham. Reproduced with permission.

30 Razor wire and CCTV: surveillance as boundary enforcement in the city. © Stephen Graham. Reproduced with permission.

31a and 31b *Cabinas pública de Internet* (public Internet cabins) on the streets of an informal district in Lima. © Ana María Fernández-Maldonado. Reproduced with permission.

32 New financial services buildings in the Pudong redevelopment area, Shanghai. © Stephen Graham. Reproduced with permission.

33 Kiosk advertising Copenhagen's 'cybercity' strategy, Denmark. © Stephen Graham. Reproduced with permission.

34 New Malaysian government buildings under construction in Putrajaya, 1999. © Tim Bunnell. Reproduced with permission.

35 Former residents of the Putrajaya plantation estates employed as gardeners and cleaners in the new 'cyber city'. © Tim Bunnell. Reproduced with permission.

36 The spatial town square metaphor as used to structure content on Amsterdam Digital City. © Geert Lovink. Reproduced with permission.

37 A 'home neighbourhood' within Amsterdam Digital City. © Geert Lovink. Reproduced with permission.

38 'Vectorial Elevation, Relational Architecture 4', 1999–2002. Large-scale interactive intervention with 18 robotic searchlights controlled by the public over the Internet at http://www.alzado.net. First installed at the Zócalo Square in Mexico City. © Martin Vargas. Reproduced with permission.

39 Internet café in central Amsterdam 2001. © Stephen Graham. Reproduced with permission.

40 Street advert on a Vienna pub, Austria, 2002. © Stephen Graham. Reproduced with permission.

FIGURES

1 Physical and virtual structures and their interelationships. Source: William Mitchell, 1999, 'The city of bits hypothesis'. In *High Technology and Low Income Communities*, p. 126. © MIT Press. Reproduced with permission.

2 Map of the Melbourne CityLink network on completion. © CityLink Melbourne Limited. Reproduced with permission.

6 Map of the expanding Amazon.com empire, as of January 2000. Information gathered from Amazon's annual reports and press releases. © Martin Dodge. Reproduced with permission.

8 Tourism space: authenticity, reality, virtuality and tourist pressure. Source: Jean Michel Dewailly, *Tourism Geographies*, 2000, p. 50. © Taylor & Francis. Reproduced with permission.

TABLES

2 Communication Alternatives. Source: William Mitchell, 1999, 'The city of bits hypothesis'. In *High Technology and Low Income Communities*, p. 109. © MIT Press. Reproduced with permission.

3 Advantages and disadvantages. Source: William Mitchell, 1999, 'The city of bits hypothesis'. In *High Technology and Low Income Communities*, p. 110. © MIT Press. Reproduced with permission.

7 Neighbourhood networks in Netville. © Keith Hampton. Reproduced with permission.

COPYRIGHT INFORMATION

Anniversary Conference of the Sociology Department, University of Essex, UK. The Reading is adapted from an earlier publication, 'The Compulsion of Proximity' in Deirdre Boden and Roger Friedland (eds), *Now/here: Time, Space and Social Theory*, Berkeley and Los Angeles: University of California Press (1993). The authors acknowledge helpful advice from Steven Clayman, Randall Collins, Carol Gardner, Anthony Giddens, Jack Katz, and Don Zimmerman. Copyright © 2004 Harvey Molotch. Reprinted by permission.

LUKE Luke, Timothy, 'The Co-existence of Cyborgs, Humachines and Environments in Postmodernity: Getting Over the End of Nature'. Preliminary versions of this Reading were the basis for a presentation at the International Studies Association, 28 March–1 April, 1994, and the Western Political Science Association, 24–6 March 1998. Brian Opie, Gary Downey, Paul Knox, Gearóid Ó Tuathail, and Stephen K. White all provided helpful comments on the initial drafts. Copyright © 2004 Timothy Luke. Reprinted by permission.

III CYBERCITIES: HYBRID FORMS AND RECOMBINANT SPACES

BOERI Boeri, Stefano, 'Eclectic Atlases'. Copyright © 2004 Stefano Boeri. Reprinted by permission.

MITCHELL Mitchell, William, 'The City of Bits Hypothesis'. From *High Technology and Low Income Communities*, 108–29. Copyright © 1999 MIT Press. Reprinted by permission.

CRANG Crang, Mike, 'Urban Morphology and the Shaping of the Transmissible City'. From *City*, 4(3), 303–14. Copyright © 2000 Taylor & Francis. Reprinted by permission.

CAREY Carey, Zac, 'Generation Txt: The Telephone Hits the Street'. Copyright © 2004 Zac Carey. Reprinted by permission.

GRAHAM Graham, Stephen, 'Excavating the Material Geographies of Cybercities'. This Reading is an adapted and updated version of S. Graham (2001), 'Information Technologies and Reconfigurations of Urban Space', *International Journal of Urban and Regional Research*, 25(2), 406–10. Copyright © 2004 Stephen Graham. Reprinted by permission.

TOWNSEND Townsend, Anthony, 'Learning from September 11th: ICT Infrastructure Collapses in a "Global" Cybercity'. Anthony Townsend would like to acknowledge the support of an NSF Urban Research Initiative grant 'Information Technologiues and the Future of Urban Environments', in completing the research that went into his Reading. Copyright © 2004 Anthony Townsend. Reprinted by permission.

IV CYBERCITY MOBILITIES

BARLEY Barley, Nick, 'People'. From Nick Barley (ed.), *Breathing Cities: The Architecture of Movement*. Copyright © 2000 Birkhäuser Publishing Ltd, 9–13. Reprinted by permission.

OHANA-PLAUT Ohana Plaut, Pnina, 'Do Telecommunications Make Transportation Obsolete?' Copyright © 2004 Pnina Ohana Plaut. Reprinted by permission.

SHELLER and URRY Sheller, Mimi and Urry, John, 'The City and the Cybercar'. This Reading is a significantly shortened and refocused version of 'The City and the Car', *International Journal of Urban and Regional Research*, 24, 737–57. Copyright © 2004 Mimi Sheller and John Urry. Reprinted by permission.

HOLMES Holmes, David, 'Cybercommuting on an Information Superhighway: The Case of Melbourne's CityLink'. Some parts of this Reading appeared previously in D. Holmes (2000) 'The Electronic Superhighway: Melbourne's CityLink Project', in *Urban Policy and Research*, 18(1), 2000. Copyright © 2004 David Holmes. Reprinted by permission.

EASTERLING Easterling, Keller, 'The New Orgman: Logistics as an Organizing Principle of Contemporary Cities'. Copyright © 2004 Keller Easterling. Reprinted by permission.

GOTTDEINER Gottdeiner, Mark, 'Deterritorialisation and the Airport'. From *Life in the Air: Surviving the New Culture of Air Travel* (Rowman and Littlefield: Oxford, 2001), 32–5. Copyright © Rowman and Littlefield. Reprinted by permission.

VII CYBERCITY PUBLIC DOMAINS AND DIGITAL DIVIDES

VIII CYBERCITY STRATEGY AND POLITICS

IX CYBERCITY FUTURES

Plate 40 Street advert on a Vienna pub, Austria, 2002. (Photograph by Stephen Graham.)

Index